DUSTY PLASMAS IN THE NEW MILLENNIUM

Related Titles from AIP Conference Proceedings

635 Atomic Processes in Plasmas: 13th APS Topical Conference on Atomic Processes in Plasmas
Edited by David R. Schultz, Fred W. Meyer, and Fay Ownby, October 2002, 0-7354-0090-3

611 Superstrong Fields in Plasmas: Second International Conference on Superstrong Fields in Plasmas
Edited by Maurizio Lontano, Gérard Mourou, Orazio Svelto, and Toshiki Tajima, April 2002, 0-7354-0057-1

606 Non-Neutral Plasma Physics IV: Workshop on Non-Neutral Plasmas
Edited by François Anderegg, Lutz Schweikhard, and Fred Driscoll, February 2002, 0-7354-0050-4

598 Solar and Galactic Composition: A Joint SOHO/ACE Workshop
Edited by Robert F. Wimmer-Schweingruber, December 2001, CD-ROM included, 0-7354-0042-3

563 Plasma Physics: IX Latin American Workshop
Edited by Hernán Chuaqui and Mario Favre, May 2001, 1-56396-999-8

537 Waves in Dusty, Solar, and Space Plasmas
Edited by F. Verheest, M. Goossens, M. A. Hellberg, and R. Bharuthram, October 2000, 1-56396-962-9

528 Acceleration and Transport of Energetic Particles Observed in the Heliosphere: ACE 2000 Symposium
Edited by Richard A. Mewaldt, J. R. Jokipii, Martin A. Lee, Eberhard Möbius, and Thomas H. Zurbuchen, July 2000, 1-56396-951-3

471 Solar Wind Nine: Proceedings of the Ninth International Solar Wind Conference
Edited by Shadia Rifai Habbal, Ruth Esser, Joseph V. Hollweg, Philip A. Isenberg, May 1999, 1-56396-865-7

457 Trapped Charged Particles and Fundamental Physics
Edited by Daniel H. E. Dubin and Dieter Schneider, January 1999, 1-56396-776-6

446 Physics of Dusty Plasmas: 7th Workshop
Edited by Mihály Horányi, Scott Robertson, and Bob Walch, October 1998, 1-56396-809-6

414 Two-Dimensional Turbulence in Plasmas and Fluids: Research Workshop
Edited by Robert L. Dewar and Ross W. Griffiths, December 1997, 1-56396-764-2

To learn more about these titles, or the AIP Conference Proceedings Series, please visit the webpage **http://proceedings.aip.org/proceedings**

DUSTY PLASMAS IN THE NEW MILLENNIUM

Third International Conference on the
Physics of Dusty Plasmas

Durban, South Africa   20–24 May 2002

SPONSORING ORGANIZATIONS
International Union of Pure and Applied
   Physics (IUPAP)
International Centre for Theoretical
   Physics (ICTP), Trieste, Italy
Department of Energy, USA
Office of Naval Research, USA
Naval Research Laboratory, USA
National Science Foundation, USA
National Research Foundation, South Africa

EDITORS
R. Bharuthram
M. A. Hellberg
*University of Natal
Durban, South Africa*
P. K. Shukla
*Ruhr-Universität
Bochum, Germany*
F. Verheest
*Universiteit Gent
Gent, Belgium*

Melville, New York, 2002
AIP CONFERENCE PROCEEDINGS ■ VOLUME 649

Editors:

R. Bharuthram
Director: Research

M. A. Hellberg
School of Pure & Applied Physics

University of Natal
King George V Avenue
Durban, 4041
SOUTH AFRICA
E-mail: bharuthramr@nu.ac.za
          hellberg@nu.ac.za

P. K. Shukla
Institut für Theoretische Physik IV
Ruhr-Universität Bochum NB7/23
D-44780 Bochum
GERMANY
E-mail: ps@tp4.ruhr-uni-bochum.de

F. Verheest
Sterrenkundig Observatorium
Universiteit Gent
Krijgslaan 281
B-9000 Gent
BELGIUM

E-mail: frank.verheest@rug.ac.be

The articles on pp. 59-65, 152-156, 157-161, and 357-360 were authored by U.S. Government employees and are not covered by the below mentioned copyright.

Authorization to photocopy items for internal or personal use, beyond the free copying permitted under the 1978 U.S. Copyright Law (see statement below), is granted by the American Institute of Physics for users registered with the Copyright Clearance Center (CCC) Transactional Reporting Service, provided that the base fee of $19.00 per copy is paid directly to CCC, 222 Rosewood Drive, Danvers, MA 01923. For those organizations that have been granted a photocopy license by CCC, a separate system of payment has been arranged. The fee code for users of the Transactional Reporting Service is: 0-7354-0106-3/02/$19.00.

© 2002 American Institute of Physics

Individual readers of this volume and nonprofit libraries, acting for them, are permitted to make fair use of the material in it, such as copying an article for use in teaching or research. Permission is granted to quote from this volume in scientific work with the customary acknowledgment of the source. To reprint a figure, table, or other excerpt requires the consent of one of the original authors and notification to AIP. Republication or systematic or multiple reproduction of any material in this volume is permitted only under license from AIP. Address inquiries to Office of Rights and Permissions, Suite 1NO1, 2 Huntington Quadrangle, Melville, N.Y. 11747-4502; phone: 516-576-2268; fax: 516-576-2450; e-mail: rights@aip.org.

L.C. Catalog Card No. 2002114936
ISBN 0-7354-0106-3
ISSN 0094-243X
Printed in the United States of America

CONTENTS

Preface... xiii

SECTION 1
OVERVIEW LECTURES

Structure and Dynamics of Strongly Non-Ideal Dusty Plasmas..................3
    V. E. Fortov
Dusty Plasmas in the Ionosphere and its Environment.......................13
    O. Havnes
Dusty Plasmas in the Solar System 22
    M. Horányi
Dusty Plasmas in Astrophysics and Cosmology 32
    E. Sedlmayr

SECTION 2
TOPICAL REVIEW LECTURES

Plasma Grown Particles: From Injected Gases to Nanoparticles
and Nanomaterials, from Injected Particles to Dust Clouds in the
PKE Experiment..45
    P. Roca i Cabarrocas, A. Fontcuberta i Morral, A. V. Kharchenko, S. Lebib,
    L. Boufendi, S. Huet, M. Mikikian, M. Jouanny,
    and *A. Bouchoule*
Micro-dynamics in 2D Dusty Plasma Liquids 53
    Y.-J. Lai, L.-W. Teng, P.-S. Tu, H.-Y. Chu, and *L. I*
Plasma Response to a Single Grain/Electrostatic Interaction
between Grains..59
    M. Lampe
Magnetic Effects in Dusty Plasmas 66
    N. Sato
Dust Crystal in the Electrode Sheath of a Gaseous Discharge...............74
    I. V. Schweigert and V. A. Schweigert
Nonlinear Waves in Dusty Plasmas 83
    P. K. Shukla and A. A. Mamun
Complex Plasmas under Microgravity Conditions: First Results from
PKE-Nefedov...91
    G. E. Morfill, *H. M. Thomas*, B. M. Annaratone, A. V. Ivlev, R. A. Quinn,
    A. P. Nefedov, V. E. Fortov, and PKE-Nefedov Team
Physics of Collective Dust-Dust Attraction and Dust
Structure Formation..110
    V. N. Tsytovich and G. E. Morfill

Italicized name indicates the author who presented the paper.

SECTION 3
ORAL PRESENTATIONS

A Nonlinear Theory of Void Formation in Colloidal Plasmas 121
    K. Avinash, *A. Bhattacharjee*, and S. Hu

Dust Growth in Astrophysical Plasmas 126
    R. Bingham and V. N. Tsytovich

Dust Particles Growth and Behavior under Microgravity Conditions 135
    M. Mikikian, *L. Boufendi*, A. Bouchoule, G. E. Morfill, H. M. Thomas,
    H. Rothermel, T. Hagl, A. P. Nefedov, V. E. Fortov, V. I. Molotkov,
    O. Petrov, A. Lipaev, Y. P. Semenov, A. I. Ivanov, V. Afanas'ev,
    C. Haigneré, and K. Kozeev

Rotation of Dust Coulomb Clusters in Axial Magnetic Field 139
    F. M. H. Cheung, A. A. Samarian, and B. W. James

Self-Consistent Dusty Sheaths 144
    Y. I. Chutov

Dynamic Phenomena in Complex Plasmas 148
    N. F. Cramer, S. V. Vladimirov, A. A. Samarian, and B. W. James

Causes of Small Particle Growth in Silane Discharges 152
    G. Bano, K. Rozsa, and *A. Gallagher*

Phase Transition in Dusty Plasmas: A Microphysical Description 157
    G. Ganguli, G. Joyce, and M. Lampe

Aerosol Phenomena in Plasma 162
    A. Ignatov

Ion Trapping within the Dust Grain Plasma Sheath 166
    D. Jovanović and P. K. Shukla

On the Modification of Powder Particles in a Process Plasma 170
    H. G. Thieme, M. Quaas, *H. Kersten*, H. Wulff, and R. Hippler

A Fluid Dynamic Approach to the Dust-Acoustic Soliton 174
    J. F. McKenzie and T. B. Doyle

Normal Mode Spectra of Thermally Excited 2D Finite
Coulomb Clusters 180
    A. Melzer

Typical Characteristics of RF Voltage Thresholds for Planar
Dusty RF Discharges 184
    S. Nonaka, Y. Nakamura, S. Ikezawa, and K. Katoh

Voids in Dust Clouds Suspended in the Plasma Sheath 188
    G. V. Paeva, W. W. Stoffels, R. P. Dahiya, E. Stoffels,
    and G. M. W. Kroesen

Geometry Induced Defects in a Confined Wigner Lattice 192
    F. M. Peeters, M. Kong, and B. Partoens

Dynamical Phenomena in Strongly Coupled Dusty Plasma under
Microgravity Conditions 196
    O. S. Vaulina, A. P. Nefedov, *O. F. Petrov*, and V. E. Fortov

Italicized name indicates the author who presented the paper.

Experiments and Simulation of Elastic Waves in a Plasma Crystal Radiated from a Point-Dipole-Source 200
    A. *Piel*, V. Nosenko, and J. Goree
Shock Wave-like Structures in Complex Plasmas: Theory and Experiments 204
    S. I. *Popel*, A. P. Golub', and T. V. Losseva
Charge Exchange Collisions and the Current to Probes and Dust Particles 208
    S. *Robertson* and Z. Sternovsky
Observation of Ion-Ion Instability in Dusty Plasmas 212
    Y. *Saitou* and Y. Nakamura
Modification of Shielding and Wake Potentials Due to Dust-Lower-Hybrid Waves in Flowing Magnetized Dusty Plasmas 216
    M. *Salimullah*
Solitons and Oscillitons in Complex Plasmas 220
    K. *Sauer*, E. Dubinin, and J. F. McKenzie
New Kinetic Variables and the Effective Temperatures in Dusty Plasmas 224
    P. P. J. M. *Schram*, S. A. Trigger, and A. G. Zagorodny
Dust-Acoustic and Shear Waves in Strongly Coupled Dusty Plasmas 231
    G. Prasad, J. Pramanik, B. M. Veeresha, A. *Sen*, and P. K. Kaw
Experimental Dust Levitation in a Plasma Sheath near a Surface 235
    A. A. *Sickafoose*, J. E. Colwell, M. Horányi, and S. Robertson
Grain Oscillations Induced by Electrode Voltage Modulation 239
    G. *Sorasio*, D. P. Resendes, and P. K. Shukla
Boundary Phenomena in RF and DC Glow Discharge Dusty Plasmas 243
    E. *Thomas Jr.*, W. E. Amatucci, and G. E. Morfill
Transport of Macroparticles in Weakly Ionized Dusty Plasma of Gas Discharges 247
    O. S. *Vaulina*, O. F. Petrov, and V. E. Fortov
Stationary Equilibria of Self-Gravitating Dusty Plasmas 251
    F. *Verheest* and V. M. Čadež
Stability of Particle Arrangements in a Complex Plasma 255
    S. V. *Vladimirov*, A. A. Samarian, J. Albrecht, B. W. James, S. A. Maiorov, and N. F. Cramer

SECTION 4
POSTER PRESENTATIONS

Experimental Study of Coulomb Crystal Formation in Hollow Cathode Discharge 261
    A. K. *Agarwal* and G. Prasad
Observation of Rotating Dust Particles 265
    A. K. *Agarwal* and G. Prasad

Italicized name indicates the author who presented the paper.

Numerical Investigation of Ponderomotive Force Effect Based
Contamination Control in Dusty Plasmas................................269
    M. Amroun, M. Djebli, and *R. Annou*

Dusty RF Discharges with Secondary Electron Emission....................273
    Y. I. Chutov, W. J. Goedheer, O. Y. Kravchenko, and O. A. Lavrov

Dusty Sheaths in Plasmas with Two-Temperature Electrons.................277
    K. Asano, *Y. I. Chutov*, O. Y. Kravchenko, N. Ohno, A. F. Pshenychnyj,
    R. D. Smirnov, and S. Takamura

Waves in Magnetized Plasmas with a Spectrum of Dust Sizes 281
    N. F. Cramer, F. Verheest, S. V. Vladimirov, and M. Wardle

Ballooning Instability in the Jovian Magnetosphere......................285
    N. Das and K. S. Goswami

Dust-Acoustic Wave Instability at the Diffuse Edge of RF Inductive
Low-Pressure Gas Discharge Plasma......................................289
    A. V. Zobnin, A. D. Usachev, and *V. E. Fortov*

Spatial Separation of Dust Particles by their Sizes at the Diffuse Edge
of RF Inductive Discharge Plasma..293
    A. V. Zobnin, A. D. Usachev, and *V. E. Fortov*

In Situ Study of Surface Modification of Nano-Particles in Reactive
Plasmas by Mie-Ellipsometry 297
    G. Gebauer, T. Galka, and J. Winter

On the Magnetosonic Wave and Instability in a Dusty Plasma..............301
    M. A. Hellberg, A. P. Matthews, and F. Verheest

Study of Particle Formation and its Applications in Ar/CH_4
and Ar/C_2H_2 Mixtures 305
    S. Hong, J. Berndt, and J. Winter

Electrostatic Discharging of Dust near the Surface of Mars 309
    C. E. Krauss, *M. Horányi*, and S. Robertson

Validity of Epicyclic Description of Saturnian Dust Grain Orbits 313
    C. J. Mitchell, *M. Horányi*, and J. E. Howard

Brownian Motion of Absorbing Dust Grains 317
    A. M. Ignatov, S. A. Trigger, S. A. Maiorov, and W. Ebeling

Influence of Dust-Ion Collisions on Waves in Self-Gravitating
Dusty Plasmas..321
    G. Jacobs, V. V. Yaroshenko, and F. Verheest

Dust Grains as a Diagnostic Tool for RF-Discharge Plasma 325
    B. W. James, A. A. Samarian, and F. M. H. Cheung

Low Frequency Acoustic Waves in Nebulae with Gravitational Field
Induced by Dust 329
    V. Kaizr, P. Kulhánek, D. Břeň, and J. Pašek

On the Interaction of a Complex Plasma with an External Ion Beam.........333
    H. G. Thieme, R. Wiese, D. Gorbov, *H. Kersten*, and R. Hippler

Characteristics of Nonlinear Dust Acoustic Waves in Planetary Ring 337
    M. Khan, S. Sarkar, T. K. Chaudhuri, S. Ghosh, and M. R. Gupta

Ion Drag in Complex Plasmas 341
    S. A. Khrapak, A. V. Ivlev, G. E. Morfill, and H. M. Thomas

Italicized name indicates the author who presented the paper.

Langmuir Probe Measurements in a Complex Plasma under Microgravity Conditions 345
    M. Klindworth, A. Melzer, A. Piel, U. Konopka, and G. E. Morfill

Complex Plasma Experiments: The Role of Negative Ions. 349
    B. A. Klumov, A. V. Ivlev, and G. Morfill

Particle-in-Cell Simulation of Helical Structure Onset in Plasma Fiber with Dust Grains. 353
    P. Kulhánek, D. Břeň, V. Kaizr, and J. Pašek

Collisional and Nonlinear Effects on Grain Charge and Intergrain Force 357
    M. Lampe, G. Ganguli, V. Gavrishchaka, R. Goswami, and G. Joyce

A Kinetic Study of Electron Density Fluctuation in a Dusty Plasma 361
    F. Li

Direct Numerical Simulation of Yukawa Systems by Particle-in-Cell Methods 365
    W.-C. Müller, A. Zeiler, and G. E. Morfill

Interaction of an Ion Beam with Dusty Plasmas 369
    Y. Nakamura and V. N. Tsytovich

Scattering of an Ion Beam by Charged Fine Particles with Coulomb Force. 373
    H. Amemiya and *Y. Nakamura*

Secondary Emission from Small Spherical Grains 378
    Z. Nemecek, J. Pavlu, J. Safrankova, I. Richterova, and I. Cermak

Charging Properties of Dust Grain Clusters 382
    J. Pavlu, J. Safrankova, Z. Nemecek, and A. Velyhan

Dust Ion-Acoustic Solitons: Role of Trapped Electrons 386
    S. I. Popel, A. P. Golub', T. V. Losseva, A. V. Ivlev, and G. Morfill

Dusty Plasma Structures under the External Influences. 390
    V. E. Fortov, V. I. Molotkov, V. P. Efremov, A. P. Nefedov, *M. Y. Poustylnik*, and V. M. Torchinsky

Dusty Plasmas in a DC Glow Discharge 394
    V. E. Fortov, A. P. Nefedov, V. I. Molotkov, O. F. Petrov, *M. Y. Poustylnik*, V. M. Torchinsky, and A. G. Khrapak

Oscillations of Few Particle Vertical Structures. 398
    N. J. Prior, L. W. Mitchell, and A. A. Samarian

Lunar and Martian Dust Charging on Surfaces 402
    Z. Sternovsky, M. Horányi, and *S. Robertson*

Dust Vortex in Complex Plasma. 406
    A. A. Samarian and O. Vaulina

Charging of Different Size Dust Particles in the Plasma Sheath. 410
    A. A. Samarian, S. V. Vladimirov, and S. A. Maiorov

Nanoscale SiO_2 Particles at High Temperatures: Size Dependent Properties 414
    I. V. Schweigert, K. E. J. Lehtinen, M. J. Carrier, and M. R. Zachariah

Dynamical Phase Transition in Dust Crystals 418
    V. A. Schweigert, I. V. Schweigert, V. Nosenko, and J. Goree

Italicized name indicates the author who presented the paper.

Electrostatic Response of a Dusty Plasma with a Grain Size Distribution to a Moving Test Charge 422
    M. Shafiq and M. A. Raadu

Chaotic Behavior of Electron-Positron Dusty Magnetoplasma with Equilibrium Flows 426
    A. M. Mirza, M. Shafiq, M. A. Raadu, and K. Khan

Effect of Grain Charging Dynamics on the Response of a Dusty Plasma to a Moving Test Charge 430
    M. A. Raadu and M. Shafiq

Two-Dimensional Strongly Coupled Plasma on a Liquid Surface 434
    T. Shoji, K. Shinohe, H. Tomita, and Y. Sakawa

Levitation and Transport of Charged Dust over Surfaces in Space 438
    J. E. Colwell, M. Horányi, S. Robertson, and A. A. Sickafoose

Dust-Acoustic Waves with a Non-Thermal Ion Velocity Distribution 442
    S. V. Singh, G. S. Lakhina, R. Bharuthram, and S. R. Pillay

Production Mechanism and Chemical Structure of Dust Particles in Fluorocarbon Plasmas 446
    K. Takahashi, K. Ono, and Y. Setsuhara

Relativistic Dust Particles Approaching the Earth as Highest Energy Cosmic Rays 450
    V. N. Tsytovich and R. Bingham

Instability Caused by Dust Drift and the Observed Polar Mesospheric Summer Echoes (PMSE's) 454
    V. N. Tsytovich and O. Havnes

Strong Damping and Universal Instability of Dust-Ion Sound Waves (DISW) and Dust-Acoustic Waves (DAW) 458
    V. N. Tsytovich and K. Watanabe

Theory of Small Atomic-Like 2D Dust Clusters 463
    S. G. Amiranashvili, N. G. Gousein-zade, and V. N. Tsytovich

Influence of Charge Variations on Dust Dynamics in a Planar rf Discharge 467
    O. S. Vaulina, A. A. Samarian, and B. W. James

Criteria for Phase-Transitions in Yukawa Systems (Dusty Plasma) 471
    O. S. Vaulina, S. V. Vladimirov, O. F. Petrov, and V. E. Fortov

Rotational Modes of Oscillation in a Chain of Rod-Shaped Particles in a Plasma 475
    S. V. Vladimirov, M. P. Hertzberg, and N. F. Cramer

Particle Flows and Ambipolar Electric Field in Dusty Partially Ionized Plasmas with Temperature Gradients 479
    V. S. Tsypin, S. V. Vladimirov, R. Galvão, I. Nascimento, M. Tendler, and A. de Assis

Nonlinear Periodic Waves in Dusty Plasma with Variable Dust Charge 483
    L. L. Yadav and R. Bharuthram

Italicized name indicates the author who presented the paper.

**Instability of Dust Lattice Waves Due to Periodically Varying Charges
of Dust Particles** 487
    V. V. *Yaroshenko* and G. E. Morfill
**Planetary Ring and Spoke Simulation Experiment in Fine
Particle Plasmas** 491
    T. Yokota
**Thermophoretic and Ion Drag Forces Acting on Free-Falling Charged
Particles in an RF-Driven Plasma** 495
    C. Zafiu, A. Melzer, and A. Piel

SECTION 5
SUMMARY AND PROGNOSIS

**A Synopsis of Recent Theoretical Developments in Dusty
Plasma Physics** 501
    D. A. Mendis
**A Synopsis of Recent Experimental Developments in Complex (Dusty)
Plasma Physics** 507
    G. E. Morfill

Author Index 513

Italicized name indicates the author who presented the paper.

Preface

This monograph is a collection of the invited oral and poster presentations at the Third International Conference on the Physics of Dusty Plasmas (ICPDP-2002) held in Durban, South Africa, 20-24 May 2002. The Conference was a follow up to the two previous conferences in the series, the First in Goa, India (1996) and the Second in Hakone, Japan (1999).

Plasmas are ubiquitous in nature and are playing an increasingly important role industrially. In most situations in which plasmas occur, they may be accompanied by relatively massive, charged particulates that are commonly called dust grains, leading to the nomenclature "dusty plasmas". Alternative names used in the literature are "colloidal plasmas" (their behaviour has much in common with that of colloids) and "complex plasmas". The interaction between the plasma and the dust has given rise to some surprising effects and as a result, this interdisciplinary research field has developed into one of the fastest growing areas of science. In the strongly coupled regime, dust-dust attraction occurs, with the usual Coulomb repulsion between similarly-charged dust grains being overcome, and Coulomb liquids and crystals being formed. Such exotic behaviour is, within this field, becoming normal, and these model systems are turning out to be very useful for developing understanding in areas such as phase transitions.

The Conference was well attended by delegates from all parts of the world, and was characterized by the high quality of presentations and discussion. The wide range of topics covered in the presentations revealed how rapidly the field of dusty plasma physics has advanced in the past three years, with increased observations of such plasmas in astrophysical, space and laboratory environments. The experimental, theoretical and computer simulation presentations were enthusiastically received by the delegates and aroused much discussion. The highlight of the meeting was the presentation of results from the microgravity experiments recently performed on the International Space Station.

The five overview lectures of this conference dealt with the role of dust in astrophysics and cosmology, in the solar system, in the mesosphere and ionosphere, and in industrial applications (unfortunately, a manuscript was not submitted), and described the properties of strongly-coupled dusty plasmas in the laboratory and under microgravity. A further eight topical reviews were devoted to current "hot topics". These discussed research areas such as the coagulation from gas phase to thin films (from nanoscale to microscale), microdynamics in dust Coulomb liquids, the response of a plasma to a grain and the resulting plasma-mediated interaction between grains, collective dust-dust attraction and its effects, dust crystal behaviour in the electrode sheath, observations of complex plasma behaviour under microgravity, various aspects of nonlinear waves in dusty plasmas, and magnetic field effects in dusty plasma experiments.

The selected manuscripts in this monograph are presented in the following order: overview lectures, topical review lectures, oral papers and posters, with the papers within each section being in alphabetical order of the presenting authors. Based on the Conference papers, the final two articles summarize aspects of the current state of dusty plasma research and attempt a look into the future. We wish to thank all the participants for their stimulating oral and poster presentations. Special thanks go to those who submitted their manuscripts for publication in this issue.

It is a pleasure to acknowledge that ICPDP-2002 would not have been the undoubted success that it was, both from the scientific and social perspectives, without the efforts and support of the international advisory committee, the scientific programme committee, and the local organising committee, and the generosity of our sponsors: International Union of Pure and Applied Physics (IUPAP); International Centre for Theoretical Physics (ICTP) in Trieste, Italy; Department of Energy (USA); Office of Naval Research and Naval Research Laboratory (USA); National Science Foundation (USA); National Research Foundation (South Africa); eThekwini Municipality; University of Durban-Westville; University of Natal; M L Sultan Technikon; Premier Bookshop; Sharp Electronics; Castle Wines and South African Breweries.

Finally, this monograph is dedicated to the memory of the late Anesh Sooklal, the youngest member of the local organising committee. With his excellent knowledge of computers and electronics, Anesh played an important role prior to and during the Conference. He was assisting in the compilation of this monograph when he met an untimely death in a freak accident.

R. Bharuthram
University of Natal, Durban, South Africa
M.A. Hellberg
University of Natal, Durban, South Africa
P.K. Shukla
Ruhr-Universität Bochum, Germany
F. Verheest
Universiteit Gent, Belgium

SECTION 1: OVERVIEW LECTURES

Structure and Dynamics of Strongly Non-Ideal Dusty Plasmas

V. E. Fortov

*Institute for High Energy Densities,
Russian Academy of Sciences, Izhorskaya 13/19, 127412, Moscow, Russia*

Abstract. Both the theoretical and experimental study of the strongly coupled plasma over a wide range of plasma pressures and temperatures were carried out. The plasma was investigated under conditions of low pressure dc and rf gas discharges (the electron temperature - in the range of 20000-50000 K, the ion and atom temperature - in the range of 300-400 K). The plasma was also formed from positively charged dust grains in the presence of a flux of ultraviolet (UV) photons. The results of ordered structure measurements and their theoretical analysis are presented. The experimental studies of self-excited oscillations of dust density were carried out in dc and rf glow discharges over a wide range of plasma parameters (dust temperatures and sizes, gas pressure etc.). The dust concentration, the pair correlation function, the dust temperatures and diffusion coefficients have been measured. The coupling parameter, characteristic dust frequency and particle charges have been obtained.

INTRODUCTION

Extremely high Coulomb interparticle interaction in low-density dusty plasmas arises from the high charge of micron-size macroparticles. To generate such nonideal plasmas in experiments we use glow DC and inductive RF discharges, ultraviolet light beams. Liquid- and solid-like structures are seen, and phase diagrams and transitions are investigated by experiment and simulation. The theoretical and experimental study of the strongly coupled dusty plasma was carried out over a wide range of plasma parameters and experimental devices. The thermophorestic effects in the dusty plasma with spatial thermal non- uniformity were investigated. The thermophorestic forces were used for dust cloud manipulations, cloud separations and for dust cloud levitation by compensating the gravitational forces. Measurements of the charge of dust grains have been performed in a quasineutral plasma in a wide range of grain sizes. A new method was established for measuring the charge on grains levitating in the striations of a dc glow discharge. The obtained dependence of dust-grain charge on its size was found to be strongly nonlinear in the experimental conditions. The dynamical processes in dusty plasmas were studied. The experimental studies of self-excited oscillations of dust density were carried out in RF inductive discharge. Dusty plasma experiments under microgravity conditions were carried out on the space station "MIR" and "International Space Station". The ionization of bronze dust particles by solar UF radiation generates Coulomb liquid. Ambipolar diffusion of electron-dust-ion plasma was analyzed and the diffusion coefficients was measured on the "kinetic" level by the visualization.

DUSTY PLASMA STRUCTURES UNDER EXTERNAL INFLUENCES

Investigations of different external influences on the dusty plasma structures are of great interest. First of all, the influences that make negligible perturbation in the background plasma may serve as diagnostic means. External influences may be also used to control the spatial position and the order of the dusty plasma structures. Moreover, external influence may be employed to drive energy into the system of charged dust grains in a plasma and observe the dynamics of the response.

This work represents the investigations of three different external influences in the dc glow discharge striations. The following influences are: the effect of focused laser light, neutral gas temperature gradient and the gas-dynamic influence.

The experiments were carried out in a vertically positioned discharge tube [1], in which the glow discharge with cold electrodes was created. The tube is filled with neon up to the pressure 0.1-2 torr. The discharge current varies from 0.1 to 4 mA. In these range of regimes the standing striations existed. The dust grains were held above the discharge area in a container with the grid bottom. When shaking the container the dust grains fall down through the grid and form the ordered structures in the striations. The dust particles were highlighted with a laser diode 50 mW power. The scattered light was observed by a CCD camera at a frame rate of 25 fps and in some cases by a FANTOM-6 camera at frame rate up to 300 fps. The observations were conducted in the first striation above the cathode.

The influence of the focused laser light on the dust grains was used for the measurements of the charge of dust grains levitating in striations [2]. The light beam from an Ar^+ laser was focused onto a single particle in the structure. The beam power was up to 200 mW, the waist thickness was about 60 μm. Under the effect of the laser light the particle moves 1.5-3 mm out of the structure, then leaves the beam and returns back into the structure. The velocity of a dust particle has a maximum on the returning trajectory. In the point of the maximum the neutral drag force equals to the radial electric force. From this condition the charge is found.

As it was shown in a number of experiments [3,4] thermophoresis may have a great influence on the dust grains levitating in the discharge. In this work this influence was used to measure the value of the radial force acting on a dust grain in the discharge. To create the gradient of the gas temperature a 16 cm long heating wire was maintained on a tube parallel to its axis (see Fig. 1).

Five thermocouples were set to measure the temperature profile on the tube circumference $T_w(\phi)$, ϕ is the polar angle. Since the length of the heater is much longer than the size of a dusty plasma structure the temperature distribution in the volume of the discharge could be considered as two-dimensional. The temperature variation in this experiment is several degrees, so the dependence of the gas heat conductivity on the temperature may be neglected. Therefore the temperature distribution is determined by the stationary linear two-dimensional heat conductivity equation and boundary condition. Solving this equation we aquire the temperature distribution, which allows to calculate the thermophoretic force in each point inside the tube.

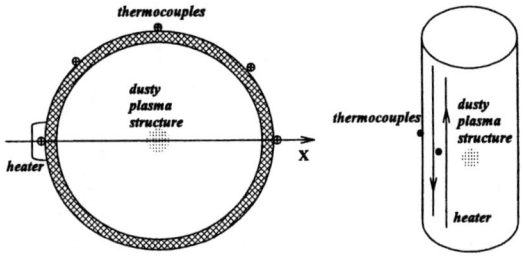

FIGURE 1. Scheme of the experiment on the thermophoretic influence.

The formula for the thermophoretic force acting onto a spherical particle with the radius a is given in [5]:

$$F_{th} = -\frac{8}{45}\sqrt{\pi}\, pa^2 l \frac{\Delta T}{T} \qquad (1)$$

where p is the neutral gas pressure and l is the mean free path of gas atoms. The experiments were done with vertical chains of dust particles. Fig. 2 shows the effect of thermophoresis on such a chain.

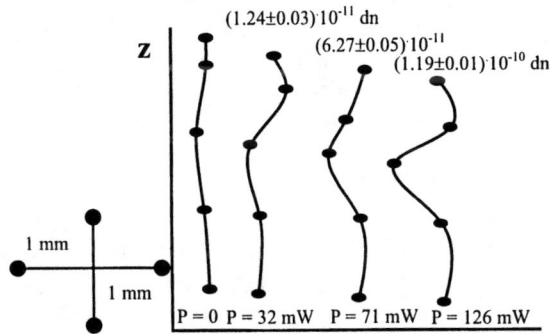

FIGURE 2. Influence of the thermophoresis on a vertical chain of dust grains. P denotes the power released in the heater, numbers above each configuration of a chain show the corresponding value of the thermophoretic force.

It is seen that the chain is distorted in such a way that the unknown interparticle force has the constituent in the radial direction for all the particles with the exception of the lowest one, which experiences only the radial trapping force and the thermophoresis. The value of thermophoresis is determined by the reconstructed temperature distribution and the position of a dust grain. In this way the radial trapping force profile is obtained.

Gas-dynamic influence was employed for the excitation of nonlinear waves in the dusty plasma structures. The cathode was made in the form of a hollow cylinder. Below the cathode a plunger was set into the tube. It was moved manually with the help of a magnet at a velocity of 30-40 cm/s. A grid kept under the floating potential was inserted 7 cm above the cathode. The discharge regime was chosen in such a way that the first striation was formed exactly below the grid (pressure 0.3 torr, current 0.1 mA). Upper edge of the dusty plasma structure was 4 mm separated from the grid. Melamine formaldehyde dust grains 1.03 μm diameter were used in these experiments.

After moving the plunger downward the structure was again for some time streaming downward, then it stopped and began moving towards its initial equilibrium and when it returned to the stable position a disturbance propagating through it appeared. The disturbance consists of three parts: first compression, rarefaction and the second compression. The amplitudes of the dust density for these three parts respectively are 1.6 and 1.3 respectively and 0.65 for the rarefaction. This means that the wave is strongly nonlinear. The velocities of the compressions are about 2.5 cm/s, but the second compression always runs a bit faster reaching the speed of 3 cm/s. The dust acoustic speed is estimated using the following formula [6]:

$$C_{da} = \sqrt{\frac{Z_d^2 T_i}{m_d} \frac{n_d}{n_i}}, \qquad (2)$$

where n_i, n_d are the ion and dust density, T_i is the ion temperature and Z_d, m_d are the dust grain charge and mass respectively. The value of the dust particle charge was calculated by the extrapolation of experimental dependences obtained in [2] and was assumed to lie in the range 400-750 electrons. The ion density is estimated to be equal to $4 \cdot 10^7$-10^8 cm^{-3}. The dust density also slightly changes inside the initial dusty plasma structure around the average value of $3 \cdot 10^4$ cm^{-3}. Thus C_{da} = 1.8-5.2 cm/s.

DUST-ACOUSTIC WAVE INSTABILITY IN INDUCTIVE GAS DISCHARGE PLASMA

A scheme of the experiment for research of a dust acoustic wave (DAW) instability in RF inductive discharge plasma and on visual diagnostics of dust wave parameters is presented on Fig. 3. The RF glow inductive discharge (100 MHz, ~1 W) was excited in a vertically oriented cylindrical glass tube of diameter 3 cm and length 65 cm with a help of two-rings inductor in neon. The measurements were performed at 5 pressures of a neon. Monodisperse melamine formaldehyde particles of 1.87±0.04 μm diameter were used. The wave parameters (phase velocity v_{DAWI}, frequency ω_{DAWI}, distribution of grain concentration $n_d(H)$ along vertical axis H) were measured with a help of the high speed CCD camera Redlake 500. The distributions of the background plasma parameters (electron density $n_e(H)$, electron temperature $T_e(H)$ and space potential $U_s(H)$) were measured by a single mobile Langmuir probe. The dust grain charge has been calculated numerically [7] for conditions of collisional plasma on the basis of measured plasma parameters. Note that all the data were measured in the region where wave amplitude was low to compare them with the DAW analytical model [8-10].

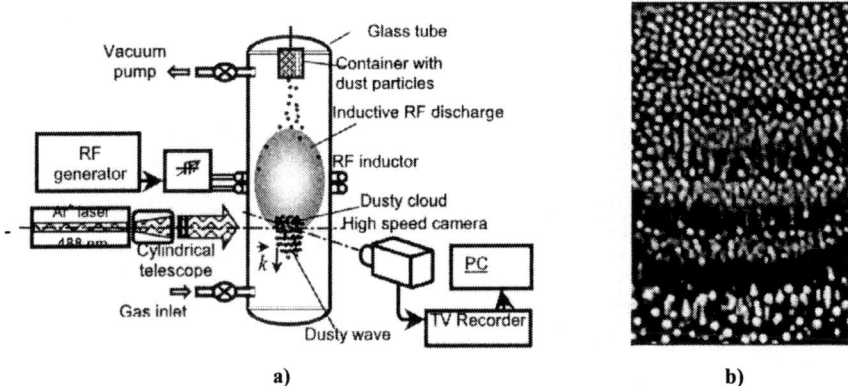

FIGURE 3. The scheme of experiment for investigation of the dust acoustic wave instability at the diffuse edge of RF inductive discharge (a); Single frame video image of DAW instability at 50 Pa of neon pressure (b).

As the observable phenomenon, DAW instability, represents a kind of traveling waves of a grain density, its theoretical interpretation was conducted within the framework of the DAW theory in a collisional plasma [8-10]. To find a dispersion relation $k = k(\omega)$ the Poisson equation was linearized with the $\delta\varphi \sim \exp(ikx-i\omega t)$ dependence for electrostatic potential:

$$k^2 \delta\varphi = 4\pi e(-\delta n_e + \delta n_i - Z_d \delta n_d - n_d \delta Z_d). \qquad (3)$$

The electron density perturbation δn_e was found in Boltzmann approximation. The ion density δn_i and dust density δn_d perturbations were found using the standard fluid approach [8-10]. In this case the dimensionless dispersion equation is

$$1 + \frac{\chi P}{1+P} + \tilde{k}^2 + \tilde{E}\tilde{k} = \frac{\tilde{k}^2 + \tilde{E}\tilde{k}(1+\chi)}{\tilde{\omega}(\tilde{\omega} + \tilde{\eta})}, \qquad (4)$$

where $\tilde{k} = kr_{Di}$, $\tilde{E} = eE_0 r_{Di}/T_i$, $\tilde{\omega} = \omega/\omega_{pd}$, $\tilde{\eta} = \eta/\omega_{pd}$, $P = Z_d n_d/n_e$, r_{Di} is the ion plasma Debye length, $E_0 \sim 4$ V/cm is the permanent electric field strength, $\omega_{pd} = 2eZ_{d0}(\pi n_d/m_d)^{-1/2}$ is the dusty plasma frequency, η is the viscosity coefficient for dust particles, $\chi(n_D/n_{e0})$ is the logarithmic derivative of Z_d on the base of n_e/n_i and $\delta Z_d = \chi Z_d$ ($\chi = 0 \div 0.33$). The numerical solutions of the Eq.4 with respect to $\tilde{k}_{Re} = \tilde{k}_{Re}(\tilde{\omega})$ and $\gamma = \gamma(\omega) = -k_{Im}(\omega)$ are presented on Fig. 4a,b.

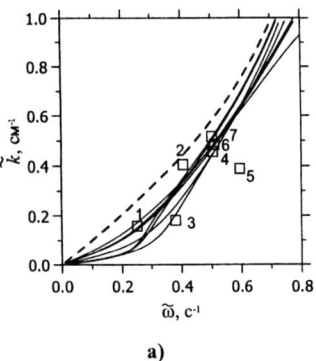
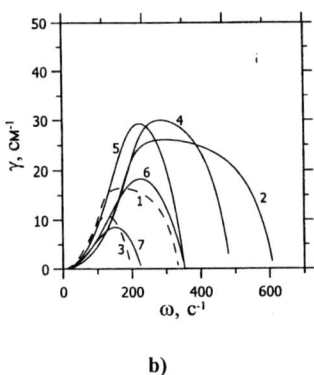

a)   b)

FIGURE 4. Theoretical (lines) and measured (squares) dispersion dependences. The digits near the squares mean the number of experiments. The dashed line is the calculated dispersion dependence for collisionless plasma. 7 solid lines are results of calculations by Eq. 4 for collisional dusty plasma for experimental data, $\chi = 0$ (a); Numerical solution of Eq. 4 relative the grow rate $\gamma(\omega)$ for the experimental conditions. The digits near the curves mean the number of experiments, $\chi = 0$ (b).

An examination of the data presented on Figs. 4 follows us to the next conclusions:
- the dispersion curves $\tilde{k}_{Re} = \tilde{k}_{Re}(\tilde{\omega})$ calculated by Eq. 4 fit the experimental data better then the "collisionless" approach [8], $\tilde{k}_{Re}(\tilde{\omega}) = \tilde{\omega}\sqrt{1 - \tilde{\omega}^2}$ (see Fig. 4);
- the frequencies ω_{max}, corresponding to maxims of calculated curves $\gamma_{max} = \gamma(\omega_{max})$, well correlate with the frequencies of instabilities measured at the experiment;
- the computational values $\gamma_{max} = \gamma(\omega_{max})$ well correlate with the DAW instabilities growing rates estimated at the experiment;
- the computational value of γ_{max} decreases as neutral gas pressure decreases; at a pressure is higher of 80 Pa the computational value γ_{max} becomes negative, as it is observed at experiment - DAW instabilities have not observed at $p > 60$ Pa;

- with other things being equal, the computational value of ω_{max} is proportional to n_d that have been observed in experiment;
- the availability of a constant electrical field E_0 is a necessary condition for a self-excitation of DAW instabilities, the forced movement of a cloud (by a passage of gas) up on 5 mm to the area with lower values E_0 ($\sim$ 1-2 V/cm) resulted in damping of DAW instabilities.

Thereby it is possible to conclude, that the considered DAW instability is the dust acoustic wave, amplified in a permanent electrical field of plasma ambipolar diffusion.

DYNAMICS OF PARTICLES UNDER MICROGRAVITY

In our work, results of an experimental study of diffusion of dust particles, charged by photoemission under microgravity are presented. The data were obtained during investigations of dusty plasma induced by solar radiation on "MIR" space station, which have shown that under the action of intensive solar radiation the micron-size particles can acquire considerable positive electric charges. The experimental study of dust diffusion was performed for bronze particles with the mean radii $a \approx 37.5$ μm in background gas (neon) at the pressure $P \approx 40$ Tor [11].

The initial dust concentration n_0 was varied from 195 to 300 cm^{-3}. The dependencies the relative dust concentration $n_p(t)/n_0$ on the time t are shown in Fig. 5. The photoemission charge of particles was obtained from the approximations of the curves $n_d(t)/n_d$ ($t=40$ s) at $t > 40$ s by the method detailed in [11] and was close to $Z \approx 4\ 10^4$ ($\pm 15\%$). Illustrations of simulation of dust transport due to the mutual Coulomb repulsion of particles by the molecular Brownian dynamic method (curve 1) and its analytical approximation (curve 2), which was used for determining dust charge [11], are presented in Fig. 5 for conditions close to experimental ones. Under the solar action, the dust motion acquired a directed motion forward the tube walls. For a time

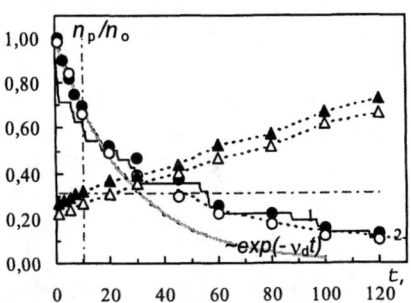

FIGURE 5. Dependencies of relative concentration n_p/n_0 (O; ●) and ratio of λ/R (Δ; ▲) versus time t for the different initial concentration n_0: (●; ▲) - 195 cm^{-3}; (O; Δ) - 300 cm^{-3}.

~ 3 s after the beginning of solar irradiation, the dust stochastic kinetic energy is increased, and the interparticle correlation changed. The pair correlation functions of particles obtained by the exclusion of interparticle distances less then $l_p/2$ are shown in Fig. 6.

Irregular fluctuations of velocity (V_x, V_y) of the separate particles on a background of their total drift motion reflect the dust temperature T, which for Maxwellian distribution can be obtained from:

$$T_{x(y)} = m_+ \{<V_{x(y)}^2> - <V_{x(y)}>^2\}, \qquad (5)$$

Determining of the dust temperature for the various experiments gives $T_x \cong 51$ eV, $T_y \cong 22$ eV, within 5%. Similar non-uniform distributions of stochastic kinetic energy

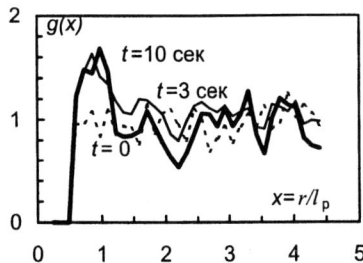

FIGURE 6. Pair correlation function $g(x)$ versus $x = r/l_p$ for several observation time t.

($T_x \neq T_y$) and "abnormal heating" were observed in a number of other dusty plasma experiments and can be related to the temporally-spatial fluctuations of dust charge, for example, due to the random nature of charging currents, or the spatial inhomogeneity of system [12-17].

As the considered system consists of the positively charged macroparticles and the photoelectrons with the density $n_e \sim Zn_d$ emitted by them, it is possible to assume, that the transport properties of this system will depend on ambipolar diffusion of the particles. Because of the large difference of mobility of electrons μ_e and dust μ_+, a negative surface charge appears on the tube walls. The incipient polarization electric field blocks further partitioning of the charged components. Therefore, the electrons and the heavy dust particles can diffuse "together" with some effective coefficient D_a of ambipolar diffusion [18]:

$$D_a = \{D_e \mu_+ + D_+ \mu_e\}/\{\mu_+ + \mu_e\}. \qquad (6)$$

Here D_e, D_+ are the free diffusion constants for electrons and particles. Then, in the case of $\mu_e \gg \mu_+$, the ratio of diffusion constants can be represented in the form:

$$D_a/D_+ \approx 1 + ZT_e/T_+ \tag{7}$$

With the measured temperatures (T_x, T_y), the ratios of diffusion constants can be estimated as $D_a/D_+^x \approx 0.8\text{-}1.6\ 10^3$, and $D_a/D_+^x \approx 1.8\text{-}3.6\ 10^3$ for $T_e = 1\text{-}2$ eV.

In a plasma with the density n, the diffusion have a ambipolar character when $\delta n = |n_e - n_+| \ll n \approx n_e \approx n_+$. For cylinder with the radius R, it is valid for $\delta n/n \approx (\lambda/R)^2 \ll 1$, where $\lambda^2 = T_e/4\pi e^2 n_e$ [18]. Taking into account that $n_e \approx Zn_p$ we have $\delta n/n \approx 2.1\text{-}6.4\ 10^{-2}$ for the considered initial conditions $n_p = n_o$ under the assumption of $T_e \cong 1\text{-}2$ eV. The dependencies of λ/R on time are shown in Fig. 5 for $T_e = 2$ eV.

Assuming that for $\delta n/n < 0.1$ the losses of charges in our experiments are connected with their ambipolar diffusion to the ampoule walls, the area of ambipolar diffusion can be determined from the mean velocity of diffusion losses of macroparticles

$$dn_p/dt = -n_d\, v_d, \tag{8}$$

where $v_d = D_a/\Lambda^2$ is the frequency of diffusion losses, and Λ is some diffusion length [18]. For cylinder with the radius R and the height $\sim 4R$, the value of $\Lambda \approx R/2$. The value of v_d can be obtained from the experimental curve $n_p(t)/n_o$ at $t < 10$ s, where the $n_p(t)/n_o$ function agrees well with the solution $n_d = n_o \exp(-v_d\, t)$ of Eq.(8) for $v_d \approx 0.0135$ s^{-1} (Fig. 5). Then, an estimate of ambipolar diffusion gives $D_a \approx 2\ 10^{-2}$ cm^2/s.

The free diffusion constants $D_+^{x(y)}$ can be retrieved from the measurements of dust temperature, and drift velocity $V_d^{x(y)}$:

$$D_+^{x(y)}(t) = \{\langle \Delta r(t)^2 \rangle - (V_d^{x(y)} t)^2\}/2t, \tag{9}$$

where $\langle \Delta r(t)^2 \rangle$ is the mean-square displacement of separate particles in the direction of axis OX (or OY). Thus, the values of free diffusion constants can be estimated as $D_+^x \approx 1.3\ 10^{-5}$ cm^2/s and $D_+^y \approx 5.7\ 10^{-6}$ cm^2/s. The value of ratio $D_+^x/D_+^y \approx 2.28$ agrees very well with the measured dust temperature, and both the measured ratios $D_a/D_+^x \approx 1538$, and $D_a/D_+^x \approx 3509$ are in agreement with the theoretical prediction.

ACKNOWLEDGMENTS

This work was supported in part by the Russian Foundation for Basic Research, Grant No. 00-02-81036 and No.00-02-17520, and by NWO Grant No. 047-008-013.

REFERENCES

1. Fortov, V.E., Nefedov, A.P., Torchinsky, V.M., et. al., *Phys. Lett. A,* **229**, 317 (1997).
2. Fortov, V.E., Nefedov, A.P, Molotkov, V.I., Poustylnik, M.Y., Torchinsky, V.M., *Phys. Rev. Lett.* **87**, 205002 (2001).
3. Jellum, G.M., Daugherty, J.E., Graves, D.B., *J. Appl. Phys.*, **69**, 6923 (1991).
4. Balabanov, V.V., Vasilyak, L.M., Vetchinin, S.P., et.al., *JET,P* **92**, 86 (2001).
5. Nitter, T., *Plasma Sources Sci. Technol.*, **5**, 93 (1996).
6. Shukla, P.K., Mamun, A.A., *Introduction to Dusty Plasma Physics*, IOP Publishing, London, 2001.
7. Zobnin, A.V., Nefedov, A,P., Sinel'shchikov, V.A., and Fortov, V.E., *JETP,* **91**, 554 (2000).
8. Rao, N.N, Shukla P.K., and Yu, M.Y., *Planet. Space Sci.*, **38**, 543 (1990).
9. Ivlev, A.V., Samsonov, D., Goree, J., and Morfill, G., *Phys. Plasmas,* **6**, 741 (1999).
10. Fortov, V.E., Khrapak, A.G., Khrapak, S.A., et al., *Phys. Plasmas,* **7**, 1374 (2000).
11. Fortov, V.E., Nefedov, A.P., Vaulina, O.S.,et al, *JETP,* **87**, 1087 (1998).
12. Vaulina, O.S., Nefedov, A.P., Petrov, O.F., and Fortov, V.E., *JETP,* **91**, 1063 (2000).
13. Vaulina, O.S., Khrapak, S.A., Nefedov, A.P., et al, *Phys. Rev. E*, **60**, 5959 (1999).
14. Quinn, R.A., and Goree, J., *Phys. Rev. E,* **61**, 3033 (2000).
15. Zhakovskii, V.V., Molotkov, V.I., Nefedov, A.P., et al, *JETP Lett.*, **66**, 392 (1997).
16. Thomas, H., and Morfill, G., *Nature*, **379**, 806 (1996).
17. Melzer, A., Homann, A.,and Piel, A., *Phys. Rev. E,* 53, 2757 (1996).
18. Raizer, Y.P., *The Physics of Gas Discharge* (Berlin: Springer Verlag, 1991).

Dusty Plasmas in the Ionosphere and its Environment

O. Havnes

Department of Physics, The Auroral Observatory, University of Tromsø, N–9037 Tromsø, Norway

Abstract. We consider dust in the upper part of the atmosphere (~70–200 km), with a special emphasis on dust in meteor trails and heads, and in the NLC (Noctilucent Clouds) and the radar phenomenon PMSE. We discuss our knowledge on the production processes of such dust and its physical state such as, e.g., its sizes, density, composition and charges.

INTRODUCTION

The presence of dust in the Earth's mesosphere and lower ionosphere observable as noctilucent clouds (NLC) and the 'smoke'-trails of meteors. While observations of the trails of meteors must be as old as mankind, the first reported observations of NLC are from 1885 [1]. It is not clear if the occurrence of strong and easily observable NLC at that occasion was a consequence of the gigantic Krakatoa volcanic eruption in 1883, or if this only led to more careful observations of the twilight sky (see [2] for a discussion of the history of NLC observations and the apparent lack of such before 1885). However, it seems clear that NLC's have increased in strength, occurrence rate and latitudinal coverage since they were first discovered [3,4,5]. This is probably mainly caused by an increased amount of water vapor in the mesosphere [4], so that more water ice particles are formed during the NLC-season. On the northern hemisphere this lasts from mid-May to mid-August. During this period the mesopause temperature can become as low as ~110°K compared to the average winter temperature of ~220°K [6].

In the same height region as the NLC appear, a radar phenomenon called Polar Mesospheric Summer Echoes (PMSE) has been observed since 1979. The PMSE occur in layers with thickness from several km and down to the radar height resolution of ~150 m and they have much internal structure and dynamics [7]. There is a strong correlation between NLC and PMSE [8].

The NLC and PMSE phenomena may be linked to the influx of meteors in several ways. Meteors probably deposit more than 100 tons of material per day into the Earth's atmosphere [9]. Part of the vaporized materials is apparently recondensing to form smoke particles [10,11] which in turn may act as seed particles for water vapor to condense on, and form NLC particles when the temperature becomes low enough. Meteor material is also deposited as free atoms and it is probable that ablation of meteoroids is the one and only source for the layers of free metals (e.g., Na, Fe, Ca, K) that exist in the atmosphere [12]. They are observed in broad permanent bands between 80 and 110 km and also in the narrow irregular layers, the so-called sporadic metal layers, which occur nearly always above 90 km (e.g., [13,14]. The metals may be accreted on to icy NLC dust particles and influence on their photoelectric properties and thereby their electric charging [15,16]. Molecules containing metals may also be part of the nucleation history for NLC particles [17].

METEOR TRAILS

More than 99% of the mass of the accreted matter from space come in sizes below one mm, of which a large portion is evaporated by frictional heating from the atmosphere. For fast meteors with incoming velocity 30 km/s, the ablation occurs approximately in the height range 90–105 km. In some cases, meteors show unexplained cometary and jet-like structures which starts to develop above 130 km and in one case as high as 196 km [18,19]. At $\sim$130 km the typical meteor trail due to ablation starts to develop. Meteors with velocities from 30 km/s and downwards ablate in the height region 75–90 km.

Meteor trails are divided into two classes: underdense and overdense. Underdense trails have electron densities that are small enough so that the local plasma frequency ω_p is less than the radar frequency ν_R (which is from 30 MHz – 50 MHz for a typical meteor radar but may be larger for other types of radars). Underdense trails are normally produced by the smallest and most numerous meteors of mass 10^{-6} kg. For larger meteors we may have $\omega_p > 2\pi\nu_R$ and the trail is overdense and the radar does not penetrate the beam. The underdense trails are being studied by radars and the Doppler shifts of the echoes from the ionized trails are used on a routine basis to measure the wind at the echo heights. By studying the weakening of the signals as the trails expand by diffusion one can find the ambipolar diffusion coefficient D [20]. For underdense trails the half-time for the decay is normally observed to be in the range 0.01 to 0.5 sec (e.g., [21]). Jones and Jones [22] and Jones [23] showed that $D \propto T/\rho$, where T is the atmospheric temperature and ρ its mass density. The measured temperatures have a large scatter but on the average they are estimated to have accuracies from 4 to 10 K in the height range $\sim$80 to 90 km [24]. Above 100 km anomalous cross-field diffusion can exceed the ambipolar diffusion [25] but below 100 km the large electron–neutral collision rate leads to equal ion and electron cross-field diffusion and no anomalous diffusion.

The possible effect of dust on the meteor trail diffusion has not been considered in any detail. It is normally assumed that ablation only leads to the release of atomic, e.g., metals and that the dust 'smoke' particles are formed as these recondences [11]. For the underdense trails, with their short lifetimes, the recondensation does not appear to be rapid enough to affect the diffusion. Rosinski and Snow [10] find that for comparatively large meteors of mass 12 mg, which would lead to overdense trails, the time to form smoke particles of $\sim$0.25 nm would be a few minutes if the trail had an initial diameter of 1 m. However, it is not unlikely that small dust particles are directly released from the meteor during its travel through the atmosphere. For the large meteors at least, visual observations show that smoke trails are immediately apparent after the passage of the meteor. Some observations [18,26] indicate that some meteors should consist of metallic or silicate grains bounded by materials of low boiling point. Such porous meteors may have a density as low as 0.7 gr/cm^3 [27].

There are some observations which may indicate that dust, when it is present in the trails, is of importance for the diffusion of the trail and for its radar scattering. Kelley et al. [28] may, during a radar and rocket campaign on PMSE in 1983, have obtained not only radar but also rocket observations of a long-lived meteor trail. Even though the meteor was estimated to have a mass of the order of 100 gr they conclude that the trail was underdense but still observable for over 6 min by a 50 MHz radar. The scatter from underdense trails could not explain the observations over such a long time and they speculate if dust could be responsible for conserving short-scale length electron turbulence in the trail by reducing the diffusion rate down close to that of the dust.

Observations of radar echoes from the head of meteors by the Arecibo 430 MHz radar [29] give the strongest echoes when the meteor is travelling down the beam, in contrast to classical meteor radar scatter which is strongest when it travels normal to the beam [30]. This unusual behavior may be the result of some instability (see also [31]) and Zhou and Kelley [29] speculate if dust may be involved.

The theory for scattering from underdense trails as observed by meteor radars at frequencies (~15 to 50 MHz), appear to be in a comparatively good shape although further work on the diffusion coefficient at different conditions should be done (e.g. [23]). At UHF frequencies the situation is not satisfactory. As discussed by Zhou and Kelley [29] the scattering cannot be described by either the underdense or the overdense scattering or the so-called rough-trail scattering theory [32]. The overdense and the rough-trail scattering is not effective enough. Incoherent scattering is also not the dominant scattering mechanism although it may play some role [33]. The alternative explanation appears to be that one or more instabilities are at work near the meteor head and in their trails. We do not know the nature of such instabilities or if dust is involved as suggested by Zhou and Kelley [29] and Kelley et al. [28].

Rosenberg [34] and Rosenberg and Shukla [35,36] consider the ability of charged dust to produce gradient-drift and Hall current instabilities. The high neutral background density will normally prevent the instabilities unless one is well above 90 km height, such as is the case for some plasma instabilities [24]. They find that the gradient-drift instability [36] does not go unstable for small enough wavelengths to influence the UHF scattering and that the growth rate for the Hall current instability [36] appears to be too long for most meteor trails.

PRODUCTION OF A GENERAL BACKGROUND OF DUST PARTICLES

There is little doubt that there must be some general background of small dust particles in the upper mesosphere and lower thermosphere where the meteors ablate and presumably recondense in many much smaller particles. There are, however, very large uncertainties both as regards the size distribution, number density and chemical composition of the dust produced by recondensation. It is generally assumed that recondensation leads to the formation of many small smoke particles. The resulting steady state number density of the background near the mesopause is estimated to be well above 10^3 cm^{-3} if their sizes are $r_d \sim 1$ nm and more than 10^4 cm^{-3} if $r_d \sim 0.5$ nm [11]. Such estimates should represent upper limits since parts of the ablated material should react with atmospheric species and not recondense as dust particles. If incoming meteors are of a fractal nature they have a much higher probability of surviving their entry into the atmosphere than a compact meteor of spherical structure and will therefore not contribute as much to the production of smoke particles [37]. Analysis of light curves of meteors indicate that at least some are not solid bodies but have a dust-ball structure

[19] with fluffy structures of densities $\sim$0.7 gr/cm^3 [18,26]. We do not know how far down in size we can have a fluffy structure but Rietmeijer et al. [38] find that the principal constituent particles of interplanetary dust particles have sizes >100 nm although condensation experiments give dust sizes 10–30 nm. Mateshvili et al. [39] may have observed a layer of dust at a height of 117 km following the Leonid meteor showers. They find sizes of $\sim$10 nm and number densities of $\sim$30 cm^{-3} for such a layer.

There may be some controversy between the theoretical production of smoke particles as a function of height [11,37], which peaks below 90 km, and the observed height distribution of meteors detected by different radars. A typical meteor radar operating at 35 MHz gives the peak of meteors at $\sim$90 km [40] while the 430 MHz Arecibo radar finds the peaks to be close to 100 km height [29]. The true height occurrence may well be somewhat between these heights which may require a different model for the bulk of the meteors than a solid body.

In situ observations of small dust in the tropical mesosphere [41] give number densities of no more than a few times 10 cm^{-3} if each small dust particle carry a charge of 1e. They observed a $\sim$5 km thick layer of positive dust from 90 km height and upwards with, apparently a thin layer of negative dust at the bottom of the layer.

NLC AND PMSE DUST

For a thorough discussion of the characteristics of the PMSE phenomenon we refer to the review of Cho and Röttger [7]. The PMSE radar phenomenon was first discovered with an MST radar at 50 MHz by Ecklund and Balsley [42] and has been the subject of intense radar studies at wavelengths up to more than a GHz and at northern latitudes from $\sim$52°N [43,44] to 78°N [45,46]. At low latitudes PMSE (or MSE) is observed $\sim$10% of the time at 53.5 MHz and only during sunlit conditions [44]. At polar latitudes of 69°N and more PMSE at 53.5 MHz is, in the midst of the season (mid-June), present most of the time [45,47]. The PMSE in the southern hemisphere is much weaker than on the northern hemisphere [48].

There is a strong frequency dependence of the backscattered power [46,49]. Although direct comparison between radars are difficult because of differences in scattering volume and since radars are seldom collocated, it appears that the backscatter at 500 MHz – 1 GHz may be 5–6 orders of magnitude less than at 50 MHz. The difference between 50 MHz and 224 MHz [49] is not so dramatic. At 224 MHz and below the PMSE signal totally dominates over the background incoherent signal. At 500 MHz and more this is not so and a weak PMSE signal may therefore often be masked by the incoherent signal [46].

The PMSE radar layers are very structured. Rocket measurements of electron structure (e.g, [50]) and dust structures [51] reveal structuring down to a meter or less. At 78°N the PMSE at 53.5 MHz have a much more pronounced double layer structure than at only 10° more south [45]. The layers have a tendency to sink, consistent with falling dust, and their overall dynamic is often dominated by wave motion (e.g., [7,45]) with periods $\sim$10–25 min to long-period and tidal waves.

PMSE was originally thought to be an indicator for the presence of strong neutral turbulence but such is not always present [52]. Kelley et al. [53] suggested that the presence of a sufficient number of charged dust particles, or large water-cluster ions, would reduce the plasma diffusion so that plasma inhomogeneities could be maintained for a sufficiently long time to explain the PMSE at longer wavelengths. According to Cho et al. [54] this requires that the charge density on dust must be comparable to that on the electrons. With such an amount of dust present there should be a corresponding reduction in the electron density, i.e. an electron bite-out. This is quite often observed and has been taken as a confirmation that dust, sometimes subvisual, was present. However, Cho et al. [54] note that on many occasions with strong PMSE there was no bite-out, indicating that little dust was present. This was confirmed by Havnes et al. [55,56] who found that even when the charge density on dust was only $\sim 5\%$ of that due to electrons, PMSE was present. It therefore seems evident that the slowed-down plasma diffusion is not a necessary condition for the PMSE phenomenon.

We do not know the nature of the NLC/PMSE dust particles. It seems overwhelmingly likely that their main constituent must be water ice but we do not know to what extent this is mixed up with other materials.

During NLC conditions, when the NLC/PMSE dust particles are large enough to be visible by lidars; 3-color lidar measurements gave an average dust size of 51 ± 21 nm [57]. With PMSE but no NLC present, so that any dust particle layer must consist of subvisual dust, rocket measurements found charge densities of up to ~ 4000 e/cm^3 probably consisting mainly of particles of radius 10–20 nm, of one to two unit charge each [51,58]. It is problematic to measure small dust particles of the order of 1 nm. They will be invisible to lidars. Also, the flow of atmospheric gas around a moving rocket will exert a drag force which may deflect them wholly or partly away from the dust probe.

A Gerdien condenser, which measures the mobility of ions, has on several occasions been flown through NLC/PMSE layers [59,60]. Mitchell et al. [59] interpret their observations of low-mobility positively charged particles to be due to small dust (aerosol) particles of masses $\sim 10^4$ amu or possibly more. This corresponds to a radius of more than 1 nm if $\rho_d \sim 1$ gr cm^{-3}. Croskey et al. [60] on another flight, on which Mitchell et al. [61] report the presence of negatively charged aerosols, measure two types of (heavy) positive ions, the one with masses ~ 2200 amu and the other group probably more than one order of mass higher. This would, again with $\rho_d \sim 1$ gr cm^{-3}, correspond to sizes of ~ 1 nm and ~ 6 nm respectively. They find that densities in the PMSE region are up to several thousand per ccm.

If the presence of such small positively charged dust particles is a general phenomenon in the mesopause region this will have implications for our understanding of the dust charging and the total dust charge density on dust there. The coagulation of dust particles will also be greatly enhanced in an environment consisting of small but numerous positive dust particles, and larger negatively charged particles [16,62]. It will be a challenge to explain the existence of small and positively charged dust particles coexisiting with larger and negatively charged particles.

On some occasions the dust near the mesopause can be dominantly positively charged [41,51,63] most likely because of a difference in dust material composition or surface composition from that of pure water ice. This can lower the workfunction W of the particles so that the photoelectric effect will dominate the charging [15,64]. Eidhammer and Havnes [16] suggests that a change of surface composition may come about as the dust temperature increase relative to the ambient gas, if the dust particle size grows [65]. At a sufficiently high temperature water vapor will no longer condense on the dust particle while other materials, e.g. metals such as Na, Fe, etc., still can, and will therefore coat the dust particles. With different assumptions on composition of dust particles, the most extreme being particles of pure Na with workfunction of $W = 2.3$ eV, Rapp and Lübken [58] found that dust charges should not exceed $\sim +4e$. However, it is not improbable that a coating with e.g. Na could lower the workfunction considerably. Qiu et al. [64] find a workfunction of only 0.8 eV when Na is codeposited with NH_3 at low temperatures. For small dust particles the photoelectric yield may also be close to 1 (e.g., [66,67]. Burtscher et al. [68] state that traces of contaminants absorbed at the surface of dust particles can alter their photoelectric yield by orders of magnitude. If we accept the positive charge density in an NLC/PMSE layer measured by Havnes et al. [51] of up to $\sim 7100e$ cm^{-3} and use the average particle density of 82 ± 52 cm^{-3} for NLC's visible by a 3-color lidar system [57], we find an average dust charge number of $Z_d \sim 87$. This value should, however, be treated with caution since the positive dust measurements of Havnes et al. [51] are modulated by secondary charging effects due to the dust impacts (e.g., [69]).

The question on how PMSE is formed is far from solved. Turbulence may play a role at lower frequencies but on many occasions the turbulence is wholly or partly absent in PMSE layers [52,70,71]. At the high frequencies from $\sim$500 MHz to 1 GHz and above, where the PMSE is weak and rare turbulence can hardly be of importance since this would require turbulence to be present at a half wavelength which is from 0.3 m to less than 0.15 m for the above frequencies. The narrow spectral widths observed in most of the PMSE's, sometimes down to the resolution limit of 1 Hz for the EISCAT 224 MHz radar also exclude turbulence as the dominant scatter mechanism. At the high frequencies it seems natural to assume that dust must play a much more direct role than just conserving electron turbulence created by the neutral gas. Scattering by 'dressed' dust particles [15,72–74] can only contribute at the highest frequencies ($\sim$1 GHz) and only in the presence of dust with large charges. Any plasma dust instability will have problems with the large damping caused by the neutral gas. A class of instabilities which has not yet been fully explored are related to the charging process themselves [75,76] although it is yet not clear to what degree these are affected by collisions with the neutral gas.

Finally, we present 3 recent observations on the PMSE phenomenon. Their consequences have not yet been fully understood but the results must certainly be clues to the nature of the PMSE phenomenon.

   i) *The effect of artificial heating of the electron gas at the PMSE height*
   Chilson et al. [77] and Belova et al. [78] observed PMSE at 224 MHz with the VHF EISCAT radar at Tromsø, Norway, while operating the EISCAT heating facility [79] which was heating the electron gas at the PMSE height. The electron temperature, which for natural PMSE conditions is identical to the neutral gas temperature of typically less than $\sim$160°K [80] was raised to $\sim$2000 to 3000°K by the heater. The reaction of the PMSE was height dependent and varied from one case to another. The height dependence may be related to differences in electron temperature enhancements with height. In one extreme case (the heater was run in a sequence with 20 sec on, followed by 20 sec off), the PMSE practically disappeared within a few seconds after the heater was turned on and reappeared within a few seconds after the heater was turned off. In other cases the decrease was much less.

ii) *Correlation between dust charge density and PMSE strength*

The first dust and plasma measurements by a rocket within a radar beam during a PMSE event revealed a close correlation between the dust charge density and the backscatter signal [55,56]. The most surprising result was, however, that strong PMSE was present even when the dust charge density was very low.

The dependence of the PMSE backscatter strength on the value of $P = N_d Z_d / n_{e0}$ (where N_d is the dust density and Z_d the dust charge number, while n_{e0} is the electron density without dust being present) was found to be a rapid increase, probably starting already as soon as dust appears, which indicate that there was no threshold for an onset of the PMSE effect in this case. There is a clear PMSE signal already at $P \sim 0.05$. At $P \sim 0.2$–0.3 there is a flattening of the curve. Since the maximum of the P value in this case was only $P_{MAX} \sim 0.3$ (there is some uncertainty because of an uncertainty in n_{e0}) we could not map the whole relationship. Results from another flight [51] must be regarded with some caution since the rocket passed the PMSE layer at a distance of ~ 20 km from the radar beam, so we cannot be sure that the PMSE structure was sufficiently similar. If they are, it appears that the radar signal becomes low when P0.8. This seems natural if electrons are essential for the scattering, since they will mainly be locked on to grains when the P is approaching 1.

iii) In one experiment by Röttger and Kubo [81] they observed a total of 4 hours with the EISCAT Sousy Svabard Radar (SSR – at 53.5 MHz) and the EISCAT Svalbard Radar (ESR – at 500 MHz) in the period 15–16 July, 2002. The radars are collocated at Longyearbyen, Svalbard at 78°N. The height resolution of ESR was 900 m with an antenna beam width of 1.6 degrees while the corresponding numbers for the SSR were 300 m and 4 degrees. In general agreement with Hall and Röttger [46] they find that PMSE is rare at the ESR frequency compared to at the SSR frequency and that there can be much and large variations in the signal strength at 53.5 MHz without any detectable signal being seen by the ESR. However, on some occasions there are events of backscatter power enhancement for the ESR lasting of the order of 10–15 min. They are not associated with especially strong, peculiar features at 53.5 MHz so there is obviously no strong correlation between the signals at the two frequencies. In one case a strong event at 500 MHz at 13:40 UT, on the 16th of July, does not seem to have a clear counterpart at 53.5 MHz.

CONCLUSION

We have discussed the two main situations where dust particles exist and may play a decisive role. These are the meteor trails, including the head of the meteor, and the other is the NLC/PMSE phenomena.

We have no well established and confirmed theories for the effect of dust on the radar scattering, from either the meteor trails or heads, or from the PMSE. While it seems reasonable that the reduced plasma diffusion due to the presence of charged dust particles, or heavy ion clusters [82], must operate in some cases it appears that this cannot be a general explanation for the PMSE. There are PMSE cases where no strong turbulence is present and also cases where very little dust appears to be present. The general view is also, in accordance with this, that there must be more than one mechanism which can give rise to PMSE.

REFERENCES

1. Backhouse, T. W., *Meteorol. Mag.* **20**, 133 (1885).
2. Gadsden, M., and Schröder, W., *Noctilucent Clouds*, Springer- Verlag, New York, 1989, pp. 165.
3. Gadsden, M., *J. Atmos. Terr. Phys.* **52**, 247–251 (1990).
4. Thomas, G. E., Olivero, J. J., Jensen, E. J., Schröder, W., and Toon, O. B., *Nature* **338**, 490–492 (1989).
5. Thomas, G. E., *Rev. Geophys.* **29**, 553 (1991).
6. von Zahn, U., and Meyer, W., *J. Geophys. Res.* **94**, 14,647–14,651 (1989).
7. Cho, J. Y. N., and Röttger, J., *J. Geophys. Res.* **102**, 2001–2020 (1997).
8. von Zahn, U., and Bremer, J., *Geophys. Res. Lett.* **26**, 1521–1524 (1999).
9. Love, S., and Brownlee, D., *Icarus* **89**, 28–43 (1990).
10. Rosinski, J., and Snow, R. H., *J. Meteor.* **18**, 736–745 (1961).
11. Hunten, D. M., Turco, R. P., and Toon, O. B., *J. Atmos. Sci.* **37**, 1342 (1980).
12. von Zahn, U., Gerding, M., Höffner, J., McNeil, W. J., and Murad, E., *Meteor. Planet. Sci.* **34**, 1017–1027 (1999).
13. Clemesha, B. R., *J. Atmos. Terr. Phys.* **57**, 725–736 (1995).
14. Gerding, M., Alpers, M., Höffner, J., and von Zahn, U., *Ann. Geophysicae* **19**, 47–59 (2001).
15. Havnes, O., de Angelis, U., Bingham, R., Goertz, C. K., Morfill, G. E., and Tsytovich, V., *J. Atmos. Terr. Phys.* **52**, 637–643 (1990).
16. Eidhammer, T., and Havnes, O. *J. Geophys. Res.* **106**, 24,831–24,841 (2001).
17. Cox, R. M., and Plane, J. M. C., *J. Geophys. Res.* **103**, 6349–6359 (1998).
18. Spurny, P., Betlem, H., Jobse, K., Koten, P, van't Leven, J., *Meteor. Planet. Sci.* **35**, 1109–1115 (2000).
19. Campbell, M. D., Brown, P. G., LeBlanc, A. G., Hawkes, R. L., Jones, J., Worden, S. P., and Correll, R. R., *Meteor. Planet. Sci.* **35**, 1259–1267 (2000).
20. Ceplecha, Z., Borovicka, J., Elford, W. G., Revelle, P. O., Hawkes, R. L., Porubcan, V., and Simek, M., *Space Sci. Rev.* **84**, 327–471 (1998).
21. Hocking, W. K., Thayaparan, T., and Jones, J., *Geophys. Res. Lett.* **24**, 2977–2980 (1997).
22. Jones, W., and Jones, J., *J. Atmos. Terr. Phys.* **52**, 185–191 (1990).
23. Jones, W., *Ann. Geophysicae* **13**, 1104–1106 (1995).
24. Hocking, W. K., *Geophys. Res. Lett.* **21**, 3297–3300 (1999).
25. Dyrud, L. P., Oppenheimer, M. M., and vom Endt, A. F., *Geophys. Res. Lett.* **28**, 2775–2778 (2001).
26. Borovicka, J., Stork, R., and Bocek, J., *Meteor. Planet. Sci.* **34**, 987–994 (1999).
27. Rietmeijer, F. J. M., in *Planetary Materials,* ed. J. J. Papike, Rev. in Mineralogy, 36, 1998, pp. 1–95.
28. Kelley, M. C., Alcala, C., and Cho, J. Y. N., *J. Atmos. Solar–Terr. Phys.* **60**, 359–369 (1998).
29. Zhou, Q. H., and Kelley, M. C., *J. Atmos. Solar–Terr. Phys.* **59**, 739–752 (1997).
30. Evans, J. V., *J. Geophys. Res.* **71**, 171–188 (1966).
31. Chapin, E., and Kudeki, E., *Geophys. Res. Lett.* **21**(22), 2433–2436 (1994).
32. McIntosh, B. A., *J. Atmos. Terr. Phys.* **24**, 311–315 (1962).
33. Pellinen–Wannberg, A., and Wannberg, G., *J. Geophys. Res.* **99**, 11,379 (1994).
34. Rosenberg, M., *IEEE Trans. Plasma Sci.* **29**, 261–266 (2001).
35. Rosenberg, M., and Shukla, P. K., *J. Geophys. Res.* **105**, 23,135–23,139 (2000).
36. Rosenberg, M., and Shukla, P. K., Manuscript (2002).
37. Kalashnikova, O., Horániy, M., Thomas, G. E., and Toon, O. B., *Geophys. Res. Lett.* **27**, 3293–3296 (2000).
38. Rietmeijer, F. J. M., Noth, J. A., and Karner, J. M., *Astrophys. J.* **527**, 395–404 (1999).
39. Mateshvili, N., Mateshvili, I., Mateshvili, G., Gheondijan, L., and Kapanadze, Z., *Earth, Moon Planets* **82–83**, 489–504 (2000).
40. Hocking, W. K., Fuller, B., and Vandepeer, B., *J. Atmos. Solar– Terr. Phys.* **63**, 155–169 (2001).
41. Gelinas, L. J., Lynch, K. A., Kelley, M. C., Collins, S., Baker, S., Zhou, Q., and Friedman, J. S., *Geophys. Res. Lett.* **25**, 4047–4050 (1998).
42. Ecklund, W. L., and Balsley, B. B., *J. Geophys. Res.* **86**, 7775–7780 (1981).
43. Reid, I. M., Czechowsky, P., Rüster, R., and Schmidt, G., *Geophys. Res. Lett.* **16**, 135–138 (1989).
44. Latteck, R., Singer, W., and Höffner, J., *Geophys. Res. Lett.* **26**, 1533–1536 (1999).

45. Rüster, R., Röttger, J., Schmidt, G., Czechowsky, P., and Klostermeyer, J., *Geophys. Res. Lett.* **28**, 1471–1474 (2001).
46. Hall, C. M., and Röttger, J., *Geophys. Res. Lett.* **28**, 131–134 (2001).
47. Hoffmann, P., Singer, W., and Bremer, J., *Geophys. Res. Lett.* **26**, 1525–1528 (1999).
48. Balsley, B. B., Woodman, R. F., Sarango, M., Rodríguez, R., Urbina, J., Ragaini, E., Carey, J., Huaman, M., and Giraldez, A., *J. Geophys. Res.* **100**, 11,685–11,693 (1995).
49. Hoppe, U.-P., Fritts, D. C., Reid, I. M., Czechowsky, P., Hall, C. M., and Hansen, T. L., *J. Atmos. Terr. Phys.* **52**, 907–926 (1990).
50. Blix, T. A., and Thrane, E. V., *Geophys. Res. Lett.* **20**, 2303–2306 (1993).
51. Havnes, O., Trøim, J., Blix, T., Mortensen, W., Næsheim, L. I., Thrane, E., and Tønnessen, T., *J. Geophys. Res.* **101**, 10,829–10,847 (1996).
52. Lübken, F.-J., Rapp, M., Blix, T., and Thrane, E., *Geophys. Res. Lett.* **25**, 893–896 (1998).
53. Kelley, M. C., Farley, D. T., and Röttger, J., *Geophys. Res. Lett.* **14**, 1031–1034 (1987).
54. Cho, J. Y. N., Hall, T. M., and Kelley, M. C., *J. Geophys. Res.* **97**, 875–886 (1992b).
55. Havnes, O., Aslaksen, T., and Brattli, A., *Phys. Scripta* **T89**, 133–137 (2001a).
56. Havnes, O., Brattli, A., Aslaksen, T., Singer, W., Latteck, R., Blix, T., Thrane, E., and Trøim, J., *Geophys. Res. Lett.* **28**, 1419–1422 (2001b).
57. von Cossart, G., Fiedler, J., and von Zahn, U., *Geophys. Res., Lett.* **26**, 1513–1516 (1999).
58. Rapp, M., and Lübken, F.-J., *Earth Planets and Space* **51**, 799–807 (1999).
59. Mitchell, J. D., Walter, D. J., Croskey, C. L., and Goldberg, R. A., *Proceedings 12th ESA Symposium on Rocket and Balloon Programmes and Related Research*, Lillehammer, Norway, 29 May – 1 June 1995, ESA SP-370, 1995, pp. 95–100.
60. Croskey, C. L., Mitchell, J. D., Friedrich, M., Torkar, K. M., Hoppe, U.-P., and Goldberg, R. A., *Geophys. Res. Lett.* **28**, 1427–1430 (2001).
61. Mitchell, J. D., Croskey, C. L., and Goldberg, R. A., *Geophys. Res. Lett.* **28**, 1423–1426 (2001).
62. Reid, G. C., *Geophys. Res. Lett.* **24**, 1095–1098 (1997).
63. Smiley et al., submitted to *J. Geophys. Res.* (2002).
64. Qiu, S. L., Lin, C. L., Jiang, L. Q., and Strongin, M., *Phys. Rev. B* **39**, 1958–1961 (1989).
65. Grams, G., and Fiocco, G., *J. Geophys. Res.* **82**, 961– 966 (1977).
66. Gail, H.-P., and Sedlmayr, E., *Astron. Astrophys.* **86**, 380 (1980).
67. Schleicher, B., Burtscher, H., and Siegmann, H. C., *Appl. Phys. Lett.* **63**, 1191–1193 (1993).
68. Burtscher, H., Schmidt–Ott, A., and Siegmann, H. C., *Z. Phys. B – Condensed Matter* **56**, 197–199 (1984).
69. Tomsic, A., Pettersson, J. B. C., Gumbel, J., and Witt, G., submitted to *J. Aerosol Sci.* (2002).
70. Lübken, F.-J., Lehmacher, G., Blix, T. A., Hoppe, U.-P., Thrane, E. V., Cho, J. Y. N., and Swartz, W. E., *Geophys. Res. Lett.* **20**, 2311– 2314 (1993).
71. Ulwick, J. C., Kelley, M. C., Alcala, C. M., Blix, T. A., and Thrane, E. V., *Geophys. Res. Lett.* **20**, 2307–2310 (1993).
72. Tsytovich, V. N., de Angelis, U., and Bingham, R. J., *J. Plasma Phys.* **42**, 429 (1989).
73. Hagfors, T., *J. Atmos. Terr., Phys.* **54**, 333–338 (1992).
74. La Hoz, C., *Phys. Scripta* **45**, 529–534 (1992).
75. Li, F., *Chinese Phys. Lett.* **19**, 214–216 (2002).
76. Tsytovich, V., and Havnes, O., this issue (2002).
77. Chilson, P. B., Belova, E., Rietveld, M. T., Kirkwood, S., and Hoppe, U. P., *Geophys. Res. Lett.* **27**, 3801–3804 (2000).
78. Belova, E., Chilson, P., Rapp, M., and Kirkwood, S., *Adv. Space Res.* **28**, 1077–1082 (2001).
79. Rietveld, M. T., Kohl, H., Kopka, H., and Stubbe, P., *J. Atmos. Terr. Phys.* **55**, 577–599 (1993).
80. Lübken, F.-J., Fricke, K.-H., and Langer, M., *J. Geophys. Res.* **101**, 9489–9508 (1996).
81. Röttger, J., and Kubo, K., Manuscript (2002).
82. Kelley, M. C., and Ulwick, J. C., *J. Geophys. Res.* **93**, 7001 (1988).

Dusty Plasmas in the Solar System

Mihály Horányi

*Laboratory for Atmospheric and Space Physics, and Department of Physics
University of Colorado, Boulder, CO 80309-0392, USA*

Abstract. Dust - plasma interactions play an important role in our solar system. Comets, planetary rings, exposed dusty surfaces, the zodiacal dust cloud are all examples where dusty plasma effects shape the size and spatial distribution of small dust particles. Simultaneously, dust is often responsible for the composition, density and temperature of its plasma environment. The dynamics of charged dust particles can be surprisingly complex, fundamentally different from the well understood limits of gravity dominated motion (vanishing charge-to-mass ratio) or the adiabatic motion of electrons and ions. Here we focus on observations that are best explained via charging effects on single particle motion and point to possible future observations where collective dusty plasma effects are expected to be important.

INTRODUCTION

Dusty plasmas represent the most general type of space plasmas. In fact, it is difficult to find any plasma environment in our Solar System that is free of dust particles - perhaps the interior of the Sun is the only exception. Comets, asteroids, interplanetary space, and planetary magnetospheres are all examples where dust particles are present, possibly forming a true 'dusty plasma' where dust particles exhibit collective behavior or a 'dust in a plasma' environment, where dust grains can be treated as test particles. Here the term dusty plasma is used to describe any environment where electrostatic charging of dust particles has physical or dynamical consequences, and only recent or rarely mentioned results are discussed. There are a number of general reviews available on this subject [1-7].

Dust particles collect electrostatic charges in many ways [8,9]. In space, electron and ion collection, secondary and photoelectron emissions, sputtering and triboelectric charging are the most common processes. At comets, for example, both positively and negatively charged dust grains coexist with electrons and ions from a large number of parent atoms and molecules. On the surfaces of asteroids or the Moon, photoelectron production dominates, resulting in a plasma environment consisting mainly of electrons and positively charged dust. The electric field generated in photoelectron sheaths above UV illuminated surfaces is thought to explain the lunar horizon glow, observed by the *Surveyor* 5, 6, and 7 spacecraft (Figure 1)[10-14].

FIGURE 1. Composite image taken by the *Surveyor* spacecraft on the surface of the Moon after sunset. The circle (S.D.) marks the position of the solar disk below the horizon. The line of light along the western horizon is likely due to forward scattered light by a cloud of characteristically micron sized dust grains, electrostatically levitated above the surface.

Charging due to dust-dust collisions or just contact between dust grains and surfaces alone can lead to significant charging. If collisions occur between particles (even if identical in composition) the larger of the two will more likely end up with a few missing electrons that are carried away on the smaller partner. Atmospheric winds can enforce size sorting due to drag, as small grains get entrained easier, leaving large ones behind. This will result in the buildup of an electric field that is limited only by the breakdown of the atmosphere, leading to electrostatic discharging (Figure 2)[15]. Sand storms and volcanic plumes on Earth are examples where triboelectric charging is thought to cause lightning and is expected to be frequent during massive sand storms on Mars [16].

Charged dust particles become coupled to electric and magnetic fields and can exhibit collective behavior. In addition to all the forces acting on uncharged dust particles (gravity, drag due to collisions and radiation pressure), charged grains respond to electromagnetic forces and are more tightly coupled to plasma flows.

FIGURE 2. Images of Sakurajima Volcano (Japan) during the day (left) and night (right) during a period of explosive eruptions in 1991. The lightning discharges are most likely due to the separation of negatively charged smaller grains from their oppositely charged bigger brothers due to drag in the gas plume.

CIRCUMPLANETARY DUST

The equation of motion of a charged dust grain (of mass m and charge Q) using an inertial coordinate system fixed to a planet's center, is

$$\ddot{\mathbf{r}} = \frac{Q}{m}(\dot{\mathbf{r}} \times \mathbf{B} + \mathbf{E_c}) - \frac{\mu}{r^3}\mathbf{r} , \qquad (1)$$

where $\mathbf{r}$ is the grain's position vector, $\mathbf{B}$ is the magnetic field, and μ equals the gravitational constant times the planet's mass. For an infinite conductivity magnetosphere that rigidly corotates with the planet with a rotation rate of Ω, $\mathbf{E_c} = (\mathbf{r} \times \Omega) \times \mathbf{B}$ is the co-rotational electric field. On the right hand side there could be other terms describing forces due to, for example, satellites, higher order terms in the planet's gravitational field, radiation pressure, and plasma and neutral drag forces. As a grain traverses the various plasma regions its charge will not stay constant, but will vary according to the current balance equation

$$\frac{dQ}{dt} = \sum_i I_i , \qquad (2)$$

where I_i represent electron and ion thermal currents, as well as the secondary and photoelectron emission currents. These are all functions of the plasma parameters, material properties, size, velocity and also the instantaneous charge of a dust particle. For very small nm sized grains, (2) has to be used with some care to recognize the quantized nature of electrostatic charges. Equations (1) and (2) can be integrated simultaneously using numerical techniques such as a variable step size Runge-Kutta method.

Stochastic charge variations due to the fluctuating nature of the charging process lead to diffusion of dust grain orbits. Systematic charge variations due to gradients in plasma parameters and/or the modulation of currents, due to the modulation of the relative speed between the grains and the plasma flow, lead to dust transport within, ejection from, and capture of grains into planetary magnetospheres.

Surprisingly, there is a great deal of new physics to be learned even with the assumption of a constant charge on a dust grain orbiting a perfectly spherical planet with a rigidly co-rotating, centered and aligned dipole magnetic field [17-22]. This is an extension of the classical Stöermer problem to include the co-rotational electric field and gravity. For example, Figure 3 shows the transition from magnetically controlled small grains to gravitationally controlled larger ones. These grains were started in the equatorial plane, where they remain confined. For negatively charged grains the transition is continuous. Increasing the particle's mass (keeping its surface potential, initial position and velocity constant) leads to increasing and deforming gyro-motion about the mean motion guiding center. However, for positively charged grains this transition is not smooth. Particles with intermediate charge to mass ratios will not stay in orbit but gain energy from the co-rotational electric field and get ejected. This is the mechanism responsible for the ejection of high-speed dust grains from the Jovian system, first discovered by the *Ulysses* spacecraft in 1992 [23].

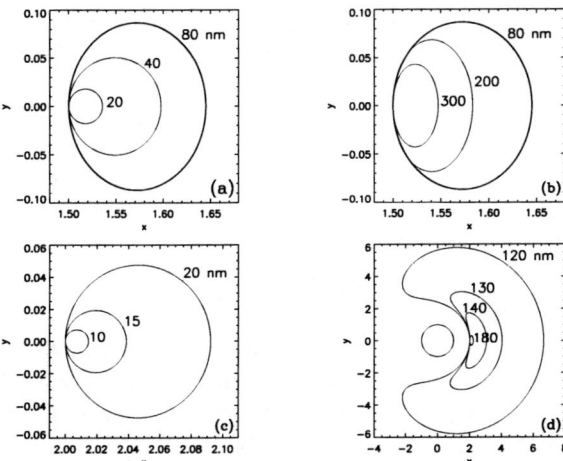

FIGURE 3. Orbits of particles with different charge-to-mass ratios are shown in their respective mean motion frame. **(a,b)** Negatively charged particles started on circular equatorial Kepler orbits with a constant surface potential of -10 V (hence their charge $Q_e \simeq -7e \times a$, where a is the grain radius in nm units). **(c,d)** Positively charged grains with a constant surface potential of +10 V started outside the synchronous orbit. Closed orbits for grains with $20 < a < 120$ nm do not exist, these grains are ejected from the magnetosphere [22].

Jupiter

Jovian dust streams: Since 1995, the *Galileo* spacecraft has made detailed observations of dust particles around Jupiter. The stream particles are now believed to have radii on the order of 10 nm and to mainly originate from the volcanic moon Io, having escape speeds from the magnetosphere in excess of 200 km/s [23-28]. It is difficult to model the orbit-to-orbit variability of these fluxes, mainly because of the time dependence of the volcanic activity on Io as well as the entire magnetosphere as it responds to variable solar wind conditions. However, the 'average' characteristics of these streams are now well explained based on our models of the Jovian magnetosphere and the detailed variations of the charging processes in this environment (Figure 4).

Dust transport: The main Jovian ring at approximately $r = 2R_J$ ($R_J = 7.1 \times 10^4$ km) is an excellent example of dust transport due to a plasma density gradient in this region. The main source of plasma is the atmosphere of Jupiter due to UV ionization. The plasma density drops with distance from the planet. The charge on a dust particle here is set by the competition between electron fluxes from Jupiter and photoelectron production from the grains. The expected equilibrium charge is positive and increases with distance from Jupiter. Small grains are continuously generated from bigger boulders of the main ring due to ongoing bombardment by interplanetary dust particles. A freshly born small grain collects charge and, due to electromagnetic perturbations, starts oscillating about its initial orbit (Figure 5). As it moves toward the planet (against the co-rotational electric field) its average (positive) charge is always above the local equilibrium due to its finite charging time. On the return path the grain is moving in the direction of the electric field with a charge that is always slightly below the local equilibrium. Hence in every oscillation period the grain loses orbital energy and migrates toward Jupiter. Small grains ($a \ll 1\mu$m) swiftly follow magnetic field lines into the atmosphere with lifetimes on the order of hours. Bigger particles drift closer to the ring plane and gain inclination due to the 'in-the-ring-plane' component of the magnetic field, possibly trapping in Lorentz resonances where the frequency of their orbital motion becomes commensurate with that of Jupiter's rotation [29]. Models based on these simple ideas have become increasingly sophisticated in recent years and are now capable of matching to a large degree the details of *Galileo*'s remote sensing observations [30].

Magnetospheric capture: Similar ideas about gradients in plasma parameters on the outskirts of the magnetosphere lead to the possible capture of interplanetary dust particles. Dust particles passing through the magnetosphere can lose enough energy and angular momentum to start orbiting Jupiter on distant highly eccentric orbits. Through consecutive visits they continue to lose energy and angular momentum and can settle into circular orbits entirely inside the magnetosphere. Depending on their initial orbits, the final state can be either prograde (just like all the regular satellites of Jupiter) or retrograde. The probability of getting into retrograde orbits is larger. The *Galileo* dust detector indeed observed micron sized dust particles on retrograde orbits. All the inner

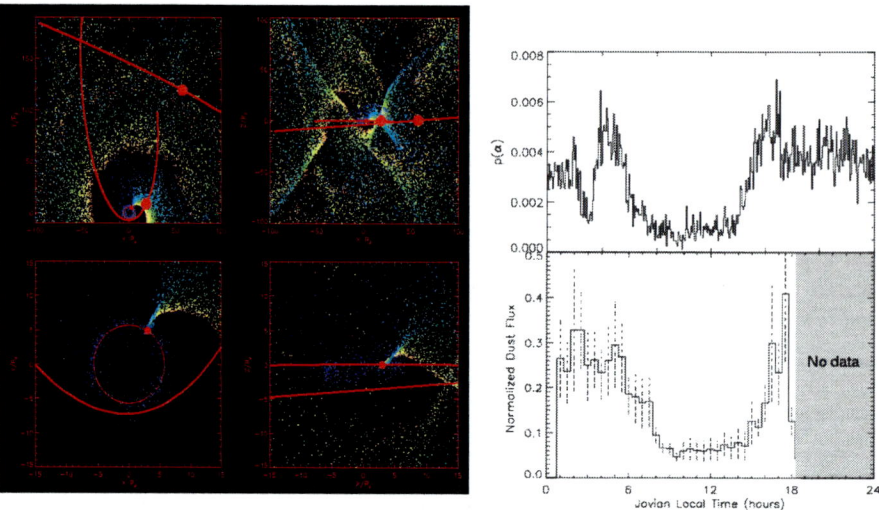

FIGURE 4. (*left:*) Snapshot of the positions of dust particles ejected from Io. The continuous lines represent the orbits of *Galileo* (curved around Jupiter) and *Cassini*. The dots show their positions at the end of 2000, at the closest approach of *Cassini* to Jupiter. The top panels show the spatial distribution on large scales, while the bottom ones only show the close vicinity of Io. The left side of this plot is a projection to the equatorial plane of Jupiter, while the right hand side shows a meridional projection. (*right:*) The flux of the Jovian stream particles as function of local time from a computer simulation and from data collected from 1995-2002 [28].

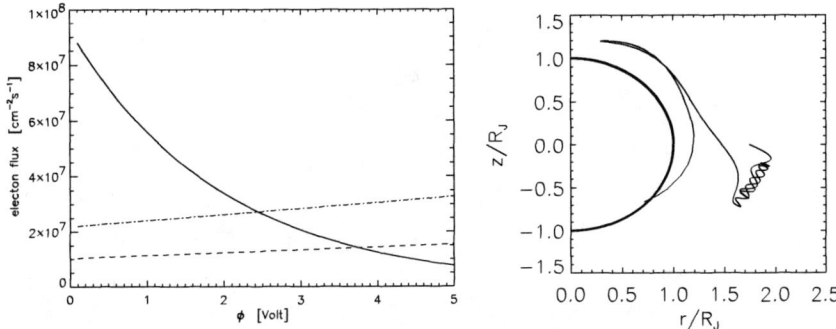

FIGURE 5. (*left:*) The flux of photoelectrons emitted from dust grains (continuous line) and the flux of electrons escaping from the atmosphere at $r = 1.8$ (dashed line) and $1.4\,R_J$ (dashed-dot line) as functions of the surface potential of a dust particle. The net flux becomes zero at the equilibrium surface potential of $\bar{\Phi}_v \simeq 3.8$ and 2.5 for $r = 1.8$ and $1.4\,R_J$, respectively. (*right:*) The trajectory of a $0.1\mu m$ dust particle ejected from the main ring shown in cylindrical coordinates (r,z) [30].

moons are sources of dust grains but these will all move on similar orbits to their sources. Hence, the detected 'backward' running population is most likely of interplanetary origin [31].

During *Galileo*'s mission at Jupiter, a number of discoveries reinforced the importance of dusty plasma effects to understand the ejection, transport and capture of dust particles. Armed with these new ideas, there is a great deal of anticipation as to which of these processes will also be observed during the *Cassini* mission, due to begin observations of Saturn in 2004.

Saturn

Before the *Voyager* discovery of Jupiter's ring, Saturn was the only planet known to have rings. Its magnificent ring structure inspired theories to explain why this is the only planet in the solar system that could sustain rings. It is a triumph of celestial mechanics that many of the fine details seen by the *Voyagers* were explained by gravitational resonances between the orbiting dust grains and various moons. However, even closer observations showed unexpected findings: periodically appearing, radially expanding dust clouds above the dense B ring, the so called spokes [32]. These clouds are dark in backscattered and bright in forward scattered light, indicating that they are comprised of sub-micron sized dust particles elevated from the main ring. Spokes appeared in consecutive images, indicating that they were born on time scales of minutes, and lasted several hours. There was no doubt that this phenomenon could not be understood by gravity alone and that dusty plasma processes can best explain the observations [33,34].

The upcoming *Cassini* observations are expected to provide high spatial and temporal resolution images of the spokes that will likely resolve several outstanding issues. These include: a) is the appearance of spokes random or a periodic phenomena? b) are spokes

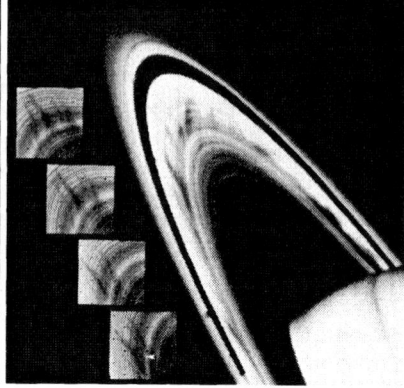

FIGURE 6. (*left:*) *Voyager* image of Saturn. Many of the fine details in the radial brightness variations can be explained by gravitational interactions between the ring particles and the moons. (*right:*) Images of spokes above the B ring. Small negatively charged dust particles are elevated above the ring plane due to a) a radially expanding plasma cloud generated by interplanetary dust impact on the ring or b) by field aligned currents from the ionosphere generated by UV electron emission from the rings. The initial radial alignment evolves into a wedge shape. The leading, slanted portion is populated by larger grains that move close to Kepler speeds, while the trailing radial edge is populated by smaller grains that are forced to co-rotate with the magnetic field lines. The vertex of the wedge is at synchronous orbit, where the Kepler rate is matches the rotation rate of Saturn.

related to the bursty and highly polarized kilometric radiation (SKR)? and/or the broad banded and unpolarized electrostatic discharges (SED)?, both discovered by the radio science experiments on *Voyagers*, and c) what is the initial radial expansion velocity?

Dynamics, transport, ejection and capture: Most of the complex dynamics of charged dust particles discovered at Jupiter are also expected at Saturn. However, contrary to Jupiter (where there is about a 10° tilt between the rotational and magnetic axes) the magnetic field of Saturn can be characterized very well as a centered aligned dipole. This highly symmetric configuration allows for the existence of 'halo' orbits, a new type of orbit for charged dust particles that does not exist for either the purely magnetic or purely gravitational cases. These are stable orbits that encircle Saturn parallel to its ring plane, never crossing the equator (Figure 7)[35]. These 'halo' particles may be found with the dust detector of the *Cassini* spacecraft towards the end of its nominal 4 year mission, when the spacecraft is planned to follow high inclination orbits.

Ejected and captured particles are expected to exist at Saturn as well. There are no volcanic moons at Saturn, but there are moons at the outskirts of the magnetosphere where the ejected grains are expected to charge positively. The co-orbital moons Dione and Helene ($r = 6.3 R_S$) and Rhea ($r = 8.7 R_S$) can act as sources of grains due to meteoroid bombardment. These grains are expected to be smaller and slower than the Jovian stream particles since the magnetic field is about 20 times weaker. The escape speeds are

$$v_{escape} \simeq \frac{3}{a_\mu} \quad \text{km/s} \quad \text{for Jupiter and} \tag{3}$$

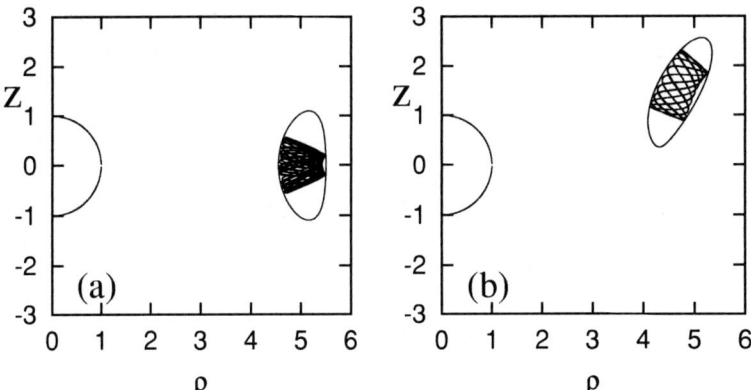

FIGURE 7. (a) The trajectory of a particle trapped in the equatorial plane shown in cylindrical coordinates. The motion is stable at this location, but it becomes unstable for starts close to the inner edge of the B ring at $r = 1.5R_S$ ($R_S = 6 \times 10^4$ km. The particle radius $a = 0.1\mu$, its charge $Q_e = 700e$, in addition to electromagnetic forces and gravity, radiation pressure was also included (this is why the grain shows excursions above and below the ring-plane). (b) A 'halo' orbit that remains stable even against radiation pressure perturbations [35].

$$v_{escape} \simeq \frac{0.6}{a_\mu} \quad \text{km/s} \quad \text{for Saturn,} \qquad (4)$$

where a_μ is the dust particle's radius measured in microns. The escaping stream particles from Saturn will remain confined close to the equatorial plane due to the high degree of symmetry of the planetary magnetic field.

COLLECTIVE EFFECTS

In recent years there has been an explosion in our understanding of the basic physics of collective phenomena in dusty plasmas. To date, however, solar system studies of dusty plasmas have focused on single particle dynamics to explain existing observations. It is only a question of time before collective dusty plasma effects will also be observed. Here we only mention two of these possibilities, dusty plasma waves and strongly coupled systems [36].

Dust acoustic waves: In a planetary ring, substantial relative speeds can be maintained between large boulders, following Kepler orbits, and small charged dust particles, because the motion of the later ones is influenced by the planetary magnetic field. This relative velocity can exceed the characteristic propagation velocity of dust-acoustic waves [37,38] and a shock wave pattern may develop [39,40]. If observed, these Mach cones could be used to remote-sense dusty plasma parameters in the main rings of Saturn using high resolution *Cassini* images.

Strong coupling: In plasmas where the dust density is significant, the average grain separation can become smaller than the plasma Debye length, and the electrostatic interaction energy between the dust particles can become comparable to the dust thermal energy. In this case, grains will no longer move independently. The stronger the electrostatic coupling between the grains, the longer the grain-grain correlation length becomes, ultimately forming a dusty plasma Coulomb lattice. The ε ring of Uranus is a candidate where strongly coupled dusty plasma effects could resolve a contradiction between the mass estimate and the optical properties of this ring [41]. A possible solution is for the micron sized dust grains to 'fly in formation' so that their effective light scattering properties imitate cm sized particles.

SUMMARY

Throughout the solar system dusty plasma effects are often responsible for shaping the size and spatial distribution of small dust grains as well as the characteristics of the plasma environment. Dusty plasma studies combine *in situ* and remote sensing observations and provide a unique tool to understand the workings of our solar system.

ACKNOWLEDGMENTS:

This work was supported by the Magnetospheric Physics Program of NASA (NAG-5-9172) and the *Cassini* project (JPL 1225700). The author is grateful for the ongoing collaboration with team members of the *Galileo* and *Cassini* dust experiments, and colleagues J.E Howard and C.J. Mitchell.

REFERENCES

1. Goertz, C.K., *Rev. Geophys., 27,* 271, 1989
2. Northrop, T.G., *Physics Scripta, 45,* 475, 1992
3. Mendis, D.A., M. Rosenberg, *Annu. Rev. Astron. Astrophys., 32,* 419, 1994
4. Bliokh, P., V. Sinitsin, V. Yaroshenko, *Astrophys. Space Sci. Lib., 193,* Kluwer Academic Publishers, 1995
5. Horányi, M., *Annu. Rev. Astron. Astrophys.,34,* 383, 1996
6. Verheest, F., *Astrophys. Space Sci. Lib., 245,* Kluwer Academic Publishers, 2000
7. Shukla, P.K., A.A. Mamun, *IOP Series in Plasma Physics,* 2002
8. Whipple, E.C., *Rep. Prog. Phys., 44,* 1197, 1981.
9. Sternovsky, Z., M. Horányi, S. Robertson, *this volume,* 2002
10. Rennilson, J.J., D.R. Criswell, *The Moon 10,* 121, 1073
11. Criswell, D.R., B.R. De, *J. Geophys. Res., 82,* 1005, 1997
12. Nitter, T., O. Havnes, *Earth, Moon and Planets, 56,* 7, 1992
13. Colwell, J.E., M. Horányi, S. Robertson, A.A. Sickafoose, *this volume,* 2002
14. Sickafoose, A.A., J.E. Colwell, M. Horányi, S. Robertson, *this volume,* 2002
15. Images taken by the Sakurajima Volcananological Observatory (SVO), Kyoto University, Japan (http://hakone.eri.u-tokyo.ac.jp/unzen/sakura/sakura.html)
16. Krauss, C.E., M. Horányi, S. Robertson, *this volume,* 2002
17. Mendis, D.A., H.L.F. Houpis, J.R. Hill, *J. Geophys. Res., 87,* 3449, 1982
18. Northrop, T.G., J.R. Hill, *J. Geophys. Res., 88,* 6102, 1983
19. Schaffer, L., J.A. Burns, *J. Geophys. Res. 99,* 17211, 1994
20. Howard J.E., M. Horányi, G.R. Stewart, *Phys. Rev. Lett., 83,* 3993, 1999
21. Howard, J.E., H. R. Dullin, and M. Horányi, *Phys. Rev. Lett. 84,* 3244, 2000
22. Mitchell, C.J., M. Horányi, J.E. Howard, *this volume,* 2002
23. Grün, E., *et al., Nature, 362* 428, 1993
24. Horányi M., E. Grün, G. Morfill, *Nature 363,* 144, 1993
25. Hamilton, D.P., J.A. Burns, *Nature 364,* 550, 1993.
26. Horányi M., E. Grün, A. Heck, *Geophys. Res. Lett. 24,* 2175, 1997
27. Graps, A., *et al., Nature 405,* 48, 2000
28. Krüger, H., M. Horányi, E. Grün, Geophys. Res. Lett., submitted, 2002
29. Schaffer, L., J.A. Burns, *J. Geophys. Res.,92,* 2264, 1987
30. Horányi, M., A. Juhasz, *J. Geophys. Res.,* submitted, 2002
31. Colwell J., M. Horányi, E. Grün, *Science 280,* 88, 1998
32. Smith, B., *et al., Science 212,* 163, 1981; and *Science 215,* 504, 1982
33. Hill, J.R., Mendis, D.A., *Moon and Planets 24,* 431,1981
34. Morfill, G.E., C.K. Goertz, *Icarus, 55,* 111, 1983
35. Howard, J.E., M. Horányi, *Geophys. Res. Lett. 28,* 1907, 2001
36. Goertz, C.K., Linhua-Shan, O. Havnes, *Geophys. Res. Lett. 15,* 84, 1988
37. Rao, N.N., P.K. Shukla, M.Y. Yu, *Planet. Space. Sci., 38,* 4, 1990
38. Melandsø, F.T., T. Aslaksen, O. Havnes, *J. Geophys. Res., 98,* 13315, 1993
39. Havnes, O., *et al., J. Geophys. Res., 100,* 1731, 1995
40. Brattli, A., O. Havnes, F. Melandsø, *Phys. Plas., 9,* 958, 2002
41. Esposito, L.W., *et al., Uranus,* U. of Arizona Press, Tucson, 1991

Dusty Plasmas in Astrophysics and Cosmology

Erwin Sedlmayr

Zentrum für Astronomie und Astrophysik, TU Berlin, Germany

Abstract. Cosmic Dust is an ubiquitous component of the interstellar medium and plays a significant role for the diagnostics and the local and global behaviour of astronomical objects, being linked both to most important microscopic and macroscopic phenomena. In this way not only the spectral appearance of dusty objects, but also their dynamical and energetical situation is essentially determined by the presence of a condensed material component. In a cosmological view the first appearance of a dust grain in the universe brings about the first 2-dimensional surface with most important consequences for the future chemical evolution of the universe.

The cosmic dust complex consists usually of three problems: microscopic dust condensation, dust processing, and dust interaction with the ambient gas and radiation which both conduct a strong feedback to the local formation and existence conditions, but also to the global structure of a dusty object.

Basically cosmic dust is intimately linked to a large variety of fundamental physical and chemical complexes. Though in astrophysics in most cases the dust component is considered to be electrically neutral, of particular interest in this context here are charged grains. It is the aim of this contribution to outline the overall situation and to address particular aspects imposed by charged particles.

INTRODUCTION

Optical and Infrared astronomy reveals cosmic dust as being a most important phase of matter ubiquitously emerging in nearly all astronomical systems, e.g. stellar environments, the interstellar medium, galaxies, galactic clusters, and even in the intergalactic space. In this sense cosmic dust turns out to be the predominant phase of matter (besides hydrogen and the noble gases) in the universe, whenever the local conditions allow solid particles to exist. For astronomers, the notion cosmic dust is essentially defined phenomenologically by its extinction, scattering, and polarization properties, as well as other specific spectroscopic features, etc., in this way suggesting typical macroscopic particles – *cosmic grains* – consisting within a wide physical, chemical and mineralogical range as "astronomical silicates", "carbonaceous species", like conducting or dielectric materials addressed as soot or graphite, and even large polyaromatic carbon hydrogens, showing an overall typical grain size of the order of the wavelength of the interacting photons.

Due to their particular extinction properties cosmic grains are assumed to have an amorphous, polycrystalline or monocrystalline mineralogical structure, the chemical nature of which may span from purely homogeneous particles to rather heterogeneous compositions. Their detailed physical morphology may show a homogeneous structure, core mantle grains of different materials, matrix inclusions, etc. and a large variety of possible geometrical shapes. In this view the cosmic dust complex phenomenologically comprises a variety of aspects of specific relevance, depending on the particular perspectives and aims of investigation: dust condensation, dust chemistry, dust processing, modelling of dust forming or dusty systems, radiative transfer and spectroscopy, individual or collective particle dynamics, role of grains in planetary physics, etc. Of particular interest in the scope of this conference are dusty plasmas, i.e. charged grains and their interaction with electromagnetic fields and with the ambient plasma. Though these aspects, without doubt, play an important role for a variety of astrophysical problems — in particular in connection with high energetic photons in the vicinity of stars or at certain regions in the interstellar medium — the dominant view of astronomers with regard to cosmic dust is either to dust blocking of stellar light or thermal dust emission and the corresponding spectroscopic signatures, and in this perspective focuses essentially on dust properties which are widely independent of electrical charge. Moreover, the bulk of cosmic dust is essentially observed in rather cool systems, e.g. in circumstellar envelopes of cool stars or in galactic clouds, where due to efficient radiation shielding only the boundary layers should be affected by photoelectric processes or by grain-plasma interactions. This supports the usual assumption that, apart from particular situations, cosmic dust may be conceived as being widely electrically neutral. On this basis it is interesting to display the cosmic dust complex and its intimate couplings to the various physical disciplines involved, as well as to the specific roles charged grains and plasma effects might play within this context.

COSMIC DUST CONDENSATION

The cosmic dust complex sketched in Fig. 1 basically is interwoven to a large variety of different fundamental problems and disciplines, like thermodynamics, hydrodynamics, particle dynamics, chemistry, radiative transfer etc., in this way addressing both microscopic or macroscopic phenomena of relevance for a reliable physical description. One most important process in this context is the question of ab initio gas phase condensation of grains, which is usually conceived as a two-step mechanism consisting of i) cluster nucleation and ii) cluster growth to macroscopic specimens, by which a gas-solid phase transition is guided which finally results in the primary condensates observed especially in cooling flows, like stellar winds of red giants and super giants, novae and supernovae, but also in hot environments of particular Wolf-Rayet stars.

This primary dust component, which essentially consists of very small high temperature stable grains ($a \ll \lambda$, with a being the typical grain size and λ the wavelength of relevant photons) injected into the interstellar medium, provides the seeds necessary for further grain growth and dust evolution.

Cluster Nucleation

The first step of dust condensation is the formation of so-called *critical clusters* out of the gas phase as consequence of the fact, that — along a cooling track within a favourable temperature and density range — molecules of increasing complexity and also small clusters are formed, which finally achieve the first thermodynamically stable

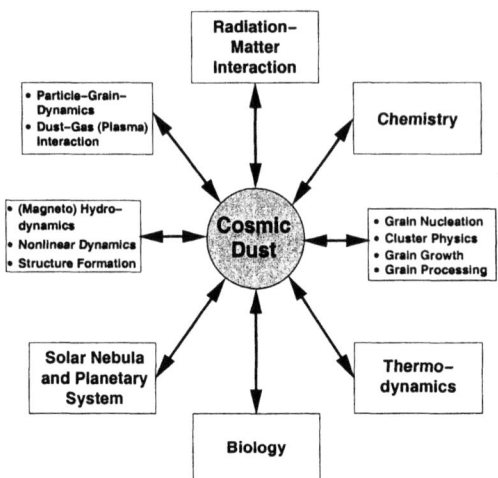

FIGURE 1. Cosmic dust is intimately linked to a large variety of fundamental physical and chemical problems.

configurations addressed to as critical clusters. These entities are the basic structures, which now in a subsequent step can grow to macroscopic grains by addition of specific admolecules. Several approaches have been developed for a reliable theoretical description of the molecule-cluster transition. Basically three conceptually different methods are applied (cf. [1]):

- *thermodynamics*: This description — called the *classical nucleation theory* — refrains from the microscopic details of the cluster formation process, but determines the first stable cluster by means of an extremum of its Gibbs function (free enthalpy of formation) as a thermodynamical limit state (smaller clusters are basically instable, larger clusters are stable), irrespective of the question in which individual way a cluster evolves from the gas phase with regard to the prevailing conditions. This approach, where only fundamental thermodynamic properties of the clusters and its molecular constituents are taken into account, yields good results for critical clusters sufficiently large to have well defined thermodynamic functions, usually split in internal energy and surface energy. This is an excellent approximation e.g. for the formation of critical water droplets, but is highly questionable for astrophysical applications, where the critical clusters are rather small inorganic (e.g. $(TiO)_N$- or $(SiC)_N$-clusters), or organic compounds, like large polyaromatic molecules or very small soot particles. Nevertheless, classical nucleation theory, sometimes modified for realistic Gibbs functions within the regime of small clusters, is widely used in astrophysical applications.
- *chemical description*: This approach conceives critical cluster formation as a result of a sequence of chemical reactions by which — starting at suitable basic molecules — clusters of increasing size and complexity are formed which finally end up in critical clusters. In this concept a critical condition is defined by the slowest

reaction step along the fastest reaction chain bridging the cluster regime, yielding the critical cluster as the product of this specific step. However, in real astrophysical situations such an approach requires large reaction networks, where usually most necessary input quantities (like individual cross sections or reaction rates) are lacking, which considerably limits its practical relevance. Moreover, the reaction network usually becomes larger and more complicated with each additional step, also introducing significant numerical problems, which explains, why no complete chemical description of this phase transition is available yet.

- *Chemical pathway*: In very special cases like the formation of carbon grains via a sequence of increasing polyaromatic molecules, the critical cluster formation can be reduced to the discussion of a few particularly effective pathways, requiring however a profound understanding of the individual molecular sequences, which seems to be at least possible for organic chemistry, in particular applying flame physics. But nevertheless, also in these restricted cases, it still remains a problem to explain quantitatively the transition from basically twodimensional to threedimensional structures, as are observed in form of soot or graphitic particles in carbon rich stellar environments.

Facing these problems, methods have been developed recently where classical nucleation theory is combined with chemical network treatments describing critical cluster formation as a result of suitable chemical equilibrium reactions ([2, 3]).

Critical cluster nucleation usually is assumed to proceed under electrically neutral conditions. One justification of this assumption is, that the primary gas-solid phase transition essentially occurs in the cooling outflows of evolved stars having rather low effective temperature and negligible UV radiation fields. For this reason these systems are governed by neutral chemistry. This is in good agreement with the observations which indicate a significant decrease of the dust formation efficiency for objects possessing pronounced UV fields due to chromospheres. A second aspect concerns the cluster formation process itself: If the condensating molecules are positively charged, significant Coulomb barriers of the charged clusters are expected, which should drastically limit the subsequent formation of larger entities. Of course the cluster charge could be reduced by adelectrons, and in this way simulate cluster neutrality. But then ion addition along a scheme

$$\ldots \longrightarrow A_N + A_1^+ \longrightarrow A_{N+1}^+ + e^- \longrightarrow A_{N+1} + A_1^+ \longrightarrow A_{N+2}^+ + e^- \longrightarrow \ldots$$

is basically similar to neutral cluster formation if $n_{A_1} + n_{A_1^+}$ is considered as the basic gas-phase monomer density. Because sufficiently small clusters behave electrically like molecules, they should be positively charged due to photoelectric effect, which again produces effective Coulomb repulsion for the positive adions. Only the larger clusters show an effective electron affinity resulting in negatively charged grains, which then, however, would efficiently attract the positive adions. This situation seems to be unlikely in the nucleation regime, but could be important in the growth regime, where it might give rise to effective heterogeneous grain growth.

Grain Growth

In the size regime beyond the critical cluster size grains grow by subsequent addition of suitable molecules, atoms and ions, the supersaturation of which is larger then unity with regard to the specific grain material. In this way a hierarchy of condensates originates, producing usually a layered grain structure, so-called core-mantel particles, having monocrystalline, polycrystalline or even amorphous mineralogical structures, depending on the timescales that dominate the formation process (capture time, hopping time, time to find a proper final lattice site, etc.). These timescales depend sensitively upon the fact, whether the grains or the adparticles are neutral or electrically charged. As adions dominantly carry a positive electric charge their capture rate can be considerably reduced (enhanced) for positively (negatively) charged grains. In a plasma, grains usually may acquire a rather significant negative electric charge due to their large electron affinity, which supports the growth process in presence of suitable positive ions. However, to my knowledge, this effect has only been taken into account by a simple increase of the corresponding ion-grain collision rate, refraining from further complications, as the change of stability with regard to the specific supersaturation ratios and the necessary modifications due to pronounced surface Coulomb potentials. Thus, in the astronomical context a consistent physical, chemical and mineralogical description of the growth of electrically charged grains is still a desideratum.

PLASMA IMPLICATIONS IN THE COSMIC DUST COMPLEX

Plasma effects play a major role in the grain-gas interaction, as well as in the individual or collective grain dynamics – especially in systems where magnetic fields are present. They provide a close coupling between the electrically charged particles and the local magnetic field configuration. In various problems with regard to dusty plasmas, very different questions arise and require specific answers within the considered context. Such problems are briefly addressed in the following, referring to different boxes sketched in Fig. 1.

Radiation-Matter-Interaction

In this context four processes seem to be of significant importance: i) efficient blocking of radiation by grain absorption and scattering, ii) charging of grains due to the photoelectric effect caused by sufficiently energetic photons, iii) momentum coupling of the dust particles and the ambient plasma, and iv) energetic coupling of the radiation field and the dust component.

i) This process, though being very important with regard to the spectral appearance of a dusty object, is essentially independent on the grain charge and hence on specific plasma effects. It has, however, an indirect influence upon the local grain charge and plasma status via spectral reddening caused by the effect that photons travelling through a dusty system become gradually less energetic. We refrain from a further discussion of this genuine radiative transfer problem.

ii) A most important effect in the context of dusty plasmas is photoelectric charging of grains due to the presence of high energetic photons. In this view the knowledge of the photoelectric yield Y_e, i.e. the production efficiency of photoelectrons per incident photon, is of significant relevance. Y_e depends on the electric properties of the grain material, the grain size, the surface potential well, and the energy of the absorbed photon. For small neutral dielectric grains Y_e is approximately unity or even larger ([4]). With increasing positive grain charge the escape probability of photoelectrons will be considerably reduced and approaches zero beyond some critical value of the grain's electron attractive Coulomb potential. Photoelectric charging of grains is important in the environment of hot stars, in the outer layers of clouds exposed to the interstellar radiation field, but also in the solar system. For an extensive and detailed discussion of photoelectric grain charging see the recent excellent contribution of Weingartner and Draine [5] and references therein, but also the review of Horányi [6], devoted to charged dust dynamics in the solar system.

iii) Due to the large extinction efficiency of dust particles, the grains acquire radiative momentum very effectively, which is transferred to the ambient gas by frictional coupling. In this way pronounced large scale motions are induced, which in circumstellar environments are responsible for massive stellar winds and heavy mass loss. The efficiency of momentum transfer to the gas depends sensitively on the microscopic coupling, which mainly depends on the gas density and the collisional cross sections of the grain-gas interaction. If charged grains and ions are involved, due to the Coulomb forces this coupling is very close, hindering drift effects, thus producing a maximum momentum transfer from the radiation field to the gas component mediated by dust.

iv) Due to its large absorption coefficient dust grains are a very effective cooling agent in an astrophysical medium, where they absorb high energetic photons and emitting predominantly infrared radiation. Though this process has only an indirect impact on the charge situation of a dusty plasma, it is most important for the energy balance between the various material components and determines in particular the spectral distribution of the local radiation field, and hence of the ionizing photons.

An important effect on the energy balance of the grain-gas system may be due to heating by photoelectronic emission, the surplus energy of which is transferred to the plasma. The heating efficiency depends on the specific grain material and on the energy distribution of the incident radiation field, as well as on the adopted grain size, and is increasing for smaller grains. Likewise, a cooling effect is conducted by accretion of ions onto a grain, causing the kinetic energy of the adions to be removed from the plasma. Both effects may play some role in the interstellar medium, in particular for situations where effective atomic or molecular heating and cooling processes are negligible.

Dust Dynamics

With regard to the dynamical processes in astronomical systems, the presence of grains has to be considered in different regards:

Particle-Grain Dynamics

In this case, the motion of a test dust particle is treated under the action of various external forces, like gravity, drag forces, radiation pressure and electromagnetic acceleration. In the context of charged dust dynamics, electric and magnetic fields play a particularly important role, as well as frictional coupling enhanced by Coulomb interaction with the ambient ions. This type of single particle description is extensively applied in investigations of dust motion in the solar and planetary magnetospheres and planetary ring dynamics, in order to identify the large scale streaming of dust grains or to explain specific dust experiments in space, like the observations of small grains around Jupiter by Ulysses (cf. [6]), but also in order to investigate possible resonance and stability situations with regard to orbital motions in planetary rings or discs.

Grain-Plasma Interaction

The frictional force acting on a grain is due to collisions with nearby gas particles. In this way in most systems a close momentum coupling between the grain and the ambient medium is generated, usually leading to a joint hydrodynamical motion of the combined dust-gas system. However, in many situations of low gas density, pronounced drift effects may occur, when the local differential acceleration of a grain, e.g. due to radiation absorption, is much larger than that of the gas system. Then the grain-gas collision rate is no more determined by the isotropy of the thermal velocity of the gas, but by the directed drift velocity of the grain, for sufficiently large values leading to a considerably increase of the collision rate, i.e. an increase of the drag force, which may finally approach the so-called *equilibrium drift* of a grain, defined by the balance of its radiative acceleration and the actual collisional momentum input from the gas. Such a situation of equilibrium drift is increasingly favoured the more the grain-gas momentum-coupling becomes tighter. This is true for sufficiently dense gases but also for an increase of the grain's momentum transfer cross section, e.g. provided by a Coulomb potential with regard to electrically charged grains and the interacting ions and electrons.

Another important aspect of grain-plasma collisions is grain charging by electron and ion currents, controlled by the local balance equation:

$$\frac{dQ}{dt} = \sum_l J_l,$$

where Q is the electrical charge of the considered grain and J_l represents the corresponding charging current carried by each of the contributing plasma components. The net source term on the r.h.s. accounts for both the direct transfer of electric charge due to sticking of electrons and ions, but also for the production of secondary electrons due to collisional ionization. The effectivity of each individual component strongly depends on the corresponding attractive or repulsive surface potential (cf. [6]).

Sticking Coefficients

A fundamental problem of grain-gas or grain-plasma interaction is the calculation of sticking coefficients. For neutral grain-gas reactions this problem is not generally solved. Only for very special cases, like silicate growth by SiO addition, quantummechanical surface mode calculations are available, indicating sticking coefficients s of the order of unity ([7]). In particular for very small inorganic grains full quantummechanical treatments of the reaction of a proper molecule (monomer) with the cluster, applying the concept of an energy surface result in high dimensional energy surfaces – even for clusters consisting of only few monomers – which usually allow for no definite conclusions about the reaction probability (e.g. [8, 9]). For larger grains, having already a well defined geometrical surface, chemisorption and physisorption arguments yield sticking probabilities in the range $0.1 < s < 1$ for astrophysically relevant inorganic materials. The situation is much better for carbonatious components, like PAHs or graphitic grains, where based on flame physics or on laboratory experiments reaction probabilities for important cases (radical reactions or reactions with dangling C-H-bonds) are known. In view of charged grains and plasma partners the sticking problem has been investigated by Weingartner and Draine [5] deriving reliable sticking coefficients both for electrons and adions.

The Shadow Force

A very interesting suggestion in the context of dusty plasmas is that of the so-called *shadow force* (cf. [10]), induced by a net momentum transfer as a result of breaking the impact isotropy of gas accretion due to grain shadow effects. Under the assumption of thermal accommodation of the adparticles at the grain's surface, this force depends on both the gas temperature and the dust surface temperature. This shadow force can be quite substantial in the outer regions of late type stellar winds, where the dust temperature is smaller than the gas temperature, and according to Bingham and Tsytovich [10] should give rise to enhanced grain growth. To my knowledge, the shadow force has not yet been considered in hitherto consistent quantitative models of dust formation in stellar outflows, but nevertheless should be included in future models of dust growth in circumstellar environments.

Hydrodynamical Approach

In a hydrodynamical description, as usually adopted for modelling dust forming objects, the gas-dust system is considered as a multicomponent continuous medium, subject to the equation of *mass conservation* and additional equations of continuity for the individual components, e.g. a *chemical reaction network*, an *equation of motion* accounting for local momentum conservation, and the *energy equation*. Of particular interest in a dust forming system are the so-called *dust moment equations*

$$\frac{\partial K_i}{\partial t} + div(K_i \mathbf{v}) = j_* \delta_{i0} + \frac{i}{d\,\tau_{gr}} K_{i-1},\ i = 0, 1, 2, 3$$

describing formation and growth of the dust components, with $\mathbf{v}$ being the hydrodynamical velocity, j_* the local nucleation rate of critical clusters, d dimension of the grains, and τ_{gr} the characteristic growth timescale, governing grain growth ([11]). The moment K_0 accounts for the local number density of grains, the moments K_i represent the mean values for the typical size ($i = 1$), surface area ($i = 2$) and volume ($i = 3$) of the grains. From the grain density equation ($i = 0$) the local grain size distribution function $f(t, \mathbf{x}, a)$ follows by simple functional derivation, with $f(t, \mathbf{x}, a)\, da$ representing the actual local number density of grains with a size in the interval $(a, a + da)$ ([12]). This description, which is applied for both homogeneous and heterogeneous cosmic dust formation, is rather flexible and allows also to incorporate Non-LTE effects, multicomponent grains, grain charge, particle drift, etc. With regard to charged grains, to my knowledge, only explorative calculations for dusty stellar winds have been performed so far ([13]) by generalizing the methods derived in [14].

Instabilities

The presence of dust in astrophysical media may induce inhomogeneities on different geometrical scales. This concerns in particular dust forming shells of pulsational variables, the shock wave pattern and dynamics of which are essentially determined by the dust formation efficiency and by radiative acceleration of the dust component and its frictional coupling to the gas. In this way a monoperiodic, multiperiodic or even chaotic dynamical behaviour of the dust shell can be produced. Grain charging may increase this effect by a tighter coupling of the dust component to the plasma. Though being present, the role of magnetic fields is not yet studied in this context, however, due to guiding effects and symmetry breaking hydrodynamical inhomogeneities, possibly can be strongly influenced in magnetohydrodynamical situations. Also various types of instabilities can be induced in a gas when dust is present: dust induced dynamical instability, thermal instability (thermal bifurcations) and radiative instability of dust condensation, leading each to pronounced structure formation (cf. [15]). Only the dynamical instability, which is connected to the strong radiative acceleration of the grains, dust induced shock waves and the enhancement of chemical reaction rates and thus to the dust formation process behind the shocks, seems to be directly influenced by charged grains, depending on whether, for sufficient plasma ions present, the grains have a negative or positive electric charge increasing or decreasing the effect. Also magnetic fields might play a role when hindering or supporting the necessary compression effect.

A global instability in a sufficiently dusty system may be due to the shadow force ([10]), yielding a Jeans-like dispersion relation for a gravitational instable situation, if gravity is replaced by the shadow force. In this way, individual clumps of masses much smaller than the corresponding Jeans mass could be formed, possibly bridging the mass regime from asteroids up to stars. However, a more realistic modelling is required in order to confirm the relevance of this process.

CONCLUSION

In the broad context of cosmic dust the problem of dusty plasmas is still not yet generally solved. The available treatments seem to be confined to special problems, where electrically charged particles and electromagnetic fields play a dominant role. This is to some extent due to the physical complexity introduced by plasma effects (Debye-shielding, correlation problems, etc.), but also to the difficulty of treating MHD-equations on a general scale. As in most systems ionizing photons and pronounced electrical and magnetical fields are present a quantitative ab initio description of the fundamental problem of dusty plasmas is an urgent need. Important questions addressing dynamical equilibria, Non-LTE-effects, charge induced instabilities, reliable electromagnetic field configurations, induced plasma waves,..., and their impact on observational effects are still open for many astrophysical situations.

ACKNOWLEDGMENTS

I thank very much U. Bolick, C. Kieschke, and Dr. A.B.C. Patzer for assistance preparing the manuscript.

REFERENCES

1. Sedlmayr, E., "From molecules to grains," in *IAU Colloquium 146: Molecules in the Stellar Environment*, edited by U. G. Jørgensen, Springer–Verlag, Berlin, 1994, pp. 163–185.
2. Patzer, A. B. C., Gauger, A., and Sedlmayr, E., *A&A*, **337**, 847–858 (1998).
3. Yamamoto, T., Chigai, T., Watanabe, S., and Kozasa, T., *A&A*, **380**, 373–383 (2001).
4. Gail, H.-P., and Sedlmayr, E., *A&A*, **86**, 380–385 (1980).
5. Weingartner, J., and Draine, B., *ApJ*, **134**, 263–281 (2001).
6. Horanyi, M., *ARAA*, **34**, 383–418 (1996).
7. Zöckler, M., *Molekulardynamische Simulation des Wachstums zirkumstellarer Silikat-Staubteilchen*, Master's thesis, Technische Universität Berlin, Germany (1997).
8. Chang, C., Patzer, A. B. C., Sedlmayr, E., , Steinke, T., and Sülzle, D., *Chem. Phys. Lett.*, **324**, 108–114 (2000).
9. Jeong, K. S., Chang, C., Sedlmayr, E., and Sülzle, D., *J. Phys. B*, **33**, 3417–3430 (2000).
10. Bingham, R., and Tsytovich, V., *A&A*, **376**, L43–L47 (2001).
11. Gail, H.-P., and Sedlmayr, E., *A&A*, **206**, 153–168 (1988).
12. Dominik, C., Gail, H.-P., and Sedlmayr, E., *A&A*, **223**, 227–236 (1989).
13. Augustin, K., *Einfluß von Ladungsprozessen auf das Staubwachstum in zirkumstellaren Hüllen*, Master's thesis, Technische Universität, Berlin, Germany (1990).
14. Gail, H.-P., and Sedlmayr, E., *A&A*, **41**, 359–366 (1975).
15. Woitke, P., *Reviews in Modern Astronomy*, **14**, 185–207 (2001).

SECTION 2: TOPICAL REVIEW LECTURES

Plasma Grown Particles : From Injected Gases to Nanoparticles and Nanomaterials, from Injected Particles to Dust Clouds in the PKE Experiment

P. Roca i Cabarrocas[*], A. Fontcuberta i Morral[*], A.V. Kharchenko[*],
S. Lebib[*], L. Boufendi[†], S. Huet[†], M. Mikikian[†], M. Jouanny[†], and
A. Bouchoule[†]

[*]*Laboratoire de Physique des Interfaces et des Couches Minces (LPICM, UMR 7647 CNRS),
Ecole Polytechnique, 91128 Palaiseau Cedex, France)*
[†]*Groupe de Recherches sur l'Energétique des Milieux Ionisés,(GREMI, UMR 6606,
Université d'Orléans, BP 6744, 45067 Orléans Cedex 2 France)*

Abstract. This paper will be focused on two specific examples related to particle growth phenomena in low pressure plasmas. The first one concerns the formation of nanometer size silicon particles and their impact on plasma processing of thin films where the deposition of these particles leads to a significant improvement of quality or to the opening of new windows in thin films technologies. Solar cells based on "polymorphous silicon" , based on nano-particles grown in Silane-Hydrogen plasmas will illustrate the first point, while nanometer crystals of silicon, deposited in SiF_4-H_2 plasmas and not embedded in an amorphous matrix, evidence attractive phototoluminescence properties.
The second illustration is related to the growth of particles in an RF argon plasmas, where externally injected polymer particles represent the primary source for dust growth. Particle growth process, as observed in preliminary laboratory studies, evidences successive growth-etching phenomena reported here. These studies lead to optimized conditions for growing dust clouds in the PKE-NEFEDOF experiments, achieved in microgravity conditions on board of ISS. In these experiments the size of grown particles was not available by ex-situ measurements and laser scattering data suggest only the existence of at a dense cloud of particles too small to be observed individually. We will show that particle size and density can be estimated from macroscopic data such as the RF discharge impedance and the propagation across the cloud of low frequency mechanical excitation driven by sheath width modulations.

INTRODUCTION

Dusty or complex plasmas are a source of rich physical and chemical processes which find applications in a wide range of fields [1].
Part I of this paper emphasizes two examples of PECVD of nanomaterials where the formation of nanoparticles in plasma plays a key role. The first one concerns nanoparticles grown in SiH_4 – H_2 plasma chemistry. The processing parameters can be controlled in such a way that densities of nanometer size particles in gas phase are significantly high but lower than threshold values opening the way to coagulation phenomena and large size dust formation. In these conditions new nanostructured thin films can be obtained, so-called polymorphous silicon, having improved transport properties with respect to hydrogenated

amorphous silicon [2]. A sensitivity of plasma tool parameters for avoiding coagulation is also evidenced. The second illustration concerns particles grown in SiF_4–H_2 plasma chemistry. A fast powder formation process is observed and it is shown that thin films including almost pure nanocrystalline particles can be obtained. Preliminary data on photoluminescence of such films appear of interest here also for devices.

The part II of the paper focuses on particles grown from solid particles previously injected in a plasma tool. This study concern the growth and behavior of dust clouds observed in the PKE RF discharge when an argon plasma is launched after studies of plasma crystals involving externally injected melamine formhaldéhyde particles. This process is at work in an experiment achieved on board of the International Space Station, in the framework of the "ANDROMEDE" taxi flight mission launched by CNES. Preliminary studies, in the laboratory, gave guidelines for the ISS experiment parameters and reveal some details in terms of particle material and sizes of injected and grown particles. In these experiments laser scattering images reveal only the presence of high density of particles too small to be evidenced individually. The paper shows that an estimation of size and density of particles in this experiment can be derived from two set of macroscopic data. The first one is the modification of the RF discharge impedance, progressively induced by the cloud formation. The second one is the wavelike propagation of fluctuations across the cloud induced by low frequency sheath width modulation.

PART I : NANOPARTICLES AND NANOMATERIALS

I 1 : Nanoparticle clouds in plasmas: critical features

The formation of nanoparticles in silane plasmas has attracted much attention and a large number of experimental studies have lead to the description of powder formation through a stepwise process [3]. A critical one, in terms of formation of large size particles, is the coagulation step of nanometric size particles. This transition from a plasma with nanoparticles towards a dusty plasma involves critical densities of nanoparticles. For a given rate of nanoparticle formation, ie plasma conditions, there is a competition between coagulation and their diffusion losses towards the walls. A criterium for coagulation to start can be derived from the condition that the diffusion losses characteristic time τ_d becomes smaller than the coagulation characteristic time τ_g. The value of τ_d is derived from the classical Brownian motion and is expressed below for a given geometry (diffusion length Λ_d) and given ambient gas conditions (index g). The parameter η (>1) takes into account transient charging effects able to enhance the residence time of nanoparticles in the plasma. The kinetic temperature of the nanoparticles (radius R_p) is assumed to be the gas temperature (T).

$$\frac{1}{\tau_d} = \eta \frac{9}{8\pi} \frac{3kT}{\Lambda^2} \frac{1}{R_p^2} \frac{1}{n_g m_g v_{gth}}$$

The coagulation characteristic time is derived from the collision rate between neutral nanoparticles :

$$\frac{1}{\tau_c} = \sqrt{2} \sqrt{\frac{8kT}{\pi M_p}} 4\pi n_p R_p^2$$

The condition for starting a coagulation process is expressed as $\tau_c < \tau_d$.
For an argon plasma at 300 K, 130 Pa, with $\Lambda = 1$ cm, $R_p = 1$ nm, this condition leads to a critical density of particles inducing coagulation : $n_p > \eta \times 6.10^{11}$ cm^{-3}. This value is consistent

with estimations derived in many experiments, showing that nanometric particle densities up 10^{12} cm^{-3} can be achieved. It shows that fairly high densities of such particles are allowed in the gas phase without inducing a coagulation process.

However it is important to note that the above equilibrium related to particles losses can be strongly affected by other forces such as thermophoresis [4] as well as by details on the geometry of the reactor. This is illustrated in figure 1 where we present the time evolution of the second harmonic of the RF current in a 3% silane in hydrogen discharge.

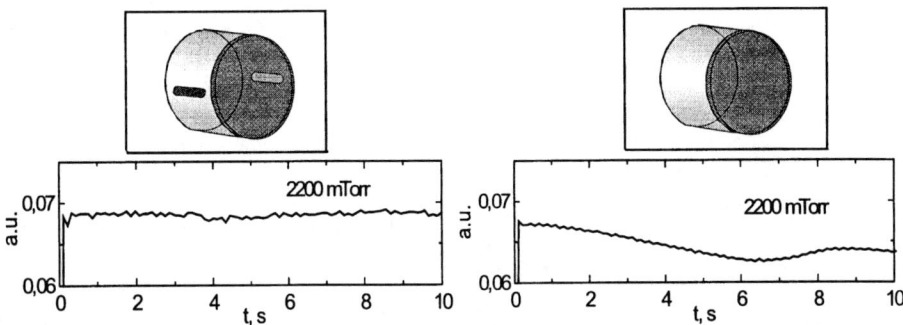

FIGURE 1. Effect of slits in the plasma box on the dynamics of powder formation. Note that for the same pressure the second harmonic of the RF current is independent of time when the slits are open while it shows a drop, characteristic of powder formation, when the slits are closed.

From nanoparticles grown in silane plasmas to polymorphous silicon films

As shown in Figure 1, the measurement of the second harmonic of the RF current can be used as a simple technique to identify the plasma conditions under which nanoparticles are formed at concentrations smaller than the critical one for agglomeration, i.e., under conditions in which film growth results from the contribution of radicals and nanoparticles. While in our previous studies we have focused on the excellent properties of polymorphous silicon films, particularly the stability of hole transport properties resulting in stable PIN solar cells [5], we want to emphasize here on the optimization of the deposition conditions and particularly on the achievement of high deposition rates. Figure 2 shows the effect of increasing the total pressure on the deposition rate at 250°C of films produced from the dissociation of a 3% silane in hydrogen mixture. Also shown in the figure is the effective power coupled to the plasma (deduced from the substractive method).

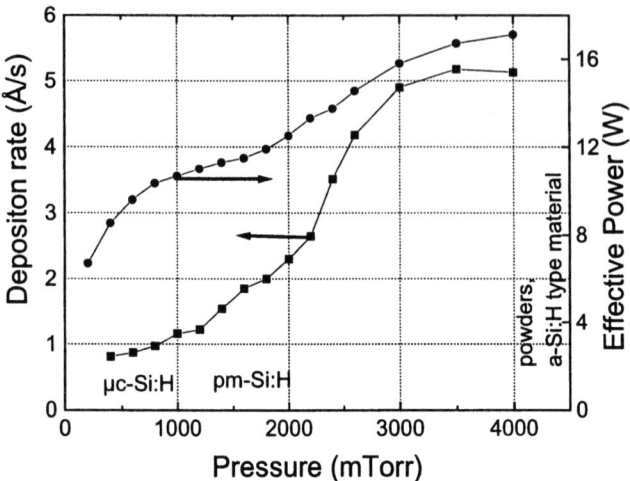

FIGURE 2. Effect of pressure on the deposition rate and the coupling of the RF power into the reactor

We can see that the increase of the total pressure results in an increase of the effective power coupled to the plasma. However, the increase in the deposition rate does not follow that of the coupled power. In particular, at low pressure we observe that the produced films are microcrystalline and that in this pressure range (400 mTorr – 1 Torr) the deposition rate is almost constant while the coupled power increases by 40%. As we increase the pressure in the range of 1 – 2 Torr the deposition rate increases by more than a factor of two while the coupled power only increases by 10%. This clearly suggest either a change of the surface reaction probability of SiH_x radicals or to a change in the nature of the species contributing to deposition. Our studies point towards an increasing contribution of clusters and nanoparticles to the growth. It is in the pressure range of 1-2 Torr that we have focused our studies of pm-Si:H films. While at high pressure (above 4 Torr in the example of Fig. 2) powders are formed and the deposition rate decreases, the nature of the films deposited in the range of 2 Torr to 3 Torr, over which the deposition rate sharply increases, remains to be determined. We suggest that the increase of the deposition rate in this pressure range must be related to an enhanced flux of nanoparticles contributing to deposition. Indeed impedance measurements unambiguously indicate that nanoparticles are present in the plasma in this pressure range and the increase in deposition rate cannot be accounted for the slight increase of the power. Further studies on the properties of the films deposited in the 2 to 3 Torr pressure range are required to determine the upper limit of the deposition rate for the obtaining of polymorphous silicon films.

Powders and nanocrystals in SiF_4 plasmas

While silane plasma chemistry and the formation powders in such plasmas have been largely studied, less is known about the formation and behavior of nanoparticles in SiF_4 mixtures. The large electronegativity of F should certainly promote the formation of powders and indeed this has been observed by the use of impedance probe measurements. Moreover in situ ellipsometry studies of the deposition from such gas mixtures have revealed:

i) a particular growth mechanism which can be interpreted as a consequence of the contribution of nanoparticles to the deposition [6]

ii) that the films are fully crystallized without any amorphous phase. This is further supported by Raman and AFM measurements which indicate that indeed the structure of the films is crystalline and that their shape is consistent with the direct deposition of the nanocrystalline silicon particles from the gas phase. This is illustrated in Figure 3 (left) where we present and AFM image of the surface of a 40 nm thick film in which agglomerated nanoparticles are observed. The crystalline nature of the nanoparticles is further demonstrated by photoluminescence studies. Figure 3 (right) gives an example of a photoluminescence measured at room temperature on a film exposed to 50 mW power from an argon ion laser (λ = 488 nm) The spectrum can be decomposed into two bands centered at 590 cm^{-1} and 660 cm^{-1}, suggesting that there is a distribution of sizes. Moreover, from the pea position of the PL we can estimate a crystallite size n the range of 2 – 3 nm [7]. While these preliminary results deserve further research, they open the way to new applications of plasma produced nanocrystals.

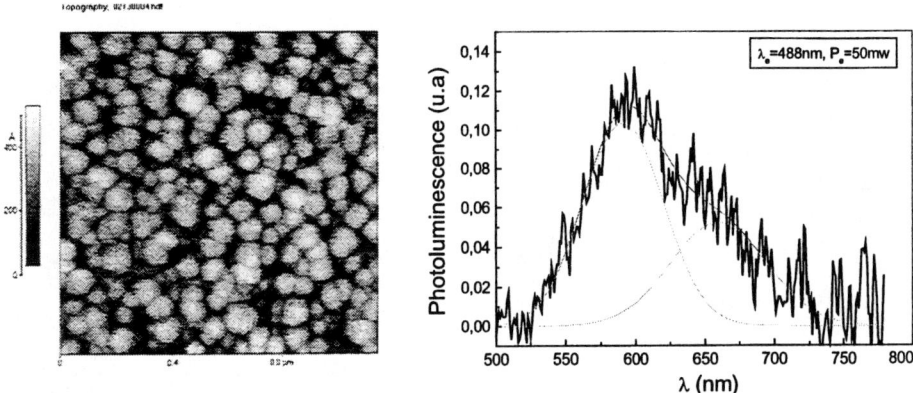

FIGURE 3. AFM image of a 40 nm thick nanocrystalline silicon films produced from SiF4 gas precursor (left) and photoluminescence spectrum of the same film measured at room temperature.

PART II : PARTICLE CLOUDS GROWN IN PKE-NEFEDOV MICROGRAVITY EXPERIMENT

Several other contributions (M. Mikikian *et al*, R.A. Quinn *et al*, H. Thomas *et al*, M. Zuzic *et al*) present the PKE-NEFEDOV experiment, both for ground studies and in microgravity conditions. This paper is focused on experiments achieved on board of ISS during the CNES taxi flight in October 2001, where the GREMI team joined the program previously launched by Max Planck für Extraterrestrial Physik (Garching, Prof Morfill) and Institut of High Energy and Devices (Moscow, Prof Fortov).

The main aim during this taxi flight mission was to study the behavior of dust clouds of small particles (< µm) grown in the PKE tool developed by MPE and previously installed on board of ISS.

Particle growth was observed in the PKE RF discharge operating with pure argon gas feeding. This growth of particles has been clearly evidenced as a consequence of plasma induced sputtering/etching of previously injected particles (mélamine formhaldehyde). Preliminary

experiment achieved in laboratory conditions evidenced several features involved in this particle growth (10) :
- when starting with a clean plasma box no particle formation is observed
- injected particles diameter is reduced from 3.4 to 3.1 μm when stored in argon plasma for 1 hour (plasma crystal studies)
- the material of grown particles is carbon while the injected particles contains C/N/O/H with a mass contribution of carbon ≈ 1/3

These studies gave guidelines for the optimal operating conditions on board of ISS i.e. argon pressure ≈ mbar, RF power ≈ 3W . In these conditions particle clouds of minute particles (few 100 nm) are typically grown in 2-5 mn.

Long term (> 20 mn) experiments reveal also that this etching/sputtering process leads to cyclic phenomena with exchange of matter between already grown particles and new generation of clouds grown in the central void generally observed in these experiments. The figure 4 below shows clearly such phenomenon : in 10 mn of plasma duration the size of particles in the initial cloud surrounding the void is reduced while another cloud is grown inside the initial void. It is interesting to note that these clouds remains well separated while the particle sizes are very close.

Figure 4 : clouds structures observed after 10 and 20 mn in PKE experiment

In the experiments on board of ISS the only available data are electrical measurements and videoframes of scattered light from the clouds illuminated by a laser sheet. This paper will be focused on particle size and concentration estimations derived from the time evolution of the discharge impedance and from experiments on wave propagation through this dusty-dense plasma.

Plasma impedance evolution during particle growth

Figure 5 shows the time evolution of the amplitude of the fundamental harmonic of the discharge current in two situations. The figures 5a concerns an argon-silane discharge where silicon dust particle are formed in an capacitive discharge with 3 cm distance between the two electrodes. Figure 5b is related to the PKE-NEFEDOV experiment on board of ISS. In these two experiments the drastic decrease of the current fundamental harmonic amplitude is due to the electron attachment on the dust particle. In both cases this drop represents 2/3 of the initial value and it means that similar area density (m^{-1}) of the dust particles are obtained. As particle size and concentration are known in the argon-silane plasma [8], this leads to an estimation of this area : $n_p R_p^2 \cong 0.1 \; m^{-1}$.

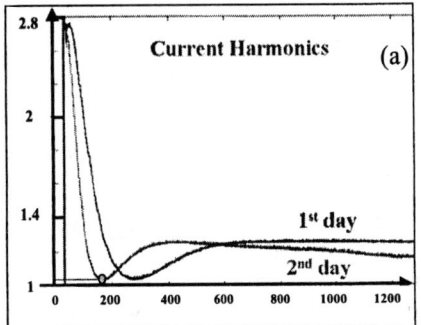
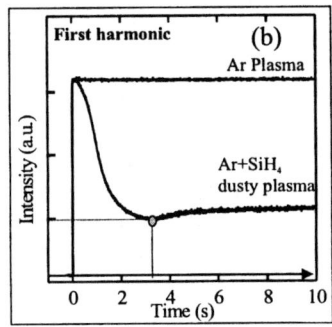

FIGURE 5. Time evolution of the amplitude of the fundamental harmonic of the discharge current for two runs on board of ISS in the PKE reactor (a) and in the argon-silane RF plasma (b).

Data derived from particle cloud dynamics

By using a low frequency voltage added to the RF power between electrodes a sheath width modulation is obtained, inducing vertical displacement of clouds edges.
Typical vertical displacements of the lower edge h_l of the dust cloud (sheath of the bottom electrode) and of its upper edge h_u (surface of the central void) are shown on figure 6.

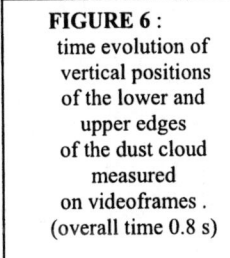
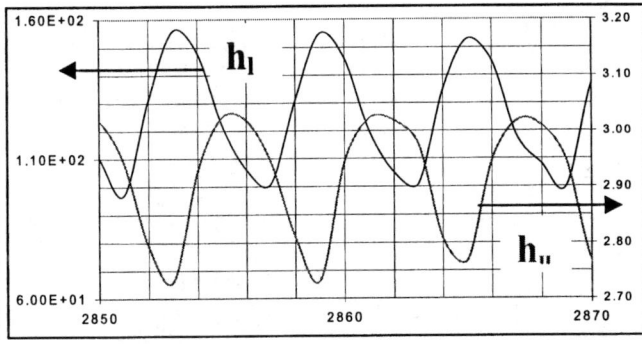

FIGURE 6: time evolution of vertical positions of the lower and upper edges of the dust cloud measured on videoframes. (overall time 0.8 s)

The time delay observed on these vertical vibrations corresponds to a propagation distance close to 2 cm. The corresponding propagation velocity is of the order of 0.3 m s^{-1}. These experimental data lead to another relation involving particles size and density, by using the measured longitudinal vibration propagation in dust acoustic wave expression (9).

This velocity is given by [9]
$$v_{daw} = \left[\frac{kT_p}{M_p} + \frac{kT_i}{M_p} \cdot \frac{\varepsilon Z^2}{1 + \frac{T_i}{T_e} \cdot (1 - \varepsilon Z)} \right]^{\frac{1}{2}}$$

where T_p, T_i and T_e are respectively the dust particle, ion and electron temperatures. M_p represents the dust mass and Z the particle charge number and ε is the ration of the particle density to the positive ion density. Knowing that the formed dust is carbon particles having a density of about ρp = 1.5 g.cm^{-3}, with $T_e \gg T_i \sim T_p$ and $n_D Z \cong n_i$ this expression is reduced to a simpler one : $v_{daw} = \left(\dfrac{3kT_g n_i}{4\pi\rho_p}\right)^{\frac{1}{2}} \left(n_p R_p^3\right)^{-\frac{1}{2}}$.

Taking into account the measured velocity a value of the volumic density of particles in the cloud is estimated as $n_p R_p^3 \cong 510^{-9}$.

Combined with the value of the surface density $n_p R_p^2$ derived from electrical behavior of RF discharge, these two set of data lead to an estimation of particle radius and particle density in the cloud. These estimations are respectively of 50 nm and 4.10^{13} m^{-3}.

These values appear to be consistent with the fact that electrical diagnostic show that a dusty-dense situation is obtained while videoframes cannot evidence individual particles. But, even if this approach in deriving microscopic particles from macroscopic data appear to be an attractive one, more precise derivations would require both a dusty plasma discharge modeling for impedance data and further experiments for the description of propagation phenomena (these are clearly not plane or spherical waves in this experimental situation).

ACKNOWLEDGMENTS

The LPICM work on polymorphous silicon films is supported by EC contract H-Alpha Solar (NNE5-1999-00133).

The GREMI work on dust clouds in PKE has been supported by the French space Agency under the CNES contract n°793/2000/CNES/8344 and data relevant of the PKE-NEFEDOV on board of the ISS have been obtained within the frame of the ANDROMEDE scientific program of the corresponding taxi flight mission (CNES, Oct 2001).

REFERENCES

1. See for a recent review : *Dusty Plasmas*, edited by A. Bouchoule, John Wiley & Sons Ltd, England (1999).
2. Roca i Cabarrocas, P., Fontcuberta i Morral, A., Lebib, S., and Poissant, Y., *Pure Appl. Chem.* **74**, 359 (2002).
3. Fridmann A., Boufendi L., Hbid T., Potapkin B., Bouchoule A., *J. A. P.* , **79** (3), 1303-1314
4. Fontcuberta i Morral A., and Roca i Cabarrocas, P., *Thin Solid Films*, **383**, 161 (2001).
5. Roca i Cabarrocas, P., Fontcuberta i Morral, A., and Poissant, Y., *Thin Solid Films*, **Vol. 403-404**, 39 (2002).
6. Kasouit, S., Kumar, S., Vanderhaghen, R., Roca i Cabarrocas, P., and French, I., *J. Non Cryst. Solids,*,**299-302**, 113 (2002).
7. Ledoux, G., Guillois, O., Porterat, D., Renaud, C., Huisken, F., Kohn, B., and Paillard, V., *Phys. Rev. B*, **62**, 15942 (2000).
8. Boufendi, L., Bouchoule A., *Plasma Sources Sci. Technol.*, **3**, 262 (1994)
9. Thomson, C., Barkan, A., D'Angelo, N., and Merlino, R.L., *Phys. Plasma*, **4**, 2331 (1997)
10. Mikikian, M., Boufendi, L., Bouchoule, A., "*First Results on Dust Formation for the PKE Experiment on ISS* ", 9th Workshop on the Physics of Dusty Plasmas, 21-23 May 2001, Iowa City (United States) and M. Mikikian, L. Boufendi, A. Bouchoule, "*Dust Particles Growth and Behavior in the PKE Experiment* ", Vth European Workshop on Dusty and Colloidal Plasmas, 23-25

Micro-dynamics in 2D Dusty Plasma Liquids

Ying-Ju Lai, Lee-Wen Teng, Pie-San Tu, Hong-Yu Chu, and Lin I

Department of Physics, National Central University, Chungli, Taiwan 32054, Republic of China

Abstract. We review the recent studies on the micro-dynamics in strongly coupled dust Coulomb liquids suspended in low pressure glow discharges. Under the interplay of stochastic thermal noise and Coulomb interaction, cooperative fast hopping strings and vortices excited from the small amplitude caged motion in the ordered lattice domains are the key cooperative excitations in the system. Their spatio-temporal statistical behaviors obey the similar generic power law scaling as sac type avalanche in coupled sub-excitable system under noise. The hopping can be further enhanced by the external stress and suppressed by the finite boundary. The liquid exhibits nonlinear visco-elastic response under high frequency AC shear drives and exhibits layering transition in a narrow gap down to a few molecular width. Travelling soliton-type micro-bubble can be formed through intense pulsed laser ablation.

INTRODUCTION

In a glow discharge system, the suspended dust particles can be negatively charged by the highly mobile electrons and turn them into the Coulomb crystal or liquid state through their strong mutual Coulomb coupling [1-3]. The proper spatial and temporal scales offer an opportunity to directly visualize the micro-structure and motions of this nonlinear many body system through video microscopy [4-12]. In this work, we concentrate on the micro-dynamics of the dust Coulomb liquid system, which is less well explored due to its complicated disordered nature In liquid, it is often thought that the stronger stochastic thermal noise excites random long time motions. Whether the micro-dynamics exhibits coherence in the time scale much shorter than the thermal relaxation time is an open question. From a more general view of modern nonlinear complex system, the liquid system can be treated as a strongly coupled sub-excitable system perturbed by noise [14-18]. Similar generic behaviors as other excitable systems should be followed. Other than the thermal excitation, the micro-dynamics under external coherent DC or AC stresses, and under tight confinement are also interesting issues [19,20]. For example, the liquid nature is drastically changed and layering transition can be observed when the confinement scale is reduced down to a few inter-particle distance [21-24]. In this paper, the above issues and their microscopic physical origins are discussed. In addition, we further report the formation of soliton-like travelling micro-bubbles generated by an intense laser pulse. The experiment is conducted in a cylindrical symmetric rf dusty plasma system as described elsewhere [7,12]. A weakly ionized discharge $n_e \sim 10^9\,\text{cm}^{-3}$ is generated in 250 mTorr Ar gas using a 14 MHz rf power system. A hollow cylindrical cell 30 mm in diameter is placed on the center of the bottom electrode to trap polystyrene particles at 7 μm diameter. It can also be replaced by rectangular cells with various widths to study the confinement induced effect. Vertically, the suspended dust particles are aligned with eight particles for each chain by the vertical ion flow induced dipoles. Particles in the same vertical chain move together horizontally. Namely, our liquid is a quasi-2D system. The particle positions in the horizontal monolayer are monitored through digital video optical microscopy.

AVALANCHE TYPE FAST PARTICLE EXCITATIONS

We use a cold liquid after melting as an example to illustrate the avalanche type cooperative excitations of fast particles. Particles alternately exhibit caged motion with small amplitude oscillations in the triangular lattice sites of the ordered domain, and then enter the fast hopping state. Spatially, we can see many hopping strings or vortices surrounding the temporally ordered domains with a few a (mean inter-particle distance) scale. The hopping particles reenter the caged state after travelling in the order of $1a$ (Fig. 1). This non-equilibrium excitation is further manifested by Fig.1 (b) which shows that the particle displacement distribution $P(\Delta R^2)$ deviates from Gaussian with a larger high speed tail at time interval τ about 1.9 s. The displacement distribution returns to Gaussian at larger τ (a few tens of a seccond). Note that the onset of fast particle motion at the time scale where the distribution is the most non-Gaussian coincides with the time scale where the diffusion switches from the caged motion dominated sub-diffusion to the hopping dominated super-diffusion [12]. We try to understand the complicated

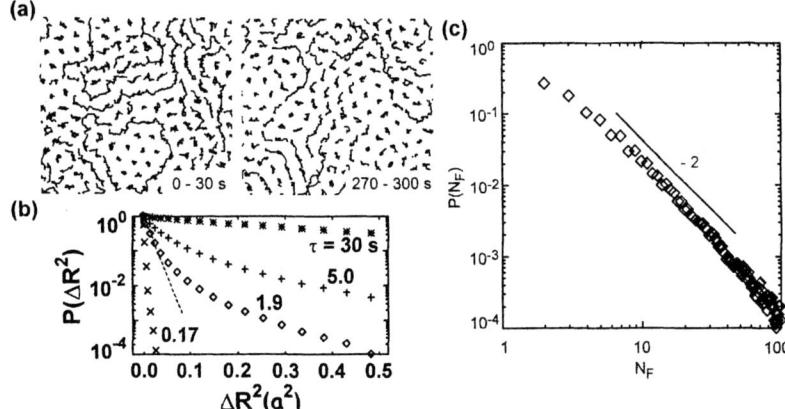

FIGURE 1. (a) The typical particle trajectories with 30 s exposure. (b) The histogram of particle displacement square at different time intervals. (c) The avalanche type power law scaling of the distribution of the cluster size of the connected sites of fast particles in the xyt-3D space.

spatio-temporal behaviors from a more general view of the recently developed nonlinear dynamics and complex systems [14]. Unlike the intuitive impression of the disruptive effect by stochastic noise, coherent state can be excited under the constructive influence of noise perturbation for coupled complex physical, chemical, and biological systems. Noise induced excitations in Diode array, sand-piles, chemical reaction systems, neural networks are the are the few good examples [14-18]. Noise plays two roles: (a) initiating the excitation through accumulation of positive perturbations, and (b) promoting and breaking the spatio-temporal propagation of excitation. For particle motion in our 2D liquid, we treat the fast particle with displacement larger than the half-width in the displacement histogram at 2 s as the excited state from the quiet caged motion. Particles start to hop cooperatively only after accumulation of sufficient constructive perturbations. Once particles hop, they can transfer energy to the neighboring particles through mutual interaction and further cascade hoppings of their neighboring particles. Their motions are slowed down if no further positive perturbations are supplied. We find that the connected sites of fast particle excitations in the xyt-3D space appear in the form of clusters [25]. Figure 1(c) shows that the cluster size (the number of the connected sites) distribution N_F follows a power law distribution with exponent about -2. The power law scaling without a typical spatio-temporal scale is similar to the photo noise excited patterns in the 2D BZ reaction [16]. Namely, the system belongs to the same class of the complex systems exhibiting SOC type behavior [14].

RESPONSE TO EXTERNAL STRESS

We shoot a horizontal narrow laser beam into the plasma and push the particle rightward through the radiation pressure [12,20]. The averaged forward velocity along the laser beam shows an S-shape nonlinear velocity response to the stress, which is proportional to the laser power [12]. Namely, the viscosity is high in the low stress regime and decreases to a constant value in the high stress regime (Fig. 2(b)). On the other hand, the transverse (to the laser beam) diffusion rate (i.e. proportional to $<\Delta R_y^2>$) increases with the stress. Many micro-vortices around both sides of the laser beam are enhanced (Fig. 2(a)). In the laser driven zone, the stress induced mechanical instability enhances the forward hopping. However, if the forward motion is frustrated through caging in the ordered domain, the particle branches off the laser zone and then induces small micro-vortices on the side. The micro-vortex mixing in turn enhances the transverse diffusion rate. At low laser power, the stress is too low to promote sufficient hopping out of the caging well. Namely, caging is the main source for the large viscosity and low transverse diffusion rate. At higher laser power, the mechanical instability is further enhanced. The dissipation of the forward momentum to excite the vortex type motion in the laser-off region are the main source for the low viscosity. Increasing temperature straighten the S-shape response curve. We further periodically chop the laser beam to investigate the physical origin of the visco-elastic response at the time scale shorter than the thermal relaxation time [20]. The averaged displacement (over 20 particles) along the laser direction does not fully return to the starting position after an on-off chopping cycle (Fig. 2(c)). Namely, unlike the macroscopic plastic deformation, the system shows both partial elastic response associated with partial plastic deformation at the micro-scale. We find that particles mayor may not exhibit backward motion when the laser is turned off. If a string or cyclic hopping is induced in or nearby the laser zone before the laser is off, the hopped particle finds a new stable position and never returns back. Hopping is the memory wash process for the viscous response and the caging is responsible for the elastic restoring.

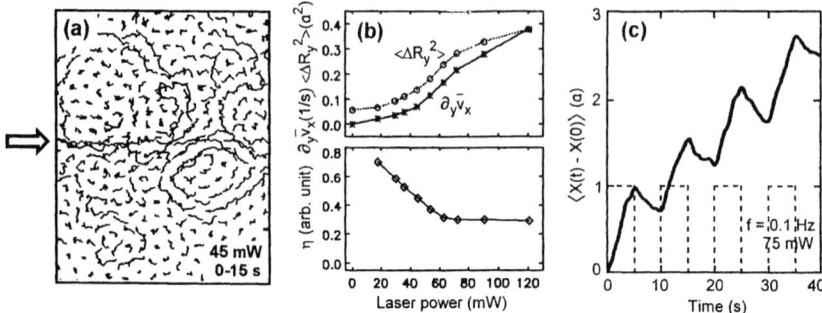

FIGURE 2. (a) Laser beam induced micro-vortices generation. The vortices have weaker intensity in the remote region. (b) The transversal mean square displacement, the shear rate and the viscosity η versus laser power. (c) The averaged particle displacement showing the visco-elastic response to the chopped laser drive. The lower dash line indicates the laser chopping cycle.

CONFINEMENT INDUCED LAYERING TRANSITION

The ultra-thin liquid confined in a small gap is an important subject in both science and technology [19,21-24]. Under confinement, liquid becomes sluggish and exhibits layering transition. Nevertheless, regardless of the numerical simulation [21-23]. Figure 3(a) shows that, similarly to their numerical simulation, the particle density distribution in our dust Coulomb liquid confined in rectangular traps with large aspect ratio also exhibits gradually damped oscillatory profile from the boundary. The straight boundary lines up the nearby particles and prohibits the nearby cyclic hopping, which not only suppress diffusion but also break the isotropy for thermal motion. Figure 3(b) shows the particle displacement histograms in which particles initially sitting in the outmost layer and in the center respectively. The motion is highly an-isotropic in the outer region. The transverse displacement histogram shows similar damped oscillatory profile. The layering in the outer region strongly cages the particle motion. Particles hop to the second layer and become trapped again. The layering trapping gradually decreases while moving away from the boundary. The displacement in the center is more isotropic with a smoother and more Gaussian like distribution profile. Reducing the gap width bellow $7d$, the hopping is also suppressed in the center region, oscillatory density profile persist through the gap. The diffusion rate in the center part drastically decreases.

LASER PULSE GENERATED TRAVELLING MICRO-BUBBLES

The bubble formation in extended media is an interesting nonlinear phenomena [26-28]. Bubbles generated by heating, ultrasonic driving, pulsed laser ablation in water, etc., are the good examples. Nevertheless, the micro-bubble with diameter down to about 10 molecular scale has never been studied. We use a pulsed Nd: YAG laser (532 nm, 50 mJ/pulse) focused at the center of a large dust liquid suspended in the plasma trap to study the micro-bubble generation [29] The pulsed laser can generate an intense plasma

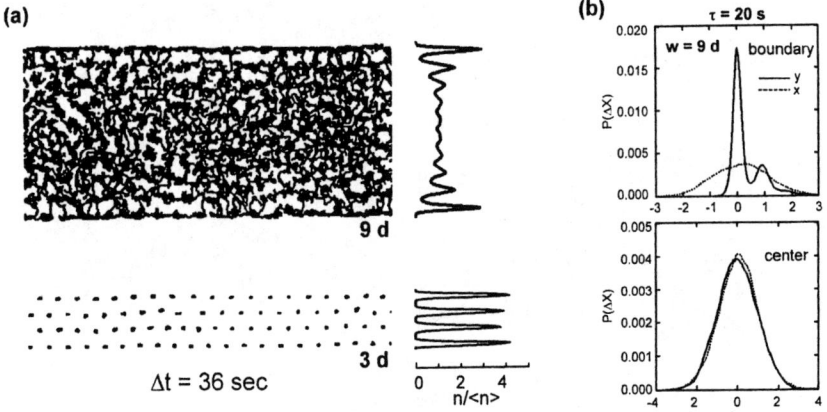

FIGURE 3. (a) The typical particle trajectories and the transverse density distributions showing the confinement induced layering. (b) The typical histograms of the transverse and longitudinal displacement at the boundary and center of the confinement gap with width 9 d, where d is the mean layer distance.

plume through the ablation of one or a few dust particles fall in the focal region, within about 0.1 to 1 microsecond after laser incidence [30]. Figure 4 shows the snap shots of the micrographs taken by our CCD camera with 17 msec exposure time at 60 Hz frame rate. The plasma plume pushes out particles at the center. The dust free bubble propagates downward as a soliton at speed about 3 cm/s and eventually disappears at the lower sheath boundary.

Since the life time of the plasma plume from the ablated particle is much shorter than the life time of the bubble, another source for maintaining the outward pressure preventing the bubble from collapsing is needed. Similarly to the spontaneous formation of great voids in the micro-gravity experiment [10], the dust particle depletion in the bubble reduces the rate of electron depletion through the electron-ion recombination on the dust particle surface. It in turn generates higher rate of electron impact ionization. The outward ion flow from the denser plasma inside the bubble provides outward momentum, which prevents the bubble from collapsing. The downward ion flow to the bottom electrode surface should be the main source for the downward motion of bubbles.

ACKNOWLEDGMENTS

This work is supported by the National Science Council of the Republic of China under contract number 90-2112-MOO8-50.

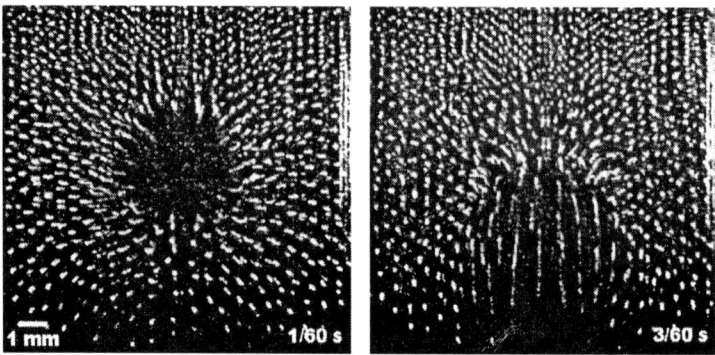

FIGURE 4. The sequential snap shots (1/30 s apart) showing the formation of a travelling micro-bubble generated by an intense laser pulse in the dusty plasma liquid

REFERENCES

1. Chu, J.H., and Lin, I., *Phys. Rev. Lett.*, **72**, 4009 (1994).
2. Thomas, G.E. *et al* , *Phys. Rev. Lett.*, **73**, 652 (1994).
3. Hayashi, Y., and Tachibana, K., *Jpn. J. Appl. Phys.*, **Part 1, 33**, 804 (1994)
4. I, Lin, Juan, W.T., and Chiang, C.H., *Science*, **272**, 1626, (1996); Juan, W.T. and I, L., *Phys. Rev. Lett.*, **80**, 3073 (1998).
5. Melzer, A.,. Homann, A., and Piel, A., *Phys. Rev. E*, **53**, 2757 (1996).
6. Pieper, J.B., and Goree, J., *Phys. Rev. Lett.*, **77**, 3137 (1996).
7. Juan, W.T., *et al*, *Phys. Rev. E*, **58**, R6947 (1998).
8. Lai, Y.J., and I, L., *Phys. Rev. E*, **64**, 015601(R) (2001).
9. Quinn, R.A., and Goree, J., *Phys. Rev. E*, **64**, 051404 (2001).
10. Morfill, G.E., *et al*, *Phys. Rev. Lett.*, **83**, 1598 (1999); Goree, J., *et al*, *Phys. Rev. E*, **59**, 7055 (1999).
11. Pramanik, J., Prasan, G., Sen, A., Kaw, P.K., *Phys. Rev. Lett*, **88**, 175001 (2002).
12. Juan, W.T., Chen, M.H., and I, L., *Phys. Rev. E*, **64**, 016402 (2001)
13. Melzer, A, Schweigert, V.A. and Piel, A, *Phys. Rev. Lett.*, **83**, 3194 (1999).
14. Bak, P., Tamg, C., and Wiesenfeld, K., *Phys. Rev. Lett.*, **59**, 381 (1987). Jensen, H, *Self-Organized Criticality*, (Cambridge University Press, Cambridge, 1998). li
15. Locher, M.., Cigna, D., and Hunt, E.R., *Phys. Rev. Lett.*, **80**, 5212 (1998).
16. Wang, J., Kadar, S., Jung, P., Showalter, K., *Phys. Rev. Lett.*, **82**, 855 (1999).
17. Jung, P ., *et al*, *J. Neurophysiol.*, **79**, 1098 (1998).
18. Van den Broeck, C, Parrondo, J.M.R., and Toral, R., *Phys. Rev. Lett.*, **73**, 3395 (1994).
19. Granick, S., *Phys. Today*, July, 26 (1999).
20. Juan, W.T., *et al*, *Physica Scripta*, **T98**, 9 (2001).
21. Bhushan, B, Israelachvili, J.N., and Landman, U., *Nature*, **374**, 607 (1995).
22. Thompson, P.A, Grest, G.S., and Robbins, M.O., *Phys. Rev. Lett.*, **68**, 3448 (1992).
23. Gao, I., Luedtke, W.D., and Landman, U., *Phys. Rev. Lett.*, **79**, 705 (1997).
24. *Chem. Phys.*, **109**, 6889 (1998).
25. Lai, Y.J., Ph.D. Thesis, Physics Department, National Central University, Taiwan (2002).
26. Walton, A., and Reynolds, G., *Adv. Phys.*, **33**, 595 (1984); Brennen, C.E., *Cavitation and Bubble Dynamics* (Oxford Univ. Press,New York, 1995).
27. Bulanov, S.V., *et al*, *Phys. Rev. Lett.*, **82**, 3440 (1999).
28. Wang, K.G., *Phys. Rev. E*, **62**, 6937 (2000).
29. Chu,. H. Y ., Master Thesis, Physics Department, National Central University, Taiwan (2001).
30. *Laser Ablation: Mechanisms and Applications -II*, edited by Miller, J.C., and Geohegan, D.B., Geohegan (American Institute of Physics, New York, (1994).

Plasma Response to a Single Grain/Electrostatic Interaction Between Grains

Martin Lampe

Plasma Physics Division, Naval Research Lab, Washington, DC 20375-5346, USA

Abstract. We review the physics of shielding around a single dust grain in a plasma, charging of the grain, and electrostatic interaction between two nearby grains. Orbital-motion-limited (OML) theory is found to be an acceptably accurate representation for the contribution of positive-energy ions to shielding, in non-flowing plasmas. However, shielding is often dominated by negative-energy trapped ions created by charge-exchange collisions. The result is usually very close to the Debye-shielded form, for $r <$ several λ_D. For larger r, $\phi(r) \sin r^{-2}$. Trapped ions also greatly increase the ion current to the grain, even in very weakly collisional cases. As a result, the grain charge can be significantly less than the OML result. For grains immersed in streaming plasma in a sheath, linear response theory is found to be a satisfactory method for calculating the wake fields, and is supported by experimental data.

INTRODUCTION

Among the most basic issues in the physics of dusty plasma are the plasma electrostatic response to a single dust grain, and the electrostatic interaction between two isolated dust grains in a plasma. What is the shielding around one dust grain? What is the charge collected by one grain in a plasma? What is the electrostatic force between two dust grains? One's first thought is that each grain generates a Debye-shielded Coulomb potential, and that this potential characterizes the interaction between grains. I hope to convince you that this is far from obvious (but nonetheless is very close to true, at least in some regimes). Even after seventy-five years of work, there are still surprises in this area of physics.

GRAIN IN NON-FLOWING PLASMA: OML THEORY

The textbook derivation of the Debye-shielded potential around a negative point charge $-Ze$, in stationary plasma, begins with an assumed Boltzmann equilibrium,

$$n_i = n_0 e^{-e\phi/T_i}, \qquad n_e = n_0 e^{+e\phi/T_e}, \qquad (1)$$

where $n_{e,i}$ are the electron and ion densities, and $\phi(r)$ is the potential. Next, Eq. (1) is linearized by assuming that $|e\phi| \ll T_e, T_i$, and inserted into the Poisson equation. The solution is then the usual Debye-shielded potential $\phi(r) = (Ze/r) \exp(-r/\lambda_D)$.

However, $\phi \to \infty$ for $r \to 0$, so that Eq. (1) indicates n_i has an essential singularity at $r = 0$. Of course, a grain is not a point particle, it has some finite radius a. But the potential at the grain surface is typically $-2T_e$, and in laboratory discharges typically $T_e > 50T_i$, so Eq. (1) would indicate that $_i(a) > e^{100}n_0$. Obviously, linearization is unjustified, and even the unlinearized Boltzmann assumption (1) is clearly wrong.

What is wrong is that the ions near the grain are not in equilibrium. In steady state, ions come in from the ambient plasma ($r = \infty$), pass by the grain and fly back out to $r = \infty$., Since the ion mean free path λ_{mfp} is typically long compared to λ_D, it seems reasonable to assume that the ion trajectories are collisionless. This is the basis of the well-known orbital-motion-limited (OML) theory [1-8]. The Vlasov equation indicates that the ion distribution function $f_i(r,v)$ is a constant along ion trajectories. Since $f_i(r,v)$ is a Maxwellian at $r = \infty$, it follows that

$$f_i(\mathbf{r},\mathbf{v}) = n_0 \left(\frac{m_i}{2\pi T_i}\right)^{3/2} \exp\left(-\frac{m_i v^2}{2T_i} - \frac{e\phi(r)}{T_i}\right) \qquad (2)$$

However, the Maxwellian is not completely filled. Ions from the ambient plasma must have positive energy. In addition, outgoing ions whose trajectories have previously hit the grain must be removed from the distribution. These two conditions require that

$$\frac{1}{2}m v^2 + e\phi(r) \geq 0, \qquad \sin\theta > \frac{a}{r}\sqrt{1 + \frac{2e[\phi(r) - \phi_a]}{m_i v^2}} \qquad (3)$$

The ion density is obtained by integrating (2) over v, with the restrictions (3). Solving the resulting nonlinear Poisson equation numerically, one finds that the potential $\phi(r)$ is fairly close to a "Debye-like" form after all, for r out to several λ_D [5,8]. However, the effective shielding length is larger than the usual Debye length, i.e. shielding is weaker. Furthermore, $\phi(r)$ transitions to a slowly decaying form for large r,

$$e\phi(r) \sim -(2ee\phi_a - T_i)\frac{T_e}{T_e + t_i}\frac{a^2}{4r^2} \sim -\frac{Ze^2 a}{2r^2} \qquad (4)$$

Equation (4) is a consequence of the absorption of ions that hit the grain. Finally, the steady state grain potential ϕ_a can be calculated by setting the ion flux to the grain equal to the electron flux. The result is that ϕ_a is given by the solution of

$$(1 - frace\phi_a T_e)\exp\left(\frac{-e\phi_a}{T_e}\right) = \left(\frac{m_i T_i}{m_e T_e}\right)^{1/2} \qquad (5)$$

Equation (5) indicates that ϕ_a is typically $2T_e$ to $3T_e$.

POTENTIAL BARRIERS AND THE VALIDITY OF OML

In writing Eq. (2) it was tacitly assumed that every positive-energy point in phase space lies on an ion trajectory that is connected to $r = \infty$. But in a spherically symmetric potential, the ion radial equation of motion can be written in terms of an effective potential $U(r)$ that includes the centrifugal force,

$$m_i \frac{dr}{dt} = \frac{d}{dr}\left(-e\phi + \frac{l^2}{2mr^2}\right) \equiv \frac{dU}{dr} \tag{6}$$

In general, $U(r)$ can have any of the forms shown in Fig. 1. If $U(r)$ has a maximum (Case 3 of Fig. 1), there can be positive-energy ion trajectories that are trapped inside the maximum, i.e. these phase space points do not connect to $r = \infty$, and the ion distribution function at these points is not given by Eqs. (2,3). However, Bernstein and Rabinowitz[2] showed that if the ambient ion distribution is taken to be monoenergetic but isotropic, then $U(r)$ has no maxima in the limit of small grain radius a; hence, one would expect no problem for dusty plasma.

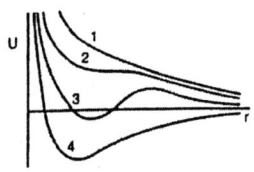

FIGURE 1

But recently, Allen et al [9] showed that this is not true when the ambient distribution is Maxwellian: for any value of a, there are always potential barriers. This has led to concern about the validity of OML theory. However, Lampe [10] subsequently showed that for typical dusty plasma conditions, specifically if $_e/T_i \ll 100\lambda_D/a$, the maximum in $U(r)$ is low and affects only a few ions. Hence it appears to be sufficiently accurate (but not exactly true) to assume that all positive-energy points in phase space are connected to $r = \infty$ by collisionless trajectories.

TRAPPED IONS

Since $\lambda_D \ll \lambda_{\text{mfp}}$, it would appear quite reasonable to assume, as OML theory does, that the ions are collisionless. However, Bernstein and Rabinowitz suggested in 1959 that negative-energy trapped ions, created by occasional collisions, could be significant. Goree [11] later made a most remarkable observation: Trapped ions, once formed, remain indefinitely in the potential well around the grain. Hence, even if the collision frequency v is very small, the density of trapped ions can build up indefinitely, limited only by other collisions that cause the ion to fall onto the grain or to escape from the well. Since both the source and the sink of trapped ions are proportional to v, the density $n_t(r)$ of trapped ions in steady state must be independent of v. Furthermore, Monte Carlo simulations [11,12] showed that $n_t(r)$ is large. Lampe et al [13] subsequently developed an analytic theory that solves for the density of trapped and untrapped ions, self-consistently with Poisson's equation, in the limit of small v and with the additional model assumption that v is independent of ion energy. The calculation proceeds by tracing self-consistently through the full lifetime of a trapped ion, from its creation as a result of a charge-exchange collision of a positive-energy ion near the grain, through any number of subsequent collisions, until the trapped ion finally falls onto the grain or escapes from the potential well. The density of trapped ions (normalized to ambient), the deviation of the untrapped ion density from ambient, and the deviation of the untrapped electron density from ambient are shown (on a semi-log scale) in Fig. 2, for the case where the neutral molecule temperature is $T = T_i = 0.01T_e$, and $a/\lambda_D = 0.015$.

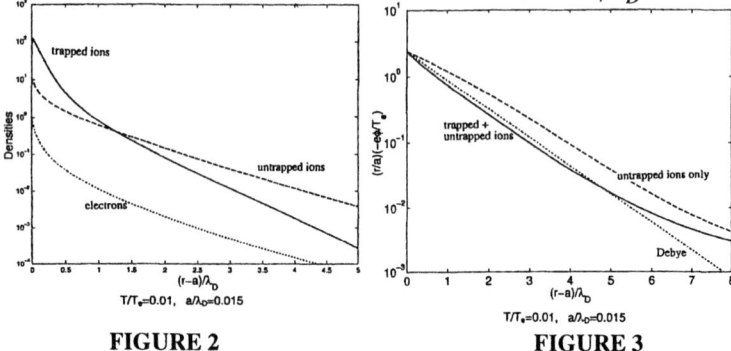

FIGURE 2   **FIGURE 3**

Note that near the grain, the density of trapped ions is an order of magnitude larger than the untrapped ion density, so OML theory is clearly in error. In Fig. 3 we show $r\phi(r)$, also on a semi-log scale. In this plot, Debye shielding is a diagonal straight line as indicated. Note that OML shielding is substantially weaker than Debye, but the complete potential, including trapped ions, is remarkably close to standard Debye shielding for $r < 5\lambda_D$. For larger r, $\phi(r)$ transitions to the r^{-2} form. It is generally true in the cases that we have calculated that $\phi(r)$ is close to Debye. It is to be expected that including trapped ions will make the shielding stronger than the OML result, but we are not aware of any simple physical explanation for why the complete calculation yields a result that is so close to standard Debye shielding. The ion density n_i is many orders of magnitude less than the nonlinear Boltzmann density, and it is invalid to linearize by assuming $e\phi \ll T_i$, but nonetheless n_i turns out to be close to the *linearized* Boltzmann density that leads to Debye shielding. Trapped ions also greatly increase the ion current to the grain F_i, which is given by

$$F_i = \frac{v}{4\pi a^2} \int_a^\infty dr 4\pi r^2 n_i(r) p(r) \qquad (7)$$

where r is the location where the ion had its last collision, and $p(r)$ is the probability that the ion's trajectory after the collision will take it onto the grain, and that it will not have another collision before reaching the grain [14]. It is possible to express $p(r)$ in terms of integrals over the ion distribution, which are too lengthy to reproduce here., Equation (7) reduces to the OML current in the limit $v \to 0$. The collisional correction to OML is first-order in v, but it has a very large coefficient. For example, we find $F_i/F_{\text{OML}} = 6$ for $v\lambda_D/c_s = 0.03$. F_i is so large because any charge-exchange collision that occurs in the potential well around the grain leads to the creation of a trapped ion, which eventually falls onto the grain. Thus the cross-section for creation of a trapped ion is very large, of order $\pi\lambda_D^2$, but this cross-section must be multiplied by the probability that an untrapped ion will have a collision while it is in the well, which is a small quantity of order $\lambda_D/\lambda_{\text{mfp}}$. Because collisions increase F_i, the grain potential ϕ_a is reduced from the OML result (5) by as much as a factor of order two. The charge on the grain is proportional to ϕ_a and is similarly reduced.

TWO GRAINS

The force exerted by one grain on another grain is not necessarily the same as the force derived from the potential around a single isolated grain. There can be nonlinear deformation of the shielding cloud around one grain due to the presence of the other grain, and if this results in extra positive charge collecting in the region between the two grains, it is possible that there could be a net attractive electrostatic force between two negatively charged grains [15,16].

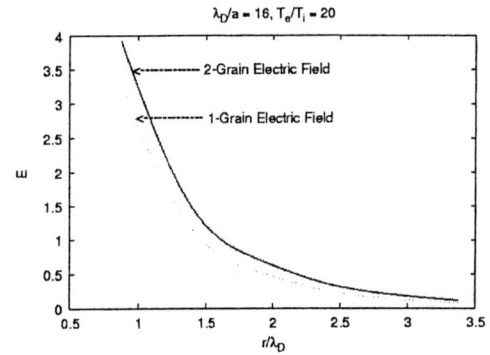

FIGURE 4

However, we have shown [14] by numerical solution of the OML equations for two grains that the nonlinear corrections actually increase the repulsion between two grains, as is shown in Fig. 4. This is in accord with analytic predictions that we made earlier for OML theory [8]. It is still possible that an attractive electrostatic force results from the overlap of the trapped ion clouds surrounding the two grains. A grain together with its trapped ion cloud is in some sense like a classical atom. When exposed to external fields, the trapped ion cloud becomes polarized, thereby shielding the grain. When two grains are nearby, Van der Waals attractive forces should result from the mutual polarization of the two trapped ion clouds. However trapped ion calculations have not yet been performed for multiple grains, and it is not clear whether Van der Waals forces can be strong enough to overcome Coulomb repulsion and lead to net attractive forces.

PLASMA STREAMING AND WAKE FIELDS

In laboratory discharges, dust normally resides at the sheath edge where ions are streaming toward the electrode at a speed u of the order of the ion sound speed c_s. Streaming greatly complicates the response of the plasma, leading to the formation of wakefields behind each charged grain. Both fluid [17] and kinetic [18] simulations have been used to calculate the wake fields around a single grain. However, streaming also simplifies the physics in one way: Ions streaming at speed c_s have kinetic energy T_e. Since the potential at the grain surface is of order $2T_e$, and the potential falls off faster than r^{-1}, the ion kinetic energy far exceeds $\phi(r)$, for all ions except those that happen to pass very close to the grain. Thus linear response theory [19-26,8] gives a very good approximation to the dynamics of the streaming ions. In Fourier space, the plasma-mediated potential associated with a dust grain of charge Z_e is thus

$$\phi(r) = \int d^3k e^{i\mathbf{k}\cdot\mathbf{r}} \frac{ze}{2\pi^2 k^2 D(\mathbf{k},\mathbf{k}\cdot\mathbf{u}_i + i v_i)}, \qquad (8)$$

$$D(\mathbf{k}, \omega) = 1 + \frac{1}{k^2 \lambda_{De}^2} - \frac{\omega_{pi}^2}{k^2} - \frac{\int d^3 v \frac{\mathbf{k} \cdot \partial f_{i0}(\mathbf{v})/\partial \mathbf{v}}{\mathbf{k} \cdot \mathbf{v} - \omega - i V_i}}{1 - i v_i \int d^3 v \frac{\mathbf{k} \cdot \partial f_{i0}(\mathbf{v})/\partial \mathbf{v}}{\mathbf{k} \cdot \mathbf{v} - \omega - i V_i}} \quad (9)$$

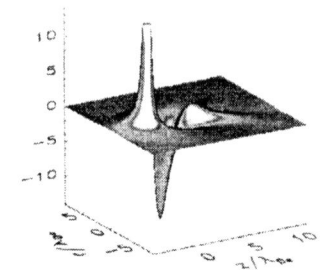

FIGURE 5

The dynamically screened potential given by the Fourier transform of (8), (9) is highly anisotropic. It is shown in Fig. 5 for a case with $T_e/T_i = 24$ and $/c_s = 1$. Upstream ($z < 0$) and transverse ($z \sim 0$) to the streaming, the potential is Debye-like, but with a screening length λ that depends on u and on the direction (because rapidly streaming ions become ineffective at screening, and λ is thus of the order of the electron Debye-length, rather than the much smaller ion Debye length).

Downstream, the negatively charged grain focuses the flow of positive ions, leading to a positively charged node behind the grain. The electron pressure overshoots at this point, giving rise to a train of ion sound oscillations behind the grain. If T_i and the gas pressure P are both small, the wake consists of a large number of nodes forming an array in the $r - z$ plane [20,8]. However, Eqs. (8), (9) include both collisional damping of the wake, which is significant at high neutral pressure, and Landau damping, which is significant at relatively high T_i/T_e; in typical dusty plasmas, there is only one strongly dominant node. The wake is strongest at $u = c_s$, but it is significant even for streaming velocities that are only a fraction of c_s. A more complete review is given in [8]. The interaction force has been measured by several methods [27,28] for grains in a sheath, confined in a plane transverse to the ion streaming. The results in all cases are Debye-like repulsive forces, consistent with the results of linear response theory.

CONCLUSIONS

OML theory appears to be satisfactory for treating the contribution to shielding of untrapped ions in non-flowing plasma. If only these ions are included in shielding, the electrostatic force between two grains is clearly repulsive. However trapped ions normally dominate the shielding around a grain. With both trapped and untrapped ions included, the shielding around a single grain is remarkably close to Debye shielding, out to several λ_D. For larger r, the potential $\phi(r)$ falls off as r^{-2}. With trapped ions included, it is not known whether an attractive electrostatic force can occur between two grains. Collisional effects (trapped ions) also greatly increase the ion current to a grain. As a result, the grain potential and charge can be reduced by a factor as large as two compared with the OML value. For dust in sheaths, where the ions stream by at speed of order c_s, linear response theory appears to satisfactory for calculating the wake fields. The interaction force transverse to ion streaming has been measured, and appears to have the Debye form, consistent with linear response theory. Attractive forces between grains have been seen only in wakefields.

In addition to electrostatic forces, nearby grains interact via shadowing forces [29,30], which result from momentum exchange with ions and neutral molecules. Space limitations have prevented discussion of these forces in the present review.

ACKNOWLEDGMENTS

My work has been in collaboration with G. Joyce, G. Ganguli, V. Gavrishchaka, R. Goswami, S. Robertson, and Z. Stemovsky. Supported by ONR and NASA.

REFERENCES

1. Mott-Smith, H. Jr., and Langmuir, I., *Phys. Rev.*, **28**, 27 (1926).
2. Bernstein, I.B., and Rabinowitz, I.N., *Phys. Fluids*, **2**, 112 (1959).
3. Laframboise, J.G., Univ. of Toronto, *Inst. For Aerospace Studies*, **Report #100**, (1966).
4. Laframboise, J.G., and Parker, L.W., *Phys. Fluids*, **16**, 629 (1973).
5. Daugherty, J.E., Porteus, R.K., Kilgore, M.D., and Graves, D.B., *J. Appl. Phys.*, **72**, 3934 (1992).
6. Allen, J.E., *Physica Scripta*, **45**, 497 (1992).
7. Boeufand J.P., Punset, C., in *Dusty Plasmas*, ed. A. Bouchoule (Wiley, New York, 1999) Chap. 1.
8. Lampe, M., G. Joyce, and G. Ganguli, *Phys. Plasmas*, **7**, 3851 (2000).
9. Allen, J.E., B. M. Annaratone, and U. deAngelis, *J. Plasma Phys.*, **63**, 299 (2000).
10. Lampe, M., *J. Plasma Phys.*, **65**, 171 (2001).
11. J. Goree, *Phys. Rev. Lett.*, **69**, 277 (1992).
12. A. Y. Zobnin, A. P. Nefedov, Y. A. Sinel'shchikov, and Y. E. Fortov, *JETP*, **91**, 483 (2000).
13. Lampe, M., Gavrishchaka, Y., Ganguli, G., and Joyce, G., *Phys. Rev. Lett.*, **86**, 5278–5281 (2001).
14. Lampe, M., Ganguli, G., and Joyce, G., Gavrishchaka, V., and Goswami, R., *in these proceedings*.
15. Tsytovich, Y.N., *Comments Plasma Phys. Controlled Fusion*, **15**, 349 (1994); Y. N. Tsytovich and D. Resendes, Plasma Physics Reports 24,65 (1998).
16. Resendes, D.P., Mendonca, J.T., and Shukla, P.K., *Phys. Lett. A*, **239**, 181 (1998).
17. Melandso, F., and Goree, J., *Phys. Rev. E*, **52**, 5312 (1995); *J. Yac. Sci. Technol. A*, **14**, 511 (1996).
18. Schweigert, Y.A., et al, *Phys. Rev. E*, **54**, 4155 (1996).
19. Shukla, P.K., *Phys. Plas.1*, 1362 (1994); Nambu , M., Vladimirov, S.Y., and Shukla, P.K., *Phys. Lett. A*, **203**, 40 (1995); Shukla, P.K. and Rao, N.N., *Phys. Plas.*, **3**, 1770 (1996).
20. Vladirnirov, S.Y., and Nambu, M., *Phys. Rev. E*, **52**, R2172 (1995); Vladirnirov S.Y., and Ishihara, O., *Phys. Plasmas*, **3**, 444 (1996), *Phys. Plasmas*, **4**, 69 (1997).
21. Rostoker, N., and Rosenbluth, M.N., *Phys. Fluids*, **3**, 1 (1960).
22. Sanmartin, J.R., and Lam, S.H., *Phys. Fluids*, **14**, 62 (1971).
23. Stangeby, P.C., and Allen, J.E., *J. Plasma Phys.*, **6**, 19 (1971).
24. Chen, L., Langdon, A.B., and Lieberman, M.A., *J. Plasma Phys.*, **9**, 311 (1973).
25. Krall, N.A., and Trivelpiece, A.W., *Principles of Plasma Physics* (McGraw-Hill, New York, 1973) Chapter 11.
26. Joyce, G., Lampe, M., and Ganguli, G., *IEEE Trans. Plasma Sci.*, **29**, 238 (2001).
27. Konopka, U., Ratke, L., and Thomas, H.M., *Phys. Rev. Lett.*, **79**, 1269 (1997); **84**, 891 (2000).
28. Hebner, G.A., et al, *Phys. Rev. Lett.*, **87**, 235001 (2001).
29. Ignatov, A.M., *Plasma Physics Reports*, **22**, 585 (1996).
30. Tsytovich, Y.N., et al, *Comments Plasma Phys. Controlled Fusion*, **17**, 249 (1996); *Physics- Uspekhi Physics*, **40**, 53 (1997).

Magnetic Effects in Dusty Plasmas

Noriyoshi Sato[1]

Graduate School of Engineering, Tohoku University Sendai 980-8579, Japan

Abstract. Here is presented a short review of our experiments on fine particles levitating in low-pressure weakly-ionized dusty plasmas under a vertical magnetic field. This includes recent results on rotation of fine-particle clouds in the azimuthal direction on the horizontal plane and preliminary measurements of spin motion of fine particles in a magnetic field. Future problems to be investigated are also pointed out, which might be important in order to clarify magnetic effects in dusty plasmas.

INTRODUCTION

Dusty plasmas including fine particles are of current interest in plasma physics and application. We have carried out a series of experiments on dynamics of fine particles in weakly-ionized plasmas at Tohoku University [1, 2]. One of our interests in dusty plasmas has been concerned with fine-particle behaviors in magnetized plasmas [3]. The magnetic field is useful for shape control of fine-particle clouds. We have also demonstrated a generation of azimuthal rotation of fine-particle clouds on the horizontal plane in the presence of vertical magnetic field. The rotation is generated even in such a weak magnetic field that positive ions are slightly magnetized. Following our observation of the rotation of fine-particle clouds, Konopka et al [4] confirmed this rotation in a rf plasma under a nonuniform magnetic field provided by permanent magnets. Here are presented some details of our experiments on fine particles in magnetized plasmas. The magnetic field, which is externally applied in the vertical direction, is varied in (1) weak (≤ 0.4 kG), (2) strong (0.4–4 kG), and (3) ultra-strong (4–40 kG) ranges. Roughly speaking, in the range of weak magnetic field, electrons are magnetized while ions are weakly magnetized. In the range of strong magnetic field, electrons and ions are magnetized while fine particles are not magnetized. In the range of ultra-strong magnetic field, fine particles are slightly magnetized in addition to electrons and ions. Even if the magnetic field is weak, shapes of fine-particle clouds can be modified by the magnetic field. Measurements demonstrate that fine-particle clouds rotate in the azimuthal direction on the horizontal plane in all ranges of the magnetic field. With an increase in the magnetic field, the rotation speed increases, saturates, and finally decreases. The rotation speed and direction are controlled by varying plasma potential and/or density profiles in the radial direction. The results might be explained by effects of modified ion motions on fine particles in the presence of the vertical magnetic field.

Kaw et al [5] have proposed a possible interpretation of the experimental results on the

---

[1] Professor Emeritus, *Kadan 4-17-113, Sendai 980-8579, Japan*

basis of the fluid equations. In this theory, the rotation is generated by the ion drag force which is balanced by ion momentum loss to neutrals in a steady state. There seems to be a good agreement between theory and experiment. Recently, on the other hand, Shukla [6] has presented a self-consistent theory for fine-particle clouds and derived a different expression for the rotation of fine-particle clouds under the same situation. Remaks are made on the two theories in this review.

The author has also been interested in spin motion of non-metal fine particles in dusty plasmas because the spin motion is accompanied by electric-current loop. Paramagnetic behaviors of fine particles are provided by the current loop in a magnetic field. The spin motion of non-spherical particles have been observed in Fukagawa et al [7] and Dahiya et at [8]. Ishihara and Sato [9] have predicted the spin motion of sperical particles in the presence of the ion velocity shear. Interaction of spinning particles with magnetic field could be quite interesting. Here, preliminary measurements of magnetic effect on the spin motion is also presented, together with future important problems to be investigated for magnetic effects on fine particles in dusty plasmas.

EXPERIMENTS

In order to observe effects of the vertical magnetic field on fine-particle clouds levitating in plasmas, two different kinds of experiments have been carried out in low-pressure weakly-ionized plasmas. One of them is concerned with a completely dc configuration, where fine particles are levitated in a diffused plasma. This experiment is for general features of fine-particle clouds in plasmas under the vertical magnetic field. The other experiment is rather specified for detailed measurements of the rotation of fine-particle clouds, where fine-particle clouds are levitated in quite simple configurations of dc and rf discharge plasmas. In both of them, a gas used for plasma production is Ar in the pressure range around a few tens of pascals and the plasma density is around $1 \times 10^8 \, cm^{-3}$, and the electron temperature is a few eV. Fine particles used are spherical with $10 \, \mu m$ in diameter. We have also used fine particles in different diameters from 0.1 to a few tens of μm. But, from a qualitative point of view, the phenomena observed do not depend on the particle diameter in this range. The vertical magnetic field is smaller than 0.4 kG in the first experiment. But, in the second experiment, we can increase the vertical magnetic field up to 40 kG. Fine particles are detected from the top and side by CCD cameras detecting laser light scattered by fine particles.

Rotation-diffused dc discharge plasma

A standard setup for investigation of fine-particle clouds in a diffused plasma is shown in Fig. 1, where a plasma is produced by a dc discharge between a mesh cathode and a ring anode. The anode, the hole of which is often covered by mesh, is electrically grounded together with the vacuum chamber. The plasma produced diffuses through the anode into a lower region. Fine particles are levitated by a levitation electrode (disc) and are confined radially by a confinement electrode (ring) in this region. We have performed various kinds of experiments on dynamics of fine-particle clouds under this dc configuration which is modified for the purposes of the experiments [1, 2].

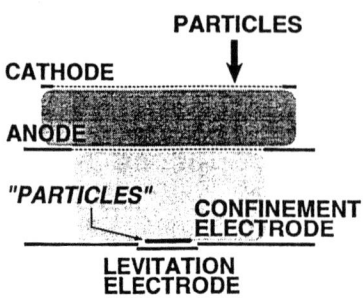

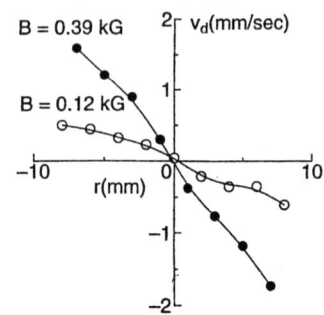

FIGURE 1. Schematic of standard set-up in our work on dusty plasmas

FIGURE 2. Rotation speed v_d as a function of radial distance r. Gas pressure $P_{Ar} \sim 30$ Pa. Discharge current = 0.5 mA.

The weak vertical magnetic field **B** up to 0.4 kG is applied in this experiment. The application of is very useful for shape control of fine-particle clouds, generating vertically column-shaped fine particles and string-shaped vertically-aligned fine particles [1-3]. Although the magnetic field applied is weak, there appears an azimuthal rotation of fine-particle clouds on the horizontal plane. The rotation is in the diamagnetic direction in this case. In figure 2, the rotation speeds are plotted as a function of the radial distance r at $B = 0.12$ and 0.39 kG. It is found that the speed is almost proportional to the radial distance. This means that the angular frequency, which is of the order of 0.1 rad/s (much larger than the fine-particle cyclotron frequency), is independent of the radial position, demonstrating a rigid-body rotation. The frequency increases with an increase in the vertical magnetic field. The interparticle distance is observed to be almost constant during this rotation.

The same rotation is generated in case of the column-shaped fine-particle clouds which is long in the vertical direction. The rotation is observed even in the case of string-shaped vertically-aligned fine particles. When the number of the strings is more than two, the vertical strings of fine particles demonstrate a clear rotation in the presence of the vertical magnetic field. The rotation is found to be accompanied by radial oscillation. The rotation depends on the particle density. When the density is extremely low, there appears no rotation. The rotation starts when the density becomes high enough to provide the strong Coulomb coupling among the particles. The rotation frequency increases with an increase in the particle density. This means that this rotation is generated in the presence of the strong Coulomb coupling among fine particles under the vertical magnetic field. In this experiment, the magnetic field is so weak that there is no direct magnetic effects on fine-particle orbits, i.e., we can neglect the cyclotron motion of the particles. But, there are magnetic effects on electron and ion orbits, which would be necessary for the rotation of fine-particles clouds.

Rotation -rf discharge plasma

Here we demonstrate some detailed features of the rotation of fine-particle clouds in the presence of the vertical magnetic field. The measurements are performed on dc and rf discharge plasmas.

An experimental configuration is quite simple, being different from our standard configuration in Fig. 1. The magnetic field is varied in a wide range from 0 to 40 kG. The experimental setup used for rf plasmas (13.56 MHz) is schematically described in Fig. 3. In the case of dc plasmas, the rf electrode is used as an anode. An upper electrode is segmented into two parts. The central part is a transparent disc electrode, to which an external voltage V_A is applied with respect to an outer ring electrode which is ground together with the vacuum chamber. Fine particles are levitated above the lower electrode in the radially central region, as shown in Fig. 3.

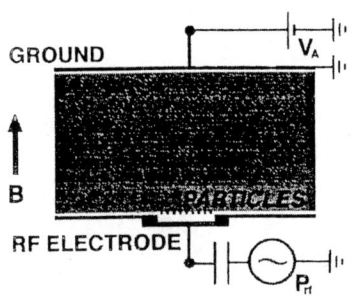

FIGURE 3. Schematic of set-up for fine-particle rotations in rf discharge plasmas.

A radial plasma potential profile is changes by varying V_A, although this is followed by a change of plasma density profile. The radial size of fine-particle clouds, however, is almost independent of V_A. This fact means that the radial potential proflle does not depend on V_A in the region of the particle levitation, being contrary to that outside this region. There is a vertical magnetic field **B** which induces the azimuthal rotation of fine-particle clouds. Almost the same results have been obtained in rf and dc plasmas. The results are presented here only in case of rf plasmas because it is more difficult to produce astable axisymmetric dc plasmas under an ultrastrong magnetic field. We can observe the rotation even if **B** is as small as 0.05 kG. In such a small magnetic field, although electrons are magnetized, positive ions are only slightly magnetized. The rotation is very sensitive to V_A. When V_A is negative, the rotation is in the diamagnetic direction, being consistent with the results for diffused dc plasmas. When V_A is positive, on the other hand, the rotation is in the paramagnetic direction. Independent of the sign of V_A, the angular frequency ω increases with an increase in the absolute value of V_A. The rotation is observed to stop around $V_A = -$a few volts. A dependence on the vertical magnetic field B is presented at $V_A = 20$ V in Fig. 4, where B is varied up to 40 kG at $V_A = 20$ V [11]. The angular frequency is found to be almost proportional to B up to 3 kG, where both electrons and positive ions are magnetized. The angular frequency approaches the maximum a little above $B = 10$ kG.

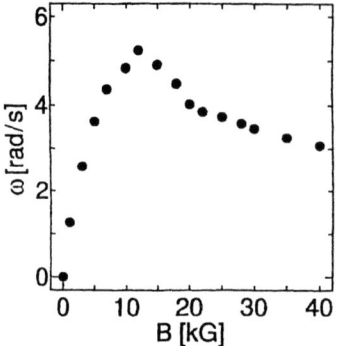
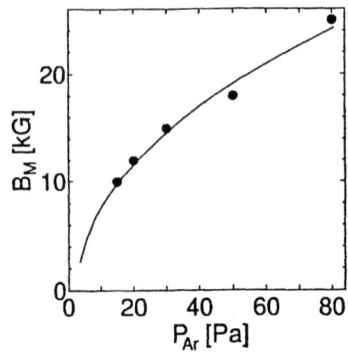

FIGURE 4. Dependence of angular frequency ω of the rotation of magnetic field B up to 40 kG at $V_A = 20$ V. Gas pressure $P_{Ar} \sim 20$ Pa. Rf power $P_{rf} = 5$ W.

FIGURE 5. Magnetic field at the maximum of ω, B_M, as a function of P_{Ar}. $V_A = 20$ V. $P_{rf} = 5$ W.

With a further increase of B, there appears a decrease in ω. Around $b = 40$ kG there should appear some direct effects of B on fine particles. The rotation, however, shows a simple decrease pressure even around this range of B. The magnetic field giving the maximum of ω, B_M is plotted as a function of gas pressure P_{Ar} in Fig. 5 and B_M is found to increase with an increase P_{Ar}. The results in Figs. 4 and 5 suggest that the rotation mechanism strongly depends on the azimuthal ion motion because the ion collision frequency is of the same order of magnitude as the ion cyclotron frequency around B_M.

The rf power P_{rf} is also varied to shnow then effect of the plasma density on the rotation. The angular frequency is found to increase with an increase in P_{rf}.

Spin Motion

It is difficult to observe the spin motion of small sperical particles. We need particles in such large diameter that we can give a spatial difference (for example, color) to the surface of particles. Then, there is a chance to observe the spin motion of spherical particles. For this purpose, we have tried a possibility of levitation of spherical particles in diameter up to $100\,\mu$m in plasmas [7]. The method employed is based on so-called "electron shower" on fine particles from downward region [1,2], which is supplied by modifying the standard setup in Fig. 1. Hollow spherical particles (glass ballon) in diameter up to 100μm are used in the experiment. But, we have not succeeded to levitate particles in diameter larger than $50\,\mu$m.

Fortunately, however, we have observed the spin motion of particles in diameter around a few tens of μm probably because some particles are not comletely spherical. In this exeriment, the measurements are made by using the motion scope whose shutter speed and recording time are variable. Dahiya *et al* [8] have also observed the spin motion of non-spherical particles. In our experiment, the spining frequency is found to be around 50 Hz. Our interest is in interaction of spinning particles with magnetic field because the loop current due to the spin motion provides a magnetic moment. In the measurements under the vertical magnetic field of 0.96 kG, more particles are observed to spin in the paramagnetic direction. This might be due to the interaction between spilling particles and magnetic field, although the magnetic moment is very small in this case.

DISCUSSIONS AND CONCLUSIONS

The vertical magnetic field is useful for generation of vertically long structures of fine-particle clouds, for example, vertically column-shaped structure and string-shaped vertically-aligned structure in our work. The vertical magnetic field provides the rotation of fine particles in the azimuthal direction on the horizontal plane in strongly-coupled dusty plasmas. This rotation is due to an effect of plasmas on fine particles, which are modified by the magnetic field. The results show that there is a drastic effect of V_A on the rotation. V_A modifies the radial potential profilee, being also accompanied by a change of the radial plasma-density profile. But, the polarity change of V_A does not necessarily means a change of the direction of the radial potential gradient. The measurements also shows that the direction of the rotation changes when the radial density gradient is reversed for a fixed relatively flat potential profile in the radial direction. The potential profile in the radial region of particle levitation is always hill-shaped, being almost independent of the polarity and absolute value of V_A. It has to be emphasized that the potential and density profiles are modified by V_A outside this region of particle levitation. A mechanism of the rotation could be explained by ion trajectories modified by the magnetic field. In general, there might be two effects of the modification of ion trajectories. One of them is due to the azimuthal ion drag force on fine particles. The other is due to a change of the potential structure around the particles, which has also an effect on fine-particle motions in the azimuthal direction. Our results up to now might be understood by the azimuthal ion drag force in the presence of the vertical magnetic field. According to the theory of Kaw *et al* [4], the rotation is generated by the ion drag force which is balanced by ion momentum loss to background neutrals. The results are compared with the typical experimental data in Fig. 6, where a good agreement is found between theory and experiment.

The theory also explains qualitatively the dependences of the rotation frequency on the rf power and the backgound gas pressure in the experiment. The prediction of Shukla [6] is based on the self-consistent fluid equations. His result looks completely different from that of Kaw *et al.* There is an interest in the difference between the two theories. As far as the magnetic field is uniform in plasmas, however, his result is found to be reduced to that of Kaw *et al.* A different theoretical approach is proposed by Ishihara *et al.*

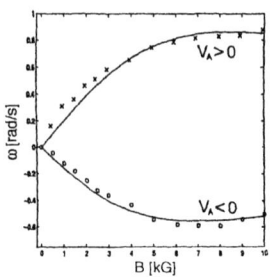

FIGURE 6. Comparison between theory (Kaw *et al* [5]) and experiment. $P_{Ar} = 10$ Pa. $P_{rf} = 2$ W.

Anyway, in the experiment, the plasma potential and density profiles are almost independent of V_A in the region where fine particles are levitated. The plasma profile outside this region depends on V_A, yielding the rotation change. From an experimental point of view, it is not clear whether some kinds of wake effect due to ion flow can explain the mechanism of this rotation in the presence of the azimuthal ion drag force.

The rotation is induced under the condition satisfying the strong Coulomb coupling among fine particles. This could be explained by the fact that a total force on fine-particle clouds is a sum up of forces on individual particles under such a condition. We must also take account of the finite ion Larmor radius because the rotation appears even when the ion Larmor radius is larger than the size of fine-particle clouds. More measurements will be performed to understand the phenomena.

The spin motion could be quite important because the loop current due to the spin provides paramagnetic phenomena of fine particles in a magnetic field. It is difficult to detect the spin motion of small spherical particles. Under the microgravity condition, however, much larger particles can be levitated and it is much easier to detect the spin motion. It has to be noted that the magnetic moment due to the spin increases with an increase in the particle diameter. In an ultrastrong magnetic field, we might have the spin and cyclotron motions of fine particles. A new orbit theory could be developed under this condition.

A magnetic field also gives an effect on charging of fine particles [12]. This problem is rather fundamental and have to be solved in connection with crystalization and other characteristic phenomena in dusty plasmas. The spin motion is interesting also from this point of view. Dusty plasmas in space are often accompanied by magnetic field. In order to understand dynamics of dusty plasmas in space environment, magnetic effects on fine particles have to be claried in basic experimental and theoretical investigations.

ACKNOWLEDGMENTS

The author thanks his many collaborators, especially G. Uchida and S. Iizuka, for their collaborations at Tohoku University. He also appreciates stimulating discussions with T. Kamimura, O. Ishihara, K. Nishikawa, P. K. Kaw, H. Ikezi, and V. H. Tsytovich.

REFERENCES

1. Sato, N., Uchida, G., Kamimura, T., and Iizuka, S., in *Physics of Dusty Plasmas*, edited by M. Horanyi *et al.* (The American Institute of Physics, New York, 1998), p.239–246.
2. Sato, N., Uchida, G., Ozaki, R., Iizuka, S., and Kamimura, T., in *Frontiers in Dusty Plasmas*, edited by Y. Nakamura *et al.* (Elsevier Science B. V., 2000), p.329 -336.
3. Sato, N., Uchida, G., Kaneko, T., Shimizu, S., and Iizuka, S., *Phys. Plasmas*, **8**, 1786 (2001).
4. Konopka, U., Samsonov, D., Ivlev, A.V., Goree, J., Steinberg, V., and Morfill, A.E., *Phys. Rev. E*, **61**, 1890 (2000).
5. Kaw, P., Nishikawa, K., and Sato, N., *Phys. Plasmas*, **9**, 387 (2002).
6. P. K. Shukla, P.K., to be published in Phys. Lett. A.
7. Fukagawa, K., Raychaudhuri, S., , Uchida, G., Iizuka, S., and Sato, N., in *Proceedings of Plasma Sci. Symp. 2001/18th Symp. on Plasma Processing* (Kyoto, Jan. 24–26, 2001), edited by N. Sato and H. Fujiyama, p.409–410.
8. Dahiya, R.P., Paeva, G., Stoffels, W.W., and Kroesen, G.M.W., in *9th Workshop on the Phsics of Dusty Plasmas*, Iowa City, May 21–23, 2001.
9. Ishihara, O., and Sato, N., *IEEE Trans Plasma Sci.* **2**, 179 (2001)
10. Ishihara, O., Kamimura, T., Hirose, K., and Sato, N., to be reported at *11th International Congress on Plasma Physics*, Sydney, July 15–19, 2002
11. Shimizu, S., Uchida, G., Iizuka, S., and Sato, N., to be published.
12. Tsytovich, V.H., and Sato, N., to be published

Dust Crystal in the Electrode Sheath of a Gaseous Discharge

I.V. Schweigert and V.A. Schweigert

Institute of Theoretical and Applied Mechanics, Novosibirsk 630090, Russia

Abstract. The phenomena observed in strongly coupled dusty plasmas in the electrode sheath of gas discharge clearly indicate that the screened Coulomb potential is not valid for inter-particle interaction. The reason why the conventional model breaks down is clear now. The strong electric field, accelerating ions toward the cathode, leads to an asymmetrical particle shielding and the appearance of an attractive component in the inter-particle force. The sheath plasma with microparticles is non Hamiltonian system because of input of energy from ion flux from the bulk plasma.
The models of interaction potential of microparticles in sheath are proposed. The first is the linear effective positive charge (EPC). On the basis of this model the stability of the dust crystal in the sheath is analyzed both analytically and in MD simulations. The scenario of crystal melting is described. The role of different types of defects in the local heating of the crystal is considered. The next non-linear model of sheath plasma with micro-particles allows to find all parameter of plasma crystal: particle charge, inter-particle distance and study the structural transition. We constructed the analytical expression for inter-particle potential and have found the mechanism acceleration of extra particle beneath the monolayer. Recently new more simple analytical kinetic approach, accounting for ion collisions, have been developed. The structural transition in the dust molecular was obtained in simulation with multipole expansion model interaction potential.

The results of study of dusty plasma behavior both in laboratory and microgravity experiments demonstrate the complex dynamics. Vortex motion of particles near the edge of the electrode, the void formation in the center of discharge chamber, void low frequency oscillations, and separation of different size particles. It is obvious that our knowledge about the spatial structure (at least two dimensional) of rf discharge even without dust as well as effect of presence of dust particles on plasma and interaction between the microparticle will impact our ability to understand and explain these phenomena.

In my talk I will emphasize on two aspects of strongly coupled dusty plasma problem. The first is the inter-particle interaction potential in ion flux which is different from usual Debye-Hukkel model. I will give the examples of application of these inter-particle potentials for modeling of particle behavior observed in the experiments. The second is the 2D simulations of ccrf discharge with new fast PIC MCC algorithms.

MODELS OF INTER-PARTICLE POTENTIAL OF INTERACTION IN ION FLUX

The dust particle interaction in plasma and, especially, the possibility of the existence of attractive forces between the charged dust particles for the explanation of the vertical alignment have been studied theoretically by means of different mechanisms. Ion bombardment of the dust particles creating a shadow behind them [1, 2, 3], or electric-field-induced dipole-dipole forces [4] give too weak forces for micron-sized particles used in the experiments. Clearly, the non-equilibrium behavior of ions due to the strong acceleration by the electric field in the electrode sheath causes deviations from the Debye-Hückel model, which is based on the Boltzmann distribution of screening electrons and ions in the particle's electric field. The deflection of the streaming ions by the particle electric field leads to focusing of ion trajectories with the enhanced ion density downstream of the particles [5, 6, 7, 8]. These positive ion clouds provide an attractive force for the negative dust particles. *Collisionless* kinetic [5] and fluid [6] approaches, which were further developed in Refs. [9, 10, 11, 12, 13, 14, 15], may not be appropriate for experimental conditions, where, generally, the ion mean free path is smaller than the inter-particle distance.

Effective positive charge (EPC) model

In [7, 8] we considered two layer crystal with hexagonal symmetry in the radial plane and aligned in the vertical direction. This layered crystal in the electrode sheath of radio-frequency discharge in helium is treated with employing the Monte-Carlo technique for ion motion accounting for ion collisions. We shift the lower layer relative the upper layer in the x direction and found the restoring force pushing the lower layer particles to their equilibrium (aligned) positions. This force is key point for further analysis of instability of the crystal. Fig. 1 illustrates the ion density distribution around the particles and forces acting on the lower particles from ion flux. In this EPC model it is assumed that i) the particles move in the plane parallel to the electrode. This simplification follows from experimental observations which show that dust particles move mainly in the horizontal (x,y)- plane. The amplitude of particle oscillations in the vertical z direction is comparably small due to the strong vertical confinement formed by the gravitational and the electrical forces; ii) the asymmetric ion distribution around the particle is replaced by an uniform ion distribution and an effective positive charge Z_c placed below the parent dust particle; iii) each ion cloud (i.e. positive point charge) is rigidly connected to the upper parent particle. This is justified since a shift of the upstream particle immediately leads to the redistribution of the ion density around this particle [7, 8], because the characteristic time of ion motion is much less than the time scale of dust particle motion.

This model is semi–empirical and demands for additional Monte-Carlo simulations to find the value of effective positive charges and their coordinates.

Instability and melting of the crystal of microparticles

In [7, 8] we have reported the effect of focusing of ion trajectories by the electric field of the dust particles result in a formation of regions with enhanced ion density below the upper particles. The attraction forces of these ion clouds on the lower layer particles leads to the vertical alignment as observed in many experiments. From these MC results

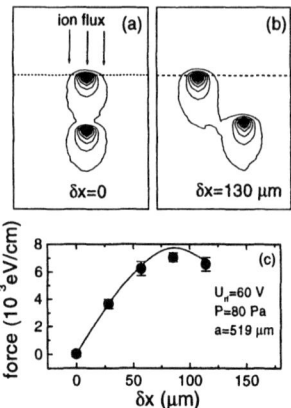

FIGURE 1. Computed ion distribution around the particles in the sheath in the equilibrium position (a) and with shift δx(b). The restoring force acting on lower particle as a function of shift from equilibrium position (c). The ions enter from the upper boudary. The dushed line denoted the upper layer position. Symbols denote the MC results, solid lines indicate the forces for positive point charges which mimic ion clouds.

an analytical model for the investigation of the stability of such aligned crystal was formulated. This model treats the many-body ion system as an effective pair potential resulting in a non–Hamiltonian system with non–reciprocal forces on lower and upper particles. The non–reciprocity is induced by ion flux asymmetrically screening the charged particles. Using the EPC interaction model we have analyzed the stability of such system by using the linear approach to the Newton equations accounting for friction [7, 8]. The vertical alignment was found to become unstable, leading to oscillations of the aligned particles below a threshold value of friction. Furthermore the characteristics of these unstable oscillations such as the wave frequency, the ratio between amplitudes of upper and lower particles, and the phase shifts between the particle oscillations have been determined. These values are in close agreement with that found in experiments. So the melting transition is explained by the onset of self–excited oscillations due to a plasma–induced instability. The oscillations are the cause for the melting transition.

Further in [16, 17], using the EPC model of inter-particle interaction, the scenario of melting transition of dust crystal was studied in the molecular dynamical simulations. The plasma crystal was shown to exhibit a nonequilibrium two-step phase transition. Note that in [15] it was found that the melting occurs in a two-step process. In Fig. 2 the mean kinetic energy of particles as function of dimensionless gas friction is shown for the crystal without defects and with defects [18]. With decreasing gas pressure, at the first critical value of friction the particle energy quickly increases due to an ion induced instability, but system retains the crystalline order. When the gas pressure becomes lower the second critical point the crystal melts. Note, that plasma crystal melts at a much higher particle energy than expected from classical model. This behavior can be explained by the fact that only high frequency modes are excited due to the instability.

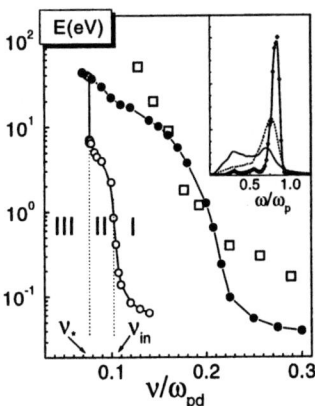

FIGURE 2. Mean kinetic energy of the dust particles as a function of the gas friction coefficient for the bilayer crystal without defects (open circles) and with extra particles above and beneath the two layer crystal (solid circles). The experimental values are also shown for comparison (solid squares). In the insert the spectrum excited phonons for crystals without defects, for the monolayer (solid line), for two layer crystal at $v = 0.1150\omega_{pd}$ (line with crosses) and at $v = 0.1225\omega_{pd}$ (dushed line). The vertical lines denote the onset of instability at v_{in} and melting at v_*.

The insert in Fig. 2 shows the spectrum of excited phonons. The low frequency modes (dangerous for crystalline order!) are absent up to melting.

Self-consistent 3D PIC MCC simulations of dust crystal in the sheath

By employing the particle-in-cell method we studied in [19] the distributions of the electric field and the electron and ion concentrations in the microparticle crystal in the electrode sheath in a radio-frequency discharge in helium. The coordinates and charge of the microparticles were found from the balance condition for the forces acting on the particles and the balance of electron and ion fluxes to the particles. With periodic boundary conditions introduced, we investigated the three-dimensional problem for the unit cell of the microparticle crystal. We examined the dependence on gas pressure and discharge voltage of the main crystal parameters: the critical particle separation at which a phase transition from a monolayer crystal to a double-layer crystal occurs, the particle potentials, and the distances between the layers in the double-layer crystal. We have obtained the critical values of the friction coefficient below which the crystal becomes unstable against the development of particle oscillations in the transverse direction. In Fig. 3 the particle charge and the critical inter-particle distance are shown as function of gas pressure. Finally, we set up an approximate model that makes it possible to calculate the main parameters of the microparticle crystal. Note, that the particle charge and the interparticle distance derived from this approximate model very well agree with PIC MCC calculation results.

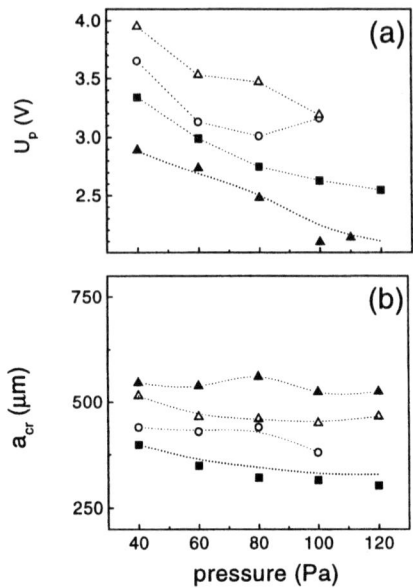

FIGURE 3. Gas pressure dependence of (a) particles surface potential and (b) critical inter-particle distance for transition from monolayer to bilayer. PIC-MCC computed results: solid triangles (40V) and solid squares (80 V), the experiment: open triangles (40V), open circles (50V).

Acceleration of fast particles beneath the monolayer

We applied the analytic expression for inter-particle potential of interaction taken from PIC MCC to study the mechanism of acceleration of the extra particle. In the experiments of Samsonov *etal.* [20], a kind of spontaneous particle motion was observed. The experiments were performed using a monolayer of particles, with a few extra particles in an incomplete second layer 200 μm below the monolayer. These extra particles moved about spontaneously, and because they were charged, they disturbed the particles in the main layer generating a Mach cone. In [21] the numerical simulations and further experiments are performed to find an explanation for the acceleration and different regimes of motion of the extra particle. The simulations take into account the ion wakefield downstream of the monolayer of particles, in the presence of ion flow. The orbit is straight at low gas pressures, but becomes crooked with increasing pressure. The extra particle is aligned with an upper particle from monolayer at higher damping. All these regimes are in agreement with experiment. This agreement supports a conclusion that the instability driven by ion flux is the cause of the particle acceleration. (For more details, see V.A. Schweigert, I.V. Schweigert, V. Nosenko and J. Goree 'Dynamical phase transition in dust crystals'.)

Linear kinetic approach and multipole expansion

Recent 3D self-consistent PIC-MCC simulations of dust crystals [19] are based on high crystal symmetry and cannot be applied, in practice, to dynamical problems such as dust crystal melting or waves. More simple modeswill be very usewful. In [22, 23, 24] the ion density and electric potential distributions around a dust particle in the electrode sheath of a gas discharge is studied using a kinetic approach. The dust particle interaction is also described with a multipole expansion including the monopole, dipole and quadrupole momenta, for which analytical expressions are derived from a linear response theory. The validity of this approach is confirmed by comparison with the results of PIC-MCC simulations. In our model [22, 23, 24], the sheath plasma is described by a kinetic approach for the ions, whereas electrons are, as usual, assumed to obey a Boltzmann distribution. According to a standard linear approach we consider small perturbations in the ion distribution function f_i and apply the Fourier transformation. After integration f_i over velocities the ion density perturbation is

$$n_k = -\frac{en_0 k^2}{Mv^2} \frac{G_1}{1-G_0} \phi_k - \frac{JG_2}{v(1-G_0)}, \quad (1)$$

where G_0 and G_1 are the integrals from function depending on the transverse ion diffusion, the ion collision frequency and the ion drift velocity [22, 23, 24].

Using linearized Poisson equation, the screened dust particle potential can be expressed as

$$\phi_k = -4\pi e \frac{Z + JG_2 v^{-1}(1-G_0)^{-1}}{k^2 \varepsilon}, \quad (2)$$

through the dielectric plasma permittivity

$$\varepsilon = 1 + 1/k^2 \lambda_e^2 + \kappa^2 G_1/(1-G_0), \quad (3)$$

where $\kappa = \omega_p/v$ describes the collisionality of the ion flow; $\omega_p = \sqrt{4\pi e^2 n_0/M}$ is the ion plasma frequency; the second and third terms in RHS of Eq. (3) describe the electron and the ion shielding, respectively. In Fig. 4 the perturbation of potential distribution along the ion flow is shown obtained within different approaches. The particle is at $z = 0$.

Dust molecule and phase transition

We have studied the phase transition with two dust particles of different mass for the experimental conditions [25]. Due to their different mass the particles are trapped at different heights in the sheath, where the electric field balances the weight of the particles. The vertical trapping potential is very strong and vertical motions of the dust particles are small in the experiment and are, therefore, neglected in the following analysis. In the horizontal plane, the particles are free to move under the influence of their mutual interaction and a weak horizontal confining potential.

At low gas pressures (55 Pa in a helium rf discharge) the two dust particles are vertically aligned, forming a bound dust molecule. By direct laser excitation of the

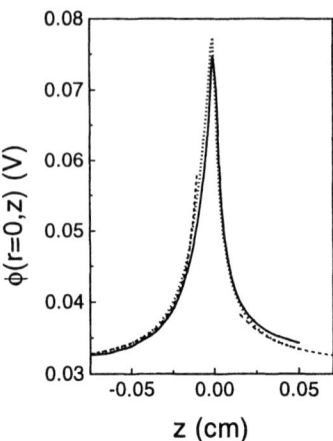

FIGURE 4. The ion cloud potential distribution along the ion flow ($\rho = 0$) obtained with PIC MCC (solid line), in the linear approximation (dotted line) and with multipole expansion (dushed line). for helium pressures at $P = 200Pa$. The particle radius $R = 1.73\mu m$, the particle charge Z=2000e, and the electrical field E=10 V/cm.

particles, the alignment is shown to be due to an asymmetric attractive force acting on the lower particle, only, without backaction on the upper particle. When the gas pressure is increased now, the aligned state is retained up to a gas pressure of 175 Pa, where suddenly the alignment breaks up and the two particles becomes separated in the horizontal plane, indicating that net repulsive forces act between the dust particles. Decreasing the gas pressure again the separated state is retained until at 55 Pa. Then the particles jump back into the aligned state, again.

Since the characteristic size of the ion cloud is much smaller than the interparticle distance in the dust molecule, $d = 750\mu m$, that allows one to describe the particle interaction by a multipole expansion. In simulation we described the inter-particle interaction with a multipole expansion with analytical expressions derived from a linear response theory. We have found a structural transition between the aligned and the separated state, which shows a hysteresis with gas pressure. In Fig. 5 the horizontal separation between two particles is shown as a function of gas pressure.

FAST ALGORITHM FOR THE 1D AND 2D SIMULATION OF CCRF DISCHARGE

Recently the Particle in Cell Monte-Carlo collision algorithm is the most advanced technique providing the complete information about kinetic processes in gas discharge. However to obtain the results within reasonable computed time (even without particles) it is necessary to enhance efficiency of this algorithm. We have modified PIC MCC

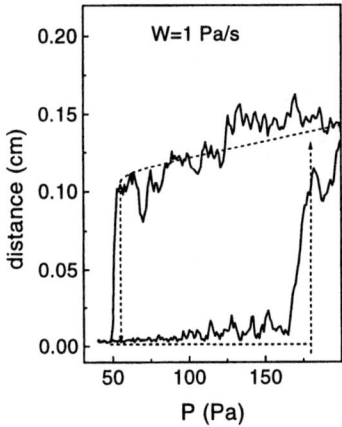

FIGURE 5. Horizontal inter-particle distances in the dust molecule for different gas pressures.

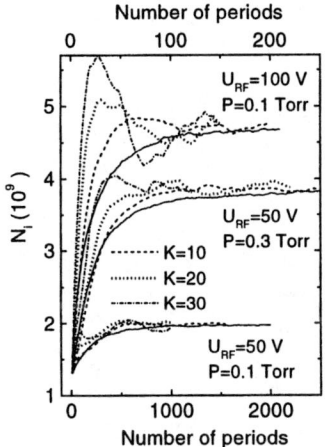

FIGURE 6. Relaxation of the total number of ions in a rf discharge simulation without (fill curves) and with (broken curves) accelaration procedure.

algorithms and considerably accelerates calculations of ccrf discharge in one and two dimensions. Two methods of enhancing the efficiency of PIC-MCC algorithms are proposed. The first is oriented on discharge under relatively high ($P = 0.1 \div 1$ $Torr$) gas pressures and the averaging of ion motion over a discharge period is used. The kinetic equation for electrons is solved once for several discharge periods, but for the rest periods the electron density distribution is only corrected in order to conserve the space charge distribution. Depending on conditions, this method allows to accelerate calculations by $K_a = 5 - 20$ times.

The second method works for low gas pressures ($P = 10^{-3} \div 10^{-1}$ $Torr$) when the non–collisional electrons heating becomes important. This method combines the PIC MCC and fluid approaches. The auxiliary fluid equations for density and flux of electrons and ions are used to find the electric field. This allows to reduce the number of simulated particles without significant increasing of the artificial electrons heating.

The efficiency of the first fast PIC MCC is illustrated in Fig. 6. The total number of ions as function of N is presented on the Fig. 6 for different gas pressures and RF-voltages. The bottom scale (solid curves) and top scale (other curves) correspond to simulations without and with described acceleration method for different chosen K_a. It is evident that our approach allows to accelerate simulation in K_a times.

REFERENCES

1. Ignatov, A. M., *Plasmas Phys.*, **22**, 648 (1996).
2. Hodataev, Ya. K.,*et al.*, *Plasmas Phys.*, **22**, 1028 (1996).
3. Fortov, V.E., and Lakhno, V. D., *Phys. Lett. A*, **250**, 149 (1998).
4. Mohideen, U., *et al.*, *Phys. Rev. Lett.*, **81**, 349 (1998).
5. Vladimirov, S. V., and Nambu, M., *Phys. Rev. E*, **52**, 2172 (1995).
6. Melandso, F., and Goree, J., *Phys. Rev. E*, **52**, 5312 (1995).
7. Schweigert, V.A., *et al.*, *Phys. Rev. E*, **54**, 4155 (1996).
8. Melzer, A., *et al.*, *Phys. Rev. E*, **54** (1996).
9. Nambu, M., and Shukla, P., *Phys. Rev. A*, **203**, 40 (1995).
10. Vladimirov, S. V., and Ishihara, O., *Phys. Plasmas*, **3**, 444 (1996).
11. Ishihara, O., and Vladimirov, S. V., *Phys. Plasmas*, **4**, 1 (1997).
12. Baisong Xie, K. H., and Huang, Z., *Phys. Lett. A*, **253**, 83 (1999).
13. Chiueh, T., and Kuo, D. M.-T., *Phys. Lett. A*, **46**, 190 (1999).
14. Melands, F., and Goree, J., *J. Vac. Sci. Technol. A*, **14**, 511 (1996).
15. Melandso, F., *J. Vac. Sci. Technol. A*, **14**, 511 (1997).
16. Schweigert, V.A., *et al.*, *Phys. Rev. Lett.*, **80**, 5345 (1998).
17. Schweigert, I.V., *et al.*, **87**, 905 (1998).
18. Schweigert, I.V., *et al.*, *Phys. Rev. E*, **62**, 1238 (2000).
19. Schweigert, V.A., *et al.*, **88**, 482 (1999).
20. Samsonov, D., *et al.*, *Phys. Rev. Lett.*, **83**, 3649 (1999).
21. Schweigert, V.A., *et al.*, submitted to *Phys. Plasmas* (2002).
22. Schweigert, V.A., *et al.*, *J.de Physique IV*, **10** (2000).
23. Schweigert, V.A., *et al.*, *Plasma Phys. Rep.*, **27**, 997 (2001).
24. Schweigert, V.A., *et al.*, submitted to *Phys. Rev. E*, **27**, 997 (2002).
25. Melzer, A., and Piel, A., *Phys. Rev. Lett.*, **83**, 3194 (1999).

Nonlinear waves in dusty plasmas

P. K. Shukla and A. A. Mamun[1]

Fakultät für Physik und Astronomie, Ruhr-Universität Bochum, D-44780 Bochum, Germany

Abstract. We present theories for dust ion-acoustic solitary and shock waves, dust-acoustic solitary and shock waves, and dust lattice solitary waves that may exist in dusty plasmas. It is found that the presence or dynamics of dust particles introduces new features to the nonlinear electrostatic waves. The implications of our theoretical results to experimental observations of solitary and shock waves are discussed.

INTRODUCTION

It is well known that dust ion-acoustic (DIA) waves [1], dust-acoustic (DA) waves [2], and dust lattice (DL) waves [3] are three normal modes of an unmagnetized dusty plasma. The linear properties of the DIA, DA, and DL waves in dusty plasmas are fully understood from both theoretical [1-3] and experimental [4-6] points of view. A large number of review articles and books have summarized the present status of waves and instabilities in dusty plasmas [7-10].

The linear theory is valid only when the wave amplitude is so small that one may neglect the nonlinearities. However, there are numerous processes via which unstable modes can saturate and attain large amplitudes. When the amplitudes of the waves are sufficiently large, nonlinearities can no longer be ignored. The effects of nonlinearities in plasmas contribute to the localization of waves, leading to different types of interesting nonlinear coherent structures (viz. solitary structures, shock waves, etc.) which are important from both theoretical and experimental points of view, and have received a great deal of attention for understanding the basic properties of localized electrostatic perturbations in space and laboratory dusty plasmas.

A number of theoretical investigations have been carried out for studying the properties of DIA solitary [11,12] and shock [13,14] waves, DA solitary [2,15,16] and shock [17,18] waves, and DL solitary waves [3,19]. The DIA solitary [20] and shock waves [21-23] and DL solitary waves [24,25] have been experimentally observed. The present topical review is aimed at systematically providing all the basic features of DIA solitary and shock waves, DA solitary and shock waves, and DL solitary waves.

---

[1] Permanent address: Department of Physics, Jahangirnagar University, Savar, Dhaka, Bangladesh

DIA SOLITARY WAVES

We consider a dusty plasma in which dust particles are stationary and provide only the background charge-neutrality. The basic equations for the one-dimensional DIA waves can be expressed in terms of normalized variables as [1]

$$\frac{\partial n_i}{\partial t} + \frac{\partial}{\partial x}(n_i u_i) = 0, \tag{1}$$

$$\frac{\partial u_i}{\partial t} + u_i \frac{\partial u_i}{\partial x} = -\frac{\partial \phi}{\partial x}, \tag{2}$$

$$\frac{\partial^2 \phi}{\partial x^2} = \mu \exp(\phi) - n_i + 1 - \mu, \tag{3}$$

where n_i is the ion number density normalized by its equilibrium value n_{i0}, u_i is the ion fluid velocity normalized by the ion-acoustic speed $C_i = k_B T_e/m_i$ (with k_B the Boltzmann constant, T_e the electron temperature, and m_i the ion mass), ϕ is the wave potential normalized by $K_B T_e/e$ (with e the magnitude of the electron charge), and $\mu = n_{e0}/n_{i0}$. The space variable x is normalized by the modified electron Debye radius $\lambda_{Dem} = (K_B T_e/4\pi n_{i0} e^2)^{1/2}$ and the time variable t is normalized by the ion plasma period $\omega_{pi}^{-1} = (m_i/4\pi n_{i0} e^2)^{1/2}$. We have assumed a Boltzmannean electron density response.

Now, introducing the stretched coordinates [26] $\xi = \varepsilon^{1/2}(x - v_0 t)$ and $\tau = \varepsilon^{3/2} t$, where ε is the expansion parameter, expanding n_i, u_i and ϕ in a power series of ε, and developing the equations in powers of ε, we can derive a Kortweg-de Vries (KdV) equation

$$\frac{\partial \phi^{(1)}}{\partial \tau} + a_i \phi^{(1)} \frac{\partial \phi^{(1)}}{\partial \xi} + b_i \frac{\partial^3 \phi^{(1)}}{\partial \xi^3} = 0, \tag{4}$$

where $a_i = (3\mu - 1)/\sqrt{2\mu}$ and $b_i = 1/2\mu^{3/2}$. In a moving frame with a speed u_0, the stationary solitary wave solution of Eq. (4) is

$$\phi^{(1)} = \left(\frac{3u_0}{a_i}\right) \operatorname{sech}^2\left[\sqrt{\frac{u_0}{4b_i}}(\xi - u_0 \tau)\right]. \tag{5}$$

It is obvious from Eq. (5) that for $\mu > (<) 1/3$, a dusty plasma supports compressive (rarefactive) DIA solitary waves which are associated with a positive (negative) potential.

DIA SHOCK WAVES

We have studied DIA solitary waves which arise because of the balance between the effects of the nonlinearity and dispersion when the effects of dissipation is negligible in comparison with that of the nonlinearity and dispersion. However, when the dissipative effect is comparable to or more dominant than the dispersive effect, one encounters shock waves. Shukla [13] presented an analytical model for the DIA shock waves in an unmagnetized dusty plasma assuming constant dust grain charge, and derived a KdV-Burgers equation, which admits shock wave solutions [27]. There is an important question regarding the source/mechanism of this dissipative term. Motivated by this question, recently, we have considered an unmagnetized dusty plasma including the dust charge fluctuation dynamics, and have proposed that dust grain charge fluctuations cause dissipation, which, in turn, is responsible for the formation of the DIA shock waves in a dusty plasma [28].

The nonlinear dynamics of the one-dimensional DIA waves in a dusty plasma including dust charge fluctuations is governed by Eqs. (1) and (2), and

$$\frac{\partial^2 \phi}{\partial x^2} = \mu \exp(\phi) - n_i + (1-\mu) z_d, \tag{6}$$

where z_d is the number of electrons residing onto the dust grain surface normalized by its equilibrium value Z_{d0}. We note that z_d is not constant but varies with space and time. Thus, Eqs. (1), (2), and (6) are completed by the normalized dust grain charging equation [28]

$$\eta \frac{\partial z_d}{\partial t} = \mu\beta \exp(\phi - \alpha z_d) - \beta_i n_i u_i \left(1 + \frac{2\alpha z_d}{u_i^2}\right), \tag{7}$$

where $\eta = \sqrt{\alpha m_e(1-\mu)/2m_i}$, $\alpha = Z_{d0} e^2/k_B T_e r_d$, $\beta = (r_d/a)^{3/2}$, $a = n_{d0}^{-1/3}$, and $\beta_i = \beta\sqrt{\pi m_e/8m_i}$. We note that at the equilibrium $\mu\beta \exp(-\alpha) = \beta_i u_{i0}(1 + 2\alpha/u_{i0}^2)$, where u_{i0} is the ion streaming speed normalized by C_i.

Now, introducing the stretched coordinates [26] $\xi = \varepsilon^{1/2}(x - v_0 t)$, $\tau = \varepsilon^{3/2} t$, and $\eta = \varepsilon^{1/2}\eta_0$, expanding n_i, u_i, ϕ and z_d in a power series of ε, and developing the equations in powers of ε, we obtain a KdV-Burgers equation

$$\frac{\partial \phi^{(1)}}{\partial \tau} + A_c \phi^{(1)} \frac{\partial \phi^{(1)}}{\partial \xi} + B_c \frac{\partial^3 \phi^{(1)}}{\partial \xi^3} = C_c \frac{\partial^2 \phi^{(1)}}{\partial \xi^2}, \tag{8}$$

where the coefficients A_c, B_c, and C_c are given in Ref. 28. An exact analytical solution of Eq. (8) is not possible. However, we can deduce some approximate analytical shock solutions [27]. The nature of these shock structures depends on the relative values between the dispersive and dissipative coefficients B_c and C_c. If the value of C_c is very small, the energy of the particle decreases very slowly and the first few oscillations at the wave front will be close to solitary waves. However, if the value of C_c is larger than a critical value, the motion of the particle will be aperiodic, and we obtain a shock wave with a monotonic structure. A monotonic shock wave solution is

$$\phi^{(1)} \simeq \psi_{sh} - \psi_{sh} \tanh\left[(\xi - U_0 \tau)/\Delta_{sh}\right], \tag{9}$$

where $\psi_{sh} = U_0/A_c$ and $\Delta_{sh} = 2C_c/U_0$ represent the amplitude and the width of the shock wave, respectively.

The DIA shock waves were experimentally excited in a dusty double plasma (DP) device by Nakamura et al. [21]. Oscillatory shock waves are excited in a plasma first without the dust and then with the dust, increasing the dust density in small steps, keeping the probe fixed from the grid. The experimental results of Nakamura et al. [21] reveal that the oscillatory wave structure behind the shock becomes less in number with increasing dust particle number density and finally disappears completely at a sufficiently high dust particle number density, leaving only the laminar shock front. The shock speed also increases with increasing the dust particle number density. It is noted that the particle density behind the shock remains constant, although the amplitude of the shock front (steepened part) seems to decrease when the dust particle number density is increased. The effect of the dust particle number density on the ion acoustic compressional pulses has also been experimentally studied by Luo et al. [22] who observed a steepening of the ion-acoustic pulses as they propagated through a dusty plasma if the percentage of the negative charge in the plasma on the dust grains was about 75% or more.

DA SOLITARY WAVES

The dynamics of low phase velocity (in comparison with electron and ion thermal speeds) one-dimensional DA solitary waves is governed by [2]

$$\frac{\partial n_d}{\partial t} + \frac{\partial}{\partial z}(n_d u_d) = 0, \tag{10}$$

$$\frac{\partial u_d}{\partial t} + u_d \frac{\partial u_d}{\partial z} = \frac{\partial \varphi}{\partial z}, \tag{11}$$

$$\frac{\partial^2 \varphi}{\partial z^2} = n_d + \mu_e \exp(\sigma_i \varphi) - \mu_i \exp(-\varphi), \tag{12}$$

where n_d is the dust particle number density normalized by its equilibrium value n_{d0}, u_d is the dust fluid velocity normalized by $C_d = (Z_{d0} k_B T_i / m_d)^{1/2}$, and φ is the electrostatic wave potential normalized by $k_B T_i / e$. The time and space variables are in units of the dust plasma period $\omega_{pd}^{-1} = (m_d / 4\pi n_{d0} Z_{d0}^2 e^2)^{1/2}$ and the Debye length $\lambda_{Dm} = (k_B T_i / 4\pi Z_{d0} n_{d0} e^2)^{1/2}$, respectively. We have denoted $\mu_e = \mu/(1-\mu)$, $\sigma_i = T_i/T_e$, and $\mu_i = 1/(1-\mu)$. We have assumed that the electrons and ions follow the Boltzmann distributions.

As before, introducing the stretched coordinates [26] $\zeta = \varepsilon^{1/2}(z - v_0 t)$ (with v_0 the soliton speed normalized by C_d), and $\tau = \varepsilon^{3/2} t$, expanding n_d, u_d, and φ in a power series of ε, and developing the equations in powers of ε, we obtain a KdV equation

$$\frac{\partial \varphi^{(1)}}{\partial \tau} + a_d \varphi^{(1)} \frac{\partial \varphi^{(1)}}{\partial \zeta} + b_d \frac{\partial^3 \varphi^{(1)}}{\partial \zeta^3} = 0, \tag{13}$$

where [15]

$$a_d = -\frac{v_0^3}{(1-\mu)^2}\left[1 + (3+\sigma_i \mu)\mu \sigma_i + \frac{1}{2}\mu(1+\sigma_i^2)\right], \tag{14}$$

and $b_d = v_0^3/2$. The stationary solution of the KdV equation, Eqs. (13) is

$$\varphi^{(1)} = \varphi_m^{(1)} \text{sech}^2[(\zeta - u_0\tau)/\Delta_d], \qquad (15)$$

where u_0 is a constant speed normalized by C_d, and $\varphi_m^{(1)} = 3u_0/a_d$ and $\Delta_d = \sqrt{4b_d/u_0}$ are the amplitude and width of the DA solitary waves. Since $v_0 > 0$, Eq. (14) reveals that a_d is always negative for all possible values of σ_i and μ. This means that a dusty plasma under consideration supports the DA solitary waves with a negative potential, but not with a positive potential.

DA SHOCK WAVES

The nonlinear propagation of the DA waves in a strongly coupled dusty plasma can be investigated by means of the generalized hydrodynamic (GH) model [29], namely Eqs. (10) and (12), and

$$(1 + \tau_m D_t)\left[n_d\left(D_t u_d + v_{dn}u_d - \frac{\partial \varphi}{\partial z}\right)\right] = \eta_d \frac{\partial^2 u_d}{\partial z^2}, \qquad (16)$$

where $D_t = \partial/\partial t + u_d \partial/\partial z$, v_{dn} is the dust-neutral collision frequency normalized by the dust plasma frequency ω_{pd}, τ_m is the viscoelastic relaxation time normalized by the dust plasma period ω_{pd}^{-1} and $\eta_d = (\tau_d/m_d n_{d0}\lambda_{Dm}^2)[\eta_b + (4/3)\zeta_b]$ is the normalized longitudinal viscosity coefficient (with η_b and ζ_b the shear and bulk transport coefficients).

To derive a dynamical equation for the DA shock waves from our basic equations (10), (12) and (16), we employ the reductive perturbation technique. Thus, as before, we introduce the stretched coordinates $\zeta = \varepsilon^{1/2}(z - u_0 t)$ and $\tau = \varepsilon^{3/2}t$, and use the expansion of n_d, u_d and φ, and finally derive a KdV-Burgers equation

$$\mathscr{A}_d^{-1}\frac{\partial \varphi^{(1)}}{\partial \tau} + \varphi^{(1)}\frac{\partial \varphi^{(1)}}{\partial \zeta} + \beta_d \frac{\partial^3 \varphi^{(1)}}{\partial \zeta^3} = \mu_d \frac{\partial^2 \varphi^{(1)}}{\partial \zeta^2}, \qquad (17)$$

where [28] $\mathscr{A}_d = (v_0^3 a_d/2)(1 + v_{dn}\tau_m/2)^{-1}$, $\beta_d = 1/a_d$, $\mu_d = \eta_{d0}/a_d v_0^3$, $a_d = (v_{dn}\tau_m - a_{\mu\sigma})/v_0^4$, and $a_{\mu\sigma} = 2v_0^4[1 + (3 + \sigma_i\mu)\sigma_i\mu + (1 + \sigma_i^2)\mu/2]/(1 - \mu)^2$. Thus, for a weakly coupled or collisionless dusty plasma ($v_{dn}\tau_m \to 0$) we have $a_{\mu\sigma} < v_{dn}\tau_m$, i.e. $a_d < 0$, i.e. all coefficients ($\mathscr{A}_d$, β_d and μ_d) are negative, whereas for a strongly coupled highly collisional dusty plasma satisfying $v_{dn}\tau_m > a_{\mu\sigma}$ we have $a_d > 0$, i.e. all coefficients ($\mathscr{A}_d$, β_d and μ_d) are positive. We note that for the usual dusty plasma parameters (viz. $\mu = 0.1$ and $\sigma_i = 1$) we have $a_{\mu\delta} \simeq 2.0$. This means that for $T_i \leq T_e$ and $n_{e0} \leq 0.1 n_{i0}$ all coefficients ($\mathscr{A}_d$, β_d and μ_d) can be positive if $v_{dn}\tau_m \geq 2$. The shock solutions of Eq. (17) and their properties, depending on the signs and values of dispersive and dissipative coefficients β_d and μ_d, can be explained in the same way as we did before.

DL SOLITARY WAVES

We consider a multi-component dusty plasma consisting of the electrons, ions, and negatively charged dust particles which interact with each other via the Debye Hückel potential energy

$$U_D = \sum_{i<j} \frac{q_d^2}{r_{ij}} \exp(-k_D r_{ij}), \tag{18}$$

where $k_D = (\lambda_{De}^{-2} + \lambda_{Di}^{-2})^{1/2}$ and $r_{ij} = |\mathbf{r}_i - \mathbf{r}_j|$, with $\mathbf{r}_k$ being the dust particle coordinate. The equation of motion for charged dust particles in a linear chain is

$$m_d \frac{\partial^2 \mathbf{r}_k}{\partial t^2} = -\frac{\partial U_D}{\partial \mathbf{r}_k}. \tag{19}$$

Now, following the perturbation technique of Melandsø [3], we expand Eq. (19) by assuming that the particle displacement $u(\mathbf{r},t)$ from the equilibrium position is small, and thereby keep the first nonlinear terms. The resulting nonlinear equation for one-dimensional DL waves is of the form

$$\frac{\partial^2 u}{\partial t^2} - C_L^2 \frac{\partial^2 u}{\partial x^2} + \frac{\gamma}{2} C_L^2 \frac{\partial}{\partial x}\left(\frac{\partial u}{\partial x}\right)^2 - C_L^2 L_d^2 \frac{\partial^4 u}{\partial x^4} = 0. \tag{20}$$

For $\kappa = k_D a \ll 1$, where a is the inter-particle spacing, we have $C_L^2 = (q_d^2/m_d a)[3 + 2\ln(\kappa^{-1})]$, $\gamma = [11 + 6\ln(\kappa^{-1})]/[3 + 2\ln(\kappa^{-1})]$ and $L_d^2 = 1/k_D^2[3 + 2\ln(\kappa^{-1})]$, while for $\kappa \gg 1$, we have $C_L^2 = (Q^2/m_d a)(\kappa^2 + 2\kappa + 2)\exp(-\kappa)$, $\gamma = (\kappa^3 + 3\kappa^2 + 6\kappa + 6)/(\kappa^2 + 2\kappa + 2)$, and $L_d^2 = a^2/12$.

The quantity $\partial u/\partial x$ and the normalized dust density perturbation $n_{d1}(\ll n_{d0})$ are related by $\partial u/\partial x \approx -n_{d1}/n_{d0} = -N$. We now seek localized solutions of Eq. (20) in the frame $\xi = x - V_L \tau$, where V_L is the constant velocity, by assuming that u is function of ξ and τ. Hence, Eq. (20) for uni-directional propagation becomes

$$\frac{\partial N}{\partial \tau} + a_L N \frac{\partial N}{\partial \xi} + b_L \frac{\partial^3 N}{\partial \xi^3} = 0, \tag{21}$$

where $a_L = \gamma C_L/2$ and $b_L = C_L L_d^2/2$. Equation (21) is a KdV equation which admits a stationary solution in the form of a compressional dust density solitary wave. The latter is of the form

$$N = \mathcal{N} \operatorname{sech}^2\left[\frac{1}{\Delta_L}(\xi - V_L \tau)\right], \tag{22}$$

where $V_L = \gamma C_L \mathcal{N}/6$ and $\Delta_L = L_d \sqrt{12/\gamma \mathcal{N}}$. It turns out that the soliton speed V_L (width δ) is directly (inversely) proportional to the soliton amplitude $\mathcal{N}$.

DISCUSSION

We have presented the basic features of different types of nonlinear electrostatic waves that may exist in weakly and strongly coupled dusty plasmas. We have studied the properties of the DIA solitary and shock waves in an unmagnetized dusty plasma whose constituents are inertial ions, Boltzmann electrons, and immobile dust grains. The DIA solitary waves with a positive (negative) potential can exist for $\mu > (<) \ 1/3$. As we increase σ_i, we need higher value of the critical Mach number (corresponding to a lower value of μ) in order to have DIA solitary waves with negative potential.

We have studied the properties of the DA solitary and shock waves in an unmagnetized dusty plasma by considering the dynamics of charged dust particles and Boltzmannean electron and ion density responses. We have shown that such a dusty plasma can support DA solitary waves with a negative potential only, corresponding to a hump in the dust number density.

A strongly coupled (correlated) dusty plasma may support the DA shock waves, instead of DA solitary structures due to the dissipative effect μ_d. When μ_d is extremely small, shock waves will have an oscillatory profile in which first few oscillations will be close to solitary structures. If μ_d is increased and if it is larger than a critical value μ_{dc}, shock waves will have a monotonic behaviour. We have finally studied the properties of the DL solitary waves in a strongly coupled dusty plasma by constructing a nonlinear evolution equation, which admits solitary dust density compressional solution.

ACKNOWLEDGMENTS

This work was partially supported by the Research Training Network entitled "Complex Plasmas: The Science of Laboratory Colloidal Plasmas and Mesospheric Charged Aerosols" of the Fifth Framework Programme of the European Commission through the contract No. HPRN-CT2000-00140. A. A. Mamun gratefully acknowledges the financial support of the Alexander von Humboldt-Stiftung (Bonn, Germany).

REFERENCES

1. Shukla, P. K. and Silin, V. P., *Phys. Scripta* **45**, 508 (1992).
2. Rao, N. N., Shukla, P. K., and Yu, M. Y., *Planet. Space Sci.* **38**, 543 (1990).
3. Melandsø, F., *Phys. Plasmas* **3**, 3890 (1996).
4. Barkan, A., D'Angelo, N., and Merlino, R. L., *Planet. Space Sci.* **44**, 239 (1996).
5. Barkan, A., Merlino, R. L., and D'Angelo, N., *Phys. Plasmas* **2**, 3563 (1995).
6. Homann, A., Melzer, A., Peters S., and Piel, A., *Phys. Rev. E* **56**, 7138 (1997).
7. Verheest, F., *Space Sci. Rev.* **77**, 267 (1996).
8. Shukla, P. K., *Phys. Plasmas* **8**, 1791 (2001).
9. Verheest, F., *Waves in Dusty Plasmas* (Kluwer Academic Publishers, Dordrecht, 2000).
10. Shukla, P. K. and Mamun, A. A., *Introduction to Dusty Plasma Physics* (Institute of Physics Publishing Ltd., Bristol, 2002).
11. Bharuthram, R. and Shukla, P. K., *Planet. Space Sci.* **40**, 973 (1992).
12. Mamun, A. A. and Shukla, P. K., *Phys. Plasmas* **9**, 1468 (2002).
13. Shukla, P. K., *Phys. Plasmas* **7**, 1044 (2000).
14. Popel, S. I., Yu, M. Y., and Tsytovich, V. N., *Phys. Plasmas* **3**, 4313 (1996).
15. Mamun, A. A., Cairns, R. A., and Shukla, P. K., Phys. Plasmas **3**, 702 (1996); Mamun, A. A., Astrophys. Space Sci. **268**, 443 (1999).
16. Ma, J. X. and Liu, J., *Phys. Plasmas* **4**, 253 (1997).
17. Melandsø, F. and Shukla, P. K., *Planet. Space Sci.* **43**, 635 (1995).
18. Shukla, P. K. and Mamun, A. A., *IEEE Trans. Plasma Sci.* **29**, 221 (2001).
19. Farokhi, B., Shukla, P. K., Tsintsadze, N. L., and Tskhakaya, D. D., *Phys. Lett. A* **264**, 318 (1999).
20. Nakamura, Y. and Sharma, A., *Phys. Plasmas* **8**, 3921 (2001).
21. Nakamura, Y., Bailung, H., and Shukla, P. K., *Phys. Rev. Lett.* **83**, 1602 (1999).
22. Luo, Q. Z., D'Angelo, N., and Merlino, R. L., *Phys. Plasmas* **6**, 3455 (1999).
23. Nakamura, Y., *Phys. Plasmas* **9**, 440 (2002).
24. Samsonov, D., Ivlev, A. V., Quinn, R. A., Morfill, G., and Zhdanov, S., *Phys. Rev Lett.* **88**, 095004.
25. Nosenko, V., Nunomura, S., and Goree, J., *Phys. Rev Lett.* **88**, 215002.
26. Washimi, H. and Taniuti, T., *Phys. Rev. Lett.* **17**, 996 (1966).
27. Karpman, V. I., *Nonlinear Waves in Dispersive Media* (Pergamon Press, Oxford, 1975).
28. Mamun, A. A. and Shukla, P. K., *IEEE Trans. Plasma Sci.*, **30**, in press (2002).
29. Kaw, P. K. and Sen, A., *Phys. Plasmas* **5**, 3552 (1998).

Complex Plasmas under Microgravity Conditions: First Results from PKE-Nefedov

G. E. Morfill*, H. M. Thomas*, B. M. Annaratone*, A. V. Ivlev*, R. A. Quinn*, A. P. Nefedov[†], V. E. Fortov[†] and PKE-Nefedov Team**

Centre for Interdisciplinary Plasma Science, Max-Planck-Institut für Extraterrestrische Physik, D-85740 Garching, Germany
[†]*Institute for High Energy Densities, 127412 Moscow, Russia*
**Quoted in Ref. [1]*

Abstract. PKE-Nefedov[1] is the first natural science experiment on the International Space Station (ISS). It is designed to study "complex plasmas" (ions, electrons, and charged microparticles) at the most elementary, the kinetic level. The PKE-Nefedov laboratory is described, the unique features of a microgravity environment are discussed and the first results of the early experiments are presented. These include an overview of the complex plasma features observed and some dedicated experiments and measurements – complex plasma boundaries, decharging of the complex plasma cloud, epitaxial-like growth of plasma crystals, and charge-enhanced coagulation – as well as an outlook regarding future planned experiments.

INTRODUCTION

"Complex plasmas" consist of ions, electrons, highly charged microparticles, and neutral gas. They cannot be considered simply as partially ionized multicomponent plasmas, because the microspheres carry a variable charge of typically $Z \sim 10^3 - 10^4$ electrons, depending on size and conditions. The charge variability of the microspheres has two origins: The first is stochastic, due to discrete collection and recombination of electrons and ions, as well as the plasma fluctuations, and the second depends on the particle number density itself, i.e., when the density increases the supply of ions and electrons per particle decreases and vice versa. This can be seen quite simply. Under quasineutrality conditions we have locally: $n_e + Zn_p = n_i$ (the subscripts stand for electrons, particles, and ions, respectively), and hence Z depends on $n_p/n_{e,i}$ as well as $n_{e,i}$.

The interest in complex plasmas has grown steadily over the last eight years. This has several reasons, all of which combine to make complex plasmas an exciting new tool for studying many-particle interacting systems at the most elementary – the kinetic level. They are:

1. One plasma component (the microparticles) is directly visible at the individual particle level.

---

[1] Named after Anatoli Nefedov, who died suddenly on February 19th, 2001, and who had done so much to make the project successful.

2. Plasma processes are slowed down, due to the heavy microparticles – the plasma frequency is typically of the order 10 Hz – and hence plasma processes can be studied at comparatively high temporal resolution.
3. Due to the large charge on the microparticles, the regime of "strongly coupled plasmas" becomes readily accessible for experimental work for the first time.
4. Complex plasmas can be easily controlled and manipulated (e.g., using electrostatic forces, laser light pressure, thermophoresis or neutral gas flows), allowing active experiments to be performed.
5. A plethora of new phenomena can be studied in unprecedented detail in these new systems. Understanding these, it is reasonable to assume will help to understand other systems – ranging from clusters, crystals, fluids to small microcanonical thermodynamical systems, etc.

In this topical review we will summarize some of the progress made in this field in recent times, in particular the first results obtained from the microgravity laboratory PKE-Nefedov, the first natural science laboratory in operation on the International Space Station (ISS), will be discussed. PKE-Nefedov is a multi-purpose fully programmable laboratory for complex plasma studies under microgravity conditions. In the next chapters we will briefly describe this laboratory and its special operations features that make it so flexible to use. We describe the overall structure and flow patterns of complex plasmas observed under microgravity conditions and discuss why microgravity is "special" for complex plasmas. Then we focus on four experiments performed with PKE-Nefedov: (i) We study the boundary between complex plasmas of different size particles and the boundary between the complex plasma and the central particle-free region – the so-called "void". (ii) We discuss the process of "decharging" – the interest here lies in the time taken for a complex plasma to give up its charge when the power is turned off. Under gravity condition the particles simply fall down and such investigations are impossible, whereas under microgravity particles remain suspended and it is possible to measure the charge decay and any rest charge (if it exists). (iii) We describe experiments on the complex plasma flow (which is similar to epitaxial growth) in the limit of high particle influx. (iv) We discuss the behavior of an overall charge-neutral particle cloud (no electrons and ions, but positively and negatively charged microparticles). We present theoretical work on the runaway charge-induced coagulation (the so-called "gelation") – a process that can be of considerable astrophysical significance, e.g., in the formation of planets. In conclusion, we summarize the first results obtained with PKE-Nefedov and outline other planned experiments.

PKE-NEFEDOV, THE FIRST NATURAL SCIENCE LABORATORY ON THE ISS

The PKE-Nefedov hardware onboard the ISS consists of two parts. The first part is a hermetically sealed container of 50 cm height and 50 cm diameter. Inside the container, a platform is mounted to the container wall using dampers to reduce jitters. On top of the platform the experimental components are mounted (see Fig. 1). This consists of the rf plasma chamber, including two microparticle dispensers with monodisperse particles of 3.4 and 6.8 μm in diameter, the rf generator, the gas control unit and the optical particle diagnostics consisting of two laser diodes with microline optics for illumination, lenses and two video cameras with different magnification of the complex plasma. The particle diagnostics is mounted on a translation stage which enables the whole system to move, so that 3D information on the complex plasma is obtained through 2D scanning. The plasma chamber is a special development for microgravity experiments. It is a symmetrically driven rf discharge in the push-pull mode, which provides the highest symmetry of the generated plasma (with respect to the horizontal midplane). A low frequency voltage ($\leq$ 100 Hz) can be applied to the electrodes to excite the complex plasma. This allows us to excite waves and measure the dispersion relation.

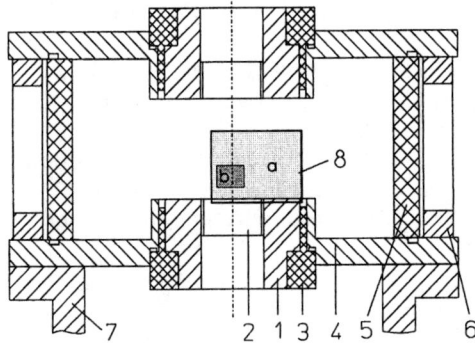

FIGURE 1. This sketch shows the basics of the experimental setup, the plasma chamber. It consists of two identical rf-electrodes, driven symmetrically in the push-pull mode (1). Each rf-electrode includes the microparticle dispenser (2). They move like a piston to inject particles into the plasma region between the electrodes. The two dispensers allow experiments with two different particle sizes and their mixture. The electrodes are insulated by a Macor ceramic (3) from the grounded shield and flanges (4). The walls of the plasma-vacuum chamber are made of glass (5). The walls have a quadratic cross-section to achieve planar geometry for the laser and viewing. The chamber is held together with spacers (6). O-rings from Viton are used for vacuum tightness of the plasma chamber. Two CCD-cameras provide two different magnifications of the complex plasmas. The overview camera (a) shows about a quarter of the field between the electrodes (28.16×21.45 mm^2), while the high resolution camera (b) is used for detailed views inside the overview field covering a field of (8.53×6.50 mm^2).

The space below the platform is used for the assembly of the electronics and the experiment computer. The computer analyzes important experimental parameters and controls the plasma generation in real time. It also runs the experimental software, which is preprogrammed and tested on Earth.

The second part of the hardware onboard ISS consists of a control computer which is used as a console for the cosmonauts and allows them to follow and manipulate, if necessary, the preprogrammed experiments. Two video recorders are connected to the computer for storage of the video signals transferred from the experimental setup.

A typical experimental run on the ISS starts with the evacuation of the plasma chamber to high vacuum for a couple of hours. For this purpose a turbomolecular pump is connected to the vacuum valve on top of the experimental block. A ventline to space is used as prevacuum for the pump. After reaching good vacuum conditions the experiment is prepared and started by the cosmonaut using the control console computer. During the experimental run, which lasts typically 90 minutes, an audio connection is available to the corresponding cosmonaut. A video connection can be established, too, for a small fraction of the experimental time. These allow the scientists on ground to be up to date with the experiment progress. After the experiment, the plasma chamber is pumped for a few hours, closed and finally filled with Argon at a few mbars.

The digital housekeeping data can be transferred electronically via e-mail down the ground, the video cassettes are brought back by the cosmonaut with the "Soyuz" rocket from the Russian side or the Space Shuttle on the American side, respectively.

COMPLEX PLASMAS UNDER MICROGRAVITY

On Earth, the need to levitate microparticles against the pull of gravity implies that the complex plasma has to be located in a strong electric field. From $ZeE \simeq Mg$ we get electric fields of typically $E \sim 100$ V/m in order to levitate micron-sized particles. The field scales as the squared particle size, so that for 3 μm particles we need $E \sim 1000$ V/m. Such fields exist, e.g., in the sheath of a rf discharge or in striations of dc glow discharges. For comparison with interparticle (Coulomb) forces, we define the "force ratio":

$$\beta = \frac{e^2 Z^2}{\Delta^2 Mg} e^{-\kappa} \equiv \beta_0 e^{-\kappa},$$

where $\kappa = \Delta/\lambda_D$ is the particle separation in units of the screening (Debye) length – the determining parameter. For micron-sized particles carrying a charge of $Z \sim 3 \times 10^3$ electrons and separated by $\Delta \simeq 150$ μm (as observed) we have $\beta_0 \sim 1$. (Note that for Mg we may write eZV_{sh}/L_{sh}, with V_{sh} the sheath potential and L_{sh} the length scale. Then β_0 is the ratio of interparticle potential to the potential change in the sheath over the particle separation, Δ). This shows that on the ground the bulk (gravity) force is comparable to (or even larger than) the interparticle Coulomb force. Hence the complex plasmas on the ground are in a highly stressed state, which in turn means that there is plenty of stored energy that can be released easily and can lead to instabilities that may mask the (possibly more subtle) effects one wishes to investigate.

In space, under microgravity conditions particles are suspended in the "quasi-isotropic" (bulk) region of the discharge. The need for confinement is still there, but the major forces are now ion drag, thermophoresis and internal (Coulomb) pressure of the complex plasma itself. The bulk forces are limited by the electric field that the plasma may sustain. This (ambipolar) field is typically very weak, of the order of $E \sim 1$ V/m, i.e., less than one percent of what is required on Earth to suspend the micron-sized particles. Under microgravity we correspondingly write the force ratio as:

$$\beta = \frac{eZ}{\Delta^2 E} e^{-\kappa},$$

which implies that externally imposed stresses are negligible compared to the interparticle forces ($\beta \gg 1$), provided that $\kappa \leq 4-5$. In an isotropic plasma the screening length, λ_D, is of the order of the ion Debye length, λ_{Di}.

It is clear from these considerations that microgravity has a profound influence on the properties, structure, dynamics, and physical state of complex plasmas, making experimental investigations possible that cannot be conducted on the Earth. Figure 2 shows the overall features of complex plasmas in the PKE-Nefedov laboratory. Unlike measurements on Earth, where the complex plasma is simply confined to a narrow layer in the plasma sheath, under microgravity it fills nearly the whole volume. The cloud contains several million particles. In this particular example two particle sizes were used, 3.4 and 6.8 μm. The features of Fig. 2 are:

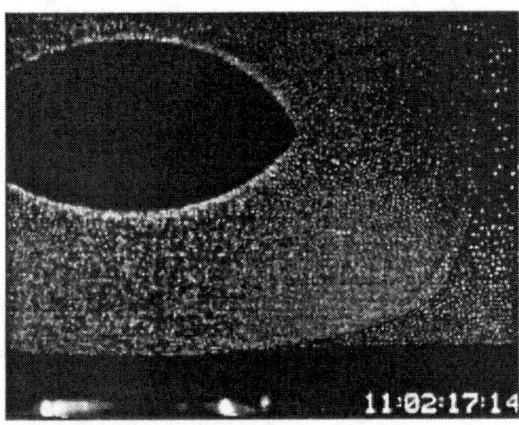

FIGURE 2. Measurements from PKE-Nefedov showing the distribution of two complex plasmas in the vertical (parallel to the axis) midplane of the cylindrical plasma chamber. The experimental parameters are: rf voltage 35.75 V, pressure 74 Pa, diameter of large particles 6.8 μm, (located outside), small particles 3.4 μm (located inside).

1. The two particle types separate, with the small particles on the inside.
2. The boundary between the two particle regimes is sharply defined, the surface roughness is of the order of the interparticle distance, possibly even less.
3. Along the chamber axis the particles form crystal structures. Regimes of around $40 \times 40 \times 20$ lattices can be obtained – between 10^4 and 10^5 particles.
4. Outside this crystalline regime there are circulating flows, the complex plasma is in a liquid state.
5. A central particle-free void exists, which has a very sharp boundary, of similar roughness as the interparticle boundary mentioned above.
6. The system is rotationally symmetric about the chamber (vertical) axis. Due to small thermophoretic effects a slight asymmetry with respect to the horizontal midplane develops when the experiment has been in operation for some time. (Figure 2, taken at the beginning of an experimental run does not show this).

It is the features described above and the controlled manipulations that make investigations into boundaries, flows, crystal growth, etc. possible. By changing the rf power, modulating the electrode voltage, changing the gas pressure, and effecting abrupt changes in some of these parameters it is possible to generate special conditions for studying complex plasmas under relatively stress-free conditions.

EXPERIMENTS

The first series of experiments, PKE-Nefedov Basic, were designed to study 3D complex plasmas in detail for the first time [1]. The mission was to investigate basic phenomena occurring in complex plasmas under microgravity conditions, such as structures and transitions in plasma crystals [1], waves and shocks [2], particle coagulation [3], etc. Complex plasmas, consisting of micron-sized charged particles embedded in rf and glow discharge plasmas, were initially studied in ground-based laboratories [4]. As the PKE-Nefedov experiments show, removing the force of gravity leads to much larger system sizes as well as a host of new, even unexpected, phenomena.

Complex plasma boundaries

The coexistence of microparticles and a quasineutral plasma has been studied only recently in microgravity experiments [5]. When the plasma sheaths surrounding the particles interact unpredictable effects arise [6, 7]. Among those the counter-intuitive effect which pushes the negative particles away from the center of the discharge, which is generally the most positive part of the plasma.

Here we investigate the microscopic and collective processes which result from the particle segregation, the self-structuring and the surprisingly sharp surfaces observed. The mechanism outlined in this paper has been suggested by the analysis of the microgravity experiments but it is of wider validity. Figure 2 represents a typical steady state side view for a plasma loaded with two different size particles during the PKE-Nefedov experiments. It is easy to identify four regions that, although not completely uniform, show internal coherence. The first region is the central void. Its extension is much larger than any screening length so that we can assume quasi-neutrality in it. Ionization does actually occur in the void but, because of symmetry, no net current crosses the void-complex plasma boundary. The second region, surrounding the void is a three-component plasma where the negative charge is distributed between the electrons and the particles to equal the ion density. Since the free electron density is lower than in the void the electron screening length is longer. Here the ionization rate is reduced with respect to the void. The third region shows larger particles with interparticle distances, hence larger screening lengths. The fourth region is dominated by the electrode rf sheath.

In this work we analyze the boundaries between the above regions assuming quasineutrality in the first three. Large discontinuities in the electrostatic potential, charge double layers (DL) will match the flow of particles from any of the above plasmas. These DL are similar to the those found at the edge of a metal or at the discontinuity between different work-function metals or also in semi-conductors. A difference with respect to the above examples is that all the three components, electrons, ions and charged particles can participate to the conduction and in general their densities are not in equilibrium.

A one dimensional fluid model is used to describe the existence of strong DL, separating void and dust volume. The ions arrive at the void-DL boundary with an averaged velocity acquired in a pre-sheath potential. Here we have assumed monoenergetic ions. In the absence of collisions in the DL the flux of ions is continuous across the sheath and energy is conserved. The void electrons enter the sheath with random velocity, we can assume, in most cases, the Boltzmann distribution law. Some correction for small DL potentials leading to non-Maxwellian distributions might be required for weak DL. A beam of electrons is leaving the complex plasma at lower potential and is accelerated by the DL field. Based on visual observations of the steady state shown in Fig. 2, the injected particles do not enter the DL but, being negatively charged, limit the number of electrons available for conduction. In complex plasmas, the ratio of the volume particle density to the free electron density defines the "Havnes parameter", P (see Ref. [7]). The above statements lead to the following net charge density:

$$\rho = \frac{en_{i0}}{\sqrt{1+V/V_0}} - \frac{J_b/\sqrt{2eV_{DL}/m_e}}{\sqrt{1-V/V_{DL}}} - en_{e0}\exp(-eV/T_e) - en_{n0}\exp(-eV/T_n),$$

where n is the density, V is the potential, J_b is the current density of the beam, T is the temperature; subscripts i,e,n refer to ions, electrons from the higher potential, and negative ions, 0 corresponds to the pre-sheath edge. Poisson equation has been integrated with quasineutrality conditions on the two sides of the DL and for a variety of currents, Havnes parameters and Maxwell stresses at the low potential side. It was found that each solution was only possible for a single triplet of the values V_0, V_{DL}, and J_b.

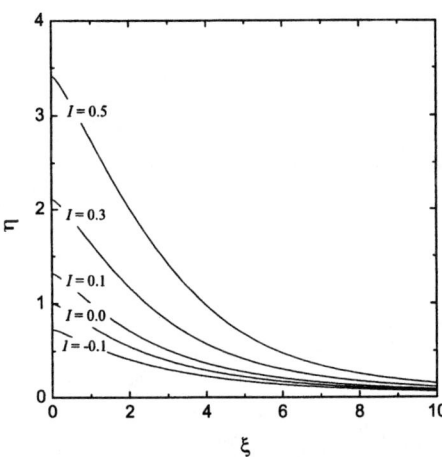

FIGURE 3. Potential profiles for several values of the current across the double layer (hypothesis of zero Maxwell stress on the "low potential" plasma edge of the double layer.

The curves for zero stress and $P = 0$ (the void-complex plasma case) are presented in Fig. 3, where the $I_b = 0$ curve applies to closed geometries, as in Fig. 2. Dimensionless potential η is measured in units of T_e/e, dimensionless distance ξ is normalized to λ_D, and the current I is normalized to the Child-Langmuir current. Matching the experimental DL thickness, as seen between the two complex plasmas in Fig. 2 with the theoretical curves allows the determination of the Havnes parameter, for the small particles complex plasma. The value found is $P \simeq 0.2$.

We conclude, that the complex plasma surface behaves almost like a wall with zero net current for closed geometries. In a microgravity environment, the microparticles arrange themselves around a void, where the ionization mainly occurs. Different complex plasmas will then self-organize and arrange themselves in order of decreasing free electron density. The double layer theory provides information on the potential structure of the observed sharp interface between the void plasma and the complex plasma. A modified equation for the DL between complex plasmas was derived taking into account the variations in the electron density. This allows us to evaluate the Havnes parameter and the coupling strength.

Decharging of a complex plasma

Here we describe a decay, or "decharging" of a complex plasma in the afterglow, after switching off the discharge power. Particles injected in a plasma are charged in a few μs and form a cloud with an ellipsoidal void in the center of the chamber. In order to study the particle charges a sinusoidal voltage produced by a function generator (FG) with amplitude ± 13 V and frequency 0.47 Hz was applied to both electrodes. This caused vertical oscillations of the cloud due to the electrode sheath modulation. The FG voltage was only applied to the outer part of the electrodes (outer rings), the particle dispensers in the center were grounded. Also, the lower ring was biased negatively (about 2-3 V) with respect to the upper one. For this experiment we used particles of 6.8 μm diameter, the neutral gas pressure was $p = 73$ Pa, a rf peak-to-peak voltage was $\simeq 90$ V, the plasma number density in the bulk is estimated as $n_0 \sim 3 \times 10^9$ cm^{-3}.

After the discharge was switched off the afterglow plasma disappeared during a few ms and the cloud started drifting upward slowly. We studied the dynamics of the cloud ("layer") below the void in detail. The vertical motion of different parts of the layer is strongly dependent on the radial (horizontal) position: Figure ?? shows the vertical position (center of masses) of the layer versus time in the center and at the "periphery" of the chamber. The first obvious feature is that the velocity of the drift increases with radial distance from the chamber axis, the second one is that the particles oscillate with the frequency of the FG voltage and with varying amplitudes which are also higher at the periphery. Since the vertical ac and dc electric fields also increase with radial distance, both observed features have a clear electric nature, i.e., the particles retained a certain negative charge after the plasma was off. We can determine this "rest charge" using the measured amplitude of oscillations, A, and the following relation:

$$A = \frac{eZ_{\text{rest}} E_{\text{FG}}}{2\gamma \omega M}. \tag{1}$$

Typical amplitudes at the periphery were $A \simeq 0.2 - 0.3$ mm, the excitation FG electric field $E_{\text{FG}} \simeq 8 - 10$ V/cm had a frequency $\omega = 2\pi f_{\text{FG}} = 2.95$ s^{-1}. Particles of mass

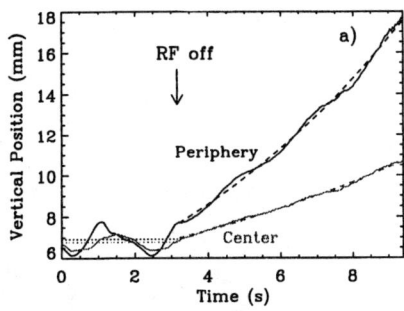 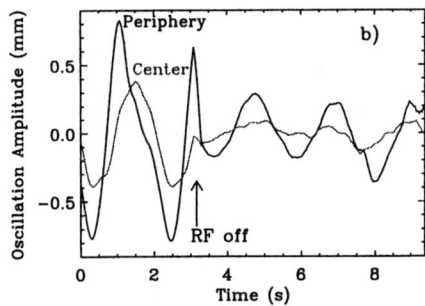

FIGURE 4. (a) Vertical position (center of masses) of the lower part of the particle cloud. The two curves show the vertical motion in the center and at the "periphery" of the experimental chamber. The motion is a combination of the quasi-steady vertical drift (dashed line) and the oscillations with the FG frequency. (b) Oscillatory part of the motion.

$M \simeq 2.5 \times 10^{-10}$ g have a neutral gas drag damping coefficient $2\gamma \simeq 140$ s^{-1}. Then from Eq. (1) we obtain that each particle carried $Z_{\text{rest}} \simeq 130 - 150$ electron charges. The rest charge is about two orders of magnitude less than the initial (plasma) value, but is still a quite significant charge. In order to explain this effect we propose a simple model of the complex plasma decharging. We study the plasma decay after the discharge power is turned off and include the self-consistent variation of the particle charge. In the model we assume that the density of particles is sufficiently low and they do not change the initial plasma charge composition significantly, i.e., the charge density of particles, Zn_p, is small compared to the initial plasma density, n_0.

The kinetics of the plasma decay is determined by a combination of the plasma recombination and the electron temperature relaxation. Recombination of a rf plasma is primarily due to diffusion on the walls of the discharge chamber [8]. (Note that the diffusional loss rate decreases as the neutral gas pressure, p, increases, whereas the time scale of another sink – recombination on the particles, does not depend on p. But even at the high pressures used in our experiment diffusion remains the fastest process). At the initial stage of the decay the diffusion is ambipolar, so that the plasma quasineutrality, $n_i \simeq n_e$, is provided. The equation for the dimensionless plasma density, $\tilde{n} = n_{i,e}/n_0$, normalized to the initial density is [8]

$$\frac{d\tilde{n}}{dt} = -\frac{\tilde{n}}{\tau_D}, \qquad (2)$$

where the diffusion loss time scale is $1/\tau_D \simeq \frac{1}{3}(v_{T_i} l_{in}/\Lambda^2)(1+\tilde{T}_e) \equiv \frac{1}{2}(1+\tilde{T}_e)/\tau_D^\infty$. Here Λ is the characteristic diffusion length (of the order of the chamber size), $v_{T_i} = \sqrt{T_i/m_i}$ is the thermal velocity of ions, l_{in} is the mean free path of ion-neutral collisions, and $\tilde{T}_e = T_e/T_n$ is the ratio of the electron to neutral temperature. We assume that the ion and neutral temperatures are equal, $T_i = T_n \equiv T$. The initial electron temperature is much higher than T (typically, $\tilde{T}_{e0} \sim 10^2$), but when the power is off it decreases due to electron-neutral collisions and tends asymptotically to T. Therefore, the diffusion time scale increases with time, from the initial value $\tau_D^0 \simeq 2\tau_D^\infty/\widetilde{T}_{e0}$ to the limit τ_D^∞. The equation of the electron temperature relaxation is [8]

$$\frac{d\widetilde{T}_e}{dt} = -\frac{\widetilde{T}_e - 1}{\tau_T}. \tag{3}$$

The time scale of the temperature relaxation is $1/\tau_T = 2\sqrt{m_e/m_i}(v_{T_i}/l_{en})\sqrt{\widetilde{T}_e} \equiv \sqrt{\widetilde{T}_e}/\tau_T^\infty$, i.e., τ_T also grows with time, varying in the range $\tau_T^\infty/\sqrt{\widetilde{T}_{e0}} \equiv \tau_T^0 \leq \tau_T \leq \tau_T^\infty$.

The particle charge number, Z, is generally a function of the electron and ion densities and the electron temperature. The kinetic equation for Z in the orbital motion limit (OML) is [9]

$$\frac{dZ}{dt} = J_e - J_i \equiv 2\sqrt{2\pi}a^2 \left[n_e v_{T_e} e^{-\gamma} - n_i v_{T_i}(1 + \widetilde{T}_e \gamma) \right], \tag{4}$$

where $J_{e,i}$ are the electron and ion fluxes on the particles, a is a particle radius, and $\gamma = e^2 Z/aT_e$ is the dimensionless surface potential of the particle. If the time scale of the charge variation is sufficiently short (less than τ_D and τ_T), then the charge is close to equilibrium, $Z \simeq Z_{eq}$. Its value is given by equation: $(n_e/n_i)\sqrt{\widetilde{T}_e}e^{-\gamma_{eq}} = \sqrt{m_e/m_i}(1 + \widetilde{T}_e \gamma_{eq})$. Note that γ_{eq} is a function of three parameters – the electron-ion density ratio, temperature ratio, and mass ratio. The dependence on $\widetilde{T}_e$ and m_e/m_i is rather weak: For quasineutral plasma of various gases, $\gamma_{eq} \simeq 2.5 - 4.5$ in the range $1 \leq \widetilde{T}_e \leq 100$. From Eq. (4) we obtain the kinetic equation for the charge fluctuations around equilibrium,

$$\frac{dZ}{dt} \simeq -\frac{Z - Z_{eq}}{\tau_Z}, \tag{5}$$

with a time scale of the variations which increases as the plasma decays, $1/\tau_Z \simeq (v_{T_i}a/\sqrt{2\pi}\lambda_{i0}^2)(1 + \gamma_{eq})\tilde{n} \equiv \tilde{n}/\tau_Z^0$, (here $\lambda_{i0} = \sqrt{T/4\pi n_0 e^2}$ the initial ion Debye length). Thus, if $\tau_Z \leq \min\{\tau_T, \tau_D\}$, then the charge fluctuates around the equilibrium value Z_{eq}, and the fluctuations obey Eq. (5). In the opposite case, the general kinetic equation for the charge, Eq. (4), should be used.

The differential equations (2), (3), and (4) along with the plasma quasineutrality condition is a complete set of equations describing decay of a complex plasma. First, let us consider the dependence of the time scales on the plasma parameters. The major parameter is the gas pressure, p (initial plasma density, n_0, is not an independent parameter, but is roughly proportional to p). Both the charging and the temperature relaxation processes accelerate with pressure, $\tau_Z, \tau_T \propto p^{-1}$, whereas the plasma density decay has the opposite tendency, slowing down as $\tau_D \propto p$. The initial charging time for micron size particles is usually much shorter than the time of the temperature relaxation: For $\widetilde{T}_{e0} \sim 10^2$ we have $\tau_Z^0/\tau_T^0 \sim 10^{-1} - 10^{-2}$. Then, we compare the temperature relaxation and diffusion processes and get the scaling $\tau_T^0/\tau_D^0 \propto p^{-2}$. Thus, initially charging is the fastest process; the plasma diffusion is the slowest process at high pressures, but it can be faster than the temperature relaxation for pressures below a certain critical value. This critical pressure,

p_{cr}, is given by the condition $\tau_T^0 \sim \tau_D^0$, which corresponds to $p_{cr} \sim 50-70$ Pa for our experiment. We consider cases of high ($p \geq p_{cr}$) and low ($p \ll p_{cr}$) pressures separately. For $p \geq p_{cr}$ we have the following hierarchy of the initial time scales: $\tau_Z^0 \ll \tau_T^0 \leq \tau_D^0$. In this case the electron temperature drops down to T ($\widetilde{T}_e \to 1$) during $t \sim \tau_T^\infty = \sqrt{\widetilde{T}_{e0}\tau_T^0}$, whereas the plasma density remains nearly unchanged [because the diffusion time scale increases to $\tau_D^\infty = (1+\widetilde{T}_{e0})\tau_D^0/2 \gg \tau_T^\infty$]. The charging time scale does not depend on $\widetilde{T}_e$, and therefore the charge is still in equilibrium and is determined by γ_{eq}. Since γ_{eq} has a very weak dependence on $\widetilde{T}_e$, we can estimate that as $t \sim \tau_T^\infty$ the particle charge is decreased by factor of $\widetilde{T}_{e0}$, i.e., $Z_{eq} \sim Z_0/\widetilde{T}_{e0}$. At $t \geq \tau_D^\infty$ the plasma density starts decreasing as $\tilde{n} \propto e^{-t/\tau_D^\infty}$ in accordance with Eq. (2) and τ_Z grows exponentially. However, as long as the plasma is quasineutral, the charge (which is a function of n_i/n_e) cannot change at this stage. If the particle volume charge, Zn_p, can be neglected, the plasma quasineutrality is violated when the Debye length becomes comparable with the chamber size, $\lambda_i(\tilde{n}_*) \sim \Lambda$, i.e., when the density drops down to $\tilde{n}_* \sim \lambda_{i0}^2/\Lambda^2$. This occurs at $t_* \sim \tau_D^\infty \ln\tilde{n}_*^{-1}$. At $t \geq t_*$ electrons and ions start diffusing independently, the ratio n_i/n_e grows and therefore the relative contribution of the ion flux in Eq. (4) increases. We can evaluate the upper limit of the relative change of the charge at $t \geq t_*$ as $|Z - Z_{eq}|/Z_{eq} \leq \tau_D^\infty/\tau_Z(t_*)$. Using the definition of $\tilde{n}_*$ we get that this change is less than a/l_{in} which is $\sim 10^{-2}$ for micron size particles and mbar pressures. Therefore, the charge variation at the last stage is negligible because the charging time exceeds the time scale of the density decay (Z is not in equilibrium anymore) – the charge cannot follow the variations in the ambient plasma and becomes "frozen". We finally conclude that for high pressures the rest particle charge, Z_{rest}, should be $\sim \widetilde{T}_{e0}$ times smaller than the initial value Z_0.

In the opposite limit of low pressures, $p \ll p_{cr}$, the temperature relaxation becomes the slowest process, $\tau_D^0 \ll \tau_T^0$. Therefore, up to $t \leq \tau_T^0$ the plasma density decreases as e^{-t/τ_D^0} (the charging time correspondingly increases), whereas the temperature (and thus the charge) remains constant. The charge can only start changing at $t \geq \tau_T^0$. However, similar to the high pressure case, the charge will be already "frozen" at this moment, if $\tau_D/\tau_Z \leq 1$. This condition can be rewritten as $(\tau_D^0/\tau_Z^0)e^{-\tau_T^0/\tau_D^0} \leq 1$. Since $\tau_D^0/\tau_Z^0 \propto p^2$ and $\tau_T^0/\tau_D^0 \propto p^{-2}$ the condition can be easily satisfied for sufficiently low p. Thus, we conclude that at low pressures the rest particle charge might be "frozen" at or near the initial "plasma level", $Z_{rest} \sim Z_0$.

In our experiment the pressure is close to the critical value (maybe, somewhat higher), and the hierarchy of the initial time scales is: $\tau_Z^0 \sim 1$ μs, $\tau_T^0 \sim \tau_D^0 \sim 30-60$ μs (initial plasma parameters are $n_0 \sim 3 \times 10^9$ cm^{-3} and $\widetilde{T}_{e0} \sim 100$, initial charge is $Z_0 \sim 10^4$). In accordance with our model we would expect the rest charge to be of the order of $Z_0/\widetilde{T}_{e0} \sim 100$, which is in very good agreement with the measured value. Thus, the proposed theoretical model for the decharging agrees quite well with the experiment and predicts the rest charge at different gas pressures.

Summarizing, we measured the rest charge which particles retained after the rf power was turned off. Microgravity allows us to perform these measurements quite precisely, because the particles remain suspended in the chamber for a sufficiently long time. A simple theoretical model for the decharging was proposed which agrees quite well with the experiment and predicts the rest charge at different gas pressures. Currently we are planning to perform a series of experiments over a wide range of pressures, in order to test the model.

Complex plasma flow

Another interesting phenomenon is a complex plasma flow induced by the particle injection, which has certain characteristics in common with the epitaxial film growth (EFG) mechanism found in thin film processing. EFG is an important technique in semiconductor and nanostructure manufacturing.

In the experiment, particles were injected into an active glow-discharge plasma device. Both 3.4 and 6.8 μm particles were initially present in the plasma, and both sizes were also injected. One of the great advantages of experimental complex plasmas is the ability to extract kinetic and structural information from direct measurements of one of the plasma components, namely the dust microspheres.

The injection process threw thousands of particles into the device. These particles, which charged nearly instantaneously, swept up the already present particles (through their repulsive interaction) and carried them up (in this microgravity context, "up" only refers to the orientation of the video camera) and into the edges of the chamber. This resulted in a highly compressed particle cloud which, after the injection event, expanded rapidly back into the central region of the chamber, seeking an equilibrium configuration. Since more particles were initially compressed into the upper part of the chamber (the particles were injected from below), the particles streamed downward around the edge of the void [5, 10] and piled up on the lower particles, as shown in Fig. 5.

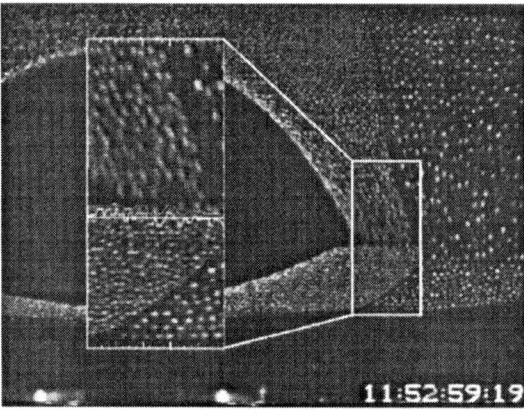

FIGURE 5. Data image showing onset of flow. Inset shows the analyzed field of view with the boundary surface position indicated.

After a brief initial period during which the lower cloud was compressed, presumably by momentum transferred from the streaming particles, the lower cloud density stayed approximately constant and the mass transfer of particles from the upper to lower cloud became essentially a surface effect. That is, particles were added at the upper surface of the lower cloud without additional compression of the lower cloud. This continued until the densities in the upper and lower clouds were equalized, a process which required a few seconds.

We can think of the process of moving particles from the upper cloud and adding them to the lower cloud at a well-defined surface as deposition on a growth front. During a significant fraction of the deposition time, the growth front was approximately horizontal, and moved upward with a slowly changing velocity. Near the end of the time period studied, the velocity of the growth front dimished because the particle flux from the upper cloud dimished as the upper cloud approached its equilibrium density. In the remainder of this chapter, we will characterize the properties of this growth front and comment on possible analogies with EFG.

In EFG, a low-density beam of atoms comes into contact with a crystal surface. When each beam atom ("adatom") contacts the surface, it is attracted to the nearest lattice cite. It then diffuses, hopping from site to site around the surface. When two adatoms meet, the sticking probability is higher and "nucleation" occurs. As more adatoms come down, other nucleation sites form, and the original ones grow. Once these nucleation "islands" are large enough, some nucleation events can take place on top of them leading to vertical structure in the growing layer. The relative amount of vertical structure is given by the surface roughness, defined as the rms variation in the surface height. Growth rate and surface roughness are two parameters of primary importance to thin film manufacturers and they are primarily determined by incident flux, desorption rate, surface area, and surface diffusion. Surface diffusion plays the primary role in the "smoothing" the surface and thus effects surface roughness (see, e.g., Ref. [11]).

From the discussion above, it is clear that in describing the complex plasma flow as a deposition phenomena, the important parameters to measure are: incident flux, desorption rate, surface growth rate, surface roughness, and surface diffusion. In addition, we must also quantify the degree of order or crystallinity of the lower particle cloud. First, we observe from the video that the particle flux is in one direction only, namely downward toward the lower lattice. In other words, there is no desorption or evaporation of particles from the boundary surface. This means that the incident flux is proportional (through the surface area) to the growth rate.

The next task is to identify the location of the boundary surface throughout the image sequence. In order to simplify this task, we focused our analysis on a small region around the boundary surface (inset, Fig. 5). An approximate boundary position was first estimated by hand such that it passed below elongated (fast) particles and also below single isolated particles near the flow region (promontories). The "actual" boundary was determined by taking the first intensity maximum within $\simeq 1.5$ interparticle spacings a below the first boundary. The boundary was then characterized by its mean height $h_0(t)$ and the standard deviation of this height $\delta h(t)$ (the surface roughness). The growth rate/incident flux is related to $h_0(t)$ through the mean density, as discussed below.

In order to obtain a as well as information about the degree of order in the lower cloud, we computed the full autocorrelation function of the image data. This is proportional to the 2D time-dependent pair correlation function $g(x,y,t)$. A slow drift of the particles in the lower cloud was also apparent from this analysis. The correlation function was

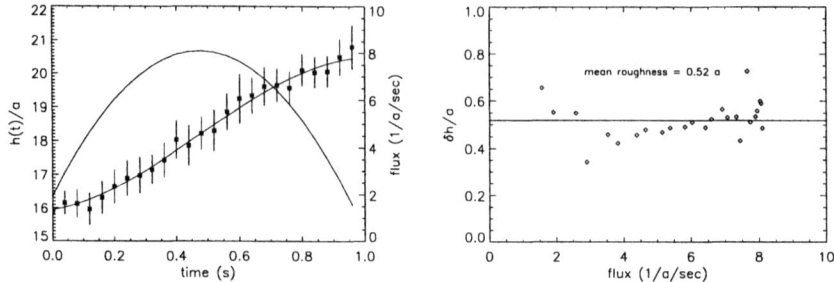

FIGURE 6. (a) Normalized height with fit and incident flux as a function of time. The height $h(t)$ has been corrected for the drift in the lower cloud, as discussed in the text. The flux is computed from the density and the time-derivative of the fit to $h(t)$. Both are normalized to the interparticle spacing a. (b) Normalized surface roughness versus incident flux. The mean value of the surface roughness is also shown for illustration. No significant correlation is evident.

computed in the two spatial dimensions and time for that part of the images below $h_0(t)$. Analysis of the correlation function yielded the interparticle spacing $a = 220$ μm and the drift velocity of the lower cloud $\mathbf{v_d} = -0.96\hat{\mathbf{x}} - 0.46\hat{\mathbf{z}}$ mm/s.

Fixing our coordinate system in the moving lower layer, the corrected height is $h(t) = h_0(t) + \mathbf{v_d} \cdot \hat{\mathbf{z}}t$. Using $n = a^{-2}$ for the density, we obtain the growth rate $\equiv$ incident flux $F(t) = n dh/dt$. Both $h(t)$ and $F(t)$ are shown in Fig. 6(a). Note that $h(t)$ has been fit to a 3rd order polynomial (shown) and $F(t)$ has been computed using the derivative of the fit. Note also that a *two-dimensional* flux has been computed, ie., just the flux within the vertical laser sheet. This could easily be translated into a three-dimensional flux by assuming symmetry in the x-y plane. Both $h(t)$ and $F(t)$ have been normalized by a in Fig. 6(a).

Given that we have a range of fluxes spanning nearly one order of magnitude, it is reasonable to look for a dependence in surface roughness δh with flux (growth rate). The surface roughness is just the standard deviation in the mean height of the surface [the error bars in Fig. 6(a)]. Roughness versus flux is shown in Fig. 6(b). No significant correlation is seen. This is instructive and suggests that even for the lowest fluxes $F \gg D$, where D is the surface diffusion coefficient, and that the surface roughness is simply due to the spatial variation in the incident flux.

The flow had certain characteristics in common with epitaxial film growth, a technique used in semiconductor manufacturing, in that it involved particles being deposited onto a growing surface. Exploring this analogy, the growth rate/incident flux were computed as a function of time, as well as the surface roughness. It was found that the growth rate was much larger than the surface diffusion rate so that it is likely that the surface roughness was dominated by variations in the incident flux.

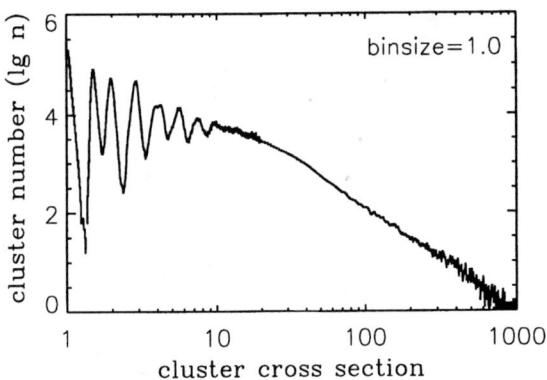

FIGURE 7. Spectrum of "small" clusters, as a function of the cluster cross section, measured $\simeq 3$ s after the injection.

Charge-enhanced coagulation

Experimental study of coagulation in a cloud of micron size particles embedded in a rarefied neutral gas is quite important for understanding the nature of the coagulation process [12]. We investigated the coagulation of micron size monodisperse particles in a neutral gas under microgravity. In several experimental runs up to $\sim 10^6$ particles of $2a = 3.4$ μm diameter were injected into the chamber at a pressure of 0.7 mbar. Figure 7 shows the mass spectrum of the resulting particle clusters – number of clusters, n, versus their geometrical cross section, s (measured in units of πa^2). Starting from certain moment $n(s)$ is not bounded at large s, but exhibits an algebraic tail, $n(s) \propto s^{-\tau}$, with $\tau \simeq 2.2$. Simultaneously, the growth of a single large agglomerate occurs, accumulating $\sim 10^4 - 10^5$ particles in a few seconds. Examples of these aggregates are shown in Fig. 8. The coagulation process develops several orders of magnitude faster than was expected. A huge agglomerate is formed while the aggregation among smaller clusters is still going on.

Further investigation of the data obtained in the experiments showed that the clusters were charged, positively or negatively (charge was measured by applying a sinusoidal voltage to the chamber electrodes, nature of the charging is discussed in [13]). The magnitude of the charges was at least a few thousand elementary charges. We believe that the enormously fast aggregation we observed is due to the particle charging. Kinetics of the charge-induced coagulation in neutral gases is very different from that in plasmas, where the charging is due to the absorption of electrons and ions and charge of the cluster is a certain function of the size (mass). In our case, however, the external sources of charging are absent and the total (initial) charge of the system is conserved.

We proposed a theory of pair clustering in a *conservative* charged system. The simplest way to describe the aggregation process is to use the so-called mean-field theory and to generalize the Smoluchowski coagulation equation [14] for the case of two independent variables – the cluster mass, m, and the charge, Q. Then the kinetic (coagulation) equation for the distribution function of clusters, $n(m,Q,t)$, can be written in the

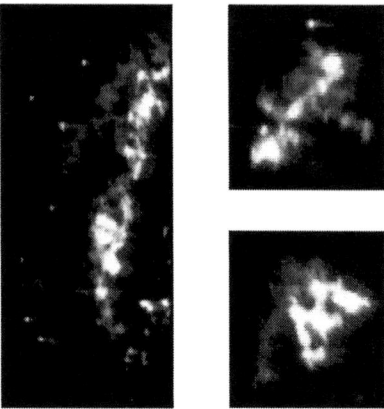

FIGURE 8. Examples of large single agglomerates. Each agglomerate accumulates $\sim 3 - 10$ % of the total amount of injected particles, horizontal size is about 1 mm.

following form:

$$\frac{\partial}{\partial t} n(m,Q,t) = \frac{1}{2} \int\limits_0^m dm' \int\limits_{-\infty}^{\infty} dQ' \, K(m',Q';m-m',Q-Q') n(m-m',Q-Q',t) n(m',Q',t)$$

$$- n(m,Q,t) \int\limits_0^{\infty} dm' \int\limits_{-\infty}^{\infty} dQ' \, K(m',Q';m,Q) n(m',Q',t),$$

where $K(m',Q';m,Q) = \langle v_r \sigma(a',Q';a,Q;v_r) \rangle$ is the coagulation rate coefficient (kernel) – the probability for a pair of clusters, (m',Q') and (m,Q), to merge. Here v_r is the relative velocity of the clusters, σ is the merger cross section, and angle brackets denote averages. The explicit relation between a and m is given by the appropriate scaling law with fractal dimension D_f [15],

$$m \propto a^{D_f}.$$

For different aggregation processes the fractal dimension can vary from $D_f \simeq 3$ (dense, or compact clusters, upper limit) down to $\simeq 1.4 - 1.5$ (fluffy aggregates) [15, 16].

It is well known that in some uncharged systems a special kind of phase transition called "gelation" is possible (see, e.g., [17, 18] and references therein). At a certain "gelation moment", t_{gel}, the ensemble becomes unstable against the formation of a single cluster of "infinite" mass. This process is also called "runaway growth". The gelation develops if the coagulation rate increases sufficiently steeply with the mass. Mathematically, this is because at $t = t_{gel}$ the distribution function for such kernels is no longer bounded exponentially at the high-mass end, but behaves algebraically, $n(m,t_{gel}) \propto m^{-\tau}$ with $2 < \tau < 3$. The total mass of the finite size clusters is not conserved in this case. There is a non-zero "mass flux" at $m \to \infty$, which causes the formation of the "infinite" gel particle. In terms of moments of the distribution function it means that the gel particle accumulate the mass comparable with the total mass of the system, so that the mass dispersion, or the second mass moment, diverges.

One can see that all major features observed in our experiment are peculiar to gelation. In order to determine the conditions when the gelation is possible we have to analyze the derived kinetic equation. Obtaining exact solutions of the coagulation equation is extremely difficult task even in the charge-independent case. However, the mathematical evidence of gelation is the divergency of the mass dispersion. Therefore, it is sufficient to analyze moments of the distribution function,

$$M_{\alpha,\beta}(t) = \int_0^\infty dm \int_{-\infty}^\infty dQ\, m^\alpha Q^\beta n(m,Q,t).$$

Equations for the moments can be derived by integrating the kinetic equation.

Major contribution to the charge-induced enhancement of coagulation is provided by the charge-charge and charge-dipole (induced) interactions. The measured charge dispersion of originally injected particles is extremely wide, and the coagulation rate for oppositely charged particles is increased by the factor $\langle Q^2/aT \rangle \sim 10^4 - 10^5$, where T is the kinetic energy of clusters. (Coagulation between particles having the same sign of charge is impossible, due to repulsion). However, each merging act decreases the charge of the resulting cluster, and therefore the charge dispersion (and thus the coagulation velocity) decreases rapidly. When the charge dispersion decays substantially (so that $\langle Q^2/a \rangle \leq T$) the charge-dipole interaction becomes more important. The corresponding coagulation rate can be written in the following algebraic form (omitting a constant factor):

$$K(m',Q';m,Q) = m^\mu m'^\nu |Q'|^\varepsilon + m'^\mu m^\nu |Q|^\varepsilon,$$
$$\mu + \nu = \lambda, \tag{6}$$

where μ and ν are the mass exponents, and ε the charge exponent. Analysis of the moment equations shows that the charge-dipole interaction with the kernel (6) allows the runaway growth, and the gelation condition is

$$\lambda + \frac{1}{2}\varepsilon > 1, \tag{7}$$

in contrast to $\lambda > 1$ for the charge-independent coagulation [17]. Criterion (7) shows that the charge-dipole interaction enhances the aggregation significantly and stimulates the gel phase transition.

Let us compare the gelation conditions for the charge-dipole and pure geometrical (charge-free) coagulation. Equation (7) shows that in the presence of charge-induced interactions the coagulation kernel need not necessarily be a steep function of the cluster mass — for the charge-dipole interaction with charge exponent $\varepsilon = 2$ the mass exponent λ is sufficient be positive. Smaller λ implies a higher value of the fractal dimension D_f of clusters at the high-mass end of the distribution. Using Eq. (7) we can get the critical values of D_f which are necessary to start the gelation for different types of coagulation:

Charge-dipole, $\varepsilon = 2$: $\lambda = D_f^{-1} - \frac{1}{2}$, gelation when $D_f < 2$.
Charge-dipole, $\varepsilon = 1$: $\lambda = \frac{3}{2}D_f^{-1} - \frac{1}{2}$, gelation when $D_f < \frac{3}{2}$.

Geometrical: $\lambda = 2D_f^{-1} - \frac{1}{2}$, gelation when $D_f < \frac{4}{3}$.

Clusters produced due to the Brownian motion are quite fragile, with an average value of the fractal dimension about $\simeq 1.8 - 2$ [16]. Thus, the gelation is most probable due to the charge-dipole interaction. The occurrence of gelation during geometrical coagulation is unlikely in this case – too low values of D_f are necessary.

In conclusion, we performed microgravity experiments on coagulation of microparticles in neutral gas. Injected particles were charged, positively or negatively, the overall charge was very small. We observed growth of single large agglomerate, accumulating $10^4 - 10^5$ particles in a few seconds. A huge agglomerate is formed while the aggregation among smaller clusters is still going on. We proposed to generalize coagulation equation taking into account the enhancement of the coagulation rate due to the charge-induced attraction. Equation allows the runaway ("gelation") solution – an aggregate of "infinite" mass is formed at a certain (gelation) moment. It is shown that the presence of charge on particles dramatically enhances the aggregation process and considerably lowers the threshold for the gelation onset. The discovered process can be of considerable astrophysical significance, e.g., in the formation of planets. The major puzzle there is rapid growth of planetesimals from dust in protoplanetary disks. We showed that the charging provides the possibility of enhanced coagulation during the early stages of dust growth, so that the rapid formation of planetesimals is possible.

SUMMARY AND OUTLOOK

The experiments described here only represent an excerpt of the research carried out so far. Other work (not described here) which is in progress at present deals with the detailed analysis of the crystal structures found in a stress-free environment for the first time, the excitation of dust-acoustic waves and the resulting diagnostics, the (possibly first) observation of dust-acoustic shock waves, the detailed examination of the potential in the central void using microspheres as test particles, experimental studies of coagulation, particle growth experiments, investigation of the global heart-beat instability, vortex flows, shear flows, boundary layers, etc.

In future experiments we plan to investigate some of these processes in more detail, over a larger parameter range, and we need to let the system settle down over a longer time scale, in particular for the crystallization experiments. The new physical insights obtained from this "first stress-free look" at the kinetic world of complex plasmas promise surprises and food for many debates.

ACKNOWLEDGMENTS

This work was supported by DLR under grant 50WM9852. The authors wish to acknowledge the excellent support from the agencies involved in making PKE-Nefedov into a success: DLR, Rosaviakosmos, Russian Control Mission Center, RSC "Energia", Kayser-Threde, Y. Gagarin Cosmonauts Training Center, IPSTC, and the Russian Basic Research Foundation.

REFERENCES

1. Nefedov, A. P. et al., submitted to *Nature*.
2. Samsonov, D. et al., submitted to *Phys. Rev. Lett.*
3. Ivlev, A. V. et al., submitted to *Phys. Rev. Lett.*
4. Thomas, H. M. et al., *Phys. Rev. Lett.* **73**, 652 (1994).
5. Morfill, G. E. et al., *Phys. Rev. Lett.* **83**, 1598 (1999).
6. Morfill, G. E., and Tsytovich, V. N., *Phys. Plasmas* **9**, 1 (2002).
7. Havnes, O. et al., *J. Geophys. Res.* **92**, 2281 (1987).
8. Raizer, Yu. P., *Gas Discharge Physics*, Springer, Berlin, 1991.
9. Melandsø, F., Aslaksen, T., and Havnes, O., *Planet. Space Sci.* **41**, 321 (1993).
10. Samsonov, D., and Goree, J., *Phys. Rev. E* **59**, 1047 (1999).
11. Jensen, P., *Rev. Mod. Phys.* **71**, 1695 (1999).
12. Blum, J. et al., *Phys. Rev. Lett.* **85**, 2426 (2000).
13. Morfill, G. et al., submitted to *Phys. Rev. Lett.*
14. Smoluchowski, M., *Z. Phys. Chemie* **92**, 129 (1917).
15. Meakin, P., *Rev. Geophys.* **29**, 317 (1991).
16. Kempf, S., Pfalzner, S., and Henning, T., *Icarus* **141**, 388 (1999).
17. Ernst, M. H., in *Fractals in Physics*, edited by L. Pietronero and E. Tosatti, North-Holland, Amsterdam, 1986, pp. 289-302.
18. Lee, M. H., *Icarus* **143**, 74 (2000).

Physics of Collective Dust-Dust Attraction and Dust Structure Formation

V. N. Tsytovich* and G. E. Morfill[†]

*General Physics Institute, Russian Academy of Science Moscow, Vavilova str. 38, 119991, Moscow, Russia.
[†]Max-Planck Institute fur Extraterrestrische Physik, 85740, Garching, Postfach 1312, Germany.

Abstract. Collective attraction forces between two test dust grain in presence of many other dust grains are related with collective plasma flux. The basic state is formed by the balance of the plasma absorption on grains and plasma ionization. In the limit where the collective interaction dominates ($a^2/\lambda_{Di}^2 \gg T_i/T_e$, a being the dust size and λ_{Di} being the ion Debye length) the amplitude of the exponentially screened part of the potential is suppressed, the screening length of it is decreased and the non-screened cosinusoidal part of the potential dominates, describing a set of attraction potential wells decreasing with inter-dust distance. The collective attraction in presence of an ion flow exceeds the pair wake attraction. The collective attraction is increased in presence of strong magnetic field $B > B_{cr}(Gauss) \approx = 10^4 (a/\lambda_{Di})\sqrt{n/10^9\,\mathrm{cm}^{-3}}(0.02 T_e/T_i)$. The dust structures created by collective flux and collective attraction are investigated for micro-gravity conditions. The collective attraction creates a large damping and universal instability of dust-acoustic waves and can cause transition of complex plasma to a crystal state.

FROM DUST CLUSTERS TO PLASMA CRYSTALS.

In matter crystal formation is deeply related to chemical bounding which is of a quantum nature (exchange interaction). In a system of dust grains the interactions are classical and first of all should be explained the presence of bounds, creating dust clusters and dust plasma crystals . The transition from clusters to crystalline structures in classical systems confined by external parabolic potential is well known [1]. The dust small number 2D clusters were investigated experimentally in [2–4], showing some features similar to the classical clusters in traps, namely the periodical shell structures, which start to be unstable with increasing numbers of dust grains in the structure. Two main difference of dust grain from Coulomb interacting charges is that: 1) the interaction between grains differs from the screened Coulomb interactions (so called Yukawa interaction), 2) the transition to the crystalline structures is related with domination of collective attraction. This transition is determined by the number of grains in the cluster or by the size of the dust cloud ($L > L_{cr}$). For $L \ll L_{cr}$ the par interaction between grains is non-collective, while for $L \gg L_{cr}$ it is collective and is determined by the density of other grains. In both collective and non-collective attraction the potential of interaction resembles the molecular type potential. It is caused by presence of plasma flux on the grains which is a non-collective flux for $L \ll L_{cr}$ and is a collective flux for $L \gg L_{cr}$. The L_{cr} plays a role of critical size of dust cluster for conversion to the dust plasma crystal state. In the presence of dust attraction the presence of an external confining parabolic potential is not necessary, i.e. the boundary free clusters and crystals can be created. Numerical investigations and theoretical estimates give [5]

$$L_{cr} = \frac{\lambda_{Di}^2}{aP} \qquad (1)$$

where λ_{Di} is the ion Debye radius, a is the dust size, and $P = n_d Z_d / n_i$ is the collective (so-called Havnes) parameter determining the relative charge on the dust grains. The experimentally observed plasma crystals satisfy the relation $L \gg L_{cr}$ while the observed dust clusters satisfy the opposite inequality. Experimental investigation of the transitions at $L \approx L_{cr}$ is of special interest.

NON- YUKAWA DUST-DUST PAIR INTERACTIONS

For $L \ll L_{cr}$ the interactions between dust grains can be considered as a sum of pair interactions of each grain with all other grains. For the parameters of the present experiments, pure Yukawa interactions are not operating due both to the interference in the charging processes and to the non-linearity in screening. The necessary condition that the non-linearity is small and that the Yukawa potential operates is that at least at the screening distance the potential of the charge ϕ is much less than the ion temperature, $e\phi \ll T_i$ or $z \ll \tau \lambda_{Di}/a$ (where $z = Z_d e^2 / aT_e$ is the dimensionless dust charge and $\tau = T_i/T_e \ll 1$ is the ion to electron temperature ratio). For the parameters of the existing experiments this requirement means $z \ll 1$ and practically gives $Z_d < 10^2$ while in experiments Z_d ranges from 10^3 up to 10^5. Therefore the screening is always non-linear. The simplest well known and widely used non-linearity (for the old problems of charging of spacecrafts in ionosphere see [6]) is related to the constrain that only ions with positive energy far from the grain can take part in screening [7]. Figure 1a demonstrates for parameters most often found in existing experiments the difference between the Yukawa exponential screening factor and the nonlinear screening factor (it is calculated by solving the non-linear Poisson equation with the mentioned restrictions). Another important feature is the presence of the plasma flux necessary to maintain the large dust charges resulting in non-screened part of potential $\propto 1/r^2$ which is described asymptotically by the screening factor $\approx a/2r$ [8]. Figure 1b shows at which distances the non-linear screening factor (which takes into account both the non-linearity and presence of plasma flux) reaches its asymptotical values.

NON-COLLECTIVE ATTRACTION AND FORMATION OF BOUNDARY-FREE DUST CLUSTERS

The presence of another dust particle shadows the bombardment flux of ions and neutrals and creates the non-collective dust attraction which was the subject of intense investigations during the last 6 years [9-12]. The results can be summarized by an expression for the ratio of attraction potential to the bare Coulomb potential $\eta_{ncoll} = V_{at}/V_{Coul}$,

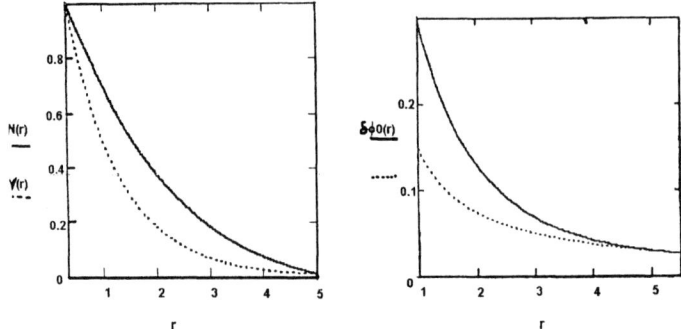

FIGURE 1. Fig.1a — Dependence of the screening factor on the distance r in units of the Debye length, ϕN is the actual non-linear screening factor, ϕY is the Yukawa exponential screening factor, Fig.1b — $\delta\phi 0$(solid line) is the correction to the screening factor due to the presence of the charging flux, dotted line is the asymptotic expression for these corrections $a/2r$, the distance r and the dust size a are given in units of the Debye length. For both figures $\tau = T_i/T_e = 0.02$, $z = 2.5$, $a \equiv a/\lambda_{Di} = 0.3$

$$V_{\text{Coul}} = Z_d^2 e^2/r$$

$$\eta_{ncoll} = -\frac{a^2}{z^2 \lambda_{Di}^2}\left(\alpha_{st,i} + \alpha_{st,n}\frac{n_n T_n T_i}{n_i T_e^2}\right) \quad (2)$$

where n_n and n_i are the concentrations of neutral atoms and ions respectively, T_n, T_i, T_e are the temperatures of neutral atoms, ions and electrons respectively and $\alpha_{st,n}$, $\alpha_{st,i}$ are the sticking coefficients of ions and neutral atoms respectively. The existing theoretical concepts and the existing experiments on dust charging indicate that the sticking coefficient of ions is close to one, while the neutral atom sticking coefficient depends much on the properties of dust surface and is the subject of intense investigation in material science. For rough dust surfaces $\alpha_{st,n}$ ranges from 0.1 up to 0.02 while for mirror type surfaces it can be very small ($\sim 10^{-4} - 10^{-6}$). An important feature of shadow attraction potential is that *it is not screened* and together with non-screened part of charging potential (Fig.1a) can forme a potential well. This occurs for $\eta_{ncoll} \ll 1$ where the molecular-type of potential well is located at distances $r_{ncoll} \approx a/\eta_{ncoll}$ larger than the non-linear screening distances. If the neutral atom flux is negligible $r_{well} \approx \lambda_{Di}^2/a$. For $\eta_{ncoll} \gg 1$ the attraction dominates at the closest dust separation and dust agglomeration should occur. Large values of η_{ncoll} can be caused in the conditions where $\alpha_{st,n}$ and the grain size is relatively large as it occurs in etching experiments. In plasma crystal experiments $\eta_{ncoll} \ll 1$ and the attraction forces provide a **classical molecular-type potential**. Molecular dynamic simulation [5] was used to illustrate that the pair dust interaction including nonlinear screening and non-collective dust attraction can form 2D and 3D dust clusters from initially random dust distributions (see Fig.2a and Fig 2b). These clusters are boundary free and are not confined by external potential. Due to (1) and estimated r_{well} the cluster formation requires $P \ll 1$.

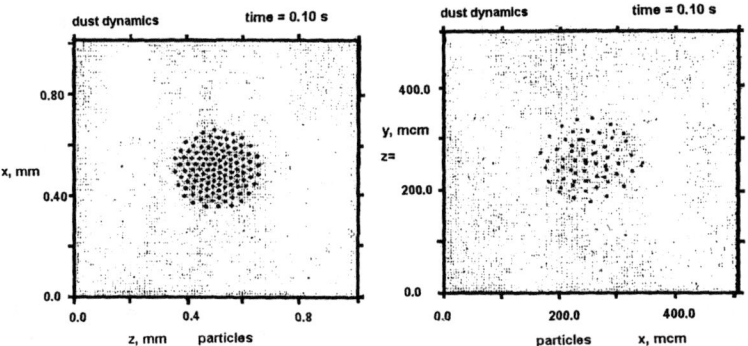

FIGURE 2. Fig.2a and Fig2b show clusters found as the final stage of evolution of initially random dust distribution for 2D and 3D cases respectively, total number of grain is 50 and 200, the grain size is $a = 2\,\mu m$ and $1\,\mu m$, the frequency of dust-neutral collision $v_{nd} = 100\,s^{-1}$ and $280\,s^{-1}$ respectively, other parameters $n_i = 3 \times 10^9$, $T_e = 3\,eV$, $T_i = 0.03\,eV$, $\lambda_D = 23.4\,\mu m$ are the same for both cases

COLLECTIVE ATTRACTION IN ABSENCE OF ION FLOW.

The first derivation of collective attraction was made in [13] using the kinetic approach. In the existing plasma crystal experiment the the relation opposite to (1) is valid and the shadows of the flux by different dust grains is overlapping and is creating a collective flux. The interactions of grains becomes collective and the interaction of any pair of grains depends on the density of other grains. Since the attraction operates often at large distances the linear static responses of the system can be used with the basic state in which both the charge quasi-neutrality and the balance of plasma ionization and absorption on dust grains is taken into account. For existing experiments the balance is established on time scales of order of of μs. Parameters of the basic state $n_{e,0}/n_{i,0} = 1 - P_0$, z_0 and the required rate of ionization are determined by a single parameter P_0. The static dielectric constant, which determines the dust-dust collective interactions at large distances, can be found from linear perturbations of the basic state using the force balance hydrodynamic equations and the continuity equations [14] and depends only on few parameters $\eta_{coll}, \tau \ll 1$ and $k^2 \lambda_{eff}^2$:

$$\varepsilon_k = 1 + \frac{1}{k^2 \lambda_{eff}^2} \frac{(k^2 \lambda_{eff}^2 - \tau \eta_{coll})}{(k^2 \lambda_{eff}^2 + \eta_{coll})}; \eta_{coll} = \frac{P_0^2 z_0^2 \alpha_{dr} \alpha_{ch} a^2}{\tau(1+z_0)\lambda_{eff}^2 \left(1 + \frac{P_0}{1+z_0}\right)^2} \quad (3)$$

where $\lambda_{eff} = \lambda_{Di}\left(1 + \frac{P_0}{1+z_0}\right)^{-1/2}$ is the effective screening length modified by collective effects, α_{dr} is the drag coefficient and α_{ch} is the charging coefficient. For Orbit Limited Model $\alpha_{ch} = 1/a\sqrt{\pi}$ and for ion multiple scattering model $\alpha_{dr} = 2\ln\Lambda/3\sqrt{\pi}$, $\ln\Lambda$ being the Coulomb logarithm. The ratio η of the collective dust-dust potential energy to the bar Coulomb $Z_d^2 e^2/r$ energy for $\eta_{coll} \ll 1$ will be [14]

$$\eta = \exp\left(-\frac{r}{\lambda_{eff}}\right) + \eta_{col} \cos\left(\sqrt{\eta_{col}\tau P_0}\frac{r}{\lambda_{eff}}\right) \qquad (4)$$

and for $\eta_{col} \gg 1$

$$\eta = \frac{1}{\eta_{col}}\exp\left(-\sqrt{\eta_{col}}\frac{r}{\lambda_{eff}}\right) + \cos\left(\sqrt{\tau P_0}\frac{r}{\lambda_{eff}}\right). \qquad (5)$$

Opposite to non-collective attraction the collective interaction has a series of attraction wells, decreasing with inter-grain distances. For $\eta_{coll} \ll 1$ the change of exponential screened interaction (see (4)) is of the order of 1 and is due to the dust charge variations, the position of the first attraction minimum is similar to the non-collective one $r_{coll} \approx \lambda_{Di}^2/aP_0^{3/2}$ but the value of the attraction well $-V_{att,coll}$ is $\approx P_0^{7/2}/\tau$ larger than $V_{att,ncoll} \approx Z_d T_e z_0 \eta_{ncoll}^2$. For the opposite limit $\eta_{coll} \gg 1$ the exponential part of the potential is much suppressed, the distance of the first attraction minimum is smaller and is close to the electron Debye length, divided by $\sqrt{P_0}$ and the value of the attraction well is much larger of the order of $Z_d T_e z_0 \sqrt{P_0} a/\lambda_{De}$. For the parameters of existing plasma crystal experiments P_0 is of the order of 1 and $1 < \eta_{coll} < 4$ where the collective attraction should be rather effective. The potential well can be comparable with the dust temperature and can provide a plasma condensation to the crystal state. The position of the attraction minimum is estimated to be close to the dust separation in the observed crystal states. Collective attraction could be the best candidate for explanation of the observed formation of dust plasma crystals. For the conditions where the collective attraction operates the crystals are self-confined by collective flux and the presence of an external confined potential is not required (the plasma crystals could be boundary-free).

COLLECTIVE ATTRACTION IN PRESENCE OF ION FLOW

Ion flow is present in most present experiments. Collective effects convert the wake attraction to an collective wake interaction [15]. The investigation of the first corrections to the collective attraction related with presence of ion flow shows that the presence of almost symmetric attraction wells survives till the velocity of the ion flow is less than the critical one u_{cr} (the ion drift velocity u is normalized on $\sqrt{2}v_{Ti}$)

$$u_{cr} = \frac{P_0^{3/2} z_0 \alpha_{ch}}{(1+z_0)\sqrt{\tau}}\frac{a}{\lambda_{Di}} = \frac{\sqrt{\alpha_{ch}(1+z_0+P_0)}}{\sqrt{\alpha_{dr}(1+z_0)}}\sqrt{\eta_{coll}P_0} \qquad (6)$$

For $\eta_{coll} \gg 1$ the critical velocity can substantially exceed the ion thermal velocity but for $\eta \ll 1$ the critical Mach number $M_{cr} = \sqrt{2\tau}u_{cr}$ is less than 1. The numerical results shows that with an increase of the flow velocity the attraction wells perpendicular to the direction of the flow are modified less than those along the flow (corrections are of the order of u^2/u_{cr}^2 in the perpendicular direction while they are of the order of u/u_{cr}

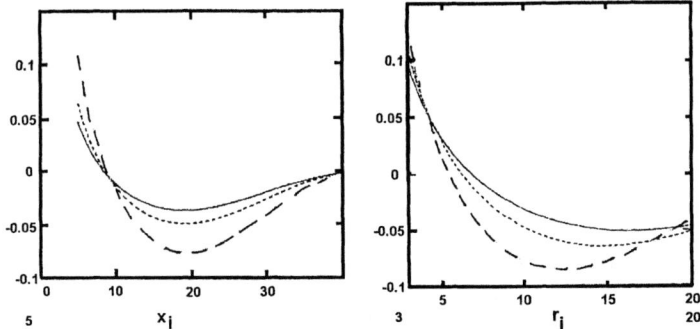

FIGURE 3. Dependence on the magnetic field strength of the normalized interaction energy (vertical axes) versus the inter-dust distance (normalized with respect to the ion Debye length); the negative values of the interaction potential correspond to dust attraction; the solid line correspond $B = 0$, the dotted line correspond $B = B_{cr}$ and the dash line correspond $B = 10B_{cr}$; Fig.3a — the grains are located along the magnetic field, Fig.3b — the grains are located perpendicular to the magnetic field

along the direction of the flow). Numerical investigation of the symmetric part of the interaction as a function of u/u_{cr} for $u \gg u_{cr}$ shows that the attraction potential wells are increasing with an increase of u/u_{cr} and are larger than the non-collective wake attraction.

COLLECTIVE ATTRACTION IN A STRONG MAGNETIC FIELD

We consider the collective attraction in presence of strong magnetic field when the Lorentz force on ions can be comparable with the ion drag force. The collective attraction was calculated at large separation between the grains using the static dielectric function modified by the magnetic field [16]

$$\varepsilon_{\mathbf{k}} = 1 + \frac{1}{k^2 \lambda_{\text{eff}}^2} \frac{\left(k^2 \lambda_{\text{eff}}^2 R_B - \tau \eta_{\text{coll}}\right)}{\left(k^2 \lambda_{\text{eff}}^2 R_B + \eta_{\text{coll}}\right)}; R_B = \frac{k_z^2}{k^2} + \frac{k_\perp^2}{k^2} \frac{B_{\text{cr}}^2}{B_{\text{cr}}^2 + B^2} \quad (7)$$

where $k_z, \mathbf{k}_\perp$ are components of the wave vector along and perpendicular to the magnetic field respectively and the critical magnetic field $B_{cr} = (acT_e/\sqrt{2}v_{Ti}\lambda_{Di})P_0^2 z_0^2 \alpha_{dr}^2$ is written in practical units form in the abstract. The dependence of the dust-dust potential energy on the inter-dust distance which includes the first attraction minimum was found numerically using the expression (7). Figure 3a shows the interaction potential along the magnetic field and Figure 3b shows the interaction potential perpendicular to the magnetic field for $\eta_{\text{coll}} = 1$, $\tau = 0.05$, $P_0 = 0.9$, $Z_0 = 2$, the value of η_{coll} is used as the one closest to the present experiments data. Numerical results where obtained also in broad range of parameters η_{coll} and B/B_{cr}. The general features of them are that 1)an increase of the magnetic field above the critical value increases the potential attraction wells both along the magnetic field and perpendicular to the magnetic field 2) with an increase of the magnetic field the position of the attraction well perpendicular to the magnetic field decreases and is not much changed in the direction along the magnetic field. This means that the structures created in strong magnetic field are contracted perpendicular to the magnetic field and the temperature of their crystallization increases.

COLLECTIVE ATTRACTION DETERMINES THE DISPERSION, DAMPING AND INSTABILITIES OF WAVES

The change by the wave disturbances of collective flux (and the change of other collective parameters of the ground state proportional to some power of the parameter P_0) causes the dispersion of waves, creates wave strong damping, strongly restricts the range of the wave existence, leads to a new modes and leads to an universal instabilities [17,18]. This is of particular importance for dust acoustic waves: 1) the dust acoustic waves exist only for $k_{cr,d} \ll k \ll 1/\lambda_{eff}$ where $k_{cr,d}\lambda_{Di} \approx P_0 a/\lambda_{Di}\sqrt{\tau}$; this restriction indicates that the dust acoustic waves cannot exist for the parameters of the present plasma crystal experiment but exist for the parameters of experiment [19] where they were experimentally investigated [19], 2) in the range $k_{cr,i} = (a/\lambda_{Di}^2)z_0 P_0^{3/2}\sqrt{\alpha_{dr}\alpha_{ch}/(1+z_0+P_0)} < k \ll k_{cr,d}$ there appears a new collective mode with almost constant frequency $\omega_{coll} = \omega_{pd}(a/\lambda_{Di})(\alpha_{ch}\alpha_{dr}P_0(1+P_0)/\tau(1+P_0+z_0))^{1/2}$ (see below), 3) for $k \leq k_{cr,i}$ an universal collective instability is present, which is related with conversion to either collective dust structure or dust crystal state 4) the dispersion of dust acoustic waves caused by the collective effects is described by expression [18]:

$$\omega^2 \approx \omega_{daw}^2 \left(1 + \frac{\alpha_{dr}\alpha_{ch}P_0 z_0(1+P_0)a^2}{k^2\lambda_{Di}^4\tau} - k^2\lambda_{Di}^2\right); \omega_{daw} = k^2 v_{daw}^2; v_{daw} = \omega_{pd}\lambda_{eff} \quad (8)$$

where ω_{pd} is the dust plasma frequency and λ_{eff} is the effective screening length taking into account the dust charge variations [see the text after equation (3)]. The expression (8) gives for the first time a correct expression for dust acoustic wave dispersion being in agreement with experiments [19].

STRUCTURES FORMED BY COLLECTIVE ATTRACTION

The collective attraction can be the major effect in formation of plasma crystals and (for a high dust temperature) can create gaseous dust structures with sharp boundaries self-confined by the collective flux. Numerical investigation [20] of **spherical structures under micro-gravity conditions,** shows (see Fig.4a) that their existence is restricted by several conditions including that the ionization rate to be less than critical value, that the ion and electron density should to be larger than critical value and that their should be present a void between the structure and the wall at floating potential where the direction of the collective flux is inverted. The total number of the dust grains self-confined in the structure cannot exceed the maximum number determined by the ratio od dust size and ion-neutral mean free path. For the parameters of present experiments it is 3×10^6 and

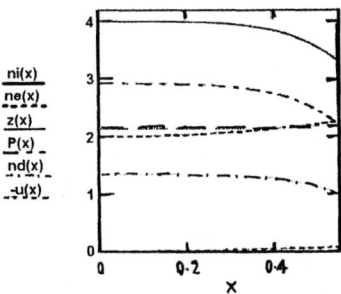
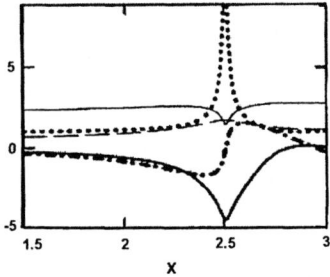

FIGURE 4. Fig4a Distribution of parameters of spherical dust structure surrounded by spherical wall at floating potential with a size larger than the ion-neutral collisions mean free path λ_{in}. The distance R (horizontal axis) is given in units of λ_{in}/τ and the end point $R = 0.55$ correspond to the surface of the dust structure where the dust density jumps to zero. Fig. 4b Distribution of the parameters of strong non-linear dissipative plane structure moving from right to left with a constant profile with Mach number $M = 1, \tau = 0.02$, the figure is obtained by exact numerical solution of non-linear balance equations assuming the dust at rest

can be reached only by fine tuning of pressure and ionization power. With lowering of the dust temperature the condensation to the crystal state is found to occur simultaneously almost in the whole structure leaving only a very thin liquid layer at the surface of the structure. The plain structure of different type can be created by collective flux. The presence of strong collective dissipation of disturbances in the whole system changes completely the formulation of shock wave problem — the dissipation at the front can be either increased or decreased. In particular [21] **the stationary dust ion-sound shocks cannot exist**, since they assume the constant ion flow velocity different on the both sides of the shock front and any flow is dissipated by collective damping if it is not supported externally. The solution of numerical problem for constant moving profiles with a constant Mach number for strong non-linear waves shows neither soliton nor shock type of structures. The strong non-linear dissipative structures found are unusual, they are asymmetric, and if the electric field strength is zero far away from one side of the structure front it reaches a constant value on the other side (after several oscillations at the front). Such a structure should be supported by an external field [21]. It can be seen from the example of numerical solution for such structure (see Fig.4b) that at the center of its first part the ion density and the ion drift velocity are largely peaked [21].

REFERENCES

1. Schweigert, V.A., and Peeters, F., *Phys. Rev. B*, **51**, 7700–13 (1995).
2. Melzer, A., Klindworth, M., Piel, A., *Phys. Rev. Lett.* **87**, 115002/1–4 (2001).
3. Juan, Wen-Tau, Huang, Z.-H., Hsu, J.-W., Ju, Y., and I., L., *Phys. Rev. E*, **58**, R6947–50 (1998).
4. Goree,. J, Samsonov, D., Ma. Z., Ghatacharjie, A., Thomas, H., Konopka, U., and Morfill, G., *Fronties in Dusty Plasmas*, Proceedings of the Second International Conference on the Physics of Dusty Plasmas-ICPDP-99, edited by Y. Nakamura, 92, Elsever, 2000.

5. Khodataev, Ya., Bingham, R., Tarakanov, V., Tsytovich, V., and Morfill, G., *Physica Scripta*, **T89**, 95 (2001).
6. Al'pert, Y., Gurevich, A., and Pitaevsky, *Space Physics with Artificial Sattelites*, Consultant Bureaut, 1965.
7. Lafranbose J., and Parker L., *Phys. Fluids*, **16**, 629 (1973).
8. Bernstein I., and Rabinovich I., *Physics of Fluids* **2**, 112 (1959).
9. Tsytovich, V. N., Khodataev, Y., and Bingham R., *Comments on Plasma Physics and Controlled Fusion, 17*, 249 (1996).
10. Ignatov *Comments P.N.lebedev Inst.A.* **58**, 1 (1996).
11. Khodataev, Ya. K., Morfill, G., and Tsytovich V. N., *J. Plasma Phys.*, **65**, 257 (2001).
12. Lampe, M., Gavrishchaka, V., Ganguli, G., Joyce, *Phys. Rev. Lett.*, **86**, 5278 (2001).
13. Tsytovich, V.N., de Angelis, U., *Phys. Plasmas*, **7**, 554 (2000).
14. Tsytovich V.N., and Morfill G., *Fhyz. Plasmy*, (in press) (2002).
15. Kompaneetz R., and Tsytovich V.N., *Fhyz. Plasmy*, (in press)(2002).
16. Tsytovich, V.N., *Contr. to Plasma Phys.*, (submitted) (2002).
17. Tsytovich, V.N., de Angelis, U., and Bingham R., *Physics of Plasmas* **10** (2002).
18. Tsytovich, V.N., and Watanabe K., *Contr. Plasma Phys.*,(accepted), preprint NIFS-720 (2002).
19. Barkan, A., Merlino, R.L., D'Angelo, *Phys. Plasmas* **2**, 3563–5 (1995).
20. Morfill, G., and Tsytovich, V.N., *Physics Plasmas*, **7**, 235 (2001).
21. Tsytovich, V.N., *Fhyz. Plasmy*, (in press) (2002).

SECTION 3: ORAL PRESENTATIONS

A Nonlinear Theory of Void Formation in Colloidal Plasmas

K. Avinash, A. Bhattacharjee, and S. Hu

Department of Physics and Astronomy, The University of Iowa, Iowa City, IA 52242

Abstract. A nonlinear time-dependent model for void formation in colloidal plasmas is proposed. For experimentally relevant initial conditions, the model describes the nonlinear evolution of a zero-frequency linear instability that grows rapidly in the nonlinear regime and subsequently saturates to form a void. A number of features of the model are consistent with experimental observations under laboratory and microgravity conditions.

Recently, a number of colloidal (or dusty) plasma experiments, in laboratory as well as under microgravity conditions, have shown the spontaneous development of voids [1-5]. A void is typically a small and stable centimeter-size region (within the plasma) that is completely free of dust particles and characterized by sharp boundaries. In the laboratory [2], the void is seen to develop from a uniform dust cloud as a consequence of an instability when the dust particle has grown to a sufficient size. It was suggested in [2], and now widely accepted, that the ion drag force plays a crucial role in causing the initial instability. This instability can be described as follows. Imagine a local depletion of negatively charged dust particles within a spatially uniform dusty plasma. The depletion will produce a positive space charge with respect to the surrounding plasma, and hence an electric field that points outward from the region of reduced dust density. This electric field will cause an inward electrical force, F_e, on negatively charged dust particles that tends to restore the dust density to its equilibrium value, and an outward force, F_d, due to the ion drag (in the direction of the ion flow) that tends to expel dust particles from the region of depletion. If $F_d > F_e$, which occurs when dust particles have grown to a sufficient size, an instability grows, deepening the initial density depletion.

Although the physical picture described above has been confirmed by several theoretical analyses, there is as yet no nonlinear time-dependent model that describes the spontaneous development of the linear instability and its subsequent saturation to produce a void. In this paper, we propose a basic, time-dependent, self-consistent nonlinear model for void formation in a dusty plasma. A basic model should contain at least three elements: (a) an initial instability caused by the ion drag, (b) a nonlinear saturation mechanism for the instability, and (c) the void as one of the possible nonlinearly saturated states, dynamically accessible from the initially unstable equilibrium. For the initial instability, we choose a simple variant of the zero-frequency mode described by D'Angelo [6] which grows when $F_d > F_e$. The saturation mechanism we adopt is relevant for collisional voids where ions achieve near-thermal velocities in the void region. In this regime, as discussed in [7], F_d initially increases with ion velocity v_i, attains a maximum for $v_i = v_{thi}$, where v_{thi} is the ion thermal velocity, and decreases for $v_i > v_{thi}$. As the linear instability grows, the ions are initially accelerated in the growing electric field, and F_d initially increases. Eventually, as the ions are accelerated to speeds larger than the ion thermal speed, F_d decreases to balance F_e and thus saturate the instability. (The reduction in F_d can also be brought about by nonlinearity in the ion mobility; however, we will not consider this mechanism here.) We demonstrate by analysis and numerical simulation that in the saturated state, a stable void is formed.

We now describe the approximations and simplifications made in reducing the fluid equations for dust, electrons, and ions to our model equations. The one-dimensional continuity and momentum equations for dust are given, respectively, by

$$\partial_t n_d = -\partial_x (n_d v_d) + D\partial_x^2 n_d \ , \tag{1}$$

and

$$m_d(\partial_t + v_d \partial_x) v_d = -ZeE + F_d - v_{dn} m_d v_d - (T_d/n_d)\partial_x n_d \ . \tag{2}$$

Here n_d is the number density of dust particles of charge $-Ze$, mass m_d, and temperature T_d, v_d is the fluid velocity of dust, D is a particle diffusion coefficient for dust, E is the electric field, and the term $v_{dn} m_d v_d$ represents the frictional drag on dust grains by neutral atoms, where v_{dn} is the dust-neutral collision frequency. We approximate the nonlinear ion drag force, F_d, by the expression $F_d = m_d v_{di} v_{thi} u/(b + u^3)$ where $u \equiv v_i/v_{thi}$, v_{di} is the ion-dust collision frequency and b is a positive constant. This nonlinear expression, with $b = 1.6$, fits well the numerically calculated F_d given in the range $0.1 \leq u \leq 5$ [7].

For electrons, each of charge $-e$, we neglect all inertial effects in their momentum equation and write $(T_e/n_e)\partial_x n_e = -eE$, where $n_e(T_e)$ is the electron density (temperature). The electric field E is determined by the self-consistent Poisson's equation $\partial_x E = 4\pi e(n_i - n_e - Zn_d)$, where $n_i(T_i)$ is the ion density (temperature). We neglect ion inertia and take the ion motion to be mobility-limited, that is, $v_i = eE/(m_i v_{in})$ where v_{in} is the ion-neutral collision frequency [2]. Although a void contains no dust particles, it contains electrons and ions as a pristine plasma (without dust particles) does. In order to sustain a continuous ion wind in the void, one generally needs to rely on a local ionization mechanism whereby ions are produced by impact ionization of electrons on neutrals. In our model, we simply assume that the ion density n_i is a constant, that is, $\partial_t n_i = 0$. This assumption ensures that a constant source of ions is available at all times, and that the ionization is rapid enough to balance convective and all other losses. Thus, in our model, ionization is assumed to be present implicitly rather than as an explicit term in the equation of ion continuity.

In order to delineate carefully the role of electrons in void formation, we will report the results of two calculations: one in which the electron density is set equal to zero in Poisson's equation, and another in which the electron density is included. In the first case, which involves only positively charged ions and negatively charged dust, we obtain the following model (in dimensionless form):

$$\partial_x E = 1 - \varepsilon_d, \tag{3}$$

$$\partial_t v_d = [-1 + a/(b + u^3)]E - \alpha_0 v_d - (\delta/\varepsilon_d)\partial_x \varepsilon_d, \quad u = \mu E, \tag{4}$$

$$\partial_t \varepsilon_d = -\partial_x(\varepsilon_d v_d) + D_0 \partial_x^2 \varepsilon_d, \tag{5}$$

where we have neglected the convective nonlinearity in the dust equation and defined $u = |u|$, $\varepsilon_d \equiv Z n_d / n_{i0}$, $a \equiv m_d v_{di}/(m_i Z v_{in})$, $\alpha_0 \equiv v_{dn}/\omega_{pd}$, $\tau_{d,i} = T_{d,i}/T_e$, $\delta = \tau_d/Z$, $\mu = \omega_{pi}/(v_{in}\tau_i)$, and $D_0 = D\tau_i^2/(\omega_{pd}\lambda_{di}^2)$. In these model equations, the distance x is normalized by λ_{di}/τ_i where λ_{di} is the ion Debye length, time t is normalized by ω_{pd}^{-1}, the densities of dust and ions are normalized by the initial ion density n_{i0}, and the electric field E is normalized by the quantity $(T_e/e\lambda_{di})$. The last terms in (4) and (5) are due to dust pressure and diffusion, respectively. These terms would generally appear to be small because the dust temperature is low, but they can be important in the nonlinear regime where spatial gradients in density are large.

We first carry out a simple equilibrium and stability analysis of equations (3)-(5). Note that these equations are satisfied identically for a homogeneous field-free equilibrium with $E = 0$, $\varepsilon_d = 1$, and $v_d = 0$. Linearizing (3)-(5) about this equilibrium and assuming for simplicity that $D_0 = 0$, it is straightforward to show that this equilibrium is unstable to a zero-frequency (that is, purely growing) mode of wave number k when $a/b > 1 + \delta k^2$. The mode with the largest growth rate has $k = 0$. Since $a/b = F_d/F_e$, the instability condition for the fastest growing mode reduces to $F_d > F_e$, which is exactly the condition discussed earlier in this paper. The linear instability saturates when F_d is reduced via the cubic nonlinearity in (4) to balance F_e. To investigate whether there is a steady-state solution containing a void that the unstable equilibrium may evolve to, we set $\partial/\partial t = 0$ in (3)-(5), and assume that a void extends from $x = 0$ to $x = x_v$ where $\varepsilon_d = 0$. Consequently, from (1), we obtain $E = x$ in the range $0 \le x < x_v$. In this region, we obtain the condition $F_d > F_e$, which causes the complete expulsion of dust particles. At the boundary $x = x_v$, determined by the relation $-1 + a/(b + \mu^3 x_v^3) = 0$, we obtain $F_d = F_e$ which continues to hold between $x = x_v$ and the dust cloud boundary $x = x_c$. In the range $x_v < x \le x_c$, we obtain $\varepsilon_d = 1$ and $E = (a - b)^{1/3}/\mu$. Although there are no dust particles in the void, by (4) there is a steady velocity profile given by $v_d = \alpha_0^{-1} x[-1 + a/(b + \mu^3 x^3)]$ which yields $v_d = 0$ at the two endpoints ($x = 0$ and $x = x_v$) of the void. For $x_v \le x \le x_c$, where there is a uniform distribution of dust particles ($\varepsilon_d = 1$) in steady state, the dust velocity is zero. Thus, we have obtained a steady nonlinear solution containing a void that the initially unstable equilibrium can evolve into. It is easy to show that this solution is linearly stable.

We now integrate (3)-(5) numerically in the range $0 \le x \le x_c$ to demonstrate the growth of the linear instability from an initially unstable equilibrium and its evolution to form a saturated void with the attributes discussed above. Experimental observations show that during the formation of the void, particles continue to escape from the cloud, so we assume that the dust cloud boundary $x = x_c$ is open (that is, $\partial_x n_d = 0$ at $x = x_c$). It is also assumed, as in the microgravity experiment[3], that the solutions are symmetric around $x = 0$ where $v_d(0,t) = E(0,t) = 0$ for all t. The equations are evolved from a homogeneous field-free equilibrium with $E = 0$, $\varepsilon_d = 1$, and $v_d = 0$, with the parameters $m_d/m_i = 5 \times 10^8$, $\tau_i = .05$, $v_{in} = 6 \times 10^6 s^{-1}$, $v_{dn} = 6 \times 10^3 s^{-1}$, and $\omega_{pd} = 3 \times 10^3 s^{-1}$. For these parameters, we obtain $a = 5$, $\alpha_0 = 2$, $\delta \approx 0.001$, $\mu = 3$, $D_0 \approx 10^{-2}$. In Fig. 1, we show various stages of the evolution of the void density. The three-dimensional plots of the dust density at $t = 0, 4,$ and 20 are obtained from the one-dimensional solution by assuming rotational symmetry about the vertical axis. Note the sharp boundary of the void region at $t = 20$. The particles originally contained in the void region are expelled in the early stages of the dynamics and escape from the cloud boundary at $x = x_c = 4$. The nonlinear growth rate of the instability, defined by the relation increases very rapidly by nearly an order of magnitude of its linear value before it reduces in the final stages of saturation. This is consistent with the near-explosive growth of the instability seen in the laboratory experiments.

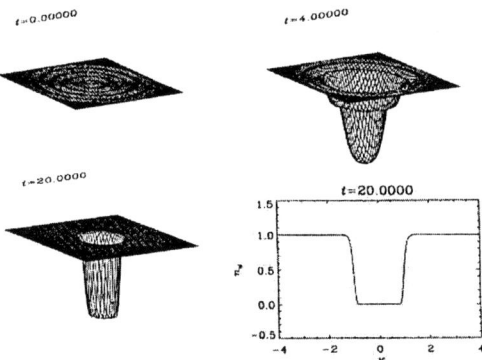

FIGURE 1 The time-evolution of the of the dust density at $t = 0, 4,$ and 20. The three-dimensional solutions are obtained from the one-dimensional solution by assuming rotational symmetry about the vertical axis. Note the sharp boundary of the void region at $t = 20$. The physical parameters are given in the text.

To examine the effect of electron depletion, we now add to our model the (dimensionless) Boltzmann relation $(1/\varepsilon)\partial_x \varepsilon = -\tau_i E$ for electrons, coupled with the

Poisson equation $\partial_x E = 1 - \varepsilon_d - \varepsilon$. We solve these two equations, coupled with (4) and (5), for the same initial conditions and parameters as that used for Figs. 1 and 2 except that we take $\varepsilon_d = 0.2$ and $\varepsilon = 0.8$ at $t = 0$. Once again, we obtain void solutions in qualitatively the same manner as shown in Fig. 1. However, as shown in Fig. 2, the void is now larger and the electric field within it is sub-linear, that is, it deviates from the strictly linear behavior seen when $\varepsilon = 0$ and is of smaller magnitude within the void.

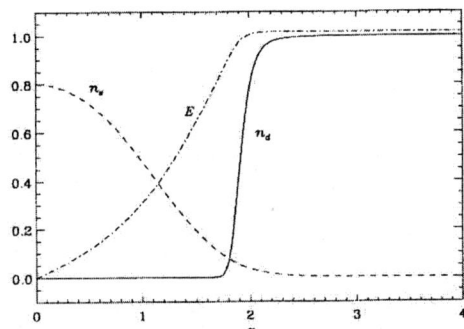

FIGURE 2 Plots of the electron density (n_e), electric field (E), and dust density (n_d) in the saturated state of the instability.

In summary, we have presented a basic nonlinear time-dependent model that describes the evolution of a zero-frequency linear instability that grows faster than exponential in the nonlinear regime and subsequently saturates to form a void. This model can provide the foundation for more complete analyses, including a more detailed account of ionization physics and effects such as charge variation and the thermophoretic force.

Acknowledgements

We thank John Goree for useful discussions. This research is supported by NASA Grant No. NAG5-2375

References

1. Prabhuram, G., and Goree, J., *Phys. Plasmas,* **3**, 1212 (1996).
2. Samsonov, D., and Goree, J., *Phys. Rev.,* **E59**, 1047 (1999).
3. Morfill, G.E., Thomas, H.M., Konopka, Rothermel, H., Zuzic, M., Ivlev, A.,and Goree, J., *Phys. Rev. Lett.,* **83**, 1598 (1999).
4. Melzer, A., Homann, A., Piel, A., Schweigert, V.A., and Schweigert, I.V., *Physics of Dusty Plasmas,* edited by M. Horanyi, S. Robertson, and B. Walch, AIP Conf. Proc. No. 446 (AIP, Woodbury, 1998), p. 167
5. Dahiya, R.P., Paeva, G.P., Stoffels, W.W., Stoffels, E., Kroesen, G.M.W., Avinash, K., and Bhattacharjee, A., to appear in Phys. Rev. Lett.
6. D'Angelo, N., *Phys. Plasmas* **5**, 3155 (1998).
7. Goree, J., Morfill, G.E.,Tsytovich, V.N.,and Vladimirov, S,V., *Phys. Rev.,* **E59**, 7055 (1999).

Dust Growth in Astrophysical Plasmas

R.Bingham* and V.N.Tsytovich[†]

Rutherford Appleton Laboratory, Chilton Didcot, OX11 0QX, UK.
[†]*General Physics Institute, Moscow, 119991, Russia.*

Abstract. Dust formation in space is important in diverse environments such as dust molecular clouds, proto-planetary nebulae, stellar outbursts, and supernova explosions. The formation of dust proceeds the formation of stellar objects and planets. In all these environments the dust particles interact with both neutral and plasma particles as well as with (ultraviolet) radiation and cosmic rays. The conventional view of grain growth is one based on accretion by the Van der Waals and chemical forces [Watson and Salpeter [14]] considered in detail both theoretically and numerically (Kempf at all [6],Meaking [7](and confirmed recently by micro-gravity experiments Blum et all [2]). The usual point of view is that the dust grow is occurring in dust molecular clouds at very low temperatures $\approx (10-30)^0$ K and is a slow process — dust grows to a size of about $0.1\,\mu$m in $10^6 - 10^9$ years. This contradicts recent observations of dust growing in winds of C-stars in about 10 years and behind the supernova SN1987A shock in about 500 days. Also recent observation of star formation at the edge of irradiated dust clouds suggests that new plasma mechanism operates in star formation. Dusty plasma mechanisms of agglomeration are analyzed as an explanation of the new astrophysical observation. New micro-gravity experiments are proposed for observing the plasma mechanisms of dust agglomeration at gas pressures substantially higher than used in ([2]). Calculations for the growth rates of dust agglomeration due to plasma mechanisms are presented. It is shown that at large neutral gas densities the dust plasma attraction provides an explanation of dust grow in about 10 days observed in H-star winds. Ionization by cosmic rays and by radioactive dust can provide the dust attraction necessary for forming dust clumping observed in molecular clouds and the fractal plasma clumping can enhance the time to reach the gravitational contraction phase operating at the final stage of star formation. A new gravitation-like dust clumping instability should operate in the dust molecular clouds at the time scale of $\sqrt{m_d/n_n T_n a^4 n_d}$ (standard notations used) and for molecular cloud conditions $n_d \approx 10^{-3}\,\text{cm}^{-3}, n_n \approx 10^3\,\text{cm}^{-3}, T_n \approx 10\,\text{K}, a \approx 3\,\mu\text{m}$ the dust agglomeration time is 3 orders of magnitude smaller than the gravitation instability time. The characteristic lengths is of the order $\approx (10^{11}-10^{12})$ cm which is 3 orders of magnitude less than the Jeans length. This new plasma mechanism can be important in star and planetesimal formation.

The subject of dust formation in space and astrophysical plasmas is important in diverse environments such as dust molecular clouds, proto-planetary nebulae, stellar outbursts, and supernova explosions. The formation of dust proceeds the formation of stellar objects and planets. And the ultimate fate of most of the dust is to be consumed in the process of making stars, planets and smaller bodies such as comets and asteroids.

In all these environments the dust particles interact with both neutral and plasma particles as well as with (ultraviolet) radiation and cosmic rays. The conventional view of grain growth is one based on accretion where condensation nuclei containing a small number of atoms formed in the process of adiabatic expansion bound together by the Van der Waals and chemical forces grow by atoms sticking and attaching themselves to the surface (Spitzer [14]). This first stage of Brownian motion-driven dust growth was considered in detail both theoretically and numerically (Weidenschilling & Cuzzi [15],

Kempf et al. [6], Meaking [7]) and confirmed recently by micro-gravity experiments (Blum et al. [2]). This model has also been used to explain the slow growth of dust grains to a size of about $0.1 \mu m$ in $10^6 - 10^9$ years. The problem in producing larger dust grains by accretion is due to a number of effects including, the sticking probability which reduces with increasing size, the ambient density and temperature of neutrals and plasma as well as the temperature of the dust grain itself. High temperatures of grains prevent further growth. Observations (Winters et al. [16]) of dust reveal typical sizes of the order of $10 - 1000 \mu m$. And even more importantly recent observations (Kempf et al. [6]) of circumstellar dust shells around long-period variables (Winters et al. [16]) demonstrate dust formation taking place on a time scale of 10 years. The winds of these C-stars are so effective at producing dust that the circumstellar dust shell totally obscures the central object. Multiple dust shells are also observed forming the circumstellar structure. Given the rate of production and multiple shell like structures we propose a new process of dust formation namely agglomeration. An important feature of the light curve from supernova SN1987A is a break at about 500 to 600 days coincident with an increase in the infrared flux and an increase in absorption of red-shifted emission lines (Chevalier[4]). These observations confirm the formation of dust within the supernova eject after about 500 days. The dust in interstellar space can be studied using telescopes. Small particles of solid matter tend to polarize light reflected and scattered by their surfaces. Particles of about a micron in size appear brighter in forward-scattered light than in back-scattered light.

Dust agglomeration has been shown to be important in laboratory etching experiments (Garscadden et al. [5]) where growth of dust is extremely rapid and is due in the final stages to dust-dust attraction by a plasma or neutral bombardment force known as the shadow effect (Tsytovich et al. [13]). Since small grains of dust are attaching themselves together the rate of growth to large sized grains of order $10 \mu m$ or larger is extremely rapid of order of hours in laboratory experiments and as we will demonstrate it is of order 10 years in C-stars.

This rapid formation of large dust grains through agglomeration can lead to a new type of instability formed as a result of the net attractive force existing between the dust grains. By replacing the force of gravity by this force we derive a linear dispersion relation describing the growth rate of a gravitation-like instability leading to clumping of the dust structures with a typical length for collapse smaller than the Jeans length for gravitational collapse. We find that the formation of clump like structures proceeds faster than gravitational collapse. The consequence is that this type of collapse is perhaps necessary before the onset of gravitational collapse resulting in a self similar or scale invariant process describing the self-organization of dust structures.

The shadow force that is responsible for this phase of collapse is due to the introduction of an anisotropic pressure gf neutrals and plasma particles on the surface of a dust grain due to the shadow or shielding by a neighboring grain. The shielding effect results in a reduction in pressure between the grains forcing them to coalesce or agglomerate. The force produced as a result of the pressure differential is proportional to the square of the ratio of dust size to separation distance. At large separation distances r this force is larger than the Van der Waals force which is proportional to r^{-8}.

In this article we estimate the rate of dust growth due to agglomeration and compare it with the normal accretion model. We will also demonstrate that it can lead to a rapid gravitational-like collapse with a length much smaller than the Jeans length for gravitational collapse. The result is the formation of dust clumps.

Our understanding of dust growth has benefited greatly from laboratory experiments (Garscadden et al. [5], Boufendi & Bouchoule [3]) carried out within the last 10 years. These experiments have been successfully explained by a combination of a series of stages of dust growth. In the first stage nucleation occurs with the chemical formation of atom clusters, this stage is well documented for both laboratory and space scenarios and it is understood that the Van der Waals and chemical binding forces play a key role in forming these nanometre sized molecular clusters. One advantage in space is that gases from outbursts or winds from stars, supernova explosions or jets are adiabatically cooled to very low temperatures forming atom clusters when the attaching (sticking) coefficient is of the order of the de-attaching (evaporation) coefficient for neutral gas atoms. Gas jet experiments also confirm the formation of atom clusters. In previous calculations of dust growth in space these atom clusters accrete single neutral atoms through collisions. Neutral gas accretion requires dust temperatures of order 10–80 K, the dust can easily maintain these temperatures by radiation cooling. In molecular clouds dust at this temperature has been detected but for the dust to grow to the size of several microns (Spitzer Jr. [10]) takes about $10^6 - 10^9$ years by the process of accretion. This may be satisfactory to explain the observations of dust in molecular clouds but in outbursts, stellar winds and supernova explosions the dust formation is extremely rapid. The main difference between these regions of dust formation and molecular clouds is the presence of plasma which is known to assist in the process of dust growth. Laboratory etching experiments are carried out in plasmas where the temperature can be as high as 3×10^4 K and the formation of atom clusters up to 10 nm are still observed, these conditions are closer to those of outbursts, stellar winds and supernova explosions than to molecular clouds, the main difference being the absence of 3×10^4 K plasma in the molecular clouds (the degree of ionization by cosmic rays or radioactivity is less than 10^{-7}) except at boundaries illuminated by UV radiation resulting in rapid growth in several stages. The presence of plasma changes the growth processes considerably.

The second stage observed in laboratory experiments carried out in the presence of plasma is rapid dust agglomeration up to grain sizes of order $0.1 - 1.0\,\mu$m. One of the main differences introduced by the plasma is the presence of charge on the grains which normally acquire a negative charge in the absence of UV radiation due to the greater electron mobility. Estimates show that for grains of 10 nm in size charges of $10 - 50$ electron charges are possible on isolated grains. At the agglomeration stage the dust density is high which has the effect of decreasing the charge on the grains to about $3 - 20$ electrons. The shadow force produced by the pressure anisotropy can easily overcome the Coulomb repulsion between the grains leading to agglomeration of more and more grains forming larger and less dense structures. The growth through agglomeration has two effects; to reduce the dust density and increase the charge on the dust to about 10^3 electrons which is large enough to overcome the shadow force through Coulomb repulsion slowing the growth (Boufendi & Bouchoule [3]).

The third stage of growth is the plasma deposition or plasma accretion stage where ions are attracted to the charged dust increasing the size of dust to about $100\,\mu$m or larger. During this stage irregular shaped dust soon becomes spherical due to a spherical converging flow of ions. The main difference between neutral accretion and ion accretion or deposition is that the ions are accelerated by the field produced by the charged dust to energies of the order of the electron temperature. The potential of the charged dust grain is of the order of the electron thermal energy in some cases as high as 3×10^4 K. This energy imparted to the ions causes ion implantation. The ion can be embedded deep within the dust in contrast to neutral particles which only sit on the surface. The presence of the plasma inevitably produces a continuous flux of ions which enhances the growth phase. For low levels of ionization the flux of neutrals can be much larger than the flux of ions. Contrary to ions, the neutrals have a very short residence time on the dust unless the temperature of the dust is extremely low of order 10 K (Spitzer Jr. [10]). At higher temperatures greater than 100 K the accretion of neutrals will not occur but accretion by ions will be rapid. In laboratory plasmas where the ion density is of order 10^9 cm^{-3} and electron ion temperature ratio $T_e/T_i \sim 10^2$, the ion deposition time for 50 μm sized particles is about 30 min i.e. rather fast, indicating that the ion implantation (bombardment) mechanism should always be considered for these conditions and needs also to be considered for in astrophysical plasmas.

The shadow forces is the main force which act in the fourth stage of dust growth. The shadow force due to neutrals and ions caused by dust is given by a simple expression which can be obtained easily qualitatively (for a more exact theory see Bingham & Tsytovich [1], Khodataev et al. [8]).

$$F = \frac{a^2}{r^2} nT\pi a^2 A \tag{1}$$

where a^2/r^2 is the solid angle, a is the dust size and r is the distance separating the dust particles, nT is the pressure of either neutrals or ions we can neglect electrons because of their smaller mass and A is a numerical coefficient of order of unity. In the calculation we have assumed that the dust particles are of equal size.

The physics of shadow forces is very simple. For a single dust particle the flux of absorbed charged and neutral plasma particles on the dust surface is symmetric resulting in no net momentum transfer. Another dust particle at distance r shadows the flux to the first dust particle with a solid angle $\approx a^2/r^2$. The net momentum transfer will be proportional to the solid angle, to the surface of dust particle πa^2 and to the pressure nT as written in (1). The numerical coefficient depends on whether the distance between the dust particle is less or larger than the mean free path in the neutral component being in both cases of the order of 1, Khodataev et al. [8]. In the case where the distance between dust particles is larger than the mean free path the physics of the force is slightly different, one dust particle creates a temperature gradient in the neutral gas component and another dust particle is attracted by thermophoretic force due to this temperature gradient (Khodataev et al. [8]). The attraction exists if the temperature of the dust surface is less than the temperature of the neutral gas which occurs almost always in astrophysical conditions. For example in optically thin dust clouds the dust surface is cooled by the radiation losses (the energy flux of ions on dust particles which heats the dust surface is usually less than the radiation losses the more than 15 orders of magnitude and the heating by other type of radiation is less than the cooling for optically thin cloud).

The shadow force is inversely proportional to r^2 same as for gravity and Coulomb forces, it depends on the neutral n_n or ion density n_i.

The fourth stage of dust growth observed under laboratory conditions but not in all situations (Garscadden et al. [5]) is the advanced stage or second stage of agglomeration which leads to dust grains of 0.1 mm in size which are observed optically. The physics of this stage is dominated by dust-dust attraction, which is proportional to the fourth power of the dust size, and becomes larger than the Coulomb repulsion of dust charges, which is proportional to the square of the dust size (Tsytovich et al. [13]).

The four stages of dust growth observed in laboratory experiments in the presence of plasma will also be important in space plasmas. For example in carbon rich star outbursts (Winters et al. [16]) dust formation takes about 10 years in regions where there is a mixture of neutral and plasmas at relatively high temperatures. The gas density in these stellar outbursts is as high as $10^{10}\,\text{cm}^{-3}$ with temperatures large enough that ionization is also present. The dust shells in these carbon rich stars form within 5 stellar radii in regions where the temperature is about 1000 K and it should be expected that the four stages of dust formation should proceed including ion implantation and agglomeration. In a gas or plasma the forces transmitted to the dust grains are the random forces transferred to the small particles during Brownian motion if we bias the collisions so that they are not isotropic for example through shielding effects results in the shadow force equation(1).

In previous studies of dust growth in space, equation (1) is not used, only the accretion of neutrals is assumed and isotropic collisions with dust grains results in no net force between the grains. This is also true if the neutral atom attaches itself and then de-attaches with the same temperature. If the neutral atom during its contact with the surface has sufficient time to thermalize to the dust temperature and is emitted with the energy corresponding to the dust temperature then the factor (Khodataev et al. [8]) $1 - \sqrt{T_{ds}/T_n}$ appears in the expression for the force given by equation (1), T_{ds} is the dust surface temperature and T_n is the neutral temperature. The force vanishes for $T_{ds} = T_n$ and is largest for $T_{ds} \ll T_n$, a condition often found in space where the dust cools by emission of radiation ata rate $\pi a^2 \sigma_{SB} T_{ds}^4$ where σ_{SB} is Stefan-Boltzmann constant. In this case the dust cloud is optically thin for dust radiation a situation found in dust-molecular clouds. In (Khodataev et al. [8]) it was shown that in optically thin dust cloud the dust temperature is always smaller than the gas temperature, the forces between dust particles due to gas particle bombardment are always attractive and that with an increase of the gas density the attraction forces "saturate" at densities of about $10^{16}\,\text{cm}^{-3}$ much larger than in any dust-molecular cloud.

The force given by equation (1) operates in the presence or absence of plasmas. In the presence of plasmas the force due to the ion and neutral flux should be compared to the Coulomb force F_c between two dust grains given by

$$F_c = \frac{Z_d^2 e^2}{r^2} = \left(\frac{Z_d e^2}{aT_c}\right)^2 \frac{n_i T_e^2 a^2}{e^2 n_i r^2} = z^2 \frac{a^2}{r^2} n_i T_e 4\pi d_e^2 \qquad (2)$$

where $z = \frac{Z_d e^2}{aT_e}$ is the dimensionless dust charge of order 1-2 and $d_e = \sqrt{\frac{T_e}{4\pi n_e e^2}}$ is the electron Debye radius. Comparison of equations (1) and (2) demonstrates that the bombardment force due to ions is $\frac{4d_e^2 z^2}{a^2}$ less than the Coulomb repulsive force for $r \ll d_e$. For $r \gg d_e$ the Coulomb force is screened while the attractive shadow force is not, this can lead to a contraction of the dust cloud until the inter-dust distances are comparable to 10 times the Debye radius. For distances larger than the Debye radius the ion accreation attraction (Tsytovich et al.[13]) can dominate the Coulomb repulsion and should be added to the effect related with gas particle bombardment attraction, but this contribution is usually small. For outbursts of carbon rich stars where $n_n \simeq 10^{10}\,\text{cm}^{-3}, n_i$ is of order n_n and $T_e \sim 1eV$ resulting in a Debye radius of between 100 and $1000\,\mu m$, attraction can occur between two equally charged grains. Using $r \sim 4d_e$ where the repulsive force can be overcome by the attractive force forming an attractive potential well, with a potential given by (the ion accretion force is usual smaller than the gas bombardment force (Khodataev et al. [8]))

$$V \simeq F_c r \approx z^2 n_i d_e a^2 T_e \tag{3}$$

we find that a "dust molecule" is formed if

$$T_d < T_e n_i d_e a^2 z^2 \tag{4}$$

where T_d is the dust kinetic temperature not the surface temperature. In environments where dust is cooled by emission of radiation condition (4) can easily be satisfied. For $n_i \sim 10^{10}\,\text{cm}^{-3}, d_e \sim 10^{-2}\,\text{cm}$ i.e. $r \sim 10^3 \mu m$ and $a \sim 1\,\mu m$ equation (4) results in $T_d < T_e z^2 \sim 10 T_e$ which is easily satisfied in carbon rich star outbursts (Winters et al. [16]).

Laboratory experiments (Thomas & Morfill [11]) demonstrate the possibility of producing regular dust structures known as dust crystals which occur for $Z_d e^2 n_d^{\frac{1}{3}} > 170 T_d$. In space this condition is more difficult to satisfy than (4) namely

$$T_d > T_e n_i a d_e^2 z^2 \left(n_d^{\frac{1}{3}} a\right) / 170 \tag{5}$$

but it may be possible which has lead to suggestions that dust crystals could exist in space environments (Morfill [9]). For $T_i \ll T_e$ numerical results (Bingham & Tsytovich [1]) show that the attraction force equation (1) can dominate the Coulomb force for $a > (8 - 60) d_i$ thus leading to dust agglomeration, d_i is the ion Debye radius.

In dust molecular clouds d_e is very large and the degree of ionization is very low, however, cosmic rays produces ionization such that $\frac{n_i}{n_d} \sim 10^{-6} - 10^{-8}$ which is only about one order of magnitude less than in laboratory experiments. In most laboratory experiments the neutral and ion shadow forces are of the same order of magnitude and give rise to the second and fourth stages of agglomeration. In dust-molecular clouds, if the dust is not radioactive, the main attraction force is determined by the neutral flux and equation (1) can be used if the dust surface temperature is lower than the gas temperature due to radiative cooling. The phenomenon of dust agglomeration related with the dust attraction due to neutral particle flux was not previously considered for dust-molecular clouds but it can operate effectively in the range of temperatures above the critical temperature range of 10–80 K for gas accretion on dust. For large temperatures this will serve as the main mechanism responsible for dust growth. An estimate of the agglomeration time is found by taking into account the relative number of particles in phase space with energies less than the attraction potential well

$$\frac{dm_d}{dt} = 2m_d v_d n_d^{\frac{1}{3}} \left(\frac{V}{T_d}\right)^{\frac{3}{2}}; \quad v_d = \sqrt{\frac{T_d}{m_d}} \tag{6}$$

Since the force acting is in the direction separating the two dust particles we can use for the time scale the time for the dust to travel the inter-dust distance which is a direct consequence of (6) given by

$$\frac{1}{\tau_{aggl.}} \simeq \sqrt{\frac{T_d}{m_d}} n_d^{\frac{1}{3}} \left(\frac{T_e n_i a^2 d_e z^2}{T_d}\right)^{3/2} \tag{7}$$

for $a \sim 1\,\mu m$, $T_d \sim 100\,K$, $n_d \sim 10^{-3}\,cm^{-3}$ we estimate to be $\tau_{aggl.}$ to be about $10^9\,sec \simeq$ 30 years. The shadow force also creates a gravitational-like instability. The dispersion relation for the agglomeration instability for the dust component can easily be written using the dispersion relation for the Jean's instability but replacing the force of gravity by the shadow force resulting in

$$\omega^2 = k^2 v_s^2 - G_{eff} n_d m_d \tag{8}$$

where according to (1)

$$G_{eff} = \frac{nT_n \pi a^4}{m_d^2} \tag{9}$$

and

$$v_s^2 = \frac{n_d T_d + n_n T_n}{n_d m_d + n_n m_m} \sim \frac{n_n T_n}{n_d m_d + n_n m_n}$$

if $n_n T_n \gg n_d T_d$. Although the attraction forces are acting only on dust particle the expression v_s will correspond to the sound speed for the frequencies much less than the dust-neutral collision frequency (Tsytovich [12]).

The corresponding Jeans type length is determined by the relation

$$L = \frac{2\pi v_s}{\sqrt{G_{eff} n_d m_d}} \simeq \frac{2\sqrt{\pi}}{n_d a^2 \sqrt{(1 + n_n m_n/n_d m_d)}} \tag{10}$$

For $a \sim 1\,\mu m$, $n_d \sim 10^{-3}\,cm^{-3}$ we have $L \sim 10^{11}\,cm$, resulting in dust clumps of order $10^{11} - 10^{12}\,cm$ in size.

These clumps are much smaller than the Jeans length for gravitational collapse and will proceed stellar and planetary formation. It can be envisaged that clumping could proceed through this mechanism forming larger and larger scales all the way to the Jeans length at each scale the structures are similar and scale invariant. These structures are a result of the self-organization found in complex physical systems, each scale could collapse forming structures from asteroids in size all the way up to star formation which still could be related with additional compression produced by shock waves. The characteristic time for the Jeans-like instability to develop is given by $\tau_{\text{ag.inst}}$ i.e. agglomeration instability time scale

$$\tau_{\text{ag.inst}} \simeq \sqrt{\frac{m_d}{n_n T_n \pi a^4 n_d}} \qquad (11)$$

which for $n_d \sim 10^{-3}\,\text{cm}^{-3}$, $n_n \sim 10^3$, $T_n \sim 50\,\text{K}$, $a \sim 1-3\,\mu\text{m}$ is

$$\tau_{\text{ag.inst}} \sim 10-100 \text{years}.$$

This is 3 orders of magnitude faster than the Jeans gravitational collapse time and the Jeans length is 10^3 orders of magnitude larger.

Recent experiments demonstrated (Blum et al. [2]) dust agglomeration under microgravity conditions. It should be emphasized that the terminology for dust growth varies from paper to paper in this paper we discuss growth of large dust grains in a high density regime whereas in (Blum et al. [2]) the growth is in a lower density regime where the term agglomeration is used for the case where Van der Waals force at small distances overcomes the repulsion causing the sticking of dust particles. For higher densities our effect dominates through equation (1) which is larger than the Van der Waals force which varies as $1/r^8$. The agglomeration discussed in (Blum et al. [2]) is in our terminology either the first chemical stage of dust clumping forming clumps of many molecules or the second stage. We emphasized here the possibility of the existence of a fourth stage in astrophysical conditions similar to that in laboratory experiments. Even in completely neutral gas the shadow force operates producing dust particles of the size 10μ or larger. In many cases the presence of ionization can also be important for dust growth. There is a possibility to check our mechanism in micro-gravity experiments. The experiments already performed deal with very low neutral gas, pressure where according to our estimate the potential V given by equation (3) is less than that due to the Van der Waals force. To check our results it is desirable to work with dust particles of larger size and with larger neutral gas pressure. In present experiments the attraction force equation (1) can help dust clumping at large distances, where the force given by equation (1) overcomes the Van der Waals force.

In this article we have described a new force of attraction giving rise to rapid dust agglomeration in astrophysical plasmas, the rate of dust formation is much more rapid than that due to the accretion of neutral particles. This new force arises due to a shadow effect between the dust particles in either a neutral or plasma atmosphere. The force of attraction that results from the shadow effect gives rise to faster and smaller scale gravitational collapse than that due to the standard Jean's instability. The structures that are then formed could easily be responsible for the formation of planetesimals.

REFERENCES

1. Bingham, R., and Tsytovich, V.N., *Special Issue Trans. IEEE Plasma Science*, in press (2000).
2. Blum, J., et al., *Phys. Rev. Lett.*, **85**, 2426 (2000).

3. Boufendi, L., Bouchoule, A., *Plasma Sources Sci. Technology*, **3**, 262 (1994).
4. Chevalier, R.A., *Nature*, **355**, 691 (1992).
5. Garscadden, A., Ganguly, B.N., Healand, P.D., and Williams, J., *Plasma Sources Sci. Technology*, **3**, 239 (1994).
6. Kempf, S., Pfalzner, S., and Henning, T., *Icarces*, bf 141, 388 (1999).
7. Meaking, P., Rev. Geophys., **29**, 317 (1991).
8. Khodataev, Ya.K., Morfill, A.G., Tsytovich, V.N., *J. Plasma Phys.*, in press 2001.
9. Morfill, G., Private communication 2000.
10. Spitzer, L. Jr., *Physical Processes in the Interstellar Medium*, John Wiley and Sons, New York, 1978.
11. Thomas, H.M., and Morfill, G.E., *Nature*, **379**, 806 (1996).
12. Tsytovich, V.N., *Physics Uspekhy*, **40**, 53 (1997).
13. Tsytovich, V.N., Rhodataev, Y., Bingham, R.,*Comments on Plasma Physics and Controlled Fusion*, **17**, 249 (1996).
    Tsytovich, V.N., Khodataev, Y., Bingham R., and Tarakanov, V., *Advances in Dusty Plasma*, World Scientific, Singapore, ed. P K Shukla, D A Mendis and T Desai 1996, page 212.
14. Watson, W.D., and Salpeter, E.E, *Ap.J.*, **174**, bf 321 (1972).
15. Weidenschilling, S.J., and Cuzzi, J.N., *Protostars and Planets III*, edited by E. Levy and J. I. Lunine, Univ. Arizona Press, Tucson 1993, page 1031.
16. Winters, J.M., Keady, J.J., Gauger, A. and Sada, P.N., *Astron.Astrophys.*, bf 359, 651 (2000).

Dust Particles Growth and Behavior under Microgravity Conditions

M. Mikikian*, L. Boufendi*, A. Bouchoule*, G.E. Morfill[†], H.M. Thomas[†],
H. Rothermel[†], T. Hagl[†], A.P. Nefedov**, V.E. Fortov**, V.I. Molotkov**,
O. Petrov**, A. Lipaev**, Yu.P. Semenov[‡], A.I. Ivanov[‡], V. Afanas'ev[§],
C. Haigneré[§] and K. Kozeev[§]

Groupe de Recherches sur l'Energétique des Milieux Ionisés, Université d'Orléans, 45067 Orléans Cedex 2, France
[†]*Max-Planck-Institut für Extraterrestrische Physik, 85741 Garching, Germany*
**Institute for High Energy Densities, Izhorskaya 13/19, Moscow, 127412, Russia*
[‡]*Russian Rocket Corporation "Energia"*
[§]*Cosmonauts of the Taxi mission on ISS*

Abstract. In this paper we present the first observations on dust particles growth and behavior under microgravity conditions obtained in the PKE-Nefedov chamber by a French-German-Russian program. These experiments have been performed on board of the International Space Station (ISS) in late october by a French-Russian cosmonauts team. Some of the results are discussed and compared with on ground experiments.

INTRODUCTION

This work is an extension of previous experiments performed by the Institute for High Energy Densities (IHED) of Russian Academy of Sciences and the Max-Planck-Institute for Extraterrestrial Physics (MPE) concerning the study of clouds of injected microparticles under microgravity conditions [1]. These conditions are required to avoid perturbations induced by gravity, revealing the real interactions between dust particles and pure plasma effects. The Research Group on Energetics of Ionized Gases (GREMI) joins IHED and MPE for an extended program on dust particles growth. Growth kinetics, spatial distribution and dynamics of grown particles have been investigated.

DESCRIPTION OF THE EXPERIMENTS

Experimental basis

The PKE-Nefedov experimental setup [1] consists in a radiofrequency discharge where an argon plasma is created in push-pull excitation mode. The parallel plate electrodes are separated by 3 cm and their diameter is 4 cm. Polymer particles (melamine formaldehyde $\sim$ 3.4 and $\sim$ 6.9 μm) can be injected in the plasma with two dispensers inserted in the electrodes. A thin laser sheet perpendicular to the electrodes crosses the chamber and the light scattered by the particles is observed at 90° by two CCD cameras with different magnification. The system laser-camera can be moved horizontally to scan the entire cloud and obtain tridimensional informations on the cloud structure. The behavior and the organization of these clouds have been investigated either on ground and in microgravity by MPE (tridimensional structure, presence of a dust-free region called "void" in the center of the discharge).

Growth procedure

On ground, after each experiment the injected dust particles fall down on the lower electrode. This deposited matter can be used to generate a new dust particles population. In the PKE-Nefedov chamber, this growth has been obtained in GREMI laboratory at high pressure (> 1 mbar) and low power ($\sim$ 1 W) with a typical appearance time (on screen) around 2 to 3 mn. After this growing step where the particles seem to be monodisperse, a constant process of sputtering and growth leads to particles of various sizes (between 0.2 and 0.8 μm in diameter) as shown by scanning electron microcospy. The X-ray fluorescence analysis reveals that these particles are principally constituted of carbon, the nitrogen present in melamine formaldehyde has disappeared. Due to their submicron size, gravity is not the predominant force acting on them and we can obtain their trapping in the whole volume of the plasma even on ground. The void region has been observed and large crystalline regions can appear as shown in figure 1.

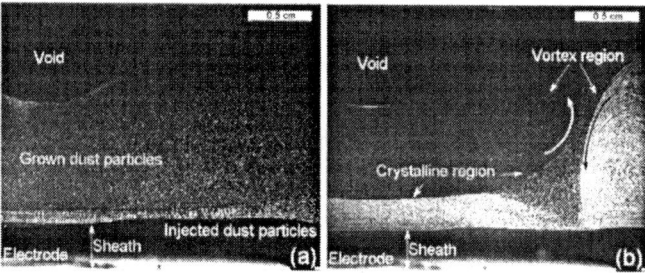

FIGURE 1. Dust clouds on ground, a) grown and injected particles (3.4 μm), b) crystalline and vortex-like regions of various sizes grown particles (15 images superimposed)

Microgravity experiments preparation

One of the unknown parameter for the microgravity experiments is the quantity of matter deposited inside the reactor from previous experiments : injected particles do not fall down directly on the electrodes when the plasma is turned off. On ground we have observed a direct correlation between the size of the grown particles and their density. Their size will remain very small while their density is very large. Another requirement for this experiment is to obtain a rather low base pressure. Our laboratory experiments have shown that the growth can be inhibited if the base pressure before an experimental run is not low enough (10^{-3} Pa), that is to say if the argon plasma is not pure (presence of O_2 or/and N_2). This aspect has been taken into account for the microgravity experiments and a turbomolecular pump has been installed on board replacing the direct vacuum connection to space previously used.

OBSERVATIONS AND RESULTS

Dust particles growth has been achieved on board the ISS leading to a situation with a huge amount of very small size particles. This configuration has been obtained in the laboratory but is not the most common one. For the moment it is not clear if it is due to microgravity or to a high purity of the plasma and a lot of matter available. In these conditions the particles can not be individualized on images and then our study is concentrated on a global observation of the cloud. We can notice that a first interesting result is the presence of the void structure even with these small size particles.

Growth evidence

As shown in recent experiments [2] performed in our laboratory in an argon-silane plasma, the dust particles growth can be followed through the evolution of the current harmonics amplitudes. Dust particles decrease the plasma electron conductivity through collisions or attachment, leading to a drop of the electron current reaching the electrodes. This effect has been observed during the on board experiments as shown on figure 2. We have plotted on the same figure the evolution of the video signal intensity recorded by the cameras (i.e. dust appearance) to show the correlation between these two quantities.

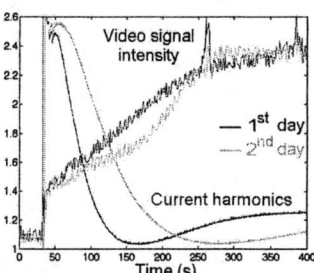

FIGURE 2. Time evolution of the current harmonics and correlation with the video signal intensities

Dynamical behavior

As observed on ground, big vortex cells appear in the on board experiment showing that this behavior is not due to thermal convection as it could be supposed but caused by pure plasma effects and more precisely by the electric field near the electrodes edges. Another interesting dynamical behavior is the response of the cloud to a low frequency excitation applied on the electrodes. On figure 3b the cloud is constituted of grown particles and injected ones (3.4 µm, down right). We have recorded the position of the caracteristic points labeled on the image in fonction of time for various frequencies contained between ∼ 0.5 and ∼ 34 Hz (figure 3a for a frequency about 1 Hz). The points (P1,P2,P3) move in phase and then this movement propagates through the cloud of grown particles which is moving in phase at low frequency but with a phase shift at higher frequency. This experiment could give informations on dust acoustic wave propagation. A more precise investigation will be conducted on ground to see if microgravity plays a role in this phenomenon.

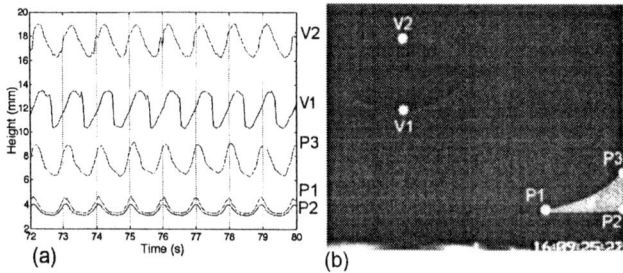

FIGURE 3. Low frequency excitation of a cloud constituted of grown and 3.4 µm particles, a) Time evolution of the height of the points labeled in b)

CONCLUSION

This work opens interesting perspectives in the field of dust particles growth. Clouds of submicron particles of various sizes can be grown in the whole plasma volume. Furthermore, this experiment offers the opportunity to study experimentally the void region even on ground. However, microgravity is needed to avoid effects due to size segregation and to study dynamical effects.

ACKNOWLEDGMENTS

The authors from GREMI wish to thank their russian and german colleagues for this successful collaboration. This work was supported by CNES under contract 793/2000/CNES/8344.

REFERENCES

1. Morfill, G. E., Thomas, H. M., Konopka, U., Rothermel, H., Zuzic, M., Ivlev, A., and Goree, J., *Phys. Rev. Lett.*, **83**, 1598–1601 (1999).
2. Boufendi, L., Gaudin, J., Huet, S., Viera, G., and Dudemaine, M., *Appl. Phys. Lett.*, **79**, 4301–4303 (2001).

Rotation of Dust Coulomb Clusters in Axial Magnetic Field

F.M.H. Cheung, A.A. Samarian, and B.W. James

School of Physics, University of Sydney, NSW 2006, Australia

Abstract. In an inductively coupled rf plasma system, Coulomb clusters of 2 to 12 particles have been rotated under the influence of an axial magnetic field. The rotation of the clusters is dependent on the particle number and its structural configuration. The threshold magnetic field strength required to initiate the rotation is inversely proportional to the square of the number of particles. The angular velocity of the clusters increased and the radius of the clusters decreased as the magnetic field strength was increased. Qualitative comparison of the driving force and the ion drag force is presented.

The rotation of Coulomb clusters with different numbers of micron-sized particles is studied experimentally in an inductively coupled magnetised dusty plasma. When the magnetic field was absent, the clusters exhibited small random fluctuation but always remained around their equilibrium position. However when the magnetic field was present, the small clusters were observed to undergo rotational motion. The direction of the rotation was in the left-handed sense with respect to the magnetic field direction. Here we report the rotation of Coulomb clusters with N number of particles, N equal from 2 to 12. To our knowledge this is the first time such small crystal systems are systematically studied to exhibit nature of rotation.

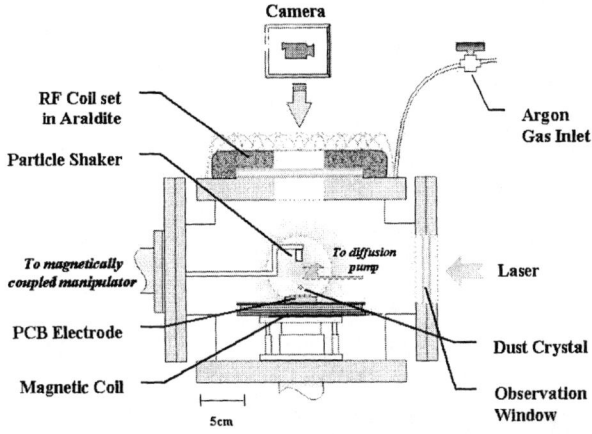

FIGURE 1. The experimental apparatus used to produce the coulomb dust clusters.

The experiment is conducted in a radio-frequency (rf) discharge with a printed-circuit board (PCB) electrode system. Fig. 1 shows the interior of the experimental apparatus. The dust cluster formed above the confining electrode are illuminated by a fully height adjustable He-Ne laser for observation. The motion of the dust crystals was observed under the microscope and from the video images on the televisions generated from the cameras. Before the images of the crystal's rotational motion were recorded, the video signals from the video camera were passed through a sinc-pulsed signal amplifier to be converted from noise signal into a clear sharp picture. The processed images of the rotational motion were then recorded on videotapes at a frame rate of 50 fps and a shutter speed of 0.008 seconds. The particles are then tracked with a software program that outputs the x and the y-coordinates of their trajectories as a function of time for analysis.

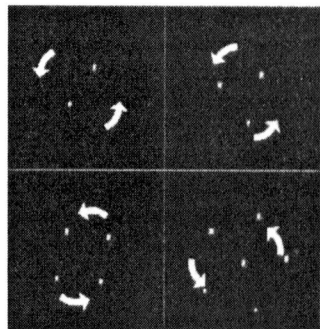

FIGURE 2. Images of planar-2, planar-3, planar-4 and planar-6 crystals

We found the cluster rotation to be dependent on the particle number and configuration. Thus clusters with smaller numbers of particles require a higher magnetic field strength in order to initiate the rotation. The threshold magnetic field strength B_{th} at which the cluster begin to rotate decreased as the number of particle N

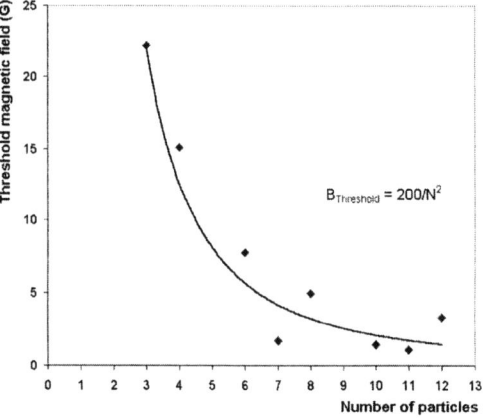

FIGURE 3. Threshold magnetic field versus number of particles.

increased and can be expressed by empirical relation $B_{th}=200/N^2$.

The angular velocity of the clusters increased and the radius of the clusters decreased as the magnetic field strength was increased. The relationship between angular velocity and magnetic field is dependent on the number of particles N in the cluster. Based on the experimental measured values the angular velocity dependence on magnetic field and number of particle can be approximated by empirical relation

$$\omega = \exp(-23/N)B^{\frac{8.3}{N^{1.5}}} - 4/N^4.$$

Let us consider the steady

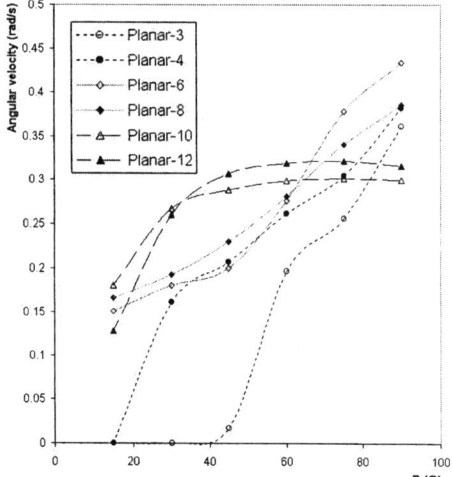

FIGURE 4. The variation of angular velocity of the dust clusters as a function of magnetic field strength.

rotation of the plasma crystal in the horizontal plane. This rotation is determined by the balance of the radial electrostatic force F_r due to the confining electric field E_r, the azimuthal driven force F_d, the azimuthal friction force due to neutral-dust collisions F_f and the force due to the interaction between the particles F_{ip}. The interparticle interaction we should take into account only in the case of spatially nonuniform system. The driven force is the force which we attribute the rotation of the plasma crystal. It can be the azimuthal component of ion drag force. The resulting equation of motion for the crystal using the cylindrical coordinates $\{r,\beta\}$, where r is the radial axis from the center of rotation and a is the azimuthal angle can be written in the form:

$$e_r m_d \omega^2 r = -e_r F_r + e_r F_{ip} + e_\beta F_d - e_\beta F_f . \tag{1}$$

where m_d is the dust particle mass, ω is the angular velocity, e_r and e_a are the unit vectors in the radial and azimuthal directions. If we consider the forces separately acting in the azimuthal direction for rigid body rotation the driven force must be equal to the friction force. One can estimate the friction force (under the assumption of complete accommodation of neutrals) using the expression[1]

$$F_f = -\frac{4}{3}\delta m_n n_n v_T \pi a^2 \omega r \tag{2}$$

where $\delta \sim 1$ is the coefficient dependent on the type of scattering, m_n is the mass of the Ar atom, $n_n = \frac{P}{kT_n}$ is the gas atom number density, P and T_n are the pressure and temperature of neutrals correspondently, $v_T = \sqrt{\frac{8kT_n}{\pi m_n}}$ is the thermal velocity of the atoms and a is the dust particle radius.

So the magnitude of the driving force can be calculated from the equation of the friction force in the azimuthal direction. The obtained value is about 10^{-17}N. Using Eq.2, along with the angular velocity dependence on magnetic field strength we measured experimentally, the driven force is linearly proportional to the magnetic field strength $F_d \sim B$. Under the assumption that the crystal rotates due to ion drag force the angular velocity can be written in the form[2]

$$\omega = \frac{\alpha BE(r)}{r} \tag{3}$$

where α is a constant. In the case of parabolic confining potential, the electric field is linearly proportional to r and the angular velocity should depend only on the magnetic field $\omega \sim B$. This fact does not agree well with experimental data, which show the dependence of the cluster rotation on the particle number.

To provide the quantitative comparison we need to estimate the value of the ion drag force, which is given by the expression[3]

$$F_{id}^{\phi} = m_i n_i u_\Sigma \pi \left(b_c^2 + 4 b_{\frac{\pi}{2}}^2 \Gamma \right) u_\phi \quad (4)$$

where m_i is the mass of the Ar$^+$ ion, n_i is the ion density, u_Σ is the mean ion velocity, b_c is the collection impact parameter, $b_{\pi/2}$ is the orbit impact parameter whose asymptotic orbit angle is $\frac{\pi}{2}$, Γ is the Coulomb logarithm integrated over the interval from b_c to λ_{De}, u_ϕ is the azimuthal component of ion drift velocity, μ_0 is the zero field mobility and α_0 is a parameter for argon.

It is easy to see that in comparison with the friction force, the ion drag force is about order less. Thus the attempt attribute the rotation of plasma crystals to the action of ion drag force gives us only explanation of linear proportionality angular velocity to the magnetic field strength. Another item that we should take into account is the dependence of radial electric field on magnetic field (see Figure 5). Obtained data show that electric linear proportional to magnetic field thus if we assume the ion drag force as driven the angular velocity was proportional to the square of magnetic field.

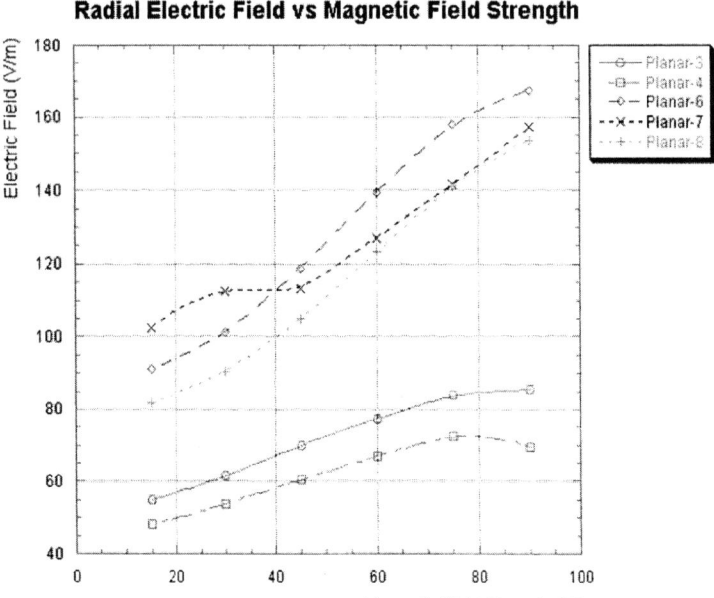

FIGURE 5. Radial electric field dependence on axial magnetic field

It was demonstrated from the experiment that the rotation of small dust coulomb clusters is possible with the application of an axial magnetic field. It is easier to initiate the rotation of the clusters with larger number of particles than smaller number of particles at very low magnetic field strength. The angular velocity of the clusters increases while the radius of the clusters decreases as the magnetic field strength increases.

ACKNOWLEDGMENTS

This work was supported by Australian Research Council, the Science Foundation for Physics within the University of Sydney. AAS was supported by University of Sydney U2000 Fellowship.

---

[1] P. S. Epstein, Phys. Rev. 23, 710 (1924)
[2] U. Konopka, D. Samsonov, A. V. Ivlev, J. Goree, V. Steinberg and G.E. Morfill, Phys. Rev. E, 61, 1890 (2000)
[3] M.S. Barnes, J.H. Keller, J.C. Forster, J.A. O'Neill and D.K. Coultas, Phys. Rev. Lett. 68(3), 313(1992).

Self-Consistent Dusty Sheaths

Yu. I. Chutov

Taras Shevchenko Kiev University, Kiev, Ukraine, yuch@univ.kiev.ua

Abstract. Self-consistent dusty sheaths are investigated by using the kinetic PIC/MCC simulation of a temporal evolution of both dusty plasmas consisting of immersed electrodes (walls) and developing dusty discharges without special boundary conditions for the sheaths. Obtained results show an essential influence of dust particles on sheath parameters. Spatial distributions of a dust particle charge are not trivial due to a self-consistent evolution of electron and ion energy distribution functions.

Electrostatic charged sheaths separate plasmas from electrodes or walls adsorbing charged particles (usually electrons and ions) from the plasmas. The sheaths are appeared due to a difference between fluxes of the adsorbed charged particles to electrodes or walls without sheaths. In the simple case of plasma without a magnetic field, the electron flux essentially exceeds the ion flux so that positively charged sheaths are created around negatively charged walls.

However the electrodes or the walls disturb the plasmas usually far from the sheaths creating here quasineutral non-uniform plasma regions with slow electric fields and slow gradients of a plasma density. The regions provide fluxes of charged particles to sheaths and it are called preasheaths [1]. Basic properties of the plasma-sheath transition were given in early works and considered in the detail review [2]. According to the works, a boundary between a sheath and a presheath can be determined only with a precision of several electron Debye lengths. Therefore according to the Bohm's theory [3] for sheaths in collisionless plasmas, a non-linear sheath around an electrode or a wall facing to plasmas consists of three distinct regions, namely: an ion (Child-Langmuir) sheath without electrons, a Debye sheath ranging over a few Debye lengths, and a plasma region where quasineutrality holds (presheath). In a case of equilibrium electrons and cold ions in collisionless plasmas, the sheath edge can be fixed conditionally as a point where a drift ion velocity u is equal to an ion sound speed $u_s = (kT_e/M)^{1/2}$ or an electric field E is equal to the characteristic electric field $Eo = kT_e/L_D$. The sheath edge fixed by the ion sound speed corresponds to the well known Bohm's boundary condition which used very often at a consideration of plasma sheaths.

However the electron equilibrium can be disturbed even in collisionless sheaths (including the sheath edge) in plasmas with single electron specie. Indeed, electron energy distribution functions in collisionless plasmas are formed in any points by opposite electron fluxes, which have to be identical for equilibrium functions. Adsorbing electrodes or walls create directional electron fluxes due to a collection of tail electrons with the energy greater than the sheath potential energy so that the electrons do not return back into the plasma. As a result, the electron energy distribution function is truncated, however as was shown earlier [4], the disturbance is not essential for plasmas with a single electron species due to the smallness of such electron fraction.In case of non-equilibrium plasma electrons, the disturbance of the electron energy distribution functions can be more essential. For example, such situation can take place in plasmas with two-temperature Maxwellian electron distributions [5].

The sheaths can consist of dust particles appeared as the product of the plasma-wall interaction in various technological devices [6,7] including controlled fusion devices [8,9] or created due to coagulation of various components in chemically active plasmas with their subsequent transport into sheaths [10]. Besides, dust particles can be immersed from outside and be trapped in sheaths creating plasma crystals [11] investigated intensively now. Dust particles can essentially influence sheath parameters due to the continuous selective collection of background electrons and ions that can cause an essential change of both electron and ion energy distribution functions [12]as well as an ion flux in sheaths.

Usually sheaths are investigated without presheaths because of essential differed space and time scales of both regions. Of course at the investigations, boundary conditions have to be formulated at a sheath edge. Bohm's boundary conditions (or their later modifications) formulated for a hydrodynamic description of collisionless sheaths a long time ago [13], are used very often. The conditions consist of assumptions about equilibrium electrons, could ions as well as the zero electric potential and field at the sheath edge where a drift ion velocity is equal to the ion sound speed in non-disturbed plasma. The conditions were modified than for cases of warm ions including ions with a given energy distribution function [2], plasmas with two-temperature electrons [14,15] and are very popular also for investigations of sheaths in dusty plasmas up to last time [16,17].

Unfortunately, the Bohm's boundary conditions used in various works very often for sheaths in collisionless plasmas are not self-consistent. Indeed, it is not possible to find in the plasmas a point where the conditions are valid because an acceleration of ions to the ion sound speed from undisturbed plasma is possible only by a self-consistent electric field causing a continuous change of an electric potential. The change can be essential in the case of non-equilibrium electrons and ions indicated above.

However there are several possibilities to consider the self-consistent sheaths without special boundary conditions for the sheaths. The consideration is based on a study of a temporal evolution of systems including sheaths as a part of the systems. Examples of the consideration can be the asymptotic behavior of rarefaction waves created by electrodes or walls immersed into plasma [18] or the general consideration of entire discharge plasmas with given boundary conditions on electrodes and walls [1].

The self-consistent description of dusty sheaths has to consist of a self-consistent behavior of dust particles including their charge depending strongly on electron and ion energy distribution functions. The kinetic simulation of the self-consistent dusty sheaths was developed in [19-22] using the PIC/MCC method (1D3V model) described in detail earlier [23-24] for computer simulations of the plasma without dust particles. The method is based on a kinetic description of the motion of positive and negative "superparticles" in phase space under an influence of a self-consistent electric field E. The field E is obtained by solving of the Poisson equation using a computational grid, which is introduced by dividing the simulation region. The Monte Carlo technique is used here to describe various elementary processes in plasmas.

The method was developed in [19-22] for dusty plasmas using the self-consistent charging of dust particles according to the Orbit Motion Limited (OML) theory [25]. The Monte Carlo technique [23,24] is used in developed method to describe interactions of electrons and ions with dust particles. The interactions include Coulomb's collisions of electrons and ions with dust particles, as well as the electron and ion collection by dust particles. The cross-sections of an electron and ion collection by dust particles are taken from [25]. The Coulomb cross-section for electron and ion scattering by immobile dust particles is taken from [26]. The simulation region size is chosen to be equal to several hundreds of the Debye length so that the region exceeds essentially a sheath size. Electrodes or walls collect a "superparticle" if its center reaches an electrode (wall) surface.

In addition to a usual PIC/MCC scheme, the weighting procedure is used in the developed method for the determination of a superparticle charge, which is interacting with a dust particle. In all cases of sheath simulations, one boundary of the simulation region is bounded by an electrode or a wall in front whose the sheath is created. The second boundary can be located in a plasma with given parameters where presheath is developed during the plasma evolution. The continuous exchange by superparticles takes place on the second boundary that has to be taking into account at the computer simulation. In this work, the original model of the exchange was developed with the self-consistent change of the electric potential as well as electron and ion energy distribution functions on the boundary.

Obtained results show essential influence of dust particles on sheath properties. The influence is caused by a selective collection of electrons and ions by dust particles with self-consistent electric charge and depends on the ratio between an ion transit time through a sheath and characteristic times of ion collection and scattering [19]. Dust particles can strongly change an ion flux so that the flux is not constant in the sheath. That causes a change of both the well known Bohm's criterion on the sheath boundary and the distributions of other sheath parameters.

In the case of dusty sheaths in non-equilibrium plasmas with two-temperature electrons [27], non-monotonic distributions of the self-consistent electric potential caused by spatial distributions of total charge of dust particles take place in sheaths. The non-monotonic self-consistent electric potential in the front of electrodes and walls with a potential minimum in dusty plasma causes a protection of the electrodes and walls from the intensive ion bombardment because the ions reach the electrodes and walls with a lower energy corresponding to the electrode and walls potentials. The protection is depending plasma parameters including the temperature and density ratios of hot and cold electrons.

In the case of sheaths in oblique magnetic fields [20], dust particles change also boundary conditions and spatial distributions of sheath parameters causing a double structure of sheaths due to an effective collection of ions by dust particles in sheaths. In both cases indicated above, the dust particle charge is non-trivial due to peculiarities of spatial distributions of electrons and ions in dusty sheaths and can even change the sign.

Dust particles influence on parameters of non-stationary dusty RF sheaths [21,22]separating the electrodes from a quasi-neutral central part of RF discharges. The dust particle charge changes non-monotonously across the interelectrode gap and has maximum close to a sheath edge, due a peculiarity of the electron energy distribution function in the quasi-neutral central part of the RF discharge and the spatial distribution of plasma parameters. The secondary electron emission from electrodes and walls can strongly influence on RF discharge parameters however dust charge spatial distributions are invariability [28] due to the conservation of the ratio of the electron and ion currents into a dust particle.

The author thanks JSPS for a support of his stay at the Nagoya University where the paper was begun.

REFERENCES

1. Lieberman M.A., Lichtenberg A.J. *Principles of Plasma Discharges and Material Processing, Wiley, New York, 1994, pp. 154-170, 327-471*.
2. Riemann K.-U. *J. Phys. D: Appl. Phys.*, **24**, 493-518 (1991).
3. Chen F.F. *Introduction to plasma physics, Plenum Press, New York and London, 1975, pp. 243-250*.
4. Self S.A. *Phys. Fluids*, **6**, 1762-1768 (1963).
5. Song S. B., Chang C. S., Choi D. *Phys. Rev.*, **55**, 1213-1216 (1997).
6. Hollenstein Ch. at al., *in Frontiers in Dusty Plasmas*, ELSEVIER, 2000, pp. 169-176.
7. Stoffels E., StoffelsW.W., Swinkels G.H.P.M., Kroesen G.M.W. *in Frontiers in Dusty Plasmas*, ELSEVIER, 2000, pp.177-184.
8. Winter J. *Plasma Phys. Control.Fusion*, **40**,1201-1210 (1998).
9. Sharpe J. P.at al. *PSI-15 Gifu, Japan, May 27 - 31, 2002. Program and Book of Abstracts*, p. P3-46.
10. Boufendi L., Bouchoute A. *Plasma Sources Sci. Technol.*, 3.262-267,(1994).
11. Lin I. at al. *in Frontiers in Dusty Plasmas*,ELSEVIER, 2000, pp.177-184.
12. Chutov Yu., Kravchenko A., Schram P.P.J.M. *Physica B*, **128**,11-20 (1996).
13. Bohm D. *The charactristic of electrical discharges in magnetic fields, MacGrow-Hill, New York, 1949, p. 77*.
14. Takamura S. *Phys. Letters*, **133**, 312-314 (1988).
15. Boswell R.W., Lichtenberg A.J., Vender D. *IEEE Trans. Plasma Sci.*, **20**, 62-65 (1992).
16. Liu D., Wang D., Wang X., Liu J. *Phys. Plasmas*, **8**, 1427-1431 (2001).
17. Mahanta M. K., Goswami K. S. *Phys. Plasmas*, **8**, 665-668 (2001).
18. Cipolla J.W. (Jr.), Silevitch M.B. *J. Plasma Phys.*, **25**, 373-389 (1981).
19. Chutov Yu., Kravchenko O., Schram P., Yakovetsky V. *Physica B*, **262**, 415-420 (1999).
20. Chutov Yu., Kravchenko O., Yakovetsky V. *J. Plasma and Fusion Research. SERIES*, 3, 558-561 (2000).
21. Chutov Yu., Goedheer W., Kravchenko O., Zuz V., Yan M. *J. Plasma and Fusion Research. SERIES*, **4**, 340-344 (2001).
22. Chutov Yu., Goedheer W., Kravchenko O., Zuz V., Yan M. *Materials Science Forum: Plasma Processing and Dusty Particles*, **382**, 69-79 (2001).
23. Birdsall C.K. *IEEE Trans. Plasma Sci.*, **19**, 65-85 (1991).
24. Vahedi V., Surendra M. *Comput. Phys. Commun.*, **87**, 179-198 (1995).
25. Allen J. *Phys. Scr.*, **45**, 497-503 (1992).
26. Trubnikov B.A. *"Particle collisions in full ionized plasmas" in Problems of plasma theory,*, Gosatomizdat, Moscow, 1963, PP.98-182.
27. Asano K., Chutov Yu.I., Kravchenko O.Yu., Ohno N., Pshenychnyj A.F., Smirnov R.D., Takamura S. *(tsis book)*.
28. Chutov Yu.I., Goedheer W.J., Kravchenko O.Yu., Lavrov O.A. *(tsis book)*.

Dynamic Phenomena in Complex Plasmas

N. F. Cramer*, S. V. Vladimirov*, A. A. Samarian* and B. W. James*

*School of Physics, The University of Sydney, N.S.W. 2006, Australia

Abstract. A complex or dusty plasma exhibits many dynamic phenomena associated with single and few-particle motions, lattice collective excitations, phase transitions, and void and cluster formation. Experiments have been performed to investigate these processes, using detailed modelling of the discharge and powerful diagnostic techniques. Observations and theory of self-excited motions of dust grains and their interaction with the background plasma are reported. Such observations can be used to elucidate the basic physics of the dynamical processes coupling the grains to the plasma, and to diagnose the parameters of the dusty plasma.

The study of collective processes involving the dynamics of dust grains in a comcomplex "dusty" plasma is an area of rapidly growing interest. The complex plasma, made of highly (usually negatively) charged particulates of micrometer size, demonstrates a wealth of dynamic phenomena associated with self-organization, phase transitions, lattice collective excitations and the formation of lattice defects, vortices tices, voids, clusters, etc. Interactions of the charged dust grains between themselves as well as with the ambient plasma can cause various instabilities influencing the propagation of plasma waves, and leading to the excitation of dust wave modes and oscillations. A distinctive feature of a dusty plasma is that the particle charge, often extremely large compared to the electron/ion charges, is not fixed and is coupled self-consistently to the parameters of the surrounding plasma. Dynamic phenomena in a complex plasma naturally involve varying dust charges, thus opening even more possibilities for plasma-dust interactions and associated instabilities. Here, recent advances in the study of dynamic phenomena and cooperative particle motions in a complex plasma are presented, including experimental results as well as theoretical modelling and numerical simulations. A natural progression is from the dynamics of single particles, through the dynamics of several-particle clusters, to the dynamics of many interacting particles in crystal-like arrays.

Experiments have been carried out at Sydney with an rf-discharge plasma in a 40 cm inner diameter cylindrical stainless steel vacuum vessel with many diagnostic ports, with the aim of verifying much of the theory described above. Argon plasmas were generated at pressures in the range 0.5–20 Pa by applying a 15 MHz signal to a disk electrode. A compensated single Langmuir probe was used for measurements of the plasma parameters. The typical plasma parameters are $n_e \simeq 0.8 - 10 \times 10^8 \text{cm}^{-3}$, $T_e \simeq 2\,\text{eV}$, and plasma potential $V_p \simeq 40 - 80\,\text{V}$ with respect to the grounded electrode [1].

We can classify the various dynamic phenomena according to the number of particles involved:

Single particles

Neglecting the mutual interactions of particles, the dynamics of individual particles levitating in the sheath region depends on their response to the external forces, which are in turn determined by the behaviour of the plasma in the discharge. The dynamics of single particles therefore provides a useful diagnostic of the parameters of the plasma, provided a good model of the plasma discharge is employed. Such a model should include such effects as ionization of the background gas, and yields profiles of the plasma potential, equilibrium charge and positions of the grain, and characteristic frequency of the grain's oscillations; an example of a fluid model of the discharge is that in Ref. [2]. The detailed interaction of the plasma with a single particle has been modelled with PIC simulations; the charging of the particle, and the transfer of kinetic energy to it, by the flowing ions and electrons in [3], and the creation of ion focussing and plasma wakes in the background plasma in [4]. The equilibrium states of single dust grains have also been used as a diagnostic tool for sheath measurements in the rf-discharge plasmas described here [5].

Few particles

Two or more close particles will be coupled together due to their interaction potential: the electrostatic Debye potential between vertically aligned particles is modified assymmetrically by the existence of a wake field behind the upper particle [6], which may also play a role in the vertical alignment of particles in a Coulomb crystal. This coupling can be shown directly in the laser excitation of a pair of vertically aligned particles [7]. An instability of the particle equilibrium may occur even for two particles, when no cooperative lattice modes are involved. Recent modelling has shown that this phenomenon depends on the vertical and horizontal external potentials, as well as on the asymmetric interaction potential between the grains [8]. Such instabilities may also have a bearing on the self-excitation of many-particle arrays described in the next section. Observations have also been made of the rotation of fine dust clusters in an axial magnetic field [9]. The clusters, of 2 to 12 particles, are observed to rotate as a rigid body in the left-hand direction. The mechanism may be ion drag, but is not well understood. Many particles Particles in the ordered array of the dust-plasma crystal can interact with each other and with the background plasma, and so support a variety of oscillations and waves. The observation of such waves can again serve as a diagnostic, and yield information about the plasma parameters. The vertical vibration modes for a single horizontal line of grains [10], two horizontal lines [11], and a vertical string of grains [12] have been investigated. All these models assume nearest neighbour interactions.

A unique feature of particles interacting in the vertical direction (the direction of flow of ions to the electrode) is that, in addition to the discharge sheath electric field and the gravitational field, the asymmetric wake potential operates between vertically aligned particles.

For a one-layer structure, it is found experimentally that when the pressure is decreased below a critical value, the dust particles begin to oscillate spontaneously in the

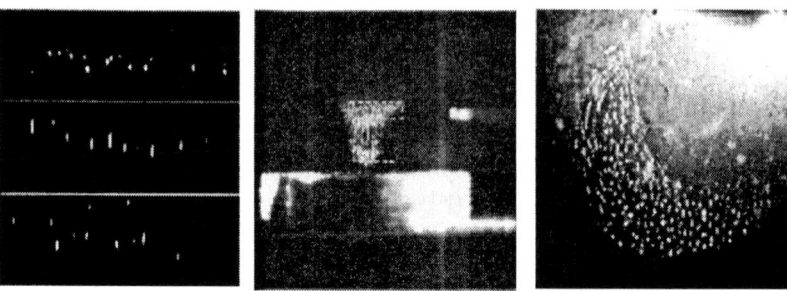

FIGURE 1. Dust motion in an rf-discharge plasma: (1) Self-excited vertical oscillations, input powers 100 W, 35 W, 15 W top to bottom, (2) Longitudinal compressive waves in 3-D structure, (3) Vortex motion in horizontal plane.

vertical direction [1]. The amplitude was several millimetres, and the frequency was greater than 10 Hz. When the rf input power was decreased, the oscillation amplitude was found to increase, with a dramatic increase for pressures below 4.5 Pa. There was similar increase of amplitude when the pressure was reduced. This is consistent with a phase transition from an ordered dust crystal state to a fluid state. The frequency and amplitude of the oscillations are independent of the number of particles for a monolayer structure. When the number of particles reaches hundreds, the particles form several layers, and self-excited oscillations are also observed, although of a more complicated nature.

To excite vertical oscillations the energy input must be sufficient to overcome the damping by friction on the background neutral gas. One possible way for a particle to gain energy is by the delayed charging effect [13]. In this case the energy gain in one oscillation cycle depends on the local electric field, the oscillation amplitude, and the variation in charge during the cycle. A detailed analysis of this energy gain, compared with the frictional energy loss, reveals that for this experiment the pressure should be no greater than 0.4 Pa for excitation to occur. However, we observe oscillations for pressures up to 5 Pa, indicating that different mechanisms will operate; one such possible mechanism is based on the spatial variation of grain charges [14], another on an instability involving a layer of charged particles considered as a virtual wall resonator [15].

Other types of self-excited motions of dust grains have been observed [16], besides the vertical oscillations (Figure 1). Using an additional ring electrode placed on the powered electrode, a 3-D crystal structure was created. On decreasing the input power or the pressure, or by increasing the number of particles in the structure, compressive density waves are observed to be excited, travelling downwards. The individual particles are found to move on an ellipse-like trajectory. Dust vortex motion in the horizontal and vertical planes was also observed, induced by a positively biased pin electrode [17]. Surface waves have also been observed to propagate on the interfaces between a dust cloud and a void. The mechanism of all these self-excited motions may be a dissipative instability due to the spatial variation of grain charges mentioned above. The charge gradient occurs due to a disturbance of quasi-neutrality or electron temperature caused by the extra electrode in the rf discharge.

In conclusion, plasma collective processes have a large influence on the configuration of crystal-like Coulomb structures, as well as the motions and vibrations of colloidal particles in the structures, the formation of clouds and voids, phase transitions and particle diffusion, and waves and vortices in the complex plasma. Various instabilities and self-excited oscillations observed in dust structures have been observed and classified. The existence of the asymmetric wake potential appears to play an important role in the dynamics of few- and many-particle systems. The self-excitation of different types of dust motions may depend on instabilities driven by gradients in the dust and plasma parameters of experimental rf discharges.

REFERENCES

1. Samarian, A.A., James, B.W., Vladimirov, S.V. and Cramer, N.F., *Phys. Rev. E*, **64**, 025402(R) (2001).
2. Vladimirov, S.V. and Cramer, N.F., *Phys. Rev. E*, **62**, 2754 (2000).
3. Vladimirov, S.V., Maiorov, S.A., and Cramer, N.F., *Phys. Rev. E*, **63**, 045401(R) (2001).
4. Maiorov, S.A., Vladimirov, S.V. and Cramer, N.F., *Phys. Rev. E*, **63**, 017401 (2001).
5. Samarian, A.A., and James, B.W., *Phys. Lett. A*, **287**, 125 (2001).
6. Ishihara, O. and Vladimirov, S.V., *Phys. Plasmas*, **4**, 69 (1997).
7. Prior, N.J., Mitchell, L.W., and Samarian, A.A., in preparation.
8. Vladimirov, S.V. and Samarian, A.A., *Phys. Rev. E*, **65**, 046416 (2002).
9. Cheung, F., Samarian, A.A., and James, B.W., *Physica Scripta*, **T98**, 143 (2002).
10. Vladimirov, S.V., Shevchenko, P.V., and Cramer, N.F., *Phys. Rev. E*, **56**, R74 (1997).
11. Vladimirov, S.V., Cramer, N.F., and Shevchenko, P.V., *Phys. Rev. E*, **60**, 7369 (1999).
12. Vladimirov, S.V., and Cramer, N.F., *Physica Scripta*, **58**, 80 (1998).
13. Nunomura, S., Misawa, T., Ohno N., and Takamura, S., *Phys. Rev. Lett.*, **83**, 1970 (1999).
14. Vaulina, O.S., Petrov, O.F., Samarian, A.A., and James, B.W., *Proc. Int. Conf. Plasma Phys.*, Quebec, 2000, p. 376.
15. Tsytovich, V.N., Vladimirov, S.V., and Benkada, S., *Phys. Plasmas*, **6**, 2972 (1999).
16. Samarian, A.A., James, B.W., Vaulina, O.S., Vladimirov, S.V., and Cramer, N.F., *Proc. XXV Int. Goof. Phenomena in Ionized Gases*, Nagoya, 2001, Vol. 1, p. 17.
17. Samarian, A.A., Vaulina, O.S., Tsang, W., and James, B.W., *Physica Scripta*, **T98**, 123 (2002).

Causes of Small Particle Growth in Silane Discharges

G. Bano, K. Rozsa, and A. Gallagher

JILA, National Institute of Technology and University of Colorado, Boulder, CO 80309-0440

Abstract. Silicon-based particles that grow in a silane radio frequency (RF) discharge are studied by light scattering. Average particle size and the corresponding particle density are measured as functions of silane density, RF voltage, and chamber temperature (T). The particle density is almost independent of T, as is the film growth rate. In contrast, the particle-size growth rate (G) decreases by a large factor as T increases from 300—500 ^oK. It is also observed that the few percent of higher silanes that are produced by the discharge cause a major increase in G.

The growth of silicon particles in silane (SiH_4) discharges has long been recognized, and many aspects of their behavior have been reported[1]. However, a full understanding of their growth, density, and escape has not been established, particularly for device production conditions. These particles cause important limitations in plasma production of amorphous silicon (a-Si:H) devices, so a detailed understanding of their causes and behavior, and their influence on the plasma and devices, has major economic significance. We report here observations of the temperature dependence of this particle growth, and evidence that higher silanes play a major role in this particle growth.

Devices are usually produced in capacitively coupled 10–80 MHz RF discharges in 0.1–1 Torr of SiH_4, often accompanied by 1–4 Torr of H_2. The particles that normally occur above the substrate are very small diameter (d_P), and those that escape to the growing devices are < 20 nm diameter [2]. As these small particles yield very small scattering signals, we use ensemble scattering and measure afterglow particle diffusion to establish d_P. The combination of Rayleigh cross-section and gravitational perturbation of free diffusion limits this technique to a $d_P = 6 - 50$ nm range. Scattered intensity then yields density [$n_P(d_P)$], and measurements versus discharge on time (t_{on}) yield the d_P growth rate (G_P). In this strongly attaching gas, $n_e/n_+ \ll 1$ [3], particle charge is much smaller than in inert-gas discharges, and many neutral particles escape from the plasma.

In most studies of particle growth in SiH_4 discharges [1,4-6], gas flow continuously replaces dissociated silane, as in industrial reactors. The gas-drag force quickly caries particles to the downstream end of the quasi-neutral plasma region, where due to their negative charge they are stopped by the sheath field. For typical industrial reactor size and gas-flow speed (v_G), gas dwell time (T_D) is $\sim$ 1 s and these particles may grow to $d_P \sim 30$ nm while traveling to the downstream end, but for experimental reactors typically $T_D \sim 0.1$ s and $d_P < 4$ nm, which is too small to observe with laser scattering. However, at the downstream end of the plasma the trapped particles grow in size due to radical reactions, sometimes accompanied by agglomeration, and more small particles continuously arrive to increase n_p. If v_G is large and the pressure (P) low enough, these particles are eventually dragged away to the pumps, often in clumps that appear to modify and overcome the sheath field. But we find that at lower v_G and higher P these trapped particles are suddenly (< 2 s) spread far upstream into the main plasma, probably due to local modification of the plasma potential caused by the particles. This can cause a delayed but sudden appearance of laser scatter within the main plasma region, as has often been reported [1,4,5]. This particle buildup and sudden spreading usually occurs many seconds after discharge initiation, so it can easily be prevented by occasionally terminating the discharge for ~ 0.1 s, allowing the trapped particles to be dragged beyond the sheath region. Although this allows the smaller particles within the main plasma region at discharge termination to incorporate into the film, the larger particles at the downstream end are dragged harmlessly away.

Although particles trapped at the plasma boundary may not directly influence the amorphous silicon (a-Si:H) film, there is still a problem in trying to use the data from a small experimental reactor to model a large industrial reactor, due to the very dif

ferent T_D. Only if T_D of an experimental reactor is increased can one hope to measure the relevant particle growth and density within the main plasma region (before trapping at the plasma edge), which is what matters most to the growing film or device. One might achieve this by slowing the gas flow, but the discharge power must also be lowered to avoid severe depletion of the silane feed gas. These two changes have not been done in most silane discharge particle studies, and we believe that when severely delayed scattering occurs it is probably the result of the upstream spreading of the trapped particles.

Particle agglomeration is also observed in a variety of silane discharge experiments (e.g., Refs. 4-6). This is accompanied by d_p growth consistent with the observed n_p if one assumes that a large fraction of particles are neutral. However, most of these experiments are in a vapor of $\sim 5\%$ SiH_4 in Ar or He, and the SiH_4 is often considerably depleted. In contrast, we observe 10^2-10^3 times smaller n_p in pure SiH_4 discharges with minor depletion, and agglomeration plays an insignificant role.

What one needs from an experimental reactor, in order to understand and perhaps control particle behavior in a large reactor, is to know the size and density distribution of particles within the main plasma region versus the time fresh gas has been in the

discharge region. This time translates into position or gas dwell time within an industrial reactor. In the experiments reported here we do this by working without gas flow, so that particles remain spread throughout the quasi-neutral plasma region. We start with fresh SiH_4 gas, and measure particles versus discharge operating time (t_{on}). We do not observe severely delayed onsets of particle scattering.

Here we report measurements of the temperature (T) dependence of G_P and n_P. A major increase in the scattering-onset delay with increasing T has been reported [5], indicating that G_P and/or n_p decreases as T increases. However, a T gradient (ΔT) yields thermophoretic forces that increase particle escape and significant ΔT probably occurred in this experiment, so it is not known if the plasma chemistry changes with T.] Here we measure the dependence of G_P and n_P on T, silane density (n_S), and RF voltage with a very small ΔT, thereby isolating the plasma chemistry. We find that there is indeed a decrease in G_P as T increases (at constant n_S), although G_P is equally sensitive to n_S and V_{RF}. Increasing ΔT beyond ~ 5 C/cm further suppresses n_p, through thermophoresis, and shifts the (size dependent) particle location within the quasi-neutral plasma. Measured d_P versus t_{on} is shown in Fig. 1 for $T = 360°K$, and G_P from similar data are shown as a function of T and P in Fig. 2. This data is plotted versus $n_S G_F$ in Fig. 3, since G_P and the film growth rate (G_F) are related as described next.

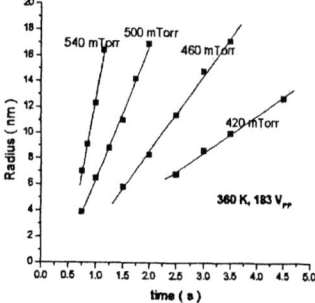

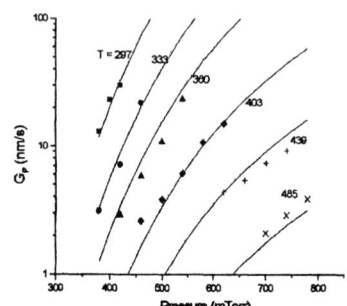

FIGURE 1. Particle radius versus discharge operating time.

FIGURE 2. Particle growth rate versus silane pressure (P) at the temperature (T) indicated. The lines are an analytic fit to this P and T dependence.

If film growth and particle growth are caused by a radical that does not react with the vapor, there is a simple relation between the radical collision rate with particles (R_P) versus film (R_F). The rate of incorporating silicon atoms into particles per collision is $R_P S_P$, where $S_P = \beta_P \varepsilon_P$, S_P is the probability of silicon incorporation per collision, β_P the reaction probability, and ε_P the probability of silicon incorporation per reaction. The rate of incorporating silicon atoms into the film is $R_F \varepsilon_F$, where ε_F is the incorporation probability per film reaction. (A radical that is non-reactive in the vapor reacts almost entirely with film, regardless of how many times it collides with the surface before reaction, so β_F does not influence the film growth rate.) The result is that $G_P/G_F \propto R_P S_P/R_F \varepsilon_F \propto n_S S_P/\varepsilon_F$, so we have plotted G_P versus $n_S G_F$ in Fig. 3. The solid lines correspond to $G_P \propto n_S G_F$, as occurs if both growths are dominated by a radical that does not react significantly with the vapor (SiH$_3$ is the primary expected specie). The data at each T converges to the solid lines at the smaller G_F, at the indicated value of $S_P \varepsilon_F$, but as T increases the S_P/ε_F of the fit decreases rapidly. This demonstrates clearly that there is a major T dependence to S_P/ε_F for SiH$_3$, or any other significant non-reactive radical such as Si2H$_5$. This is quite surprising, since the particles and film should be made of the same material, and the particle charge does not add a significant polarization potential for a radical at its surface. If $\beta_F = \beta_P$ and $\varepsilon_F = \varepsilon_P$, then $S_P/\varepsilon_F = \beta_F$, and measurements show that $\beta_F \cong 0.3$ from $T = 300 - 500°K$ (ε_F is less certain, but found to be ~ 0.3) [7]. The 300°K data in Fig. 3 reasonably matches $S_P/\varepsilon_F \cong 0.3$, but $S_P/\varepsilon_F \ll \beta_F$ for the higher T. This surprise is even more mysterious since G_F is nearly independent of T for constant plasma conditions (n_S and RF voltage). In contrast to this T dependence of G_P, the data (not shown here) indicates that $n_P(d_P)$ does not vary significantly with T.

The points at larger $n_S G_F$ in Fig. 3 rise above the solid lines, suggesting that an additional species (not a radical that is non-reactive in the vapor) contributes to particle growth but not film growth under these conditions. The data in Fig. 4 indicates that higher silanes (HS) produced by the discharge cause the appearance of this additional species.

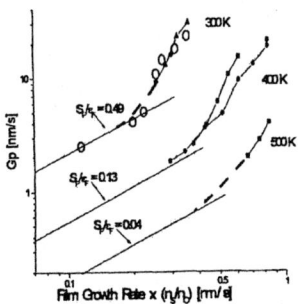 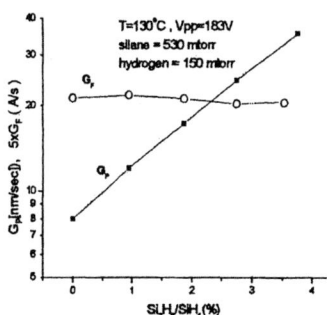

FIGURE 3. Particle growth rate versus film growth rate times normalized silane density, N_s, at three temperatures.

FIGURE 4. Particle ( ) and film (o) growth rates versus disilane pressure, as a percentage of silane density.

To obtain the data in Fig. 4, the discharge was repeatedly operated without introducing fresh silane, allowing HS to buildup in the chamber. (The Si_2H_6 density, measured by mass spectroscopy, was used as the indicator of all HS.) As shown in Ref. 8, HS builds up to $\sim 5\%$ of n_S in a silane discharge, and the data in Fig. 4 shows that this percentage will cause an $\sim 8\times$ increase in G_P but negligible change in G_f. The $\sim 0.3\,s$ delayed onset of the final pa
rticle growth rate, evident in Fig. 1, is also attributed to the delayed buildup of HS and thereby to this additional species. We have no direct evidence regarding the character of this specie, but it is interesting that its contribution to G_P at elevated T goes down in nearly the same manner as the low-GF data (solid lines) that is primarily due to SiH_3. This indicates that the decrease in S_P as T increases is a property of the particle surface and that the additional species is not an ion or small particle for which $S_P \sim 1$ is expected.

In summary, we have measured the temperature and higher silane dependence of particle growth in silane discharges. We observe that $n_P(d_P)$ is almost independent of T, but G_P decreases rapidly as T increases. We also observe that only a few percent of HS in the vapor have a very major effect on G_P, but no effect on G_F.

ACKNOWLEDGMENTS

This work was supported in part by the National Renewable Energy Laboratory.

REFERENCES

1. J. Perrin and Ch. Hollenstein, "Sources and Growth of Particles" in Dusty Plasmas, edited by A. Bouchoule, Wiley, Chichester, UK, 1999.
2. D. M. Tanenbaum, A. L. Laracuente, and A. Gallagher, *Appl. Phys. Lett.* **68**, 1705 (1996).

3. C. Courteille, Ch. Hollenstein, J.-L. Dorier, P. Gay, W. Schwarzenbach, E. Betran, G. Viera, R. Martens, and A. Macarico, *J. Appl. Phys.* **80** 2069 (1996).
4. L. Boufendi and A. Bouchoule, *Plasma Sources Sci. Technol.* **3**, 262 (1994); L. Boufendi, J. Hermann, A. Bouchoule, B. Dubrueil, E. Stoffels, W.W. Stoffels, and M.L. Giorgi, *J. Appl. Phys.* **76**, 148 (1994).
5. M. Shiratani, H. Kawasaki, T. Fukuzawa, T. Yoshioka, Y. Ueda, S. Singh, and Y. Watanabe,*J. Appl. Phys.* **79**, 104 (1996).
6. J. Perrin, M. Shiritani, P. Kae-Nune, H. Videlot, J. Jolly, and J. Guillon, *J. Vac. Sci. Technol.* A **16**, 278 (1998).
7. J.R. Doyle, D. A. Doughty, and A. Gallagher, *J. Appl. Phys.* **68**, 4375 (1990).

Phase Transition in Dusty Plasmas: A Microphysical Description

Gurudas Ganguli*, Glenn Joyce* and Martin Lampe*

*Plasma Physics Division, Naval Research Laboratory, Washington, DC 20375-5346, USA

Abstract. We report on analytical and simulation studies of microphysical processes that trigger phase transitions in a dusty plasma subject to ion streaming. For pressures below the critical pressure P_c for condensation, the grains acquire a large random kinetic energy and form a weakly coupled fluid. If P is increased to greater than P_c, the grains lose their kinetic energy and reach a strongly coupled crystalline state. The dust heating in the fluid phase is due to an ion-dust two-stream instability, which is stabilized at $P > P_c$ by the combined effect of ion-neutral and dust-neutral collisions. If one starts from the crystalline state and decreases the pressure to below the critical pressure P_m for melting, transverse phonons are destabilized by ion streaming, which destroys the short range ordering of the dust grains and triggers melting. It is found that $P_m < P_c$. For $P_m < P < P_c$ mixed phase states can exist.

INTRODUCTION

Dust grains immersed in plasma discharges acquire a large negative charge and settle into a dust cloud at the edge of the sheath. In this region, the plasma ions stream toward the electrode at a velocity $\mathbf{u} \sim c_s = (T_e \, m_i/2$. It is found experimentally that at sufficiently high gas pressure P, the random kinetic energy of the grains is damped by gas friction, and the grains are strongly coupled and self-organize into a crystalline configuration [1-3]. For lower pressures a very different behavior is seen. Despite the dissipation of grain kinetic energy to gas friction, the dust grains reach a steady-state kinetic temperature T_d which is much larger than the temperature of any other component in the plasma and T_d is so large that the dust acts like a fluid [1-3].

We have used the dynamically shielded dust (DSD) model [4] to simulate these physical processes. We fmd that the known experimental features are nicely reproduced in the simulations, that additional features are revealed, and that the simulations are particularly suitable for critically examining the phenomena and developing physics-based models. In both the experiments and the simulations, the grains typically line up directly behind each other along the streaming direction, but transverse to the streaming form a hexagonal lattice [1-3]. It has been understood for some time that this is because the ion flow creates an electrostatic wake downstream of each grain, and there are positively charged points in the wake structure behind each grain which attract other grains. However, there are other features that have not been understood. For example, there has been no satisfactory explanation of the grain heating mechanism that leads to such a large T_d at low pressure. In addition, the simulations reveal features of the evolution that have not previously been reported in the experimental literature. In Fig. 1

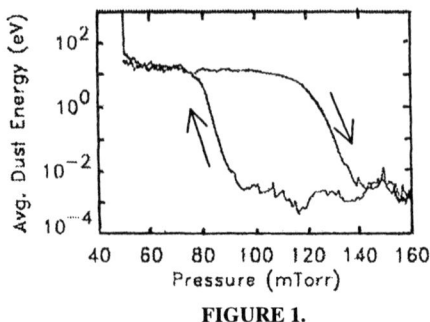

FIGURE 1.

we plot the variation of T_d as P is continuously varied in a DSD code run. A marked difference is evident between the critical pressure P_m for the solid-to-fluid melting) transition, which occurs as P is decreased, and the critical pressure P_c for the inverse fluid-to-solid (condensation or freezing) transition which occurs as P is increased. For $P_m < P < P_c$, mixed phase states are seen. As we shall see, this hysteresis occurs because the instability which triggers melting is different from the instability which heats the dust in the fluid phase, and thereby inhibits freezing.

FLUID-TO-SOLID TRANSITION (CONDENSATION)

At low pressure, the dust is subject to a two-stream instability with the ions. In situations of interest, it is easily seen that the phase velocity of the instability does not overlap either the ion or the dust velocity distribution, and therefore both species can be treated as cold. The dispersion relation simplifies to [4],

$$1 + \frac{1}{(k\lambda_{De})^2} - \frac{\omega_d^2}{\omega(\omega + i\nu_d)} - \frac{\omega_i^2}{(\omega - \mathbf{k}\cdot\mathbf{u})(\omega - \mathbf{k}\cdot\mathbf{u} + i\nu_i)} = 0 \qquad (1)$$

where ω_i and ω_d are ion and dust plasma frequencies, λ_{De} is the electron Debye length, and ν_d and ν_i are dust- and ion-neutral collision frequencies. This instability is responsible for the high temperature of the dust at low pressure. Although the nonlinear evolution of the instability is not yet understood in detail, some features are clear. Because of collisions, the instability is convective in the dust frame. Therefore, the extent of dust heating increases with the dust cloud thickness. However, even thin dust clouds exhibit sufficient dust heating to maintain the dust in a fluid state. Even for dust homogeneously filling an infmite region, the instability saturates before reaching an amplitude sufficient for trapping. T_d ranges typically from several tens of eV in thin dust clouds to hundreds of eV in thick clouds.

We have shown [4] that at pressures above a critical pressure P_c, the instability is stabilized by the combined effects of ion-neutral and dust-neutral collisions. Numerical solutions of Eq. 1), Fig. 1 of Ref [4], show that all modes are stable if both ν_d/ω_d

and v_i/ω_i are large enough. Since the ion-dust streaming instability is the heat source which keeps the dust in the weakly-coupled fluid phase, its elimination is the trigger for condensation. It is found that the critical pressure P_c determined from the solution of Eq. 1) agrees quantitatively with the pressure at which simulations indicate the threshold for excitation of waves, and also with the pressure at which simulations show the fluid-to-solid phase transition. The results also agree with experimental determinations of the freezing pressure, to the qualitative extent compatible with uncertainties in the local experimental parameters. For further details we refer the reader to Ref [4].

SOLID-TO-FLUID TRANSITION (MELTING)

The basic physics underlying the melting transition has been elucidated in a series of papers by Melzer, Piel, V. and A. Schweigert, and collaborators [2,5], and by Melandso [6]. In essence, instabilities of the crystal structure arise from the asymmetry of the dynamically screened Coulomb potential in the presence of flowing ions: upstream and to the side, the potential is similar to static Debye screening, and repels neighboring dust grains, but downstream there is an ion wake, which includes some points of positive charge accumulation. Neighboring dust grains are attracted to these points. While this feature is responsible for the ordering of the dust into crystals with grains lined up directly behind other grains, it also renders the resulting crystal susceptible to instabilities, essentially a form of ion streaming instability specific to the crystal phase.

In Refs [2,5], a phenomenological model for the asymmetric interaction between grains is constructed. The ion wake is represented as a single fictitious positively charged particle, rigidly attached behind each negative grain. The charge on the fictitious particle, and its separation from the grain, are fit to simulations of ion flow. Although this model is a reasonable heuristic representation of the actual physics, it does not provide analytic scalings for the dependence of the forces on physical parameters. Additionally, the two-parameter fit is sufficient to treat small excursions of the grains from their equilibrium crystal sites, but it is not apparent that it accurately represents the forces at large deviations from crystalline order.

We have been developing a first-principles analytic approach to the melting transition, which embodies the same physics that is present in the DSD simulation code. In addition, we are in the process of extending the work of Melzer *et al*, which concentrates on two-layer crystals, to many-layer crystals. In our work, the intergrain potential is the dynamically shielded Coulomb interaction, given in k-space by

$$\phi(k) = \sum \frac{z_i e}{2\pi^2 k^2 D(k, -\mathbf{k}\cdot\mathbf{u} + iv_i)}, \quad (2)$$

where $D(k,\omega)$ is the linear plasma response function defined in Refs [4,7]. Crystal instabilities can be represented at various levels of detail. Both longitudinal and shear modes can be driven unstable, as well as obliquely propagating mixed modes, and indeed we believe that several types of modes are involved in the later stages of the melting process. However, experiments and DSD simulations indicate that the mode which initiates the melting process is a shear mode with propagation vector **k** parallel to the ion streaming, i.e. a mode in which the horizontal grain layers slide rigidly resulting in a misalignment with respect to the adjoining layers. In the linear stage of such a mode, the equation of motion of a grain in the j layer, due to the forces exerted by the grains directly above and directly below, is

$$m\ddot{x}_j = C^+(\Delta x_{j+1} - \Delta x_j) + C^-(\Delta x_{j-1} - \Delta x_j) - mv_d \dot{x} \qquad (3)$$

where

$$C^{\pm} = \left[\frac{\partial^2 Ze\phi(\Delta x, z = \pm d)}{\partial(\Delta x)^2}\right]_{\Delta x = 0}, \qquad (4)$$

$\phi(x,z)$ is the Fourier transform of Eq. (2), d is the separation between grain layers, and is the displacement of the grain from its equilibrium position in the x direction (transverse to the propagation direction z). It is straightforward to include next nearest-neighbors in the same way. But the key point is that, unlike the situation in ordinary crystals, the force constants $C^{\pm}$ are such that $C^+ \neq C^-$. For a transverse mode in an infinite crystal, $\Delta x_j \sin\exp[i(jkd - \omega t)]$, Eq. (3) leads to a dispersion relation

$$\omega = -\frac{iv_d}{2}\left\{1 \pm \sqrt{C^{\pm} = 1 - \frac{4}{mv_d^2}\left[(C^+ + C^-)(1 - \cos kd) - i(C^+ - C^-)\sin kd\right]}\right\}. \qquad (5)$$

Similar analyses can be performed for longitudinal phonons, and for crystals with free surfaces. Equation (5) is similar to the dispersion relation for phonons in an ordinary crystal, except that here instability occurs because $C^+ \neq C^-$. However the instability can be stabilized at high pressure due to the dust collisionality v_d, which is visible explicitly in Eq. (5), and/or the ion collisionality v_i, which appears as damping of the attractive wake force C^- through the dielectric D_{in} Eq. (2). The phonon streaming instability is in this sense similar to the two-stream instability that occurs in the fluid-dust phase. However, the stabilization pressure P_m for the phonon instability (i.e. the critical pressure for melting) is lower than the critical pressure P_c for Buneman instability (i.e. for freezing), as indicated by the DSD code results. Hence, there is a range of pressures where both the solid and fluid phases of the dust are stable, which allows a mixed-phase system to exist, as observed [2]. A detailed comparison of these results with simulation and experiments will be the subject of a future article.

ACKNOWLEDGMENTS

This work was supported by NASA and ONR.

REFERENCES

1. Melzer, A., *et al*, *Phys. Rev.*, **E54**, R46, 1996; *Phys. Rev. E*, **53**, 2757 (1996); *Phys. Lett. A*, **191**, 301 (1994); *Phys. Rev. E*, **54**, R46, 1996; *Phys. Rev. E*, **53**, 2757 (1996); *Phys. Lett. A*, **191**, 301 (1994)
2. Thomas, H,. et al, *Phys. Rev. Lett.*, **73**, 652 (1994); *Nature (London)*, **379**, 806 (1996); *J. Vac. Sci. Technol. A*, **14**, 501 (1996); *Phys. Rev. Lett.*, **73**, 652 (1994).

3. Chu, J. H. and Lin I, *Phys. Rev. Lett.*, **72**, 4009 (1994).
4. Joyce, G., Lampe, M., and Ganguli, G., *Phys. Rev. Lett.*, **88**, 095006-1,2002
5. Schweigert, V. A., et al., *Phys. Rev. E*, **54**, 4155 (1996); *Phys. Rev. Lett.*, **80**, 5345 (1998).
6. Melandsø, F., *Phys. Rev. E*, **55**, 7495 (1997).
7. Lampe, M., Joyce, G., and Ganguli, G., *Phys. Script*, **T89**, 106 (2001); *IEEE Trans. Plasma Science*, **29**, 238 (2001); *Proceedings ISSS*, 2002 (to appear).

Aerosol phenomena in plasma

Alexander Ignatov

General Physics Institute, 38 Vavilova str., 119991, Moscow, Russia

Abstract. Various forces acting upon a charged dust grain in a plasma with dissipative flows are discussed. In particular, it is shown that the thermophoretic force is directed against the ion heat flow.

INTRODUCTION

Although the dusty plasma physics studies aerosols in plasmas, the main attention until now was paid solely to plasma effects. A dozen papers at the most addressing aerosol phenomena in plasmas were published during the recent decade, and many of them leave out the presence of plasma itself.

Various phenomena in neutral aerosols are largely due to the energy and mass exchange between a particulate and an ambient medium. Even a small amount of ionized matter in a gas can drastically increase heat and mass flows at a grain surface. There are two reasons for this. First, nearly each ion hitting the grain recombinates and releases considerable energy that results in heating the grain surface. The amount of released heat may be estimated as $\varepsilon = E_i - E_{wf}$, where E_i is the ionization energy and E_{wf} is the electron work function, that is, $\varepsilon > 10$ eV under typical conditions. Second, due to the large negative charge a grain collects ions from a large area that also increases the net ion flow at its surface. Therefore, even with the small ionization level, the plasma provides the heat flow towards the grain surface that may exceed the heat flow endowed by the neutral gas.

Heating of grain surface and, particularly, equilibrium temperature of a grain were addressed in several papers [1]. If the temperature of a grain is below the temperature of an ambient gas, then converging heat flow is formed that may result in effective attraction of grains due to the thermophoretic force [2]. Seemingly, for the first time the problem of thermophoresis in a plasma was brought up in [3], later a more systematic theory applicable to large grains was developed [4].

In the present paper, the energy exchange between slightly anisotropic plasma and a dust grain as long as various forces arising due to the grain heating are discussed. The plasma anisotropy is provided by dissipative flows, which are formed by the large-scale processes. Two particular examples are discussed: the current-carrying plasma and the plasma with a heat flow. The size of a grain, a, is small: $a \ll \lambda_D \ll \lambda_0$, where λ_D is the electron Debye length and λ_0 is the ion-neutral collision mean free path.

HEAT BALANCE

Let a spherical grain of radius a be situated at the origin. In the vicinity of the grain plasma is described by the Vlasov equation supplemented with two boundary conditions. First, the plasma recombinates at the grain surface, and there is no plasma particles leaving the grain: $f_\alpha(\mathbf{p},\mathbf{r})|_{r=a, p_r>0} = 0$ ($\alpha = e, i$). Second, distribution functions at infinity tend to perturbed Maxwellian distributions:

$$f_\alpha(\mathbf{p},\mathbf{r})|_{r\to\infty} = F_{M\alpha}(E_\alpha)[1 + p_z g_\alpha(E_\alpha)], \tag{1}$$

where $E_\alpha = p^2/2m_\alpha$, and it assumed that a dissipative flow is aligned along the z axis, $p_z = p\cos\theta$. The deviation from Maxwellian distribution is given by

$$g_\alpha(E_\alpha) = \begin{cases} \frac{Q_\alpha}{n_\alpha T_\alpha^2}, & \text{convective flow} \\ \frac{Q_\alpha}{n_\alpha T_\alpha^2}\left(\frac{2E_\alpha}{5T_\alpha} - 1\right), & \text{heat flow,} \end{cases} \tag{2}$$

where Q_α is the magnitude of an energy flow. In particular, if velocity of a convective flow is w_α, then $Q_\alpha = n_\alpha T_\alpha w_\alpha$.

To simplify the problem, the self-consistent potential in is replaced by the Coulomb potential, i.e., $\varphi(\mathbf{r}) = \varphi_0 a/r$. Then the Vlasov equation may be explicitly solved that eventually results in distribution functions of the plasma species at the grain surface, which are too cumbersome to be written here. Of major importance for our purposes is the ion energy flow at the grain. Assuming that the energy released by an ion hitting the grain is $\varepsilon + p^2/2M$, the heat flow is given by

$$q_\alpha^{(p)}(\theta) = -n_i v_{Ti}(\varepsilon\chi + T_i\chi^2) + \gamma Q_i \cos\theta \tag{3}$$

where θ is the spherical angle of a point at the grain surface, $v_{T\alpha} = \sqrt{T_\alpha/2\pi m_\alpha}$, and $\chi = |e\varphi_0|/T_i$ is the normalized grain potential. In what follows it is taken into account that χ is usually rather large, $\chi \sim 10^2 - 10^3$. The nondimensional coefficient γ depends essentially on the kind of the dissipative flow in the bulk of the plasma. For the case of a convective flow

$$\gamma = \gamma^{(c)} = \frac{16}{15}\sqrt{\frac{\chi}{\pi}}\left(\frac{\varepsilon}{T_i} + \chi\right), \text{ while a heat flow yields } \gamma = \gamma^{(h)} = -\frac{1}{5}\gamma^{(c)}. \tag{4}$$

It should be stressed that different kinds of dissipative flows in the bulk of the plasma result in coefficients, γ, of different signs in Eq. (4). In other words, with the heat flow propagating from top downward, the lower side of the grain is hotter. This is pictorialized in the figure, where the dipole part of the net energy flow in the vicinity of a grain is shown. In a sense, electric field of a grain affects the heat flow like a focusing lens.

The net energy balance at the grain surface includes also the heat exchange with neutrals and radiation cooling. The neutral heat flow, $q^{(n)}(\theta)$, is obtained assuming the free-molecular regime and the complete energy accomodation. The radiation cooling is modeled by $q^{(r)}(\theta) = \sigma T_s(\theta)^4$, where $T_s(\theta)$ is the grain surface temperature and the value of σ generally depends on grain optical properties as long as on its size.

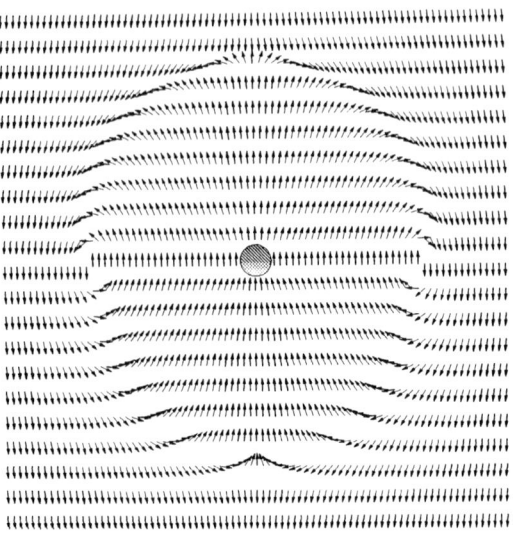

FIGURE 1. Heat flow distribution around a grain.

SURFACE TEMPERATURE

In order to determine the grain surface temperature we have to solve the heat conduction equation inside the grain, $\Delta T(\mathbf{r}) = 0$, supplemented with boundary conditions $T(\mathbf{r})|_{r=a} = T_s(\theta)$ and $-\kappa \frac{\partial T}{\partial r}\big|_{r=a} = q^{(p)}(\theta) + q^{(n)}(\theta) + q^{(r)}(\theta)$, where κ is the heat conductivity of the grain material. This equation is readily solved expanding $T(\mathbf{r})$ in powers of $\cos\theta$ that results in the grain surface temperature, $T_s(\theta) = T_{s0}(1 - \zeta\cos\theta)$. The average surface temperature, T_{s0}, which was studied in details in [1], is obtained by equating the net heat flow to zero. The uneven grain heating is characterized by the nondimensional coefficient, ζ, looking like

$$\zeta = \frac{\gamma Q_i}{p + \kappa T_{s0}/a}, \text{ where } p = 2T_{s0}(n_n v_{Tn} + n_i v_{Ti}\chi) + 4\sigma T_{s0}^4. \tag{5}$$

The structure of this expression is typical for the problem of heat exchange in aerosols. The numerator corresponds to the uneven external heating, while two terms in the denominator describe smoothing the temperature distribution due to the heat conduction ($\kappa T_{s0}/a$) and due to the cooling by the ambient medium. Usually, the ratio $\eta = pa/T_{s0} \ll 1$, however, with sufficiently small κ or large a both smoothing factors may be comparable. Numerical estimations show that the value of ζ varies from 10^{-4} to 10^{-2}, that is, the temperature drop over the grain surface may be up to several degrees.

FORCE

The force acting upon a grain is the net momentum flux integrated over the grain surface. It is conveniently represented as a sum of two terms, $F = F^{(rm)} + F^{(d)}$, where

$$F^{(rm)} = \zeta \frac{\pi a^2}{3} \sqrt{T_n T_{s0}} \left[n_n + n_i \chi \sqrt{\frac{T_i}{T_n}} \right], \tag{6}$$

and

$$F^{(d)} = \lambda \frac{\pi a^2}{3} \frac{Q_i \chi^{3/2}}{v_{Ti} \sqrt{\pi}}. \tag{7}$$

The force given by Eq. (6) is the radiometric force provided by neutral atoms bombarding the unevenly heated grain. The correction to the neutral density, *i.e.*, the second term in the brackets, arises due to the recombinated ions. The small anisotropic correction to the bulk distribution (1) results in the second force (7). For the case of the bulk convective flow this is the drag force, and the coefficient $\lambda = 32/15$. If the bulk distribution is distorted by the heat flow (2), then Eq. (7) corresponds to the thermophoretic force with the negative coefficient $\lambda = -64/75$. The relation between the two forces (6,7) depends on numerous parameters. Qualitatively, if $\eta \ll 1$, then $F^{(rm)} < F^{(d)}$, otherwise $F^{(rm)} \geq F^{(d)}$.

Finally, it should be pointed out that both radiometric and thermophoretic forces are directed against the plasma heat flow. On the other hand, the thermophoresis in a neutral gas drives a grain parallel to the heat flow. Under the conditions of the gas discharge, the relation between heat flows in the ion component and in the neutral component is determined by numerous factors that are beyond the scope of the present paper. It is fairly probable that the reverse thermophoresis discussed above is relevant in a plasma with the sufficiently high level of ionization, *e.g.*, in the near-wall regions of large tokamaks.

ACKNOWLEDGMENTS

This work was supported by NWO (project 02-02-16439) and RFBR (project 047-008-013).

REFERENCES

1. Daugherty, J.E., Graves, D.B. *J. Vac. Sci. Technol. A*, **11**, 1126 (1993); Graves, D.B., Daugherty, J.E., Kilgore, M.D., Porteous, R.K. *Plasma Sources Sci. Technol.* **3**, 433 (1994); Swinkels, G.H.P.M., Kersten, H., Deutch, H., Kroesen, G.M.W. *J. Appl. Phys.*, **88**, 1747 (2000); Kozyrev, A.V., Shishkov, A.N., *JTP Lett.*, **28**, #12 (2002).
2. Brattli, A., Havnes, O. *J. Vac. Sci. Technol. A* **14** 644 (1996); Tsytovich V.N.,Khodataev Ya.K., Morfill G.E., Bingham R., Winter D.J. *Comm. Plasma Phys.*, **18**, 281 (1998); Fortov, V.E., Nefedov, A.P., Petrov, O.F. *et al.,JETP,* **116**, 1601 (1999)
3. Brock, J.R., *Nature*, **207**, 69 (1964)
4. Chen X. *Plasma Chem. and Plasma Proc.*, **19**, 33 (1999)

Ion Trapping within the Dust Grain Plasma Sheath

D. Jovanović* and P.K. Shukla[+]

*Institute of Physics, P. O. Box 57, Yu-11001 Belgrade, Yugoslavia
[+]Institut für Theoretische Physik IV, Ruhr-Universität Bochum, D-44780 Bochum, Germany

Abstract. One of the most important and still unresolved problems in the physics of dusty plasmas is the determination of the dust charge. The grains are not directly accessible to measurements and it is necessary to have a reliable theoretical model of the electron and ion dynamics inside the Debye sphere for the interpretation of the relevant experimental data, which include also the effects of the surrounding electron and ion clouds. Recent computer simulations [6] and laboratory experiments [9] indicate that the plasma sheath is dominated by trapped ions, orbiting the grain on closed trajectories at distances smaller than the Debye radius, that cannot be accounted for by the classical theories. We present the first analytical, fully self-consistent, calculations of the electrostatic shielding of a charged dust grain in a collisional plasma. In the regime when the mean free path for the ion-dust collisions is larger than that for the ion-neutral collisions, we solve the kinetic equation for the ions, coupled with Boltzmann distributed electrons and Poisson's equation. The ion velocity distribution function, in the form of a spherically symmetric ion hole, is found to be anisotropic in the presence of charge-exchange collisions. The number of trapped ions and their spatial distribution are determined from the interplay between the collective plasma interaction and the collisional trapping/de-trapping. The stationary state results from the self-tuning of the trapped ion density by the feedback based on the nonlocality of the collisional integral, and on the ion mixing in the radial direction along elongated orbits. Our results confirm the existence of a strong Debye shielding of the dust charge, allowing also the over-population of the trapped ion distribution (ion hump).

Systematic measurement of the dust charge started only recently [1]. As the grains are not directly accessible, the data include the effects of the electron and ion clouds, and their interpretation depends on the theoretical model of the electron and ion dynamics inside the Debye sphere. The standard model for the dynamics close to the grain is the orbit-motion-limited (OML) theory [2, 3, 4]. It regards the steady state in which collisionless electrons and ions move along ballistic trajectories, determined by the electrostatic potential of the space charge, including the dust and the surrounding electrons and ions. A major shortcoming of the OML theory comes from neglecting the low energy ions orbiting the grain on closed trajectories, that emerge due to energy loss in collisions [2]. Recently, a simple analytical model for the effects of trapped ions on the dust charging was proposed [5], and the self-consistent Monte-Carlo simulations [6] showed that the OML is inappropriate if the ion mean free path is much larger than the Debye radius. Consider an unmagnetized, spatially homogeneous, dusty plasma consisting of electrons, ions, spherical dust particulates and neutral atoms (denoted by the subscripts e, i, d and n) whose charges are $-e$, $z_i e$, $-z_d e$, and 0, respectively. The dust radius r_d is much smaller than the Debye radius, $r_d \ll \lambda_D$, the plasma is weakly ionized, $n_n \gg n_i^{(0)} \sim N_e^{(0)} \sim z_d n_d$ and weakly collisional, $\lambda_{mfp}^{(n)} \gg \lambda_D$ is the mean free path for ion-neutral collisions), and for small grains the ion-dust collisions are less frequent than ion-neutral. For example, in a typical hot electron glow discharge [$T_e \geq 10$ eV, weakly ionized plasma $n_n/N_i^{(0)} \sim 10^3 - 10^4$, dust potential $e\phi(r_d)/T_e \sim 2.5$, sub-micron grains of carbon, $r_d \sim 10^{-7}$ m] the ion dust collision mean free path is $\sim 10^3$ times larger than that for the ion-neutral collisions. The ions are described by the kinetic equation

$$\frac{\partial f_i}{\partial t} + \vec{v} \cdot \nabla f_i - \frac{q_i}{m_i} \nabla \phi \cdot \frac{\partial f_i}{\partial \vec{v}} = \left(\frac{\partial f_i}{\partial t}\right)_{\text{coll}}. \quad (1)$$

We assume that they are subjected only to charge-exchange collisions. The main contribution to the collision integral $(\partial f_i/\partial t)_{\text{coll}}$ comes from the asymmetric events in which i) an energetic (free) ion exchanges its charge with a slow neutral and disappears, while the newly born ion is trapped in the potential minimum (trapping), or ii) a trapped ion collides with a fast neutral, and the newly born ion escapes from the well (detrapping). The collision integral may be written as the sum of the sources and sinks for the free and trapped ions denoted by the superscripts F and T, respectively,

$$(\partial f_i^F(\vec{v})/\partial t)_{\text{coll}} = \int d^3\vec{v}' \sigma(|\vec{v}-\vec{v}'|) [f_i^T(\vec{v}')|\vec{v}|f_n(\vec{v}) - f_i^F(\vec{v}')|\vec{v}'|f_n(\vec{v}')] \quad (2)$$

$$(\partial f_i^T(\vec{v})/\partial t)_{\text{coll}} = \int d^3\vec{v}' \sigma(|\vec{v}-\vec{v}'|) [f_i^F(\vec{v}')|\vec{v}|f_n(\vec{v}) - f_i^T(\vec{v}')|\vec{v}'|f_n(\vec{v}')] \quad (3)$$

We adopt $\sigma = $ const. In a weakly ionized plasma, the distribution function of the neutrals, f_n, is Maxwellian in $f_n(\vec{v} = n_n/(2\pi v_{T,i}^2)^{3/2} \exp(-v^2/2v_{T,i}^2)$. For weak collisions, $(\partial f_i/\partial t)_{\text{coll}} \ll \vec{v} \cdot \nabla f_i$. we expand the ion distribution function as $f_i = f_{i,0} + \delta f_i$. The leading component $f_{i,0}$ and the perturbation δf_i; are found from,

$$\vec{v} \cdot \nabla f_{i,0} - \frac{z_i e}{m_i} \nabla \phi_0 \cdot \frac{\partial f_{i,0}}{\partial \vec{v}} = 0, \quad (4)$$

$$\vec{v} \cdot \nabla \delta f_i - \frac{z_i e}{m_i} \left(\nabla \phi_0 \cdot \frac{\partial \delta f_{i,0}}{\partial \vec{v}} + \nabla \delta_0 \cdot \frac{\partial f_{i,0}}{\partial \vec{v}}\right) = \left(\frac{\partial f_{i,0}}{\partial t}\right)_{\text{coll}} \quad (5)$$

where the potentials ϕ_0 and $\delta\phi$ are calculated self-consistently from Poisson's equation.

As the grains are strongly shielded by orbiting ions, we regard them as mutually isolated. Then, the potential ϕ is spherically symmetric, the Vlasov equation (4) has two conserved quantities, the energy $W_o = m_i v^2/2v_{T,i}^2 + ez_i\phi_0(r)/T_i$ and the angular momentum $\vec{l} = \vec{v} \times \vec{r}/v_{T,i}\lambda_D$, and $f_{i,0}$ can be expressed as a function of W_0 and $\vec{L}$.

The free and trapped particles correspond to $W_0 > 0$ and $W_0 < 0$, respectively. The angular momentum effects on the free ions reduce the shielding by $\sim 10^{-2}(T_e/T_i)(r_d/\lambda_D)$, [4], and the deviation from the OML is negligible if $(T_e/T_i)(r_d/\lambda_D) \ll 30$. For $r_d \to 0$ the OML result reduces to the Boltzmann distribution, and we adopt for the free particles

$$f_{i,0}^F(\vec{v}) = n_i^0 (2\pi v_{T,i}^2)^{-3/2} e^{-W_0} \quad \text{for} \quad W_0 > 0. \quad (6)$$

The trapped particle density is sharply peaked close to the grain. Thus, their creation and destruction rates, Eq. (3), are also strongly spatially dependent, and the trapped ion population on elongated orbits is different from that on circular orbits with the same energy, which we model by multiplying Schamel's distribution [7] by a form factor the angle between r and v, we integrate Eqs. (6) and (7) and obtain the leading-order densities of the free and trapped ions in the form

$$f_{i,0}^T(\vec{v}) = n_i^0 (2\pi v_{T,i}^2)^{-3/2} e^{-\beta W_0} \exp(\gamma L \sqrt{-W_0}) \quad \text{for} \quad W_0 < 0. \tag{7}$$

Using $\int d^3\vec{v} = 4\pi \int_0^\infty v^2 dv \int_0^{\pi/2} \sin\theta d\theta$ and $L \equiv |\vec{L}| = (vr/V_{T,i}\lambda_D)\sin\theta$, where θ is the angle between $\vec{r}$ and $\vec{v}$, we integrate equations (6) and (7) and obtain the leading-order densities of the free and trapped ions in the form

$$n_{i,0}^F = n_i^0 e^{-W_m} \left[1 + 2e^{W_m} \sqrt{-W_m/\pi} - \text{erf}(\sqrt{W_m}) \right], \tag{8}$$

$$n_{i,0}^F = \frac{n_i^0 e^{-\beta W_m}}{2\beta^{3/2}} \pi \left[I_1(-\xi) - L_{-1}(\xi) \right] \left[2e^{\beta W_m} \sqrt{-\beta W_m/\pi} - \text{erf}(\sqrt{\beta W_m}) \right]. \tag{9}$$

Here $W - m(= z_i e\phi_0(r)/T_i)$, $\text{erf}(\xi) = (2/\sqrt{\pi}) \int_0^\xi \exp(-t^2) dt$, $\xi = \gamma r W_m / \sqrt{2}\lambda_D$ and I_1 and L_{-1} are the modified Bessel and Struve functions of the order 1 and -1. Using (8), (9) and Boltzmannean electrons, $n_{e,0} = n_e^{(O)} \exp[-e\phi_0(r)/T_e]$, Poisson's equation yields

$$\frac{\varepsilon_0}{e} \left(\frac{\partial^2}{\partial r^2} + \frac{2}{r} \frac{\partial}{\partial r} \right) \phi_0(r) = z_d n_d + n_{e,0} - z_i (n_{i,0}^F + n_{i,0}^T). \tag{10}$$

We proceed by deriving an analogous equation for the potential δ_0. For δf_i that is vanishing at $r \to \infty$, the kinetic equation (5) is integrated along the characteristics as

$$\delta f_i = z_i e \frac{\delta\phi}{T_i} \frac{\partial f_{i,0}}{\partial W_0} - \int_r^\infty \frac{dr'}{V_r(r')} \left(\frac{\partial f_{i,0}}{\partial t} \right)_{\text{coll}}, \tag{11}$$

where $V_r = \vec{r} \cdot \vec{r}/r$ is the radial ion velocity. Integration is performed by expressing the kernel of the integral as a function of r', W_0 and $\vec{L}$. Poisson's equation for $\delta\phi$ is then

$$\frac{\varepsilon_0}{e} \left(\frac{\partial^2}{\partial r^2} + \frac{2}{r} \frac{\partial}{\partial r} \right) \delta\phi_0(r) = \frac{en_{e,0}}{T_e} \delta\phi(r) - z_i 4\pi \int_0^\infty v^2 dv \int_0^{\pi/2} \delta f_i \sin\theta d\theta. \tag{12}$$

The coupled nonlinear Eqs. (10), (12) are solved numerically using at $r = r_d$ the dust floating potential $\phi_0(r_d)$ [8] and the charge coefficient $S = [\partial \phi_0(r_d)/\partial r_d]/[r_d/\phi_0(r_d)]$ (where $1.01 \leq S \leq 1.15$) found in the numerical simulations [6]. The parameters of the trapped particle distribution function, $\beta 3$ and γ, are fitted so as to produce a localized potential at $r \to \infty$. The calculations are performed for a nonisothermal ($T_e/T_i = 10$) hydrogen plasma, with $z_d n_d / n_i^{(0)} = 0.2$. The dust radius is adopted as $r_d/\lambda_D = 0.01$. As expected, the solution is strongly dependent on the value of the charging parameter S. A localized solution does not exist for the unshielded Coulomb potential, $S = 1$.

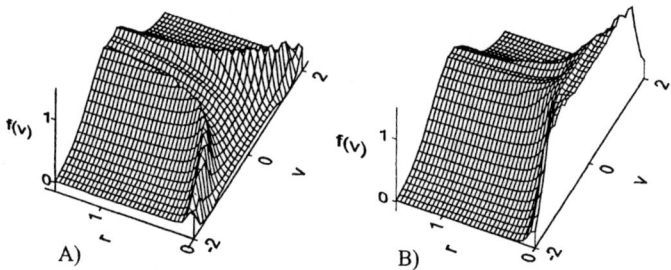

FIGURE 1. Ion distribution function (in arbitrary units), as a function of the particle velocity $v/\sqrt{2}v_{T,1}$; and the distance from the grain, $r/\lambda_{D,i}$; in the case of a hole (A) and hump (B).

For its minimum value found in [6], $S = 1.011$, we find a relatively small population of the trapped ions, $\beta = -1.5$. The effect of the ion hole is enhanced by its radial broadening due to the strong anisotropy, $\gamma = -5$. For a larger charge parameter, $S = 1.15$, an overpopulation of the trapped ions is found, corresponding to an ion hump with $\beta = 0.577$, $\gamma = 3$. The shielding of the potential in both cases is almost the same (less than 1% difference). The ion distribution functions and the radial profile of the potential are displayed in Figs. 1 and 2, respectively. Our results confirm the existence of a strong Debye shielding of the dust charge, and provide a theoretical explanation for such behavior of the electron and ion clouds. They are in a good agreement both with the simulations [6] and the recent laboratory experiments [9], where the total charge of trapped ions of the order of 20–30% of the dust charge was found, corresponding to $\sim 10^5$ trapped particles.

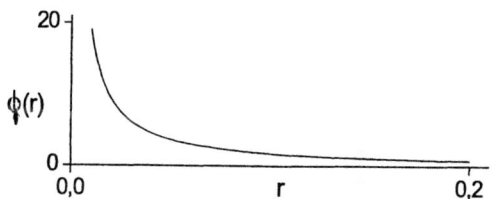

FIGURE 2. Radial profile of the potential $e\phi(r)/T_i$, r is normalized by $\lambda_{D,i}$. The grain size is $0.01\lambda_{D,i}$.

REFERENCES

1. Walch, B., Horányi, M., and Robertson, S., *Phys. Rev. Lett.*, **75**, 838 (1995).
2. Bernstein, I.B., and Rabinovich, I.N., *Phys. Fluids*, **2**, 112 (1959).
3. Laframboise, J.G., and Parker, L.W., *Phys. Fluids*, **16**, 629 (1973).
4. Lampe, M., *J. Plasma Phys.*, **72**, 171 (2001).
5. Lampe, M., Gavrishchaka, V., Ganguli, G., and Joyce, G., *Phys. Rev. Lett.*, **86**, 5278 (2001).
6. Zobnin, A.V., Nefedov, A.P., Sinel'shchikov, V.A., and Fortov, V.E., *JETP*, **91**, 483 (2000).
7. Schamel, H., *Phys. Plasmas*, **7**, 4373 (2000).
8. Shukla, P.K., and Mamun, A.A., *Introduction to Dusty Plasma Physics*, Institute of Physics Publ., Bristol, 2001.
9. Kingrey, D., Robertson, S., and Sternovsky, Z., *Bull. Am. Phys. Soc.*, **46**, 120 (2001).

On the Modification of Powder Particles in a Process Plasma

H. G. Thieme*, M. Quaas†, H. Kersten*, H. Wulff* and R. Hippler†

*E.-M.-Ardt-University, Institute for Physics, 17487 Greifswald, Germany
†E.-M.-Ardt-University, Institute of Chemistry, 17487 Greifswald, Germany

Abstract. SiO_2 particles have been coated in an acetylene plasma. Depending on the process conditions, also small carbon particles were synthesized. The decomposition of C_2H_2 and particle growth has been monitored by laser absorption. In a second experiment, the stability of luminphore particles could be improved by coating with protective Al_2O_3 films which are deposited by a PECVD process using a metal-organic precursor gas ATI.
PACS 5225Zb,5275Rx,8240Ra

INTRODUCTION

Powders which are produced or modified by using plasma technology have interesting and potentially useful properties, e.g. very small sizes (nanometer to micrometer range), uniform size distribution, and chemical activity. Size, structure and composition can be tailored to the specific requirements, dependent on the desired application [1-3]. The large variety of plasma-powder technology includes :

- treatment of soot and aerosols for environmental protection,
- particle synthesis in high pressure and low pressure plasmas,
- enhancement of adhesive, mechanical and protective properties of powder particles for sintering processes in metallurgy,
- fragmentation of powder mixtures in order to sort them,
- improvement of thin film properties by incorporation of nanocrystallites for amorphous solar cells and hard coatings,
- coating of lubricant particles,
- functionalization of micro-particles for pharmaceutic and medical application,
- production of colour pigments for paints,
- surface protection of fluorescent particles (luminophores),
- tailoring of optical surface properties of toner particles,
- application of tailored powder particles for chemical catalysis.

For example, an approach for coating of externally injected toner particles has been demonstrated in [4], where an argon rf-plasma was employed to charge and confine particles, while a metal coating has been performed by means of a separate dc-magnetron sputter source.

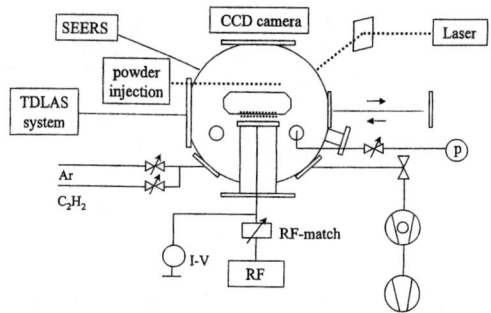

FIGURE 1. Scheme of the experimental set-up PULV A1.

PARTICLE SYNTHESIS AND COATING IN AN C_2H_2-PLASMA

In comparison to the particle coating process by magnetron sputtering [4], we perform the deposition of thin amorphous carbon (a-C:H) films onto SiO_2 grains ($\sim 1\,\mu m$) in a acetylene process plasma. The experiments have been carried out in a reactor PUL V A1 which is schematically drawn in Fig. 1. The rf-electrode as well as different diagnostics (video, TDLAS) are mounted in a spherically shaped vessel. Typical discharge conditions are as follows: power 5...100 W, pressure 1...10 Pa, gas composition Ar : C_2H_2 ...5.

Under relevant experimental conditions not only a-C:H deposition onto the externally injected SiO_2 particles but also generation of small carbon dust particles occur. After examination of the collected particles by electron microscopy (SEM) one can observe a rather small amount of large coated SiO_2 grains and a huge amount of small carbon dust particles (~ 100 nm). Surprisingly, the carbon dust particles show almost the same size and form clusters, see Fig. 2. In order to quantify the consumption of the precursor gas for film deposition onto SiO_2 grains and C-dust formation the acetylene molecules have been monitored by IR laser diode absorption spectroscopy (TDLAS) [5]. Already after a very short process duration (about 20 s) the TDLAS absorption decreases which indicates a fast and efficient decomposition of the C_2H_2 molecules. Simultaneously, the laser light intensity through the plasma drops remarkably due to the formation of carbon dust in the course of a- C:H deposition (Fig. 3).

DEPOSITION OF PROTECTIVE COATINGS

As another example for technological powder treatment in a process plasma, fluorescent particles have been coated by an alumina layer in a metal-organic discharge. The deposited layers shall protect the particles against degradation and aging during plasma and UV irradiation.

The metal-organic precursor ATI has been dissociated in an rf-discharge containing an ATJ/Ar or ATJ/air mixture, respectively. The rf-power has been varied between 10 and 100 W and the gas flow rate between 1 and 10 sccm. Changes in the electron density

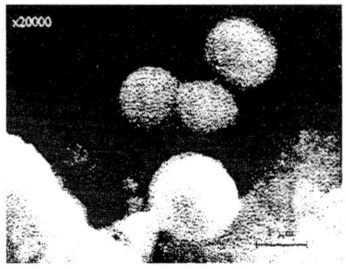

FIGURE 2. a-C:H coated SiO_2 particles (middle)

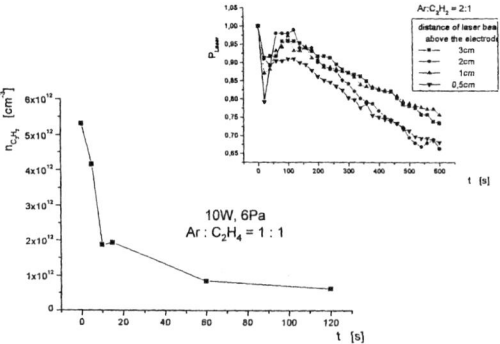

FIGURE 3. Decrease of the C_2H_2 absorption by and synthesized small carbon particles which decomposition of the precursor gas (bottom) form large clusters (left and right). and decrease of the laser light due to particle formation.

have been measured during the process by self-excited electron resonance spectroscopy (SEERS) [6]. Since the ATI fragments tend to form negative ions the electron density shows only a weak variation with the power, whereas in a pure argon plasma a strong increase could be observed. This observation is due to the electron attachment by the radicals which contribute to the film growth at the luminophore particles. The original fluorescent particles which are not coated by a protective alumina layer show a remarkable decrease in their light intensity after Ar plasma treatment which simulates the process conditions in a lamp. In comparison to the non-treated particles, PECVD by decomposition of ATI in the rf-plasma for obtaining transparent Al_2O_3 films onto fluorescent particles results in a much higher stability against plasma irradiation, e.g. against UV radiation and particle bombardment at low energies. Whereas the light intensity of non-treated luminophores decreases at plasma irradiation at high power, the light intensity of coated luminophores stays stable even for high plasma power, see Fig. 4.

Since the fluorescent properties of the grains should be preserved, it is important that there is no change in the emission spectra by the Al_2O_3. It could be shown that the luminophore particles are not influenced by the protective layers. An additional

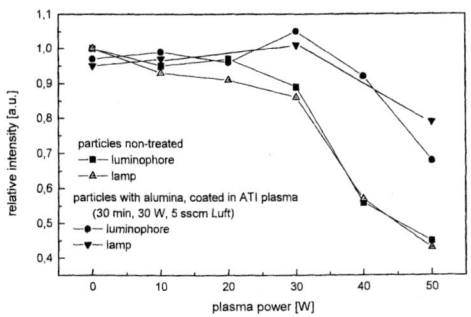

FIGURE 4. Comparison of light intensities of non-coated and coated luminophore particles after plasma irradiation.

advantage of plasma treatment plasma treatment under optimized conditions (30 W, 5 sccm air/ATL 30 min) is the decomposition of the glue material, which makes an annealing process unnecessary.

ACKNOWLEDGMENTS

This work has been supported by the Deutsche Forschungsgemeinschaft (DFG) under SFB 198/A14. The authors gratefully acknowledge Dr. J. Roepcke (INP Greifswald) for his support in TDLAS measurements.

REFERENCES

1. Bouchoule, A. (Ed.), *"Dusty Plasmas"*, J. Wiley & Sons, (1999).
2. Stoffels, E., Stoffels, W. W., Kersten, H., Swinkels, G. H. P. M., Kroesen, G. M. W., *Physica Scripta*, **T89**, 168, (2001).
3. Kersten, H., Deutsch, H., Stotfels, E., Stoffels, W. W., Kroesen, G. M. W., Hippler, R., *Contrib. Plasma Phys.*, **41**, 598 (2001).
4. Kersten, H., Schmetz, H., Kroesen, G. M. W., *Sulj:Coat.Technol.*, 108–109, 507 (1998)
5. Ropcke, J., Mechold, L, Kaning, M., Anders, J., Wienhold, P. G., Zahniser, M., *Rev. Sci. Instrum.*, **71**, 3706 (2000)
6. Klick, M., *J. Appl. Phys.*, **79**, 3445 (1996)

A Fluid Dynamic Approach to the Dust-Acoustic Soliton

J.F. McKenzie* and T. B. Doyle[†]

Max-Planck Institut für Aeronomie, 37191 Katlenburg-Lindau, Germany, and School of Pure and Applied Physics, University of Natal, Durban, South Africa
[†]*School of Pure and Applied Physics, University of Natal, Durban, South Africa*

Abstract. The properties of dust-acoustic solitons are derived from a fluid dynamic viewpoint in which conservation of total momentum, combined with the Bernoulli-like energy equations for each species, yields the structure equation for the heavy (or dust) speed in the stationary wave. This fully nonlinear approach reveals the crucial role played by the heavy sonic point in limiting the collective dust-acoustic Mach number, above which solitons cannot exist. An exact solution illustrates that the cold heavy species is compressed and this implies concomitant constraints on the potential and on the flow speed of the electrons and protons in the wave.

INTRODUCTION

The properties of linear and weakly non-linear "dust-acoustic-waves" (DAW) were first investigated by Rao, Shukla and Yu [1] (see also Shukla [2]) within a multi-fluid framework. In these waves, changes in the dynamic pressure of the cold heavy ions and the thermal pressures of the hot electrons and protons are balanced by electric stresses (McKenzie [3]). In this paper we adopt a fluid dynamic approach, in which total momentum conservation, combined with the Bernoulli energy equations for each species, is cast as the structure equation for the flow speed of the cold heavy ion component in a stationary wave. A soliton can be constructed provided the collective "dust-acoustic" Mach number exceeds unity and is less than a certain critical value (which equals two in the special case where $\gamma_e = \gamma_e = 2$). The wave is characterised by a compression in the heavy ions accompanied by a potential hump (dip), a rarefaction (compression) in the protons and a compression (rarefaction) in the electrons if the heavies are positively (negatively) charged. A fully non-linear exact solution in terms of elementary transcendental functions highlights these features.

THE STRUCTURE EQUATION FOR THE SOLITON

We formulate the problem of stationary waves in a plasma consisting of electrons, protons and a heavier species (e.g. dust ions or alpha particles) in the wave frame in which the flow speeds appear steady ($\partial/\partial t = 0$) and stream along the x-axis with speed U at $x = -\infty$. The mass flux, $m_i n_i u_i$, of each species is conserved and is equal to $m_i n_{io} U$, where n_{io} is the unperturbed density and $i = e$ (electrons), p (protons), h (heavies).

Conservation of total momentum flux along the direction of propagation (x-axis) may be written:

$$U^2 \sum_{i=h,p,e} m_i n_{io} P_i(u_i) = \frac{\varepsilon_o E^2}{2} = \frac{\varepsilon_o}{2}\left(\frac{d\phi}{dx}\right)^2 \qquad (1)$$

in which E is the electric field and ϕ is its potential. The normalised particle momentum function, $P_i(u_i)$, for each species, (with u_i normalised to the wave speed U) is given by

$$P_i(u_i) = u_i - 1 + \frac{1}{\gamma_i M_i^2}\left(\frac{1}{u_i^{\gamma_i}} - 1\right) \qquad (2)$$

in which we have assumed adiabatic flow ($p_i \propto n_i^{\gamma_i} \propto u_i^{-\gamma_i}$) and M_i is the species Mach number given by

$$M_i^2 = \frac{U^2}{c_{io}^2}, \qquad c_{io}^2 = \gamma_i p_{io}/m_i n_{io}. \qquad (3)$$

where c_{io} is the species acoustic speed.

The energy equation for each species may be written

$$\varepsilon_i(u_i) = \frac{1}{2}(u_i^2 - 1) + \frac{1}{(\gamma_i - 1)M_i^2}\left(\frac{1}{u_i^{\gamma_i - 1}} - 1\right) = -\frac{eZ_i \phi}{m_i U^2} \qquad (4)$$

which of course implies

$$\frac{m_h}{Z_h}\varepsilon_h(u_h) = m_p \varepsilon_p(u_p) = -m_e \varepsilon_e(u_e). \qquad (5)$$

The energy, ε_i, and momentum, P_i, functions are of the Bernoulli type, exhibiting a minimum at the species sonic point ($u_i = M_i^{-2(\gamma_i+1)}$). The behaviour of the flow speeds of each species in a dust-acoustic soliton follows directly from Eq. (4), which is given a graphical interpretation in Fig. 1(a) for subsonic ($M_i < 1$) and supersonic ($M_i > 1$) conditions. As we shall see subsequently, the heavier species is "supersonic" ($M_h > 1$) and is compressed ($u_h < 1$) in the wave, while the protons and electrons are "subsonic" (M_e and $M_p < 1$). If $Z_h > 0$ the electrons are decelerated (compressed) while the protons are accelerated (rarefied) and driven towards their sonic point, and the potential has a positive hump shape. On the other hand if $Z_h < 0$ the compression requirement on the heavies necessitates a (negative) potential dip with the result that the protons are compressed and the electrons are rarefied. These interesting features follow directly from Fig. 1(a), which is a graphical interpretation of the Bernoulli equation [Eq. (4)].

The structure equation for the heavy flow speed u_h in the wave follows immediately from momentum conservation [Eq. (1)] in which we eliminate the electric field E in favour of u_h from the equation of motion, i.e.

$$E = \frac{m_i U^2}{eZ_i}\left(1 - \frac{1}{M_i^2 u_i^{\gamma_i+1}}\right) u_i \frac{du_i}{dx}. \qquad (6)$$

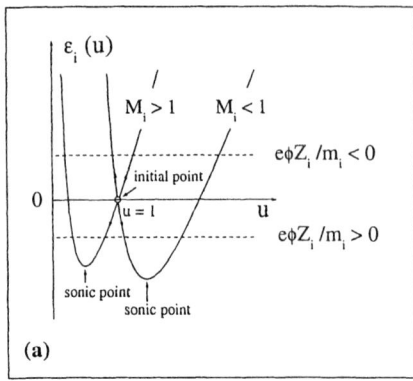
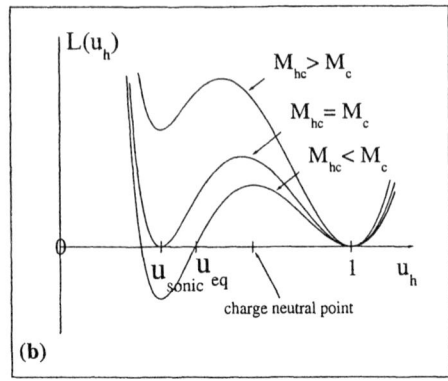

FIGURE 1. (a) The behaviour of the energy function $\varepsilon(u_i)$, as a function of flow speed u, for supersonic ($M_i > 1$) and subsonic ($M_i < 1$) initial conditions. If $M_i < 1$, positive (negative) charged particles are decelerated (accelerated) in a potential hill ($\phi > 0$), which is depicted as intersections between horizontal straight lines and the energy function curve. The opposite obtains for $M_i > 1$. Note that the energy function attains a minimum where the flow speed becomes sonic. (b) The heavy species momentum function $L_{(u_h)}$ as a function of u_h for three different values of the heavy-acoustic Mach number including the critical case. The sub-critical case yields an equilibrium point between the charge neutral point and the heavy sonic point. Solitons exist if the flow is super heavy-acoustic and sub-critical.

This equation along with the relationships between the electron and proton speeds in terms of the heavier dust speed, implicit in Eq. (4), casts the momentum conservation as a highly nonlinear differential equation for the heavy speed u_h. The strength, or centre, of the wave is determined by the zero of the total particle momentum function on the LHS of Eq. (1), (denoted by $L(u_h)$), nearest to the initial point ($u_h = 1$)). In general this equilibrium point ($du_h/dx = 0$) must be attained while the heavy flow speed remains supersonic for otherwise the flow becomes choked ($du_h/dx = \infty$) at the local sonic point of the heavies, namely, $u_h = 1/M_h^{2/(\gamma_h+1)}$. It is, in fact, this transonic constraint which places an upper limit on the speed (or collective Mach number) of the wave.

THE SOLITON STRUCTURE AND AN EXACT SOLUTION

Constraint of space rules out a complete discussion of the full structure equation for the heavy flow speed. However the general points made above can be highlighted in certain approximations, which assume that the pressure (and enthalpy) of the electrons and protons dominates their dynamic pressure (and kinetic energy). This is equivalent to their flows being highly subsonic $M_{e,p} \ll 1$ - so that Equations. (4) and (5) solved for u_i ($i = p, e$) as a function of u_h, yield

$$u_i = [1 \mp (\gamma_i - 1)M_{ih}^2 \varepsilon(u_h)/Z_h]^{-1/(\gamma_i-1)}, \tag{7}$$

where the upper (lower) sign is chosen if $i = e$ ($i = p$) and the combined Mach number M_{ih} is based on the electron (proton) pressure and the heavy mass

$$M_{ih}^2 = \frac{U^2 m_h n_{io}}{\gamma_i p_{io}}. \tag{8}$$

Near the initial point (the beginning of the wave) where $\delta \equiv u_h - 1 \ll 1$ the structure equation approximates to

$$\frac{d\delta}{dx} = \pm\sqrt{\kappa^2}\delta, \tag{9}$$

in which

$$\kappa^2 = M_{hc}^2 - \frac{1}{(1 - M_h^{-2})}, \tag{10}$$

$$M_{hc}^2 = \frac{(n_{po} M_{ph}^2 + n_{eo} M_{eh}^2)}{Z_h^2 n_{ho}}. \tag{11}$$

Equations. (9)-(11) describe the dispersion equation for linear, stationary, dust-acoustic waves. The spatial co-ordinate has been normalized to the natural length scale l_h, given by

$$l_h = \frac{U}{\omega_{ph}}, \tag{12}$$

where ω_{ph} is the plasma frequency of the heavy species.

Hence evanescent solutions near the initial point require $\kappa^2 > 0$, which implies that the "collective" Mach number M_{hc} must be "supersonic dust-acoustic", i.e.

$$M_{hc}^2 > \frac{1}{(1 - M_h^{-2})}. \tag{13}$$

Fig. 1(b) shows the behaviour of $L(u_h)$ for three values of the collective Mach number. In general the total momentum function $L(u_h)$, the LHS of Eq. (1), in which we use the "subsonic" expressions for u_e and u_p [Eq. (7)], has a double zero at the initial point $u_h = 1$ (as witnessed by Eq. (9)) and, provided the wave (flow) speed is super dust-acoustic, as in Eq. (10), and two other simple zeros in the compressive range $u_h < 1$, the first of which is interlaced by a maximum at the charge neutral point ($Z_h n_h = n_e - n_p$) and a minimum at the dust sonic point. The critical value, above which solitons cannot be constructed, corresponds to where these two simple zeros coalesce at the heavy dust sonic point. These properties follow from the derivative of $L(u_h)$, namely,

$$\frac{\partial L(u_h)}{\partial u_h} = \left(1 - \frac{1}{M_h^2 u_h^{\gamma_h+1}}\right)\left(1 - \frac{n_{eo}}{Z_h n_{ho}}\frac{u_h}{u_e} + \frac{n_{po}}{Z_h n_{ho}}\frac{u_h}{u_p}\right), \tag{14}$$

in which the zero of the first bracket corresponds to the heavy sonic point and the zero of the second bracket is the charge neutral point ($Z_d n_d = n_e - n_p$) where the electric field E maximizes.

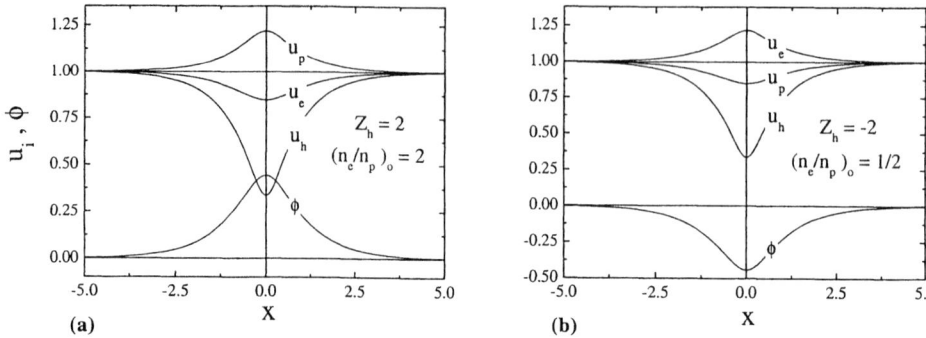

FIGURE 2. The spatial structure for the dust (heavy) soliton for the case of (a) positively, and (b) negatively charged dust (heavies). Note that in both cases the heavy species undergoes compression.

There is no simple expression for this critical Mach number and consequently the parameter regimes in which solitons or double layers exist are sometimes evaluated numerically (see e.g. Bharuthram and Shukla [4], Mace and Hellberg [5]). However, in the special case of $\gamma_e = \gamma_p = 2$, and a cold heavy species ($M_c = \infty$), the structure equation takes on the transparent form

$$u_h \frac{du_h}{dx} = \pm(u_h - 1)\sqrt{\left(\frac{M_{hc}}{2}\right)^2 (u_h + 1)^2 - 1}, \qquad (15)$$

which is readily integrated in terms of elementary transcendental functions (McKenzie [3]). The equilibrium point (or centre of the wave) occurs at the compression

$$u_{heq} = \frac{2}{|M_{hc}|} - 1, \qquad (16)$$

which relates the amplitude of the wave to its collective Mach number M_{hc} and, through the requirement $u_{heq} > 0$, yields the value two as the upper critical Mach number. Hence solitons exist in the range

$$1 < M_{hc} < 2. \qquad (17)$$

In this, the case of cold heavies, the heavy Mach number remains "supersonic" throughout, and at the upper critical collective Mach number of two gives rise to an infinite compression ($u_{heq} \to 0$) which, in a degenerate sense, does indeed correspond to the sonic point of the heavies.

The general signature of the dust-acoustic soliton is shown in Figs. 2(a) and (b) for the cases of positively and negatively charged heavies. In both cases the heavy species undergoes compression, while the potential is a positive hump shape if $Z_h > 0$ and a negative dip if $Z_h < 0$ with correspondingly reverse behaviour for the electrons and protons, as already noted, and as indicated in Figs. 2(a) and (b).

ACKNOWLEDGMENTS

The authors wish to thank the Max-Planck Institut für Aeronomie, the University of Natal, and the Foundation for Research and Development of South Africa for supporting this work.

REFERENCES

1. Rao, N.N., Shukla, P.K. and Yu, M.Y., Dust-acoustic waves in dusty plasmas. *Solar and Space Plasmas*, **38**, 543 (2000).
2. Shukla, P.K., Collective Processes in Complex Plasmas. Waves in Dusty Plasmas. *Solar and Space Plasmas*, (Ed. F. Verheest *et al*), AIP Conference Proceedings, Melville, New York, Vol. 537, 3 (2000).
3. McKenzie, J.F., The Ion-Acoustic Soliton: A Gas Dynamic Viewpoint, *Physics of Plasmas*, **9**, pp 800-05 (2002).
4. Bharuthram, R. and Shukla, P.K., Large Amplitude Double Layers in Dusty Plasmas, *Planetary Space Science*, **40**, 405 (1992).
5. Mace, R.L. and Hellberg, M.A., Dust-Acoustic Double Layers: Ion Inertial Effects, *Planetary Space Science*, **41**, 235 (1993).

Normal Mode Spectra of Thermally Excited 2D Finite Coulomb Clusters

A. Melzer

Institut für Experimentelle und Angewandte Physik, Christian-Albrechts-Universität Kiel, 24098 Kiel, Germany

Abstract. The normal mode spectra of 2D dust Coulomb clusters trapped in the sheath of an rf discharge have been measured from the thermal motion of the microspheres. With this technique, the full physical properties of the clusters like the interaction strength between the particles and the mode energies have been revealed. It is shown that the clusters are in thermal equilibrium with an effective temperature close to room temperature.

INTRODUCTION

The study of 2D finite Coulomb dust clusters allows a detailed insight into the dynamic properties of complex plasmas. These clusters consist of a small number of charged microspheres ($N = 1\ldots100$) trapped in the sheath of an rf discharge at low gas pressure. The particles are confined vertically in a strong potential well due to gravity and electric field forces, thereby forming pure two-dimensional systems. Horizontally (radially) the microspheres are confined in a shallow parabolic potential. Due to the interplay of the (screened) Coulomb repulsion of the microspheres and the radial confinement the clusters arrange in concentric rings ("shells") establishing a "periodic table" of 2D finite clusters [1, 2, 3, 4].

The dynamical properties of finite clusters are described in terms of normal mode oscillations. In recent experiments, certain modes have been selectively excited by active external disturbances applied to the system. Klindworth et al. [4] have stimulated the intershell rotation of $N = 19$ and $N = 20$ clusters by a pair of laser beams, whereas Melzer et al. [5] excited so-called "breathing" and "antisymmetric" modes in clusters with $N = 3, 4$ and 7 particles by a pulse-like modulation of the discharge power.

Here, simultaneous measurements of *all* the possible normal modes of finite clusters are presented. The normal mode spectra are derived purely from the thermal motion of the microspheres in the cluster. This technique, previously used in experiments by Nunomura et al. [6] on waves in "infinite" 2D plasma crystals, has been adapted here for finite clusters. The normal mode spectra of different clusters under various plasma conditions have been measured.

The description of two-dimensional finite clusters of N particles in a complex plasmas starts from the total energy E of the system

$$E = \frac{1}{2}m\omega_0^2 \sum_{i=1}^{N} r_i^2 + \frac{Z^2 e^2}{4\pi\varepsilon_0} \sum_{i>j}^{N} \frac{1}{r_{ij}} \exp\left(-\frac{r_{ij}}{\lambda_D}\right) \tag{1}$$

which is the sum of the radial confining potential energy and the (screened) Coulomb interaction between the particles. Here, m denotes the mass of the particles and Z is their charge number. The strength of the confining potential is denoted by ω_0. In addition, $r_i = (x_i^2 + y_i^2)^{1/2}$ is the radial coordinate of the ith particle and $r_{ij} = |\vec{r}_i - \vec{r}_j|$ is the distance between particle i and j.

Using normalized units, $r/r_0 \to r$ and $E/E_0 \to E$ as well as the screening strength κ, with

$$r_0 = \left[\frac{Z^2 e^2}{4\pi\varepsilon_0} \frac{2}{m\omega_0^2}\right]^{1/3} \quad E_0 = \left[\left(\frac{Z^2 e^2}{4\pi\varepsilon_0}\right)^2 \frac{m\omega_0^2}{2}\right]^{1/3} \quad \kappa = \frac{r_0}{\lambda_D} \qquad (2)$$

the total energy simplifies to

$$E = \sum_{i=1}^{N} r_i^2 + \sum_{i>j}^{N} \frac{1}{r_{ij}} \exp\left(-\kappa r_{ij}\right) \quad . \qquad (3)$$

The normal modes of the 2D clusters are then calculated from the dynamical matrix

$$E_{\alpha\beta,ij} = \frac{\partial^2 E}{\partial r_{\alpha,i} \partial r_{\beta,j}} \qquad (4)$$

with α and $\beta = x, y$ and i, j denoting the particle number. The normal mode frequencies ω_ℓ of the $2N$ modes are the eigenvalues of the dynamical matrix (in units of $\omega_0/\sqrt{2}$) and its eigenvectors describe the cluster oscillation mode pattern [2].

For the case of pure Coulomb interaction ($\kappa = 0$) there are three normal modes that are independent of the particle number N [2]: (i) $\omega = 0$ for the rotation of the entire cluster around the center of the confinement, (ii) $\omega = \omega_0$ (twofold degenerate) oscillation of the center-of-mass of the cluster in the horizontal potential well, and (iii) $\omega = \sqrt{3}\,\omega_0$ corresponds to a coherent radial oscillation of all particles (breathing mode). For screened interaction ($\kappa > 0$), the frequency of the first two modes is unaffected since they do not involve a relative particle motion. In contrast, the frequency of the breathing mode and all other modes becomes dependent on κ and on the particle number N.

EXPERIMENT, RESULTS AND DISCUSSIONS

The experiments have been performed in a parallel plate rf discharge operated in argon at 13.56 MHz with gas pressures of 1.5 to 2 Pa and discharge powers between 3 and 40 W (see Fig. 1a). A few ($N = 3$ to 20) melamine-formaldehyde (MF) particles of 9.55 μm diameter are immersed into the plasma and are illuminated by a laser fan (at 690 nm, 40 mW). The particles form 2D finite Coulomb clusters above the lower electrode (see Fig. 1b). The horizontal confinement for the particles is realized by a shallow circular parabolic trough in the electrode. The particle motion is viewed from top and from the side with video cameras and stored into the computer for further processing.

In the experiment the particle motion was recorded for 1 minute (corresponding to 1500 frames at 25 frames per second). No excitation in any form was applied, only the

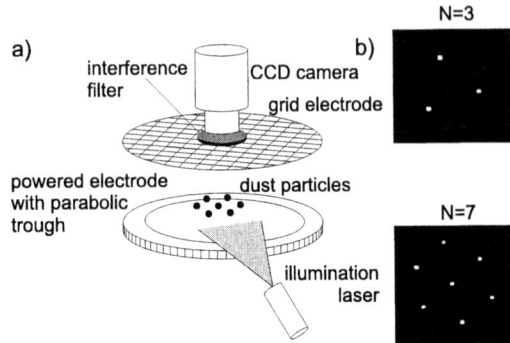

FIGURE 1. a) Scheme of the experimental setup, b) snap shots of the clusters with $N = 3$ and $N = 7$.

thermal Brownian motion of the particles around their equilibrium positions $\vec{r}_i(t)$ was recorded. These thermal fluctuations were used to obtain the normal mode spectra in the following manner: First, the time series of the particle velocities $\vec{v}_i(t) = d\vec{r}_i(t)/dt$ projected onto the direction of the normal mode vectors was determined for each mode number $\ell = 1\ldots 2N$, i.e. the quantity $f_\ell(t) = \sum_{i=1}^{N} \vec{v}_i(t) \cdot \vec{e}_{i,\ell}$ was calculated. Here, $\vec{e}_{i,\ell}$ is the eigenvector of particle i for mode number ℓ describing its oscillation amplitude and direction of oscillation. Thus, $f_\ell(t)$ is the contribution of the thermal fluctuations to mode number ℓ in the time domain. Finally, the normal mode spectra are obtained in form of the spectral power density $S_\ell(\omega) \propto |\int f_\ell(t)\exp(-i\omega t)dt|^2$

A result of such a measurement is shown in Fig. 2. There, the mode spectrum of a $N = 3$ cluster is shown for a discharge power of 18 W. One can see that the thermal fluctuations of the microspheres around their equilibrium are small (Fig. 2a), but they are nevertheless sufficient to determine the mode spectrum. The 6 eigenmodes of this clusters are depicted in Fig. 2b, with the breathing mode ($\ell = 1$), rotation of the entire cluster ($\ell = 2$), a twofold degenerate "kink" mode ($\ell = 3, 4$) and the two sloshing modes

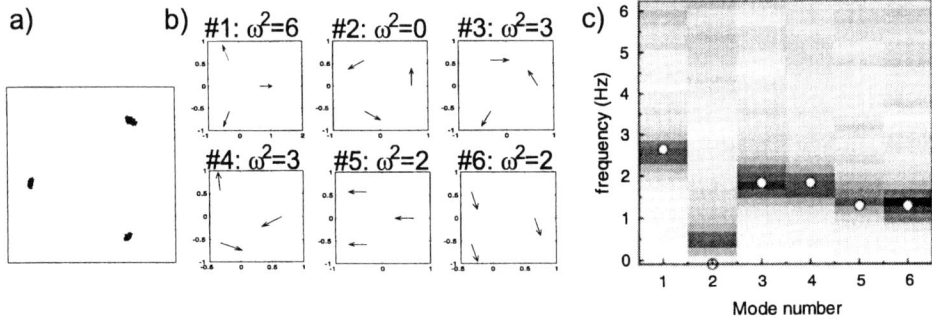

FIGURE 2. Normal mode spectrum of a $N = 3$ cluster. a) Particle trajectories over 1 minute. b) Normal modes of the cluster, the indicated mode frequencies ω_ℓ^2 are normalized to $\omega_0^2/2$. c) Measured mode spectrum of the 6 modes. The spectral power density is shown in grey scale. The circles correspond to the calculated mode frequencies.

($\ell = 5, 6$). The mode frequencies ω_ℓ are also indicated for $\kappa = 0$. For $\kappa > 0$ the oscillation pattern of the eigenmodes is unchanged, their frequencies, however, decisively depend on κ.

In Fig. 2c, the measured mode spectrum is shown as a grey scale plot. Dark regions correspond to large power densities. For comparison, the calculated mode frequencies are also indicated by the circles. One can see the very good agreement between the measured power spectrum and the calculated mode frequencies. Thus, the measured spectrum very well reflects the expected values. The calculated mode frequencies contain the adjustable parameters ω_0 and κ. Best agreement with the measured spectrum is found in the range $0 \leq \kappa \leq 2$ and $\omega_0/(2\pi) = 1.3 \pm 0.2$ Hz resulting in a dust charge of $Z = 11\,000 \pm 1200$. These values are in excellent agreement with those obtained from excitation techniques [5].

Finally, the effective temperature of the cluster can be extracted from these data. It is seen from the intensities of the grey-scale plot that the different modes seem to have comparable power densities. Indeed, the stored energy $E_\ell = \int S_\ell(\omega)d\omega = 47$ meV is the same for each mode (within 5 % error). This energy corresponds to a temperature of the dust particles of $T = 540$ K which is slightly above room temperature. This means that the cluster modes are in thermal equilibrium and the principle of equipartition holds, here. These findings are substantiated by similar experiments on clusters with $N = 5, 7, 12, 16, 19$ and 20 particles.

In conclusion, the normal mode spectra of finite clusters have been determined from the thermal motion of the microspheres around their equilibrium positions. This technique, adapted for clusters from wave experiments of Nunomura et al. [6], allows to measure the energy stored in the modes and the corresponding mode frequencies of all modes of cluster. The cluster modes are in thermal equilibrium with a temperature close to room temperature. The particle interaction can be described by a screening strength κ of the order of one and a dust charge of $Z = 11\,000$.

Helpful discussions with M. Klindworth, A. Piel and I.V. Schweigert are gratefully acknowledged.

REFERENCES

1. Bedanov V. M. and Peeters F. *Phys. Rev. B* **49**, 2667 (1994)
2. Schweigert V. A. and Peeters F. *Phys. Rev. B* **51**, 7700 (1995)
3. Juan W.-T., Huang Z.-H., Hsu J.-W., Lai Y.-J. and I L. *Phys. Rev. E* **58**, 6947 (1998)
4. Klindworth M., Melzer A., Piel A. and Schweigert V. *Phys. Rev. B* **61**, 8404 (2000)
5. Melzer A., Klindworth M. and Piel A. *Phys. Rev. Lett.* **87**, 115002 (2001)
6. Nunomura S., Goree J., Hu S., Wang X. and Bhattacharjee A. *9th Workshop on the Physics of Dusty Plasmas* Iowa City, 2001

Typical Characteristics of RF Voltage Threshold for Planar Dusty RF Discharges

S. Nonaka*, Y. Nakamura[†], S. Ikezawa[+] and K. Katoh[#]

*Toyota Technological Institute, Tempaku, Nagoya 468-8511, Japan
[†]Institute of Space and Astronautical Science, Kanagawa 229-8510, Japan
[+]Department of Electronic Engineering, Chubu University, Aichi 487-8511, Japan
[#]Densoh K.K., Kariya-city, Aichi, 448-0029, Japan

Abstract. Effects of dielectric dust grains on capacitive planar RF discharges are investigated analytically. Typical characteristics of RF voltage thresholds for maintaining the dusty RF plasma and for jumping between two different discharge states are described.

INTRODUCTION

In capacitive RF discharges for material processing by plasmas, growing dust particles have been reported [1]. We have recently considered effects of dielectric dust grains on the planar RF discharges [2]. This paper deals with the RF voltage thresholds for maintaining the plasma and for jumping two different discharge states. The results obtained will shed some light on dusty RF plasma applications for material processing in "International Microgravity Plasma Facility (IMPF).

MODELLING OF PLANAR DUSTY RF DISCHARGES

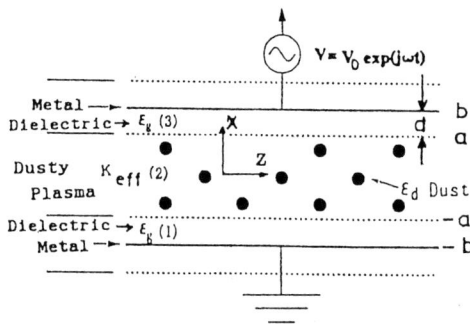

FIGURE 1. Model of dusty RF discharges.

In Fig. 1, each electrode ($x = \pm b$) is covered by a lossless dielectric material with relative permittivity ε_g and thickness $d(\equiv b - a)$. Subscripts, g and p, denote layers of dielectric ($a \leq |x| \leq b$) and uniform plasma ($-a \leq x \leq a$). Dusts are assumed to be uniformly dispersed with the density N_D. Neglecting motions of dusts and ions, the effective macroscopic permittivity K_{eff} of the dusty plasmas is expressed, applying

Wagner's electrostatic theory [3], as:

$$K_{eff} = \frac{\delta_0(\kappa_0 - X^* - j\eta X^*)(\kappa_N - X^* - j\eta X^*)}{\kappa_1 - X^* - j\eta X^*)} \quad (1)$$

where $\delta \equiv N_D(4\pi/3)(D/R)^3$, $\delta_0 \equiv 2(1-\delta)/(2+\delta)$, $j = \sqrt{-1}$, $\kappa_0 \equiv 1$, $\kappa_1 \equiv 1 + \{(1-\delta)/(2+\delta)\}\varepsilon_d(\geq 1)$, $\kappa_N \equiv 1 + \{(1+2\delta)/(1-\delta)\}\varepsilon_d(\geq 1)$, $\kappa_0 < \kappa_1 < \kappa_N$), $\varepsilon_p = 1 - X^* - j\eta X^*$, $\eta = v/\omega$, $X^* = X/(1+\eta^2)$, $X = (\omega_p/\omega)^2$, $\omega_p^2 = n_e e^2/m\varepsilon_0$.

Other notations are standards; δ is the volume ratio of all the dusts with radius D inside a virtual volume R^3; ε_p and ε_d are permittivities of cold plasma and dusts, respectively. Equation (1) denotes dust contribution to K_{eff} via δ and ε_d. The density N_D is about 10^{10} cm^{-3} when $\delta = 0.01$ and $D = 1 \mu$m, for example.

DETECTION OF HYSTERESIS PHENOMENA

For power balance, the RF power P_{ab} absorbed by electrons per unit electrode-area is derived [3], whereas all the averaged electron-energy loss Q_{loss} per unit time and area is proportional to electron density n_e as: $Q_{loss} = 2an_eQ \equiv \omega\varepsilon_0 aQ_0 X$, and so $Q_0 = 2\omega(m/e^2)Q$ [4]. Therefore, the power balance for normalized plasma density X to maintain RF plasmas is obtained by $1/P_{ab} = 1/Q_{loss}$ as follows:

$$\frac{\delta_0(1+\eta^2)}{\eta} \frac{Y^{*2} + \eta^2 X^{*2}(2X^* - \kappa_0 - \kappa_N - K_{G0})^2}{(\kappa_1 - X^*)^2 + (\kappa_N - \kappa_1)(\kappa_1 - \kappa_0) + \eta^2 X^*} = \left(\frac{U}{\sqrt{Q_0}}\right)^2, \quad (2)$$

where

$Y^* \equiv \kappa_N \kappa_0 + \kappa_1 K_{G0} - (\kappa_0 + \kappa_N K_{G0})X^* + (1-\eta^2)X^{*2}$
$K_{G0} \equiv K_G/\delta_0 \qquad K_G \equiv (a/d)\varepsilon_g$
$Q = v_{en}\frac{3m_e}{m_i}\kappa T_e + \langle v_i\rangle\varepsilon_i + \Sigma_j\langle v_j\rangle\varepsilon_j + v_i(2 + \ln(m_i/m_e)^{1/2})\kappa T_e \quad (>0)$

It is noted here that Q includes all the losses of electron energy dissipated in the plasma, for examples, energy losses for sustaining sheath potentials around nanoparticles and in front of the two electrodes, and energy losses collisional with neutral gas-molecules (or atoms) for momentum transfer and ionizations, so that dust surface electron charges are not essential in this theory.

In the lossless case ($\eta = 0$) yields the following two resonance solutions of X:

$$X_{\pm} = (1/2)\left\{\kappa_0 + \kappa_N + K_{G0} \pm \sqrt{(\kappa_0 + \kappa_N + K_{G0})^2 - 4(\kappa_N \kappa_0 + \kappa_1 K_{G0})}\right\} \quad (3)$$

As δ approaches zero, X_+ and X_- approach the geometrical (X_{G0}) and dust-dipole (X_{D0}) resonance, respectively; i.e., $X_+ \to X_{G0} = 1 + K_G$, and $X_- \to X_{D0} = 1 + (1/2)e_d$, where subscript "0" denotes the case without energy loss ($\eta = 0$). We calculate the square root of left hand term of Eq. (2) as a function of X, and then interchange the vertical and horizontal axes. The typical three results are shown in Fig. 2; curves (A), (B), and

(C) are for $X_{G0} < X_{D0}$, $X_{G0} = X_{D0}$, and $X_{G0} > X_{D0}$, respectively. Each curve has two dip points and an intermediate peak point of $U/\sqrt{Q_0}$. The two dip points correspond to the G- or D-resonance and also to the starting or stopping the RF voltage under a fixed Q_0 for discharges, whereas an intermediate peak point corresponds to a jumping RF voltage, as indicated by arrows in Fig. 1. This is because stable solutions for X are found at the upper curve than the dips, namely in the range of $dX/d(U/\sqrt{Q}) \geq 0$. Thus, only in the case of type (A), when the RF voltage gradually increases from zero, the so-called "hysteresis phenomena" will occur with two-step up (or, down) jumping at the two dip points of $dX/d(U/\sqrt{Q}) = \infty$.

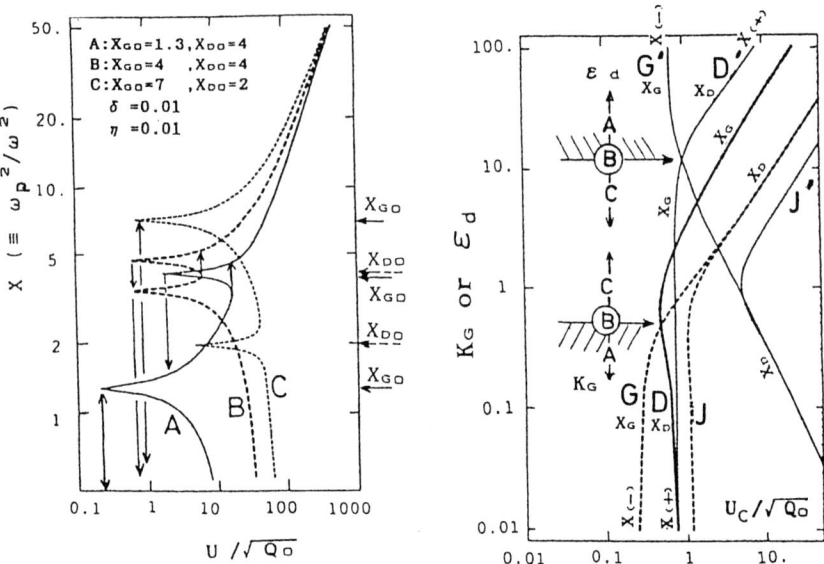

FIGURE 2. X versus $U/\sqrt{Q_0}$

FIGURE 3. Variation of RF thresholds.

RF VOLTAGE THRESHOLDS FOR DISCHARGES

When $0 < \eta < 1$, the first order approximation with respect to η is applicable to Eq. (2), so that all the terms with η^2 in (2) are negligible. Then, the RF voltage thresholds for X_{G0} and X_{D0} resonance are obtainable approximately from the condition that the first term of numerator of Eq. (2) equal zero; i.e., $Y^* = 0$. Therefore, the thresholds U_C/Q_0 are obtained by substituting Eq. (3) into Eq. (2). Similarly, the RF voltage threshold for the up-jumping point is also obtainable approximately by putting the first term of denominator of Eq. (2) zero; i.e., at $X = \kappa_1(1 + \eta^2)$. All these thresholds will vary depending on K_G or e_d. The results are shown in Fig. 3 for these thresholds in the horizontal axis as a function of the variable K_G (or e_d) in the vertical axis. Curves (G) and (D) are the case for K_G varying under fixed e_d $(= 1)$, $\delta = 0.05$ and $\eta = 0.05$, whereas curves (G') and (D') are the case for e_d varying under fixed K_G (= 9), $\delta = 0.05$ and $\eta = 0.005$. Curves (G) and (G') correspond to the dip values at $X = X_{(-)}$, and curves (E) and (E') correspond to dip values at $X = X_{(+)}$. The mode names, X_G and X_D, that are indicated along the curves, are interchanged at the degenerating vertical levels (B) of $K_G = e_d/2$.

As inserted in Fig. 3, the shaded regions (A) correspond to the regions for the hysteresis of type (A) in Fig. 2 under the condition $X_{G0} < X_{D0}$. Similarly, the levels (B) are for type (B), and the white regions (C) correspond to no hysteresis regions of type (C), respectively. When the RF voltage increases (or decreases), the discharge always starts (or stops) at the minimum threshold of all the curves (G), (D), (G') and (D') by the X_{G0} resonance. On the other hand, the down jumping for type (A) will occur at the curve (D) or (D') by the X_D resonance only in the cases of regions (A). The levels (B) denote transition between the types (A) and (C). Similarly, the results for up jumping are added by curves (J) and (J') in Fig. 3 under the condition of all the same parameters as the corresponding cases with and without the "dash", respectively. These thresholds are valid only in the shaded regions (A), not in the regions (C).

SUMMARY

The dipole resonance of the plasma around individual dust grains can produce the plasma. Typical properties of several RF voltage thresholds for starting, stopping and up (or, down) jumping between different discharge states are obtained approximately. These results will shed some light on dusty RF plasma applications, for example, material processing in space in the IMPF project.

REFERENCES

1. "Dusty plasmas"; *J. Plasma and Fusion Res.*, **73**, no.11, 1220–1274 (1997)[in Japanese].
2. Nonaka, S., Katoh, K., Nakamura, Y., Ikezawa, S., and Takamura, S., Proc. of ICPDP1999, Hakone, Japan, May 24–28, p.131, 1999
3. "Phenomena on dielectrics", ed. Inst. Electri. Engin. Jpn. (1973) pp. 143–146)[in Japanese].
4. Zhelyazkov, I., and Atanassov, V., "Axial Structure of Low-Pressure High-Frequency Discharges Sustained by Traveling Electromagnetic Surface Waves", *Physics Reports*, **Vol. 255**, No.2 and 3, April (1995) pp. 143–146.

Voids in Dust Clouds Suspended in the Plasma Sheath

G. V. Paeva, W. W. Stoffels, R. P. Dahiya,
E. Stoffels, and G. M. W. Kroesen

Eindhoven University of Technology, Eindhoven, The Netherlands

Abstract. Voids in dusty plasma are a new phenomenon, which is still not understood. In this work we have studied experimentally for first time voids in the sheath of a radio-frequency (RF) dusty plasma. Injecting big dust particles into the plasma, we form a dust cloud in the sheath. The behaviour of the cloud as a function of RF power and gas pressure is investigated using video imaging. Both dependencies show a threshold for the void formation. This threshold is characterised by a sudden decrease in the inter-particle distance, while in the non-void mode the distance increases with power and pressure. We have performed Langmuir probe measurements of the floating potential in the bulk plasma close to the sheath in order to estimate the form of the potential well trapping the dust grains.

INTRODUCTION

The interest in dusty plasmas arose fifteen years ago with the development of semiconductor manufacturing, as dust contamination was proved to be harmful for the semiconductor devices. To be able to control a process we need to understand it in detail. For this we need a lot of fundamental knowledge.

In 1996, Praburam and Goree first observed a new phenomenon which they called "great void" [1]. In their work, the void was observed in the bulk plasma as the dust particles were nanometrer sized and were grown by sputtering graphite electrodes. This work was continued by Samsonov and Goree [2]. In 1999, voids were again observed [3]. In this experiment, the dust particles were micrometre sized, but, due to the microgravity conditions, the cloud was again in the bulk plasma.

There are several attempts at finding theoretical explanation of the void formation [4, 5, 6, 7, 8].

EXPERIMENTAL PROCEDURE

Here, we describe the experimental set-up designed for this work. The argon pressures are in the range of 0.06–0.21 mbar. In the bottom of the vessel, an electrode is powered at 13.56 Mhz. The power is in the range of 3 to 60 W. We have injected spherical melamine formaldehyde(MF) particles with diameter of 9.8 mm in the plasma. The electrode is designed with a special form to create a trap for the particles — in the centre there is a circular groove 3 cm in diameter and 3 mm deep. To illuminate the particles, we

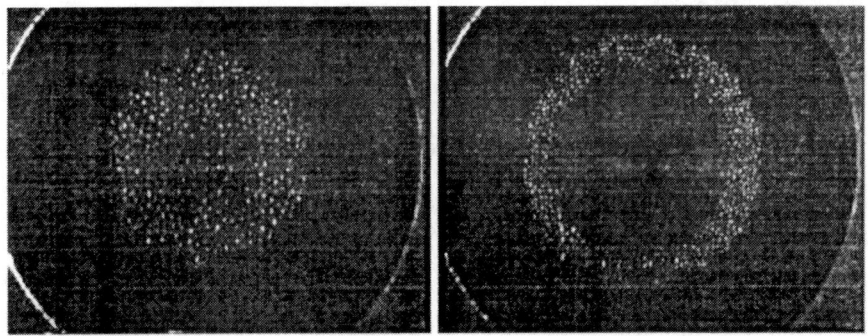

FIGURE 1. Images of the dust cloud at pressure 0.11 mbar at low and high power. The illuminated ring around the cloud is the edge of the groove in the electrode. At high power we can see in the centre a dust free region — void.

have used a horizontal laser light sheet. The particles are large enough to be imaged separately. The imaging is performed by a video camera positioned above the reactor. A Langmuir probe has been used to measure the floating potential of the discharge.

EXPERIMENTAL RESULTS AND DISCUSSION

In the argon plasma the dust particles become negatively charged. This keeps them apart. On the other hand, the trap doesn't allow them to spread into the whole vessel. The particles form a cloud positioned above the centre of the groove in the electrode.

Figure 1a shows the particle cloud. The cloud has a diameter of approximately 1 cm. By changing the conditions of the plasma, we have been able to reach a situation, in which a dust free region(void) has been formed in the centre of the cloud(Fig. 1b). We have made a series of measurements at constant pressure and varying power, and, at constant power and varying pressure. Typical graphs are shown in figures 2 and 3.

In the case of 0.11 mbar pressure the cloud has been uniformly dense up to 25 W. Above 25 W a void forms in the centre of the cloud. If we compare the area of the cloud at different powers, we see that, after forming the void, this area decreases(Fig. 2b). Figure 2 gives the impression that the void shrinks and the cloud surface increases above power of 40 W. The reason for this is the very strong glow of the plasma at high powers. This decreases the contrast of the picture, which, in turn, results in less precise measurements. The decrease of the cloud area after void formation in the centre may be due to decrease in the inter-particle distance or to shift of the particles in vertical direction.

A similar series of measurements have been performed under constant power and varying gas pressure. A typical graph(at 10 W) is shown on Fig. 3a. In this case, the cloud goes from a circular shape to an annular one at 0.1 mbar. The measurement has been performed in the range of 0.06 to 0.21 mbar, when the cloud has appeared as a ring of single particles. As it has been in the case of power dependence, we can see that up to the

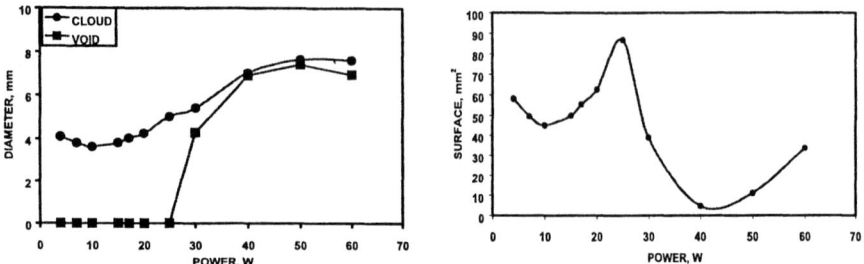

FIGURE 2. (a) Cloud and void diameter as a function of the RF power. The pressure is 0.11 mbar. At this pressure the void appears at 25W. (b) Surface of the mono-layer cloud as a function of the RF power.

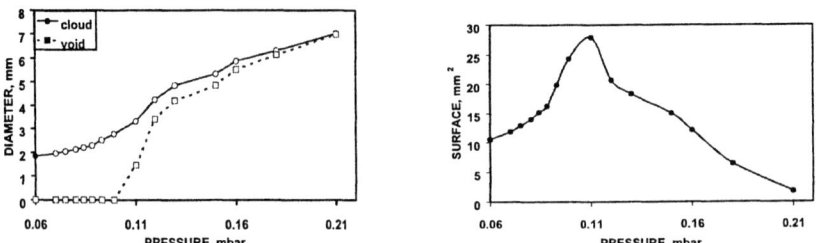

FIGURE 3. Cloud and void diameter as a function of the gas pressure. The power is 10 W. At this power the void appears at 0.1 mbar. (b) Area of the mono-layer cloud as function of the gas pressure. After the void is formed the cloud area decreases.

threshold the area of the cloud increases and after the formation of the void it decreases. To try to establish the reason for the area change, we have estimated the volume taken by a single particle. As the measurement is based on 2D images and the laser light sheet is thicker than the diameter of the particles, it is not possible to fully consider the vertical movement of the particles. However, as there is no change in the number of particles in the cloud, the estimation has showed that the inter-particle distance is the basis of the change in the cloud area.

We aim to understand the phenomenon of voids in dust clouds on the basis of the dynamics of dusty plasmas. The main forces acting on a dust grain in the plasma are electrostatic, ion drag, neutral drag, thermophoretic and gravitational. The particles levitate in the plasma in a position where the gravitational force is in equilibrium with the vertical component of the electrostatic force.

The ion drag force is due to the momentum transfer from the ions to the dust particles. The ions move in the electrostatic field and transfer momentum to the dust particles in the opposite direction to the electrostatic force.

To find out if the horizontal component of the electrostatic force is directed inwards or outwards, we need additional information. For this reason we have performed Langmuir probe measurements. As the probe measurements are not relevant in the sheath, we have measured the floating potential directly above the sheath. The results are shown in figure 4. For technical reasons, the measurements were performed at a fixed level

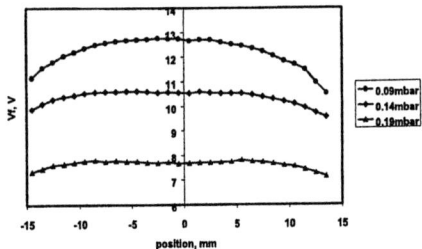

FIGURE 4. Floating potential measured at 1.5 cm above the electrode at different pressures. The measurements are performed above the groove in the electrode. Position 0 corresponds to its centre.

above the electrode. As the sheath thickness changes in response to the changing plasma parameters, our measurements are not at a constant distance from the dust cloud. By increasing the pressure, we decrease the sheath thickness and the particles also move a little lower. In this case, the measurement of the floating potential is closer to the plasma bulk and it should be less sensitive to changes in the sheath profile. We see that as the pressure increases the profile flattens and at high pressures there is even a slight drop in the centre. This drop, even though it's within the measurement error and therefore it's not reliable, already gives an idea of one possible reason for the void formation.

CONCLUSIONS

We have reported the results of laboratory observations of voids in a RF dusty plasma. The formation of voids has been investigated for the first time as a result of increased pressure or RF power. To discover whether the electrostatic or the ion drag force is responsible for this behaviour of the dust cloud, we have performed Langmuir probe measurements of the floating potential. The precision of the measurements is not enough for definitive conclusions. To obtain clear evidence that the form of the potential is causing the annular form of the cloud at higher pressures or powers, we need additional measurement techniques.

REFERENCES

1. Praburam, G., and Goree, I., *Phys. Plasmas*, **3(4)**, 1212 (1996).
2. Samsonov, D., and Goree, I., *Phys. Rev. E*, **59(I)**, 1047 (1999).
3. Morfill, G.E., Thomas, H.M., Konopka,U., Rothermel, H., Zuzic, M., Ivlev, A., and Goree, I., *Phys.Rev. Lett.*, **Vol. 83**, No.8, 1598 (1999).
4. Goree, I., Morfill, G.E., Tsytovich, V.N., and Vladimirov, S.V., *Phys. Rev. E*, **59 (6)**, 7055 (1999).
5. Ostrikov, K.N., Vladimirov, S.V., Yu M. Y., and Morfill, G.E., *Phys. Rev. E*, **61(4)**, 4315 (2000).
6. Tsytovich,V.N., *Physica Scripta*, **Vol. T89**, 89 (2001).
7. Avinash, K., *Physics of Plasmas*, **8(1)**, 351 (2001).
8. Avinash, K., *Physics of Plasmas*, **8(6)**, 2601 (2001).

Geometry induced defects in a confined Wigner lattice

F. M. Peeters*, Minghui Kong* and B. Partoens*

Departement Natuurkunde, Universiteit Antwerpen (UIA), Universiteitsplein 1, B-2610 Antwerpen, Belgium

Abstract. The configurational and melting properties of large two-dimensional (2D) clusters of charged classical particles interacting with each other via the Coulomb potential are investigated using Monte Carlo (MC) simulations. The particles are confined by a harmonic potential. For a large number of particles in the cluster (N > 150) the configuration is determined by two competing effects, namely the formation of a hexagonal lattice in the center which is the groundstate for an infinite 2D system, and the effect of the confinement which wants to impose its circular symmetry on the outer edge. In the transition region defects appear at the six corners of the hexagonal-shaped inner domain. The melting of this cluster is found to be strongly related to the local topological structure. Our results clearly show that the melting starts near the geometry induced defects and that three melting temperatures can be obtained.

There has been recently considerable theoretical and experimental progress in the study of classical systems consisting of a finite number of charged particles which are confined into an artificial circular symmetric potential. Typical experimental model systems for the study of this system are electrons on the surface of liquid helium [1], colloidal suspensions [2] and confined plasma crystals [3]. Colloidal particles dissolved in water [4] and placed between two glass plates are another example of an experimental system where classical particles exhibit Wigner crystallization. Recently, macroscopic 2D Wigner islands, consisting of charged metallic balls above a plane conductor were studied and ground state, metastable states and saddle point configurations were found experimentally [5].

Such a system with a finite number of particles, has been extensively studied during the past few years. For a small number of particles (typically $N < 100$) they are arranged in rings and a Mendeleev-type of table was constructed in Ref. [6] which gives the distribution of those particles over the different rings. Moreover, the configurations of the ground state, the metastable states and saddle point states were obtained, from which the transition path and the geometric properties of the energy landscape were given in Ref. [7]. The spectral properties of the ground state configurations were presented in Ref. [8] and generalized to screened Coulomb [9, 10, 11] and logarithmic [12, 13, 11] interactions. The melting properties of this system have been studied by several experimental studies [4] and MC [14, 15] and molecular dynamics [16, 17] simulations.

In this paper we study topological defects which are induced by the confinement potential, i. e. which are a result of the finite size of the system. We discuss the configuration and the properties of the topological defects at zero temperature. Next we investigate how these defects influence the melting of the mesoscopic 2D island.

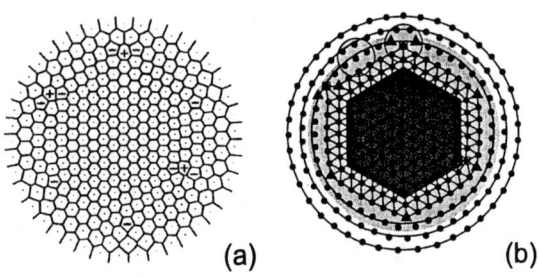

FIGURE 1. The ground state configuration for N=291 particles. (a) The Voronoi construction (b) The configuration consists of three rings at the border, an inner hexagonal region, and the defects are situated in the transition region, located at the six corners of a hexagon. The +1 and −1 topological defects are represented by the open squares and triangles, repectively.

The model system was defined elsewhere [6]. By using MC simulations together with the Newton optimization technique [8], the ground state and the metastable states can be obtained. The structure and potential energy of the system at $T \neq 0$ are found by the standard Metropolis algorithm in which at some temperature the next simulation state of the system is obtained by a random displacement of one of the particles [14].

It is well known that the hexagonal lattice is the most energetically favored structure for classical point charges in a two-dimensional infinite plane at low temperature. For a system consisting of a finite number of repelling particles restricted to 2D, which are held together by a circular harmonic potential, the cluster patterns are determined by the need to balance the tendency to form a triangular lattice against the formation of a compact circular shape. This competition leads to intrinsic defects in the 2D circular Coulomb cluster which are induced by the geometry (of the confinement potential).

We investigated the form and position of these defects in large clusters using the Voronoi constructions. An example (the groundstate for $N = 291$) is shown in Fig. 1(a). One can see the disclinations, i.e. orientational defects with five (indicated by "-") or seven (indicated by "+") fold coordination number. The total number of 5-fold N_- and 7-fold N_+ disclinations depends on the particular configuration. However, the net topological charge $N_- - N_+$ is always equal to six as was already demonstrated in Refs. [10, 11, 17]. We considered the $N = 291$ system as it minimizes the number of defects. The reason is that for this particle number the configuration has 42 particles in the outer ring, which is a multiple of the topological charge.

In these large clusters, we found a hexagonal structure for the location of the defects, as can be seen in Fig. 1(b) for $N = 291$. These defects are always situated approximately around the six corners of a hexagon, each corner with a net topological charge of "−1". Notice that a single 5-fold disclination can appear, but never a single 7-fold disclination. The configuration consists of three rings at the border with an equal number ($N = 42$) of particles (the 1D Wigner lattice), the central hexagonal structure (the 2D Wigner lattice) and the defect-region in between.

We also found that the groundstate with the same number of particles but a different interaction potential forms a different configuration and shows a different defect structure [11]. When the interaction potential changes from long range to short range, the

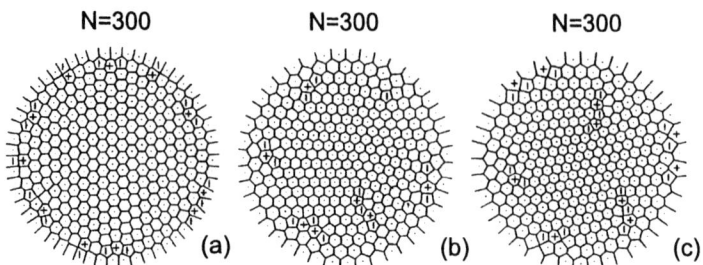

FIGURE 2. The ground state configurations and the Voronoi construction for N=300 particles with different interaction potentials: (a) for the logarithmic potential($-\ln(r_{ij})$), (b) for the Coulomb potential ($1/r_{ij}$) and (c) for the dipole potential ($1/r_{ij}^3$).

defect region will move from the system edge to the center, which is clearly shown in Fig. 2 for $N = 300$: Fig. 2(a) corresponds to the logarithmic potential, the defects only appear at the edge; Fig. 2(b) corresponds to the pure Coulomb potential, the defects appear in the transition part between the outer rings and the center; Fig. 2(c) corresponds to the dipole potential, some defects are situated near the center.

To better describe the melting process in this large-scale system, we separate the configuration of $N = 291$ into three regions as shown in Fig. 1(b). Region I (dark grey colored hexagonal area) is comprised of the defect-free hexagonal center; region II is a transition region with the defects (light grey colored area), and region III consists of the outmost two rings. In order to determine the melting transition point we calculated for each region the mean square displacement $\langle u_R^2 \rangle$, which was introduced in Ref [6].

Fig. 3(a) shows the $\langle u_R^2 \rangle$ as a function of the reduced temperature T/T_0 for the three different regions. At low temperatures the particles exhibit harmonic oscillations around their $T = 0$ equilibrium position, and the oscillation amplitude increases linearly and slowly with temperature: the particles are well localized and display still an ordered structure. Melting occurs when $\langle u_R^2 \rangle$ increases very sharply with T. After the melting point, the particles exhibit liquid-like behavior. Fig. 3(a) exhibits three different melting temperatures corresponding to the three different regions. Firstly region II, i.e. the transition region with the defects, starts to melt, then the outmost two rings melt, and finally the hexagonal region I melts.

In order to investigate the melting in the defect region in further detail we consider two new small regions as showed in Fig1 (b). One region is around a defect, the other doesn't contain a defect. In Fig. 3(b), the $\langle u_R^2 \rangle$ of these two different regions show again a different melting temperature: the melting clearly starts first around the defect. The particles around defect regions are less well interlocked and have a larger diffusion constant than the undistorted lattice regions, their thermal motions are easier to be excited [18].

In contrast to bulk systems, the melting scenario of small laterally confined 2D system was found earlier to be a two step process [6]. Upon increasing the temperature, first intershell rotation becomes possible where orientational order between adjacent shells is lost while retaining their internal order. At even higher temperatures, the growth of

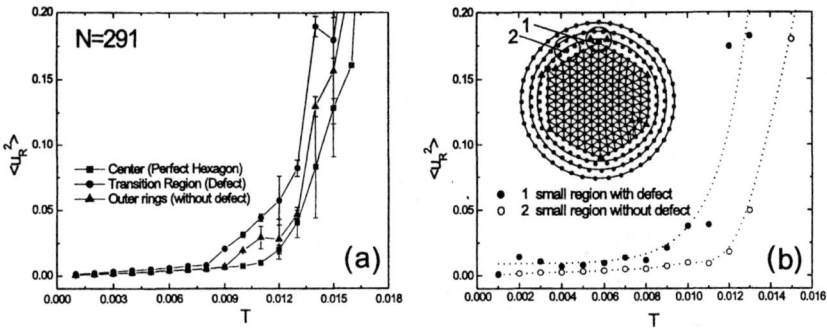

FIGURE 3. The mean square displacements as function of the temperature, (a) for the three regions defined in Fig. 1(a), and (b) for the small defect-free and defect regions.

thermal fluctuations leads to radial diffusion between the shells, which finally destroys the positional order. We also investigated the angular displacements in these large systems and we found that the radial and angular displacements change at approximately the same temperature. Thus for large clusters the intershell rotation will not be favoured.

ACKNOWLEDGMENTS

We are very grateful to I. V. Schweigert for helpful discussions. This work is supported by the Flemish Science Foundation (FWO-Vl), the Belgian Inter-University Attraction Poles (IUAP-V), the "Onderzoeksraad van de Universiteit Antwerpen"(GOA), and the EU Research Training Network on "Surface Electrons on Mesoscopic Structures".

REFERENCES

1. Grimes, C. C., and Adams, G., *Phys. Rev. Lett.*, **42**, 795 (1979).
2. S. Neser, C. B., Leiderer, P., and Palberg, T., *Phys. Rev. Lett.*, **79**, 2931 (1997).
3. Chu, J. H., and I, L., *Phys. Rev. Lett.*, **72**, 4009 (1994).
4. Bubeck, R., Bechinger, C., Neser, S., and Leiderer, P., *Phys. Rev. Lett.*, **82**, 3364 (1999).
5. Jean, M. S., Even, C., and Guthmann, C., *Europhys. Lett.*, **55**, 45 (2001).
6. Bedanov, V. M., and Peeters, F. M., *Phys. Rev. B*, **49**, 2667 (1994).
7. Kong, M., Partoens, B., and Peeters, F. M., *Phys. Rev. E*, **65**, 046602 (2002).
8. Schweigert, V. A., and Peeters, F. M., *Phys. Rev. B*, **51**, 7700 (1995).
9. Candido, L., Rino, J. P., Studart, N., and Peeters, F. M., *J. Phys.: Condens. Matter*, **10**, 11627 (1998).
10. Koulakov, A. A., and Shklovskii, B. I., *Phys. Rev. B*, **57**, 2352 (1998).
11. Lai, Y. J., and I, L., *Phys. Rev. E*, **60**, 4743 (1999).
12. Lozovik, Y. E., and Rakoch, E. A., *Phys. Rev. B*, **57**, 1214 (1998).
13. Partoens, B., and Peeters, F. M., *J. Phys.: Condens. Matter*, **9**, 5383 (1997).
14. Schweigert, I. V., Schweigert, V. A., and Peeters, F. M., *Phys. Rev. Lett.*, **82**, 5293 (1999).
15. Filinov, A. V., Bonitz, M., and Lozovik, Y. E., *Phys. Rev. Lett.*, **86**, 3851 (2001).
16. Schweigert, I. V., Schweigert, V. A., and Peeters, F. M., *Phys. Rev. Lett.*, **84**, 4381 (2000).
17. Lai, Y. J., and I, L., *Phys. Rev. E*, **64**, 015601 (2001).
18. Moore, M. A., and Pérez-Garrido, A., *Phys. Rev. Lett.*, **82**, 4078 (1999).

Dynamical Phenomena in Strongly Coupled Dusty Plasma Under Microgravity Conditions

O. S. Vaulina, A. P. Nefedov, O. F. Petrov, V. E. Fortov

*Institute for High Energy Densities,
Russian Academy of Sciences, Izhorskaya 13/19, 127412, Moscow, Russia*

Abstract. Diffusion of macroparticles, charged by the solar radiation in microgravity are studied by analyzing of experimental data obtained on the MIR space station. Temperature, velocity distributions, friction coefficient and diffusion constants were obtained for bronze particles.

One of basic transport phenomena is diffusion. Diffusion occurs in various regimes, for example, the Brownian diffusion of macroparticles suspended in a background gas. In our work, results of an experimental study of diffusion of dust particles, charged by photoemission under microgravity are presented. The data were obtained during investigations of dusty plasma induced by solar radiation on MIR space station, which have shown that under the action of intensive solar radiation the micron-size particles can acquire considerable positive electric charges [1]. The experimental study of dust diffusion was performed for bronze particles with the mean radii $a \cong 37.5$ μm in background gas (neon) at the pressure $P \cong 40$ Tor. The particles were contained in a cylindrical glass tube, the bottom of which was the uviol window intended for the solar irradiating of dust cloud. Extra irradiating of particles by a laser beam was used for improved diagnostics. The image was registered by a videocamera with the field of view ~ 8x9 mm (see Fig. 1).

The experiments were carried out under the following plan: (1) dynamic action (jolt) on the system with the closed window; (2) exposure in darkness ~ 4 s >> v_+^{-1} (v_+ is the frequency of collision of dust with the gas molecules) to reduce of initial dust velocities; (3) irradiating of the tube by solar radiation; (4) relaxation of the particles to the initial state for the time ~ 3-5 min.

The initial dust concentration n_o was varied from 195 to 300 cm^{-3}. The dependencies the relative dust concentration $n_p(t)/n_o$ on the time t are shown in Fig. 2. The photoemission charge of particles was obtained from the approximations of the curves $n_p(t)/n_p(t=40s)$ at $t > 40$ s by the method detailed in [1] and was close to $Z \approx 4$ 10^4 (±15%). Illustrations of simulation of dust transport due to the mutual Coulomb repulsion of particles by the molecular Brownian dynamic method (curve 1) and its analytical approximation (curve 2), which was used for determining dust charge [1], are presented in Fig. 2 for conditions close to experimental ones. Under the solar action, the dust motion acquired a directed motion forward the tube walls. For a time

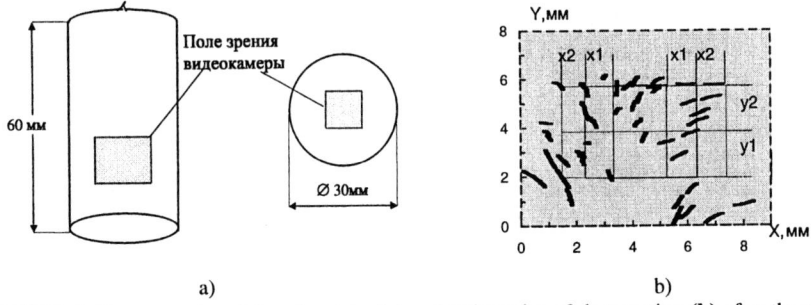

FIGURE 1. The geometry of the glass tube (**a**) and trajectories of dust motion (**b**) after the action of solar radiation.

~ 3 s after the beginning of solar irradiation, the dust stochastic kinetic energy is increased, and the interparticle correlation changed. The pair correlation functions of particles obtained by the exclusion of interparticle distances less then $l_p/2$ are shown in Fig. 3. These functions may not be suitable for quantitative analyses of the phase state of dust structure but they are reflected the qualitative changes in the system.

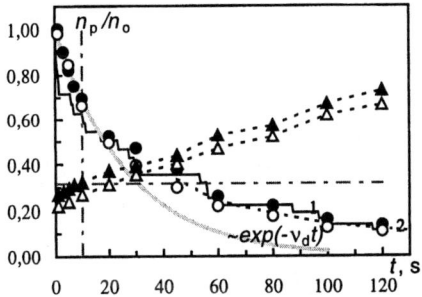

FIGURE 2. Dependencies of relative concentration n_p/n_0 (○; ●) and ratio of λ/R (△; ▲) versus time t for the different initial concentration n_0: (●; ▲) - 195 cm^{-3}; (○; △) - 300 cm^{-3}.

Trajectories of 40 particles (for 5 sec after the beginning of solar irradiation) are shown in Fig. 1b. Irregular fluctuations of velocity (V_x, V_y) of the separate particles on a background of their total drift motion reflect the dust temperature T, which for Maxwellian distribution can be obtained from:

$$T_{x(y)} = m_+ \{<V_{x(y)}^2> - <V_{x(y)}>^2\}, \tag{1}$$

Here $<>$ is the averaging of velocities on the time, and $<V_{x(y)}> = V_d^{x(y)}$ is the regular drift velocity. Determining of the dust temperature from Eq.(1) for the various experiments gives $T_x \cong 51$ eV, $T_y \cong 22$ eV, within 5%. Similar non-uniform

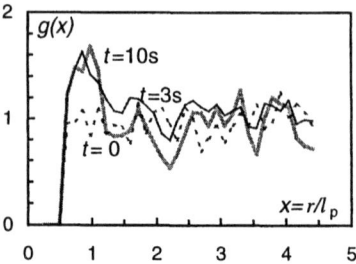

FIGURE 3. Pair correlation function $g(x)$ versus $x = r/l_p$ for several observation time t.

distributions of stochastic kinetic energy ($T_x \neq T_y$) and "abnormal heating" were observed in a number of other dusty plasma experiments and can be related to the temporally-spatial fluctuations of dust charge, for example, due to the random nature of charging currents, or the spatial inhomogeneity of system [2-7].

As the considered system consists of the positively charged macroparticles and the photoelectrons with the density $n_e \sim Zn_p$ emitted by them, it is possible to assume, that the transport properties of this system will depend on ambipolar diffusion of the particles. Because of the large difference of mobility of electrons μ_e and dust μ_+, a negative surface charge appears on the tube walls. The incipient polarization electric field blocks further partitioning of the charged components. Therefore, the electrons and the heavy dust particles can diffuse "together" with some effective coefficient D_a of ambipolar diffusion [8]:

$$D_a = \{D_e \mu_+ + D_+ \mu_e\}/\{\mu_+ + \mu_e\}. \qquad (2)$$

Here D_e, D_+ are the free diffusion constants for electrons and particles:

$$D_{e(+)} = T_{e(+)}/\nu_{e(+)} m_{e(+)}, \qquad (3)$$

where $T_{e(+)}$, $m_{e(+)}$ and $\nu_{e(+)}$ are the temperature, the mass and the frequency of collisions with the neutral gas molecules for electrons and dust, respectively. Then, in the case of $\mu_e \gg \mu_+$, the ratio of diffusion constants can be represented in the form:

$$D_a/D_+ \approx 1 + ZT_e/T_+ \qquad (4)$$

With the measured temperatures (T_x, T_y), the ratios of diffusion constants can be estimated as $D_a/D_+^x \approx 0.8\text{-}1.6\ 10^3$, and $D_a/D_+^x \approx 1.8\text{-}3.6\ 10^3$ for $T_e = 1\text{-}2$ eV.

In a plasma with the density n, the diffusion have a ambipolar character when $\delta n = |n_e - n_+| \ll n \approx n_e \approx n_+$. For cylinder with the radius R, it is valid for $\delta n/n \approx (\lambda/R)^2 \ll 1$, where $\lambda^2 = T_e/4\pi e^2 n_e$ [8]. Taking into account that $n_e \approx Zn_p$ we have $\delta n/n \approx 2.1\text{-}6.4$

10^{-2} for the considered initial conditions $n_p = n_o$ under the assumption of $T_e \cong 1\text{-}2$ eV. The dependencies of λ/R on time are shown in Fig. 2 for $T_e = 2$ eV.

Assuming that for $\delta n/n < 0.1$ the losses of charges in our experiments are connected with their ambipolar diffusion to the ampoule walls, the area of ambipolar diffusion can be determined from the mean velocity of diffusion losses of macroparticles

$$dn_p/dt = - n_p \, v_d, \qquad (5)$$

where $v_d = D_a/\Lambda^2$ is the frequency of diffusion losses, and Λ is some diffusion length [8]. For cylinder with the radius R and the height $\sim 4R$, the value of $\Lambda \approx R/2$. The value of v_d can be obtained from the experimental curve $n_p(t)/n_o$ at $t < 10$ s, where the $n_p(t)/n_o$ function agrees well with the solution $n_p = n_o \exp(-v_d t)$ of Eq.(5) for $v_d \approx 0.0135$ s^{-1} (Fig. 2). Then, an estimate of ambipolar diffusion gives $D_a \approx 2 \cdot 10^{-2}$ cm^2/s.

The free diffusion constants $D_+^{x(y)}$ can be retrieved from the measurements of dust temperature, and drift velocity $V_d^{x(y)}$:

$$D_+^{x(y)}(t) = \{<\Delta r(t)^2> - (V_d^{x(y)} t)^2\}/2t, \qquad (6)$$

where $<\Delta r(t)^2>$ is the mean-square displacement of separate particles in the direction of axis OX (or OY). Thus, the values of free diffusion constants can be estimated as $D_+^x \approx 1.3 \cdot 10^{-5}$ cm^2/s and $D_+^y \approx 5.7 \cdot 10^{-6}$ cm^2/s. The value of ratio $D_+^x/D_+^y \approx 2.28$ agrees very well with the measured dust temperature, and both the measured ratios $D_a/D_+^x \approx 1538$, and $D_a/D_+^x \approx 3509$ are in agreement with the theoretical prediction (4).

ACKNOWLEDGMENTS

This work was supported in part by the Russian Foundation for Basic Research, Grant No. 00-02-81036 and No.00-02-17520, and by NWO Grant No. 047-008-013.

REFERENCES

1. Fortov, V.E., Nefedov, A.P., Vaulina O.S.,*et al, JETP*, **87**, 1087 (1998).
2. Vaulina, O.S., Nefedov, A.P., Petrov, O.F., and Fortov, V.E., *JETP*, **91**, 1063 (2000).
3. Vaulina, O.S., Khrapak, S.A., Nefedov, A.P., et al, *Phys. Rev. E*, **60**, 5959 (1999).
4. Quinn, R.A., and Goree, J., *Phys. Rev. E*, **61**, 3033 (2000).
5. Zhakovskii, V.V., Molotkov, V.I., Nefedov, A.P. et al, *JETP Lett.*, **66**, 392 (1997).
6. Thomas, H., and Morfill, G., *Nature,* **379**, 806 (1996).
7. Melzer, A., Homann, A., and Piel, A., *Phys. Rev. E,* 53, 2757 (1996).
8. Raizer, Y.P., *The Physics of Gas Discharge* (Berlin: Springer Verlag, 1991).

Experiments and Simulation of Elastic Waves in a Plasma Crystal Radiated from a Point-Dipole-Source

A. Piel[*], V. Nosenko[†] and J. Goree[†]

[*]IEAP, Christian-Albrechts-University, D-24098 Kiel, Germany
[†]Department of Physics and Astronomy, The University of Iowa, Iowa City, IA52242, USA

Abstract. A localized elastic deformation in the plane of a 2D plasma crystal is generated by a short laser pulse. The perturbed region simultaneously radiates compressional and shear waves. The decomposition of the complex wave pattern into the fundamental wave types is achieved by calculating the divergence and vorticity of the instantaneous vector velocity map. The shear waves form two vortex-antivortex pairs, which are known as the lowest modes of excitation in finite Yukawa clusters. The higher dispersion of the compressional waves leads to the formation of a wave train. The angular intensity distribution of the two wave types corresponds to two orthogonal dipole sources. Molecular dynamics simulations closely confirm these experimental results.

INTRODUCTION

A two-dimensional lattice of particles interacting by repulsive Yukawa-type potentials has two fundamental wave types, compressional and shear waves [1]. The radiation pressure of a focused laser was used before to excite compressional waves [2, 3] and plane shear waves [4] in plasma crystals. Compressional [5] and shear wave Mach cones [6] were generated by a moving laser spot. The different wave speeds of compressional and shear waves are convenient diagnostic tools, e.g., to determine the shielding length [1]. Different from earlier investigations in the present experiments the near-field of a small antenna is studied. The radiation from of a small elastic dipole is a fundamental problem, which - like the Hertzian dipole in electrodynamics - can be used as a Green function for more complex situations.

EXPERIMENTAL ARRANGEMENT

The experimental arrangement is similar to the one used in [5]. A single-layer plasma crystal is formed of spherical plastic particles with 8.1 μm diameter. The plasma is generated in argon gas at 15 mTorr pressure and 18 W rf power. The field of view, 24 mm (x) × 18 mm (y), contains 990 particles with an average particle distance of 0.66 mm. The expanded beam of an argon-ion laser hits the plasma crystal at a small angle of incidence (10°) (Fig. 1a) and illuminates a nearly square spot of 4 mm by 4 mm size, where the laser force causes a localised shear stress that elastically deforms the crystal.

Laser power (0.5 W) and laser pulse duration (250 ms) are chosen to avoid breaking of bonds in the plasma crystal.

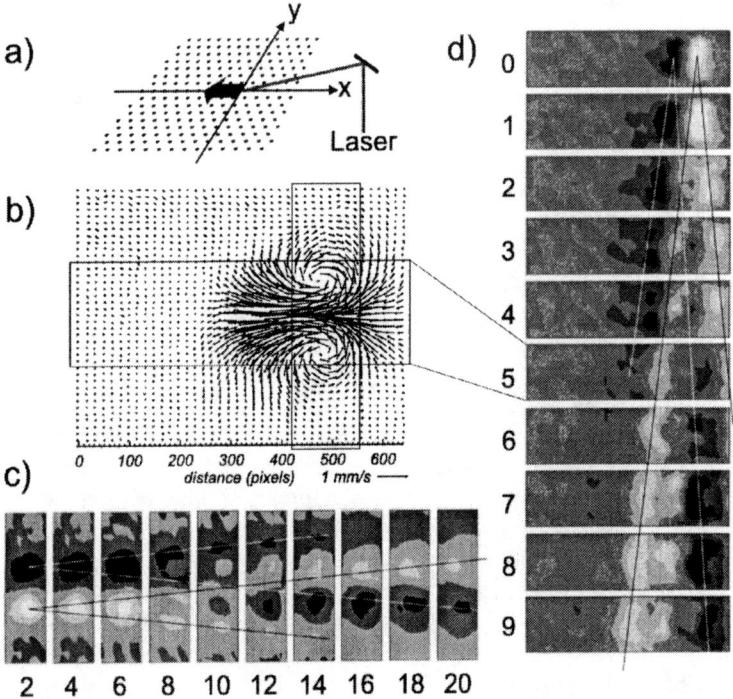

FIGURE 1. (a) Geometry for localized elastic deformation of a 2D plasma crystal by a laser force. (b) Vector velocity map of particle motion during the laser pulse. (c) Visualisation of the shear wave in terms of the vorticity of the velocity field (black=cw, white=ccw). The stripes are labeled with their frame number. (d) Visualisation of the compressional wave by the divergence of the velocity field (black=compression, white=rarefaction).

The initial perturbation is a small elastic dipole. In the x-direction, the deformation mainly consists of a compression / rarefaction pair. In y-direction a symmetric shear stress is generated. The motion of the individual particles is recorded with a video camera with 640 × 480 pixel resolution at 30 frames per second and a total of 32 frames. The experiment was repeated 100 times after 5.9 s recovery time each. The velocity field is calculated from particle positions with subpixel resolution in subsequent video frames, and interpolated to a fixed square grid for averaging. Already during the laser pulse a complicated velocity field develops, which is shown for frame 5 in Fig. 1b. This velocity field is governed by a pronounced pair of elastic vortices of opposite sign. Although this velocity field resembles streamlines in fluids, there is no macroscopic flow but only a momentary elastic deformation. After the end of the laser pulse, this elastic deformation relaxes and expands over a larger region than originally illuminated and is eventually radiated from the source region as a complex pattern of compressional and shear waves.

RESULTS AND CONCLUSIONS

The complex wave pattern is decomposed into the compressional and shear wave by calculating the divergence map $\nabla \cdot \vec{v}$ and the vorticity map $\nabla \times \vec{v}$. The vorticity was found insensitive to compressional wave activity and the divergence of a shear wave was likewise negligible. The rectangular stripes in Fig. 1b are used for displaying appropriate sequences of images of the shear wave (Fig. 1c) and compressional wave (Fig. 1d) as grey-scale contour plots.

The initial situation for the shear wave in frame 2 of Fig. 1c shows a pair of vortices with opposite sign of vorticity. The positive and negative structure show a most pronounced evolution in $\pm y$ direction. This is the expected propagation direction for a plane shear wave, if the excitation region were extended in x-direction. Two pairs of lines are superimposed to guide the eye in tracing the propagation of the individual structures. In frame 8, the splitting of the structures into an outward and inward going wave pulse becomes evident. After a cross-over of the inward going waves in frame 10, these waves appear as outward going waves in subsequent frames. The wave field eventually consists of two vortex-antivortex pairs propagating outwards in $+y$ and $-y$ direction each. The leading vortex shows a stronger damping than the subsequent antivortex.

For the compressional wave (Fig. 1d) the situation is similar, but not identical. There, the initial pair of compression and rarefaction in frame 0 is found to propagate in $\pm x$ direction. Again, the initial perturbation splits into ingoing and outgoing pulses. The ingoing pulses have a cross-over in frame 2 and reappear as outward going pulses in subsequent frames. Obviously the propagation speed for the compressional wave is higher than for the shear wave. While the wave field of the shear wave only consists of two pairs of vortices, the compressional wave develops additional wave humps, as becomes evident from frames 7-9.

The sound speeds of compressional and shear wave are different for 2D-crystals with Yukawa-type interaction potentials of the microparticles [1]. The ratio of the two sound speeds is a convenient means to determine the shielding strength $\kappa = b/\lambda_D$, in which b is the mean interparticle spacing and λ_D the Debye shielding length [5]. Here we find sound speeds of $(C_L = 24 \pm 2)$ mm/s for the compressional wave and $(C_T = 7 \pm 1)$ mm/s for the shear wave. From tabulated values of the sound speed ratio [7] we obtain $\kappa = 2.2 \pm 0.9$.

For comparison, we have performed molecular dynamics (MD) simulations of localized wave excitation. The simulation starts with a regular triangular 2D crystal with $b = 0.75$ mm. The simulation area of 24 mm (x) by 19.486 mm (y) corresponds to 32 by 30 particles. The static lattice is continued with periodic boundary conditions to confine the particles in the simulation box. Particles in the neighboring boxes are fixed, to avoid wave excitation in the exterior region. This approach allows to study the wave field until the faster wave reaches the boundary of the box. The interparticle distance, the total number of particles (960) and the particle charge $Q = 14,000e$ were chosen similar to the experimental conditions. The shielding factor $\kappa = 1$ was chosen for comparison with the theoretical dispersion curve in [8] and is lower than the experimental value. This difference does not affect the topology of the different wave modes. The shear wave (Fig. 2), develops spherical wavefronts with an approximately $\sin^2(\alpha)$ intensity distribution. The observation in the experiment, that the original vortex-antivortex pair splits into ingoing and outgoing pulses which, after crossing in the center, ultimately form two outgoing vortex-antivortex pairs, is confirmed in every detail by the MD-simulation. The compressional wave also shows pulse-splitting and crossover but eventually develops into a wave train because of the stronger dispersion.

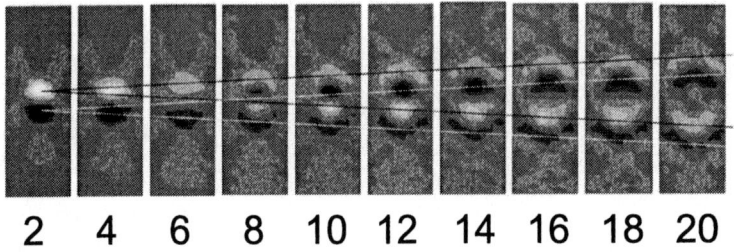

FIGURE 2. Molecular dynamics simulation showing the shear wave radiated from a localized dipole source. The vorticity of the velocity field is shown on a grey scale similar to Fig. 1c.

In conclusion, the complex wave pattern in the near field of a small elastic dipole can be decomposed into outward going compressional and shear waves, which eventually form spherical wave fronts. The finite size of the excitation region generates additional inward going waves, which cross at the center of the source region and reappear on the opposite side. The intensity distribution of the two wave types is found dipole-like with orthogonal orientation. The shear wave is found to be non-dispersive whereas the pronounced dispersion of the compressional wave leads to the formation of a wave train. This finding corresponds to the pronounced interference structures found inside the Mach cone of a compressional wave [5, 8], which is absent in shear Mach cones [6]. Because of the weak dispersion, the initial shear stress in the excitation region remains confined to the wave front of the shear wave by forming a vortex-antivortex pair. Theoretical studies of large Yukawa clusters [9] had shown before that vortex-antivortex pairs are the lowest mode of excitation. Here, we observe that vortex-antivortex pairs are the building-blocks of the wave front.

ACKNOWLEDGMENTS

AP thanks DFG for financial support under grant Pi185/22-1. Work at Iowa was supported by NASA and DOE.

REFERENCES

1. Peeters, F. M., and Wu, X., *Phys. Rev. A*, **35**, 3109–3114 (1987).
2. Homann, A., Melzer, A., Peters, S., Madani, R., and Piel, A., *Phys. Rev. E*, **56**, 7138–7141 (1997).
3. Homann, A., Melzer, A., Peters, S., Madani, R., and Piel, A., *Phys. Lett. A*, **242**, 173–180 (1998).
4. Nunomura, S., Samsonov, D., and Goree, J., *Phys. Rev. Lett.*, **84**, 5141–5144 (2000).
5. Melzer, A., Nunomura, S., Samsonov, D., and Goree, J., *Phys. Rev. E*, **62**, 4162–4176 (2000).
6. Nosenko, V., Goree, J., Ma, Z. W., and Piel, A., *Phys. Rev. Lett.*, **88**, 135001 (2002).
7. Wang, X., Bhattacharjee, A., and Hu, S., *Phys. Rev. Lett.*, **86**, 2569–2572 (2001).
8. Dubin, D. H. E., *Phys. Plasmas*, **7**, 3895–3903 (2000).
9. Candido, L., Rino, J.-P., Studart, N., and Peeters, F., *J. Phys.: Condens. Matter*, **10**, 11627 (1998).

Shock wave–like structures in complex plasmas: Theory and experiments

S. I. Popel, A. P. Golub', and T. V. Losseva

Institute for Dynamics of Geospheres RAS, Leninsky pr. 38, bld. 6, 117334 Moscow, Russia

Abstract. For the description of dust ion–acoustic shocks we have forwarded the so–called ionization source model. We show that this model is appropriate for the description of laboratory experiments in complex plasmas.

The problem of shock waves occupies an important place in present–day physics of complex (dusty) plasmas. Shock waves in a dusty plasma have specific features that distinguish them from ordinary collisional and collisionless shock waves and are attributed, in particular, to the anomalous dissipation originating from dust grain charging. That dust ion–acoustic shock waves associated with anomalous dissipation can actually exist was proved analytically in [1]. Dust ion–acoustic shock waves were observed for the first time in laboratory experiments at the University of Iowa (USA) [2] and at the Institute of Space and Astronautical Science (Japan) [3]. There is a possibility of observation of shocks related to the dust charging process in active rocket experiments, which involve the release of some gaseous substance in near–Earth space in the form of high–speed plasma jet [4]. Here, we present a theoretical model for describing dust ion–acoustic shock waves and compare theoretical conclusions with the data of laboratory experiments [2, 3].

The experiments carried out by Luo et al. [2] with the Q–machine showed that:

(i) Dust ion–acoustic shock waves are generated at sufficiently high dust densities (under the experimental conditions of [2], at dust densities such that $\varepsilon Z_{d0} \equiv n_{d0} Z_{d0}/n_{i0} \geq 0.75$, where $q_d = -Z_d e$ is the grain charge, $-e$ is the electron charge, n_d is the dust density, n_i is the ion density, and the subscript 0 stands for the unperturbed plasma parameters). In [2], the conclusion about the formation of a shock wave was drawn from the fact that the perturbation front steepens as time elapses. At sufficiently low dust densities, the perturbation front does not steepen but instead widens.

(ii) The velocity of the dust ion–acoustic waves increases considerably with increasing εZ_{d0}.

In experiments with a double plasma device, Nakamura *et al.* [3] revealed that:

(iii) The most important feature of ion–acoustic waves in a dusty plasma is the following. In the absence of dust, the effect of the electron and ion charge separation gives rise to oscillations in the shock wave profile in the vicinity of the shock front, while the presence of dust suppresses these oscillations.

The evolution of a perturbation and its transformation to a nonlinear wave structure are described by the hydrodynamical equations of the ionization source model [5, 6], which are derived on the basis of a classical *kinetic* approach to describing complex plasmas (see, e.g., [7]). The equations are:

(a) The evolutionary equations for the ions,

$$\partial_t n_i + \partial_x (n_i v_i) = -v_{ch} n_i + S_i, \qquad (1)$$

$$\partial_t (n_i v_i) + \partial_x (n_i v_i^2) = -\frac{e n_i}{m_i}\partial_x \varphi - \frac{T_i}{m_i}\partial_x n_i - \tilde{v} n_{0i} v_i, \qquad (2)$$

and a Boltzmann distribution for the electrons

$$n_e = n_{e0} \exp\left(\frac{e\varphi}{T_e}\right). \qquad (3)$$

Here, v_i is the ion velocity, m_i is the mass of an ion, n_e is the electron density, $T_{e(i)}$ is the electron (ion) temperature, S_i is the ionization source intensity (its value is chosen so that it exactly cancels the term describing the absorption of ions by dust grains in an unperturbed dusty plasma), φ is the electrostatic potential, the rate v_{ch} at which the ions are absorbed by the dust grains is equal to

$$v_{ch} = v_q \frac{Z_{d0} d}{1 + Z_{d0} d} \frac{(T_i/T_e + z_0)}{z_0 (1 + T_i/T_e + z_0)}, \qquad (4)$$

$d = n_{d0}/n_{e0}$, $v_q = \omega_{pi}^2 a (1 + z_0 + T_i/T_e)/\sqrt{2\pi} v_{Ti}$ is the grain charging rate, ω_{pi} is the ion plasma frequency, a is the grain radius, $z = Z_d e^2/aT_e$, v_{Ti} is the ion thermal velocity, the rate $\tilde{v}$ at which the ions lose their momentum as a result of their absorption on the grain surfaces and their Coulomb collisions with the grains has the form

$$\tilde{v} = v_q \frac{Z_{d0} d}{(1 + Z_{d0} d) z_0 (1 + T_i/T_e + z_0)} \left(z_0 + \frac{4T_i}{3T_e} + \frac{2z_0^2 T_e}{3T_i}\Lambda\right), \qquad (5)$$

$\Lambda = \ln(\lambda_{Di}/\max\{a,b\})$ is the Coulomb logarithm, λ_{Di} is the ion Debye radius, and $b = Z_{d0} e^2/T_i$. Expressions (5) and (6) are valid in the range $v_i/c_s < 1$, where c_s is the ion–acoustic speed;

(b) Poisson's equation for the electrostatic potential,

$$\partial_{xx}^2 \varphi = 4\pi e (n_e + Z_d n_d - n_i); \qquad (6)$$

(c) The evolutionary equation for the dust grain charge,

$$\partial_t q_d = I_e(q_d) + I_i(q_d), \qquad (7)$$

Here, the electron and ion microscopic currents to the grain surface (see, e.g., [6]) are obtained in accordance with the orbit–limited probe model.

We have taken into account the fact that in the laboratory experiments of [2] cesium ions in the plasma were produced through ionization of cesium atoms at the hot plate surface. In the experiments of [3], the electron mean free paths were so long that the neutrals were ionized presumably in collisions with the wall [6]. Consequently, under the experimental conditions of [2, 3], the ionization source term in the evolutionary equation for the ion density should be independent on the electron density and can be chosen as a constant.

We test our theoretical model against the experimental results (i)–(ii), which were obtained in [2]. To do this, we use Eqs. (1)–(7) to trace the evolution of a rectangular initial perturbation in the ion density profile under the conditions of those experiments. The conditions correspond to those of [2]. In particular, calculations were performed for the following values of the plasma parameters: the electron and ion temperatures were equal to one another, $T_e = T_i = 0.2$ eV; the background ion density $n_{i0} = 1.024 \cdot 10^7$ cm^{-3} was the same for all series of simulations; the grain radius was $a = 0.1$ μm; the width of the rectangular initial perturbation was $\Delta x = 25$ cm; and the excess initial perturbed ion density above the background ion density in the remaining unperturbed plasma was $\Delta n_i/n_{i0} = 2$ (see Fig. 2 in [2]). In the experiments of [2], a cesium vapor plasma (containing Cs$^+$ ions) was created through surface ionization. In other words, a cesium atom striking the hot plate becomes ionized. The newly produced cesium ion flies away from the plate at a certain directed velocity. Hence, we can expect the ion flux to be generated in the immediate vicinity of the plate. The intensity of the ion flux and its density are strongly sensitive to the plate temperature. In calculations, this dependence was modeled [6] by imposing the corresponding boundary condition at the surface of the hot plate (analogous to the related boundary condition in the surface evaporation problem [9]).

We show (see Fig. 1 in [6]) that there is widening of the wave front (at $\varepsilon Z_{d0} = 0$) and its steepening (at $\varepsilon Z_{d0} = 0.75$). This agrees with the experimental data from [2]. Note that it is the above boundary condition that allowed us to numerically capture the effect of the widening of the wave front in the absence of dust. The extent to which the shock front widens was calculated to be $\Delta \xi/Mc_s \sim 0.3$ ms, which corresponds to that observed experimentally (see Fig. 2b in [2]).

Figure 1 shows the dependence of the perturbation front velocity (normalized to its value in the absence of dust, $\varepsilon = 0$) on the parameter εZ_{d0}. For comparison, we also plot the experimental points (crosses) taken from Fig. 5 in [2]. The calculated results are represented by circles. The agreement between theory and experiment is quite good.

Now, we test our theoretical model against the experimental result (iii), which was obtained in [3]. The experiments described in that paper were carried out with a double plasma device. The calculations were carried out for different dust densities and for the following parameter values corresponding to the experimental those: $n_{i0} = 2.3 \cdot 10^8$ cm^{-3} (the ion background density was the same for all series of simulations), and $a = 4.4$ μm. The width of the perturbation ($\Delta x \approx 20$ cm) and its shape were determined self-consistently, in accordance with the method for exciting a shock wave. In accordance with the conclusions made in [8], in the calculations the values of the electron and ion temperatures were taken to be the same $T_e = T_i = 1.5$ eV. We show (see Fig. 3 in [6]) that the electron and ion charge separation gives rise to oscillations in the shock wave profile and that the dust suppresses these oscillations, as is the case in the experiments of [3]. The theoretically calculated rise time of the shock front is about 5 μs, which corresponds to the experimental data.

Thus the ionization source theoretical model developed here makes it possible to describe all the main results of the laboratory experiments on dust ion–acoustic shocks.

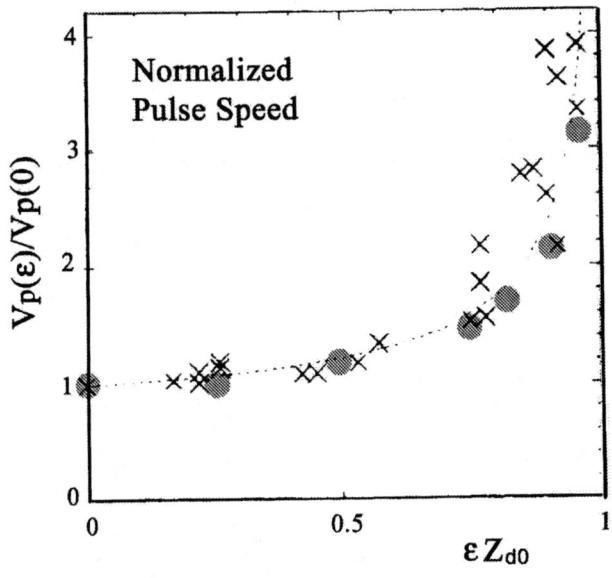

FIGURE 1.

ACKNOWLEDGMENTS

This work was supported by INTAS (grant no. 97–2149), INTAS–RFBR (grant no. IR–97–775), and RFBR (grant no. 02–02–26678).

REFERENCES

1. Popel, S. I., Yu, M. Y., and Tsytovich, V. N., *Phys. Plasmas*, **3**, 4313 (1996).
2. Luo, Q.-Z., D'Angelo, N., and Merlino, R. L., *Phys. Plasmas*, **6**, 3455 (1999).
3. Nakamura, Y., Bailung, H., and Shukla, P. K., *Phys. Rev. Lett.*, **83**, 1602 (1999).
4. Popel, S. I. and Tsytovich, V. N., *Astrophys. Space Sci.*, **264**, 219 (1999).
5. Popel, S. I., Golub', A. P., Losseva, T. V., Bingham, R., and Benkadda, S., *Phys. Plasmas*, **8**, 1497 (2001).
6. Popel, S. I., Golub', A. P., and Losseva, T. V., *JETP Letters*, **74**, 362 (2001).
7. Benkadda, S., Gabbai, P., Tsytovich, V. N., and Verga, A., *Phys. Rev. E*, **53**, 2717 (1996).
8. Nakamura, Y. and Bailung, H., *Rev. Sci. Instrum.*, **70**, 2345 (1999).
9. Knacke, O. and Stranski, I. N., *Prog. Met. Phys.*, **6**, 181 (1956).

Charge Exchange Collisions and the Current to Probes and Dust Particles

Scott Robertson and Zoltan Sternovsky

Department of Physics, University of Colorado, Boulder, CO 80309-0390 USA

Abstract. Recent theoretical work has shown that the current of charge exchange ions to dust particles can be significant even when the mean free path greatly exceeds the Debye length. We have performed an experimental test of the theory, reformulated for cylindrical geometry, using cylindrical probes in partially ionized low-density argon plasma. The charge exchange current, found by subtracting the calculated OML current from the total current, agrees with the calculated value and has the expected dependences upon electron density, neutral pressure, and probe radius.

INTRODUCTION

Charge exchange collisions affect the Debye shielding of dust particles as well as the ion current they collect [1]. Lampe et al. [2,3] have recently considered self-consistently the shielding of dust particles by ions. A large fraction of orbiting ions in the shielding cloud has been found and also an increase in the ion current. The calculations showed that the ion current arising from charge exchange can be significant, even when the charge exchange mean free path is orders of magnitude larger than the Debye length. Diagnostic methods for dust particles are not sufficiently developed for an experimental test of the theory. The cylindrical probe, however, is a standard diagnostic tool and the theory for the charge exchange current to spherical dust particles should apply to probes if the theory is reformulated for cylindrical geometry. We present a brief outline of the theory for the charge exchange current to a probe and experimental results which show good agreement.

THEORETICAL MODEL

The current of charge exchange ions collected by a probe is found from the volume integral of the charge exchange ion generation rate. The ions are collected only if their angular momentum is sufficiently small, thus the volume integral must be weighted by the collected fraction. Collection occurs if the angle between the velocity vector and the radius vector is smaller than a critical value [2]

$$\theta_c(v_1, r_1, a, V_B) = \sin^{-1}\left(\frac{a}{r_1}\right)\left[1 + \frac{2e(\Phi(r_1) - V_B)}{m_i v_1^2}\right]^{1/2}$$

where a is the probe radius, the ion is created at r_1 with velocity v_1, V_B is the probe bias potential, $\Phi(r)$ is the space potential, e is the elementary charge, and m_i is the ion mass. In cylindrical geometry, this angle defines a pair of wedges in velocity space within which the ion is collected. If the ion is trapped in the potential well, $m_i v^2 < 2e\Phi(r)$, wedges for both ascending and descending particles are collected and the collected fraction is $P(v_1,r_1,a,V_B) = 2\theta_c(v_1,r_1,a,V_B)/\pi$. In cases where the argument of the sine is greater than unity, there is no angle that escapes capture, θ_c is replaced by $\pi/2$, and the collected fraction is replaced by unity.

The probe current from charge exchange ions, I_{CX}, is obtained from the volume integral of the rate of generation $R_{cx}(r) = n_n n_0 \sigma v_c(r)$, where n_n is the neutral gas density and σ is the charge exchange cross section. The velocity of collisions v_c is determined by the potential well depth for $-e\Phi(r) > T_i$, thus $v_c(r) \cong \sqrt{-2e\Phi(r)/m_i}$. The charge exchange volume is modeled as a cylinder surrounding the probe with the same length l as the probe. The charge exchange current for a probe bias V_B is then

$$I_{CX}(V_B) = l e n_n n_0 \sigma \int_a^s 2\pi r dr \int_0^\infty 2\pi v dv \left(\frac{-2e\Phi(r)}{m_i} \right)^{1/2} \frac{2\theta_c(r,a,V_B)}{\pi^2 v_t^2}.$$

The integral extends from the probe radius a to the sheath thickness s defined by $-e\Phi(s) = T_i$. It is possible to show [4] that the integral from s to infinity is equivalent to the orbit motion limited (OML) contribution [5] to the ion current.

Evaluation of I_{CX} requires knowledge of the sheath potential profile $\Phi(r)$. This is found in a separate calculation by integration of Poisson's equation. The electron density, given by the Boltzmann factor, affects the potential profile and has a weak influence on I_{CX}. In cylindrical geometry, the ion density is constant in the sheath. The ion trajectories converge toward a negatively biased probe thus increasing the ion current density; however, this is exactly balanced by the increase in the ion velocity [6]. Except for the smallest size of probe used in the experiments, the density of orbiting ions is relatively small and has not yet been considered in the calculations.

EXPERIMENTS

The experiments are performed in the apparatus shown in Fig. 1. Argon plasma is generated by 60 eV electrons from a filament in a cylindrical chamber 31 cm in diameter and 38 cm in height. An electron emission of 2 – 40 mA in 0.15 – 0.8 mTorr of Ar gas results in plasma densities from 5×10^6 to 1×10^8 cm^{-3}. Cylindrical probes, placed in the center of the chamber, have radii from 0.063 to 0.32 mm and a length of 85 mm. The chamber also operates as an electromagnetic cavity and the volume-averaged plasma density is obtained from the shift in the frequency of a microwave mode near 2.11 GHz [7]. The electron temperature from the electron part of the probe characteristic is 1 – 3 eV. The density obtained from the electron saturation current is 0.5 – 1.0 times the microwave value and is not used in calculation. The ion part of the characteristic is found iteratively by subtraction of the electron part.

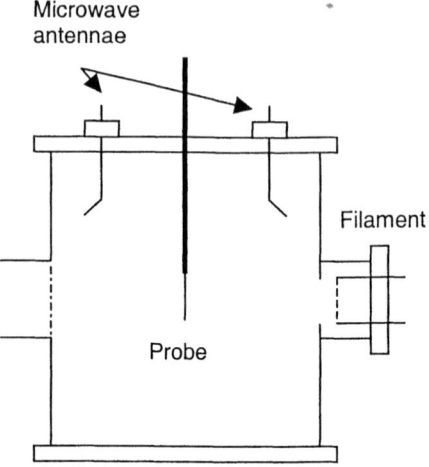

Figure 1. Schematic diagram of the experimental apparatus. The electron density is measured by the microwave cavity method rather than from the electron saturation current for greater accuracy.

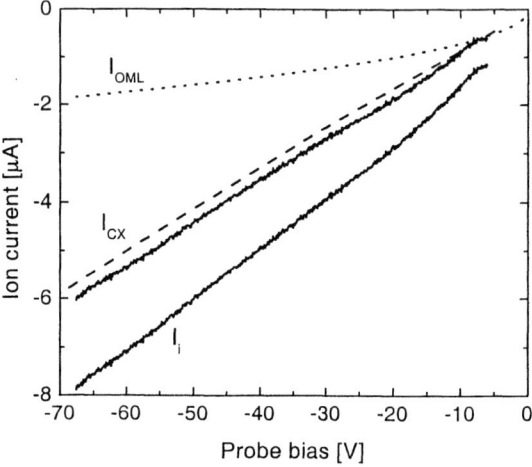

Figure 2. The measured net ion current I_i (lower line) as a function of the bias voltage of a probe with radius 125 μm. The plasma density, neutral Ar pressure and electron temperature are 3×10^7 cm^{-3}, 0.47 mTorr, and 1.8 eV, respectively. The electron Debye length is 1.82 mm.

The data in Figure 2 show the calculated and measured probe currents for typical experimental conditions. The charge exchange current (upper solid line) is found by subtracting the calculated OML current (dotted line) from the measured net current (lower solid line). The calculated charge exchange current (dashed line) is in close agreement. The OML current increases as the square root of the bias voltage and the charge-exchange current increases approximately linearly. For these conditions, the charge exchange current is about twice the current from OML theory. Figure 3 shows, for a variety of conditions, that the charge exchange current has the expected linear variation with pressure. The deviation from linearity arises in part from a weak dependence upon the electron temperature. Additional data (not shown) have verified the dependence upon probe radius for $a = 0.06 - 0.32$ mm which is 0.2–1.0 ion Debye lengths.

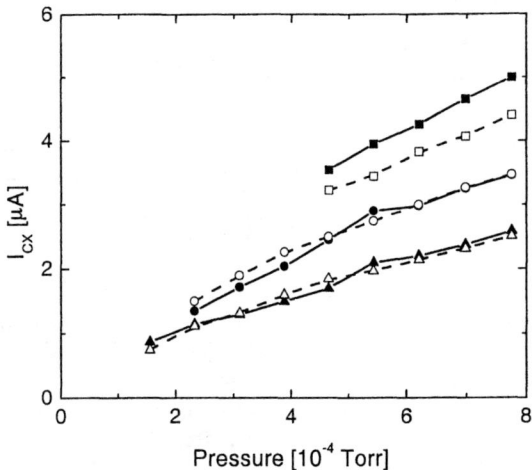

Figure 3. The charge exchange current as a function of the neutral Ar pressure for three different plasma densities: 3×10^7 cm^{-3} (solid squares), 1.4×10^7 cm^{-3} (solid circles), 0.6×10^7 cm^{-3} (solid triangles), and fixed probe bias -30V. The open symbols with dashed lines are the charge exchange currents calculated with the measured electron temperature. These data are for probe radius 188 μm.

ACKNOWLEDGMENTS

The authors acknowledge valuable discussion with Dr. Martin Lampe and support from the United States Department of Energy.

REFERENCES

1. A. V. Zobnin, A. P. Nefedov, V. A. Sinel'shchikov and V.E. Fortov, Zh. Eksp. Teor. Fiz. **118**, 554 (2000) [JETP **91**, 483 (2000).
2. M. Lampe, V. Gavrishchaka, G. Ganguli and G. Joyce, *Phys. Rev. Lett.* **86**, 5278 (2001).
3. M. Lampe, V. Gavrishchaka, G. Ganguli and G. Joyce, *Physica Scripta* **T98**, 91 (2002).
4. Z. Sternovsky, S. Robertson and M. Lampe, submitted for publication.
5. F. F. Chen in *Plasma Diagnostic Techniques*, edited by R. H. Huddlestone and S. L. Leonard (Academic, New York, 1965), Chap. 4.
6. J. G. Laframboise and L. W. Parker, *Phys. Fluids* **16**, 629 (1973).
7. M. A. Heald and C. B. Wharton, *Plasma Diagnostics with Microwaves*, Wiley, New York, 1965, Ch. 5.

Observation of Ion-Ion Instability in Dusty Plasmas

Y. Saitou* and Y. Nakamura[†]

Utsunomiya University, Tochigi 321-8585, Japan
[†]*The Institute of Space and Astronautical Science, Kanagawa 229-8510, Japan*

Abstract. Experiments on the ion-ion instability are performed in a dusty double plasma device. The spatial growth rate of the instability is reduced when glass beads are mixed in the plasma. From measured growth rates, the effective collision frequency for the instability is estimated.

INTRODUCTION

Various kinds of instability are observed in plasmas and varieties are attracting many researchers' interests. The ion-ion instability is one of such instabilities exited in an ion beam-plasma system. Basic characters of the ion-ion instability have investigated in dust-less plasmas [1] but not in dusty plasmas. In this study, we will excite the ion-ion instability in a dusty plasma and investigate an interaction between the dust and the instability. Especially, we will concentrate on an effect of the dust on the spatial growth rate of the instability. Experimental results will be compared with numerical ones.

EXPERIMENTAL SETUP

Experiments are performed in a homogeneous multi-dipole dusty double-plasma device [2]. Schematic drawing of the device is shown in Fig. 1. The device is 90 cm in length and 50 cm in diameter, and separated into a source and a target chamber by a floating mesh grid. An ion beam is injected from the source to the target by applying a potential difference between the two sections. To observe the ion-ion instability and to measure plasma parameters and noise exited in the target chamber, a Langmuir probe which is movable along the chamber axis (x-axis), is inserted in it. A dust dispersing apparatus called "dust dropper" whose cross section 5 cm × 20 cm is suitably mounted in the target chamber. The apparatus consists of a metal box whose bottom is a fine mesh grid of 300 lines per inch, which is coupled to an ultrasonic vibrator. When an alternative-current voltage V_{dd} of 30.7 kHz is applied to the vibrator, dust grains fall from the dust dropper to the plasma. The dust density is controlled by adjusting V_{dd}. The density is measured by observing the reduced intensity of the laser light passing through the dust region, and by using the relation

$$I_1 = I_0 \exp(-an_d L), \quad (1)$$

where I_0 and I_1 are the laser intensity before and after passing through the dust region, a a cross section of a dust grain, n_d the dust density, and L the length of the dust region along the laser light. Glass beads are used as dust grains, whose averaged diameter is 8.9 µm.

Argon gas is fed into the chambers at a pressure of $(2-4) \times 10^{-4}$ Torr. Plasmas in the both sections are produced independently by discharges between tungsten filaments and magnetic cages. Typical plasma parameters are as follows: the electron temperature $T_e \simeq 1.5$ eV, and the electron density $n_e \sim 10^8$ cm^{-3}. The dust density is ranged from 0 to approximately 2.5×10^4 cm^{-3}. The dust has negative charge and its charge is estimated from the quasi-neutral condition. Each dust grain has approximately $(0.5-1) \times 10^4 e$, where e is the elementary electric charge.

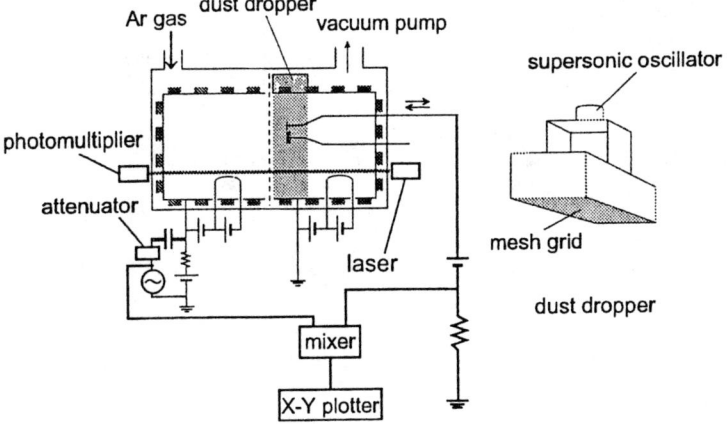

FIGURE 1. Schematic drawings of the experimental device (left) and the dust dropper (right).

EXPERIMENTAL RESULTS

When the ion beam whose normalized velocity U_b/C_s is approximately from 1 to 2 is injected to the target plasma, the ion-ion instability is excited as shown on the top of the left-hand side of Fig. 2. The frequency which gives the maximum amplitude when $x \approx 3-4$ (cm) is around 300 kHz. By applying the voltage V_{dd}, the dust is fed into the plasma. Then dust grains are charged negatively due to the electron impact. As a result, the electron density is reduced by the dust feeding. At the same time, it is found from the left-hand side figures of Fig. 2 that power of the noise corresponding to the instability decreases with increasing the dust density. In fact, as shown in right-hand side figures of Fig. 2, the spatial growth rate of the instability becomes small for a larger dust density.

DISCUSSION

In order to investigate why the growth rate of the instability is suppressed in a dusty plasma, we will compare the growth rate obtained by the experiment and by the numer-

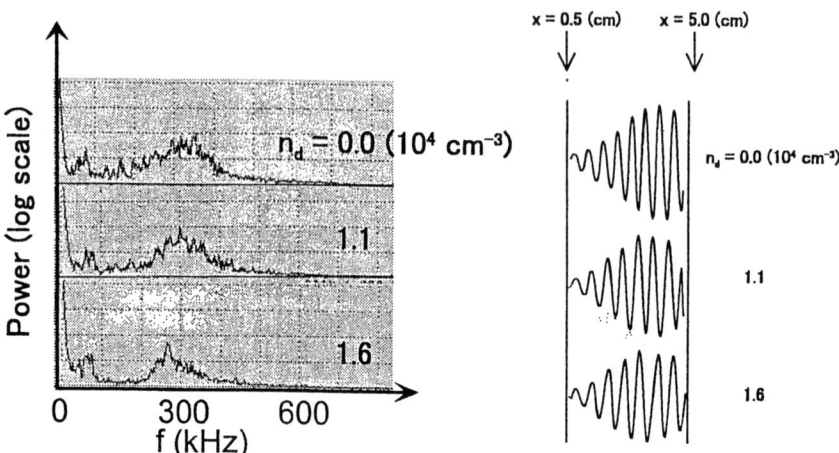

FIGURE 2. Typical examples of spectra (left) and the waveforms (right) of the ion-ion instability.

ical calculations. The dispersion relation of the ion-ion instability is given by

$$1 = \frac{\omega_{pe}^2}{k^2 v_{th,e}^2} Z'\left(\frac{\omega}{k v_{th,e}}\right) + \frac{omega_{pi}^2}{k^2 v_{th,i}^2} Z'\left(\frac{\omega}{k v_{th,i}}\right) \quad (2)$$

where ω_{pj} and $V_{th,j}$ are the angular plasma frequency and the thermal velocity of the j-th component, U_b the ion beam velocity and Z is the dispersion function [3]. When we substitute a real ω into eq. (2), we can obtain the spatial growth rate k_i. The results are shown in Fig. 3 by a solid and dotted lines for $T_e/T_i = 10$ and 25, respectively. The calculated growth rate is constant and larger than the one obtained in the experiment. The disagreement is thought to be the geometrical divergence of the beam and experimental ambiguity of the measured beam density. However, that the difference $\Delta k_i = k_{i,0} - k_{i,m}$, where $k_{i,0}$ and $k_{i,m}$ are measured spatial growth rates without and with dust, respectively, is due to collisions between the plasma and dust particles. Then,

$$\Delta k_i = \frac{v}{2 v_g}, \quad (3)$$

where v is a collision frequency and v_g is the group velocity. In the case of the ion acoustic wave, $v_g \simeq C_s$ around the unstable region, where C_s is the ion acoustic velocity. As a result, the collision frequency is approximately estimated to be as follows:

$$\frac{v}{\omega_{pi}} \simeq 9.4 \times 10^{-7} n_d, \quad (4)$$

where the dust density n_d is measured in cm^{-3}. From eq. (4), we can obtain $0 \leq \nu/\omega_{pi} \leq 0.02$ for $0 \leq n_d \leq 2.5 \times 10^4$ (cm^{-3}).

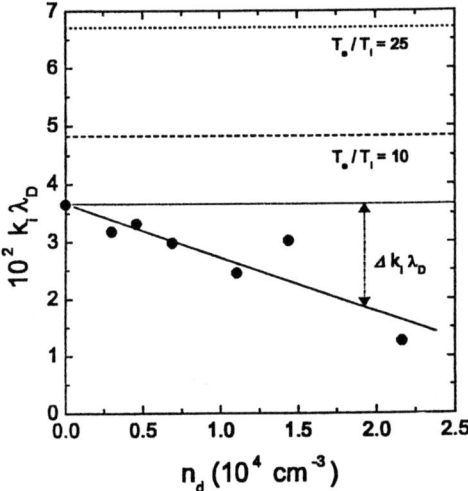

FIGURE 3. Spatial growth rate of the ion-ion instability as a function of the dust density. Closed circles are experimentally obtained with the interferometer method, and a solid line is its linear approximation. The broken ($T_e/T_i = 10$) and the dotted ($T_e/T_i = 25$) lines are obtained by numerical calculations of the plasma dispersion relation, where $T_e/T_b = 100$ and $n_i/n_b = 10$.

SUMMARY

An ion beam-dusty plasma system was realized in a dusty double-plasma device. The ion-ion instability was excited in the dusty plasma and the spatial growth rate of the instability was measured using the interferometer method. When dust particles are mixed into the plasma, the growth rate is reduced. From the reduced growth of the instability, the collision frequency is estimated.

REFERENCES

1. Saitou, Y., Nakamura, Y., et al., *Plasma Phys. Control. Fusion*, **35**, 1755 (1993).
2. Nakamura, Y., and Bailung, H., *Rev. Sci. Instrum.*, **70**, 2345 (1999).
3. Fried, B. D., and Conte, S. D., *The Plasma Dispersion Function*, Academic Pess, New York,

Modification of Shielding and Wake Potentials due to Dust-Lower-Hybrid Waves in Flowing Magnetized Dusty Plasmas

M. Salimullah

Department of Physics, Jahangirnagar University, Savar, Dhaka-1342, Bangladesh

Abstract. The effects of an external magnetic field through the ion polarization drift and the dust-modes and the role of dust dynamics on the shielding Debye-Hückel and dynamical wake potentials have been examined theoretically in a dusty plasma in the presence of a continuous ion flow and slow motion of dust grains between the bulk plasma and the sheath. The effects of the ions polarization drift and the dust-lower-hybrid waves on the strength and the effective lengths of the new static Debye shielding potential and the oscillatory dynamical potential are clearly demonstrated. It is found that for the super-sonic ion flow, the modified static Debye-Hückel screening length and the effective length, $L_\parallel$ of the oscillatory wake potential due to the dust-lower-hybrid wave [cf. Eqs.(4) and (10)] become larger by a factor of ω_{pi}/ω_{ci} than those due to the dust-acoustic wave.

In a recent paper, Shukla et al. [1] have studied the role of external magnetic field causing ion polarization drift on the dust-Coulomb crystal formation in a dusty plasma in the presence of a continous flow of ions from the bulk plasma to the crystal forming region in the sheath. In their study, the resonant interaction between the sub-sonic ion flow and the parallel phase velocity of a nearly perpendicular modified dust-ion-acoustic wave ignoring the dust dynamics was considered to be the mechanism of formation of the wake potential which would cause the robust dust crystal formation. They have shown that both the repulsive Debye-Hückel and attractive dynamical wake potentials are significantly affected by the ion polarization effect.

However, the presence of the constant electric field in the sheath region will cause slow equilibrium streaming motion of the charged dust particles ($\underline{v}_o \parallel \hat{z}$) which may resonate with the parallel phase velocity of the nearly perpendicular dust-mode, viz., the dust-lower-hybrid $[dlh]$ wave [2]. The dlh–mode involves the hybrid dynamics of cold unmagnetized and massive dust grains and the strongly magnetized ions in uniformly magnetized dusty plasmas. Moreover, the Bohm sheath criterion requires that the ion flow velocity must exceed the ion-acoustic velocity ($u_{io} \geq C_s$) in order to maintain the sheath. Therefore, the study of the static repulsive and oscillatory wake potentials in the presence of the dust-lower-hybrid wave in a magnetized dusty plasma with supersonic flow of ions should be more appropriate and relevant for the dust-Coulomb crystal formation.

We consider a homogeneous electron-ion-dust plasma in the sheath region of an rf-discharge dust crystal experiment. In the sheath plasma the relatively massive and highly charged micron/submicron sized dust grains are assumed to be streaming with constant velocities $\underline{u}_{io}$ and $\underline{v}_o$, respectively parallel to the external uniform magnetic field

($\hat{z}B_0 \parallel \underline{u}_{i0} \parallel \underline{v}_o$). The relatively hot electrons thermalize along $\hat{z}$ and follow a Boltzmann distribution, which corresponds to a shielding (to ions) response. The ions are assumed to be drifting, cold, and magnetized. The dust grains are taken to be cold, streaming, and unmagnetized. The resonant interaction of the test dust grains with the Doppler-shifted dlh-wave may significantly influence the dynamical potential due to the polarization drift term in the dielectric response function for the magnetized ions.

We consider the low-frequency electrostatic wave propagating nearly transverse to the external magnetic field ($k_\perp^2 \gg k_\parallel^2$, $\omega \ll k_\parallel u_{io}$), where the symbol $\parallel$ ($\perp$) denotes a quantity parallel (perpendicular) to the ion flow which is perpendicular to the sheath plane.

In the earlier paper [1], the assumption $\omega \approx k_\parallel u_{io}$ was made for the existence of the dust-ion-acoustic wave. Here, since the dust-modes, viz., dust-lower-hybrid and dust-acoustic waves have frequencies lower than the ion-acoustic/ion-cyclotron frequency, we assume $\omega \ll k_\parallel u_{io}$ corresponding to the lower parallel phase velocity of the dust-mode compared to the ion-streaming speed. Thus, the dielectric response function reduces to

$$\varepsilon(\omega,\underline{k}) \simeq 1 + f + \frac{1}{k^2 \lambda_{De}^2} - \frac{\omega_{pi}^2}{k^2 u_{i0}^2} - \frac{\omega_{pd}^2}{(\omega - k_\parallel v_o)^2}, \qquad (1)$$

where $f = \omega_{pi}^2/\omega_{ci}^2$, $\lambda_{De} = (T_e/4\pi e^2 n_{e0})^{1/2}$ is the electron Debye radius, T_e is the electron temperature, $n_{e0} = n_{i0} + Z_d n_{d0}$ is the electron number density, Z_d is the number of charges residing on the dust grain surface, and n_{d0} is the dust number density.

By setting $\varepsilon(\omega,\underline{k}) = 0$, one immediately obtains the Doppler-shifted dust-mode frequency [2] in a magnetized dusty plasma as

$$(\omega - k_\parallel v_o)^2 = \omega_k^2 \equiv \frac{k^2 C_d^2}{(1 - C_s^2/u_{io}^2) + (1+f)k^2 \lambda_{De}^2}, \qquad (2)$$

where $C_s = \omega_{pi}\lambda_{De} \equiv (n_{i0}/n_{e0})^{1/2}(T_e/m_i)^{1/2}$ is the modified ion sound speed and $C_d = \omega_{pd}\lambda_{De}$ is the dust-acoustic speed. It is noticed from Eq.(2) that for the low-density plasma, $\omega_{pi}^2 \ll \omega_{ci}^2$ and $C_s \ll u_{io}$, one readily obtains that the Doppler-shifted dust-acoustic mode, $(\omega - k_\parallel v_o)^2 \approx k^2 C_d^2$. However, for the high-density plasma, $\omega_{pi}^2 \gg \omega_{ci}^2$ and $C_s \ll u_{io}$, one obtains the Doppler-shifted lower-hybrid frequency, $(\omega - k_\parallel v_o)^2 \approx \omega_{pd}^2 \omega_{ci}^2/\omega_{pi}^2 \equiv \omega_{dlh}^2$.

The electrostatic potential around a test dust particulate in the presence of electrostatic modes $(\omega,\underline{k})$ in a uniform dusty magnetoplasma, whose dielectric response function is given by Eq. (1), is of the form

$$\Phi(\underline{x},t) = \int \frac{q_t}{2\pi^2 k^2} \frac{\delta(\omega - \underline{k}\cdot\underline{v}_t)}{\varepsilon(\omega,\underline{k})} \exp(i\underline{k}\cdot\underline{r}) d\underline{k} d\omega, \qquad (3)$$

where $\underline{r} = \underline{x} - \underline{v}_t t$, $\underline{v}_t$ is the velocity vector of a test dust particulate, and q_t is its charge.

Substituting Eq. (1) into Eq. (3) and performing the integration following standard procedures [1], we obtain $\Phi = \Phi_I + \Phi_{II}$, where the Debye-Hückel interaction potential is

$$\Phi_I(r) = \frac{q_t}{(1+f)r} \exp\{-r/[\sqrt{1+f}\lambda_{De}/\sqrt{1-C_s^2/u_{io}^2}]\}. \tag{4}$$

It is interesting to note that a new Debye shielding radius $\sqrt{f}\lambda_{De} = \rho_s$ for $f \gg 1$ and $u_{io} \gg C_s$ becomes operative in the nearly perpendicular direction. On the other hand, the non-Coulombian interaction potential Φ_{II}, which involves the collective interaction between the dlh-waves and a moving test dust particulate with arbitrary $\underline{v}_t$, is obtained from

$$\Phi_{II}(\underline{x},t) = \left(\frac{q_t \lambda_{De}^2}{2\pi^2}\right) \int \frac{\delta(\omega - \underline{k}\cdot\underline{v}_t)\,\omega_k^2 \exp(i\underline{k}\cdot\underline{r})}{\{(1+f)k^2\lambda_{De}^2 + (1-C_s^2/u_{io}^2)\}[(\omega - k_\parallel v_0)^2 - \omega_k^2]} d\underline{k}\,d\omega. \tag{5}$$

On performing the θ- and ω- integrations for the usual laboratory condition $k_\parallel v_t \ll \omega$, we obtain from Eq. (9)

$$\Phi_{II}(\rho,\xi) = \left(\frac{q_t c_d^2}{\pi(1+f)}\right) \int \frac{J_0(k_\perp \rho)}{(k^2 + k_D^2)} \frac{k^2 \exp(ik_\parallel \xi)\, k_\perp\, dk_\perp\, dk_\parallel}{[k_\parallel^2 \lambda_{De}^2(v_{tz} - v_0)^2(1+f)(k_D^2 + k^2) - k^2 C_d^2]}, \tag{6}$$

where $\xi = z - v_{tz}t$ and $k_D^2 = (1 - C_s^2/u_{io}^2)/(1+f)\lambda_{De}^2$. Here, ρ and z are the cylindrical coordinates of the field point.

Introducing the dimensionless notation $K = k/k_D = k\lambda_{De}\alpha$ where $\alpha = [(1+f)/(1 - C_s^2/u_{io}^2)]^{1/2}$, we rewrite Eq. (6) as

$$\Phi_{II}(\rho,\xi) = \left[\frac{q_t C_d^2}{\pi\alpha^3(v_{tz} - v_0)^2 \lambda_{De}(1 - C_s^2/u_{io})^2}\right] I, \tag{7}$$

where

$$I = \int \frac{J_0(K_\perp \rho/\alpha\lambda_{De}) K^2 \exp(iK_\parallel \xi/\alpha\lambda_{De})\, K_\perp\, dK_\perp\, dK_\parallel}{(1+K^2)[K_\parallel^2(k_\parallel^2 + K_\perp^2 + 1) - K^2\beta]}, \tag{8}$$

and $\beta = C_d^2/(v_{tz} - v_0)^2(1 - C_s^2/u_{io}^2)$.

Integrating $K_\parallel$-integration with approximation $K_\perp \rho \ll \alpha\lambda_{De}$, we obtain from Eq.(8)

$$I \approx -\left(\frac{2\pi}{\sqrt{\beta}}\right)(1+\beta) \int_0^\infty K_\perp^2 \sin\left(\frac{\sqrt{\beta}\xi K_\perp}{\alpha\lambda_{De}}\right) dK_\perp. \tag{9}$$

Taking the upper limit of $K_\perp$-integration as 1 for the wavelength of the nearly transverse dust-wave $\geq \rho_s$, We finally obtain the oscillatory dynamical wake potential as

$$\Phi_{II}(\rho = 0,\xi) \approx A_o \frac{\cos(\xi/L_\parallel)}{|\xi|}, \tag{10}$$

where $A_o = 2q_t(1+\beta)/(1+f)$ and $L_\parallel = \sqrt{1+f}\,\lambda_{De}\,|v_{tz}-v_o|/C_d$.

The derivation of Eq.(10) is valid for both sub-sonic and super-sonic ion flow and $|v_{tz}-v_o|$ greater or less than C_d. For the usual laboratory conditions, $u_{io} \gg C_s$, $\omega_{pi}^2 \gg \omega_{ci}^2$ and $\beta \ll 1$ (for $C_d^2 \ll |v_{tz}-v_o|^2$), $A_o = 2q_t/f$ and $L_\parallel \approx \sqrt{f}\,\lambda_{De}\,|v_{tz}-v_o|/C_d = \rho_s\,|v_{tz}-v_o|/C_d$ for the dlh–wave.

For the dust-acoustic wave propagating nearly perpendicular to the magnetic field, the oscillatory wake potential is given by $\Phi_{II}(\rho=0,\xi) = (2q_t/|\xi|)\cos(\xi/L_\parallel)$ with $L_\parallel = \lambda_{De}\,|v_{tz}-v_o|/C_d$ for $\omega_{pi}^2 \ll \omega_{ci}^2$, $u_{io} \gg C_s$ and $\beta \ll 1$. These conditions may be applicable to some space and astrophysical plasma systems. We note that the effective length $L_\parallel$ in the perpendicular direction to the magnetic field and the strength of the wake potential are independent of the magnetic field.

To conclude, we have studied the effect of the external magnetic field on the formation of the oscillatory wake field due to the dust-lower-hybrid wave in a dusty plasma with streaming ions and dust particles. The situation is appropriate for a sheath region of an rf-discharge plasma where a constant electric field is generated due to the high mobility of the thermal electrons compared to ions. Both repulsive Debye-Hückel screening and attractive dynamical wake potentials are drastically affected by the magnetic field. In the presence of the external static magnetic field, the symmetry in the usual Debye-Hückel repulsive potential is broken down and a new Debye shielding radius (ρ_s) appears in the direction perpendicular to the external magnetic field for $u_{io} \gg C_s$ and $\omega_{pi}^2 \gg \omega_{ci}^2$ corresponding to the high-density plasma. It may be mentioned here that the present study is valid for both the sub-sonic and super-sonic flow of ions, dust flow velocity greater or smaller than the dust-acoustic speed, and the presence of dust-lower-hybrid or dust-acoustic modes in the high- or low-density magnetized dusty plasma. However, for the super-sonic ion flow, the modified static Debye-Hückel screening length and the effective length, $L_\parallel$ of the oscillatory wake potential due to the dust-lower-hybrid wave [cf. Eqs.(4) and (10)] become larger by a factor of ω_{pi}/ω_{ci} than those due to the dust-acoustic wave.

REFERENCES

1. Shukla, P., Nambu, M., and Salimullah, M., *Phys. Lett.*, **A 291** (2001).
2. Salimullah, M., Amin, M., Salahuddin, M., and Chowdhury, A. R., *Physica Scripta*, **58** (1998).

Solitons and Oscillitons in Complex Plasmas

K. Sauer*, E. Dubinin* and J. F. McKenzie[†]

Max-Planck-Institut für Aeronomie, Katlenburg-Lindau, Germany
[†]*Max-Planck-Institut für Aeronomie, Katlenburg-Lindau, Germany, and also School of Pure and Applied Physics, University of Natal, Durban, South Africa*

Abstract. The properties of electromagnetic solitons and oscillitons in a plasma containing a massive ion (or dusty) species are discussed within a multi-fluid framework in which the magnetic field is frozen into the electrons. Oscillitons may be viewed as a generalization of a soliton in that they exhibit spatial oscillations superimposed on the classical 'hump of water' shape which characterizes classical solitons. These waves can arise in certain wave speed regimes where the linear phase and group velocities are equal. A fully nonlinear computation reveals the rich structure imbedded in large amplitude coherent waves. An application to space plasmas where coherent waves are observed is briefly discussed.

INTRODUCTION

The concept of an 'oscilliton' was first introduced by Sauer et al. [1] within the framework of nonlinear stationary waves in a bi-ion plasma, s.a. [2, 3, 4]. In this multi-fluid approach the coupling between different wave modes near cross-over frequencies leads to a stationary wave dispersion equation which, in certain wave speed regimes, yields complex values of the wave number k ($k_r + ik_i$). The implication here is that soliton-like solutions, associated with k_i, can be constructed with a superimposed spatial oscillation $2\pi/k_r$ to form a well developed nonlinear structure. A similar phenomenon arises in a cold proton-electron plasma in the whistler mode where there exists a critical speed at which the phase velocity equals the group velocity which occurs at approximately one half of the electron gyrofrequency. At wave speeds in excess of this critical speed whistler oscillitons can be constructed [5]. Here we investigate the formation of electromagnetic dust solitons and oscillitons. The latter are formed in the neighbourhood of the cut-off frequency of the coupled modes which may lie in the proton frequency range even if the dusty species is massive. We also briefly discuss how oscillitons may be the origin of large coherent waves frequently observed in multi-ion and dusty plasmas.

OUTLINE OF THE SYSTEM AND ITS STATIONARY WAVES

Within the framework of bi-ion Hall MHD equations [6, 7] stationary waves propagating along the x-axis with speed U are governed by eight conservation laws (mass, momentum, energy, B_x and the transverse electric field are constants) which, combined with quasi-charge neutrality and two of the equations of motion for the transverse deflections of the protons, complete the system for the ten physical quantities: n_i number density

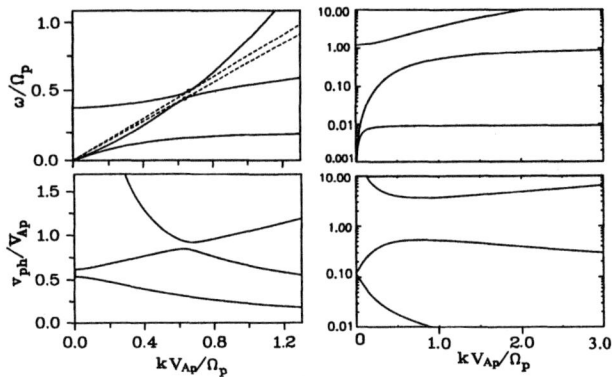

FIGURE 1. Dispersion of LF electromagnetic waves in a complex plasma. (a) Bi-ion plasma: $\alpha = n_h/n_p = 0.4, \mu = m_h/m_p = 4, Z_h = 1, \theta = 30^0$; (b) dusty plasma: $\alpha = 10^{-4}, \mu = 6 \cdot 10^5, Z_h = -5.5 \cdot 10^3, \theta = 30^0$; Ω_p: proton cyclotron frequency, V_{Ap}: proton Alfvèn velocity. The dashed lines in Fig. 1a are tangential to the curves marking two points at which phase- and group-standing waves exist.

($i = p, h$), $\mathbf{v}_i$ species velocities, B_y, B_z the transverse wave magnetic field. The linearized version of these equations yields the dispersion relation, $D(\omega = k \cdot U, k) = 0$, for stationary waves giving the wave number k as a function of the wave speed U. With $U = \omega/k$ one gets the related dispersion characteristics $\omega = \omega(k)$. As previously noted [1, 2, 3, 4] in the case of a bi-ion plasma, mode coupling gives rise to a 'throat' in the diagnostic diagram where the phase and group velocities are equal at finite values of k. In a dust electromagnetic wave the frequency at which this occurs is in the neighbourhood of the cut-off frequency

$$\omega_{cut-off} = \frac{Z_h n_h \Omega_p + n_p \Omega_h}{Z_h n_h + n_p} \sim \frac{Z_h n_h}{n_p} \Omega_p, \qquad (1)$$

where the approximation holds for massive ($m_h/m_p \gg 1$) and tenuous dust $Z_h n_h \ll n_p$ (Z_h: charge number).

Figure 1a indicates the points in the (ω, k) curves where the phase and group velocities are equal for a bi-ion plasma whose minor ion population has the mass $m_h = 4m_p$. The same features occur in a (proton) plasma containing massive, tenuous, highly charged dust (Fig. 1b). Figure 2 shows the corresponding solutions of the stationary wave dispersion equation for k as a function of the normalized wave speed U. Of particular note is the existence of a speed regime ($0.5 < U < 1.25$) in which k is complex ($k_r + ik_i$), followed by another speed regime ($1.25 < U < 1.54$) in which k is purely imaginary. In the former regime, oscillitons can be formed with a spatial oscillation ($2\pi/k_r$) superimposed on an exponentially growing (k_i) soliton-like structure, whereas in the latter, higher speed regime, a classical soliton can be constructed.

The fully nonlinear signature of an oscilliton is shown in Figure 3. The upper panel displays the longitudinal component of the dust and proton speeds in the wave frame. Although it is not clearly evident in the panel, because of scaling, the dust speed does undergo very small variations in speed ($O(m_h)$) of the same form as the protons,

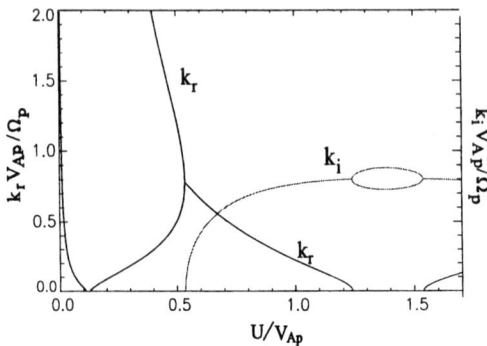

FIGURE 2. Stationary wave dispersion, k=k(U). The parameters are the same as in Fig. 1b.

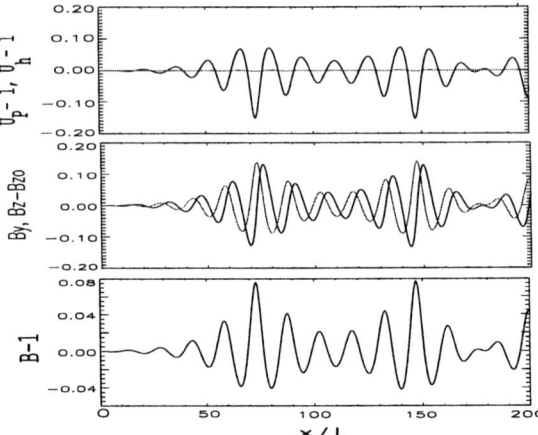

FIGURE 3. Spatial profile of a dust-electromagnetic oscilliton for the parameters of Fig. 2. The oscilliton velocity is $U/V_{A_p} = 0.53$. From top to bottom: Longitudinal velocities (U_p, U_h), transverse magnetic field (B_y, B_{z-z0}) and magnitude B.

indicating that their momentum flux comprises an essential ingredient of the wave. The second panel shows the behaviour of the transverse wave magnetic field which exhibits a remarkably coherent structure as does in the third panel for the magnitude of B.

The nonlinear structure of the soliton, formed in the upper speed regime of Figure 2, is shown in Figure 4 for two different values of the proton Mach number. Note how an increase in the Mach number brings about a substantial change from a rather smooth soliton to a strongly crested one in which $B_{max} < 2.5$ and the protons go through a deep depression. This case is similar to the oblique soliton in a cold magnetized plasma [8] in which the magnetic field attains a maximum value of three accompanied by an infinite compression in the protons.

In recent papers [1, 2, 3, 4, 5] the importance of the concept of stationary waves

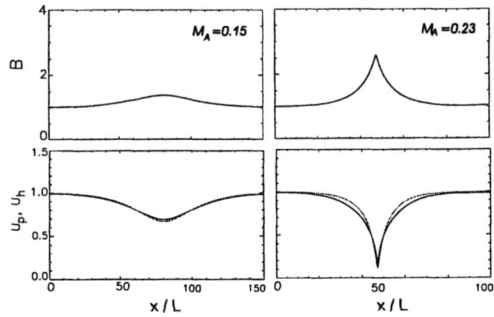

FIGURE 4. Spatial profile of a transverse dust-electromagnetic soliton for the parameters of Fig. 2, but $\theta = 90^0$. The soliton velocity (proton Mach number M_A) is given in the Figure. The magnetic field magnitude (B) and the longitudinal velocities (U_p, U_h) are shown.

(solitons, oscillitons) for the explanation of observed coherent structures in multi-ion space plasmas has been discussed. A similar situation may occur in dusty plasmas. Russel [8, 9], e.g., has observed soliton-like disturbances in the interplanetary magnetic field that were referred to the solar wind interaction with dusty debris of the asteroid Oljato. The profiles of the observed magnetic field enhancements and their characteristic width resemble structures like that shown in Figs. 3, 4. Galileo and Cassini data from Jupiter and Saturn 'dusty magnetospheres' will almost certainly yield data motivating more detailed studies of nonlinear waves in dusty plasmas.

ACKNOWLEDGMENTS

E. Dubinin thanks the Deutsche Forschungsgemeinschaft and the Max-Planck-Gesellschaft for supporting this work by grants and J.F. McKenzie the University of Natal, Durban, South Africa.

REFERENCES

1. Sauer, K., Dubinin, E., and McKenzie, J., *Geophys. Res. Lett.*, **28** (2000).
2. Sauer, K., Dubinin, E., and McKenzie, J., *Physica Scripta*, **52** (2002).
3. Dubinin, E., Sauer, K., and McKenzie, J., *Nonl. Processes Geophys.*, **9** (2002).
4. Sauer, K., Dubinin, E., and McKenzie, J., *Nonl. Processes Geophys.*, **9** (2002).
5. Sauer, K., Dubinin, E., and McKenzie, J., *Geophys. Res. Lett.*, **29** (2002).
6. Sauer, K., Bogdanov, A., and Baumgärtel, K., *Geophys. Res. Lett.*, **21** (1994).
7. Rao, N. N., *Advances in Dusty Plasmas* (1997).
8. McKenzie, J. F., and Doyle, T., *Phys. Plasma*, **8** (2001).
9. Russel, C. T., *Geophys. Research Lett.*, **14** (1987).
10. Russel, C. T., *Adv. Space Res.*, **3-4** (1990).

New Kinetic Variables and the Effective Temperatures in Dusty Plasmas

Pieter P.J.M. Schram*, Sergey A. Trigger[†] and Anatoly G. Zagorodny**

*Eindhoven University of Technology, P.O. Box 513, MB 5600 Eindhoven, The Netherlands
[†] Eindhoven University of Technology, P.O. Box 513, MB 5600 Eindhoven, The Netherlands
Institute for High Temperatures, Russian Academy of Sciences, 13/19 Izhorskaia Str., Moscow 127412, Russia
**Eindhoven University of Technology, P.O. Box 513, MB 5600 Eindhoven, The Netherlands
Bogolyubov Institute for Theoretical Physics, National Academy of Sciences of Ukraine, 14 B Metrolohichna Str., Kiev 03143, Ukraine

Abstract. The kinetic theory of dusty plasmas is derived from first principles with regard to absorption of electrons and ions by dust particles. The charging process is shown to imply a considerable modification of properties in comparison with usual plasmas. Not only the electric charge, but also the mass, the angular momentum and the inner energy of the dust particles are new dynamic variables. Their influence on the kinetic behaviour of dusty plasmas is also considered.

INTRODUCTION

For the understanding of dusty plasma experiments [1] - [4] and astrophysical phenomena [5] it is essential to derive a kinetic theory in which new dynamic variables of the dust particles play a dominant role. In the first place the electric charge. In [6], [7] the charging process was taken into full account in the derivation of kinetic equations from non-equilibrium statistical mechanics. The mass was supposed to be constant. This situation may be realistic when plasma particles are absorbed by dust particles and also neutral gas regeneration occurs due to recombination of ions with electrons on the surface of the grains. In order to describe this neutral gas regeneration correctly it seems necessary to take into account the grain momentum change due to evaporation of neutrals from the grain and to modify the kinetic equations accordingly. If the distribution of the regenerated neutrals at the instant of evaporation is isotropic, however, then this effect can be expected to leave the kinetic equations approximately unaltered.

In Section 2 this situation is considered and the derivation of the kinetic equations is sketched for the case that absorption of plasma particles is the dominant process. Approximate stationary solutions are obtained which show the existence of different effective temperatures for the velocity and charge distributions.

In Section 3 Coulomb collisions and the collisions of grains with neutral atoms or molecules are taken into account especially with regard to the influence on the effective temperatures. It is shown that the experimental observation of dusty crystals melting as a consequence of decreasing gas pressure [1], [2] is in qualitative agreement with the theory developed.

In Section 4 we consider the mass of the grains as the essential new dynamic variable. In order to simplify the theory and to concentrate on the effect of changing mass we restrict ourselves to the situation where neutral gas particles collide with neutral grains and are absorbed by them. The problem of mass transfer is typical for the conditions of experiments on grain synthesis [4] and the formation of new materials [3]. The (non-stationary) distribution function of the grains is determined. One of the main findings is that the effective temperature of the grains is lower than the temperature of the gas.

In Section 5 the inner energy of grains is introduced as a new dynamic variable. Also the angular momentum should be treated as a new dynamic variable; no attention, however, is paid to the angular momentum, since it is the main topic of a paper at this conference [8]. The evolution of the average inner energy is connected with the evolution of the average mass. Section 6 contains some conclusions.

STATISTICAL AND KINETIC THEORY OF DUSTY PLASMAS INCLUDING THE CHARGING PROCESS

Considering a plasma consisting of electrons, ions and dust particles and restricting ourselves to charging collisions in which plasma particles are absorbed by grains we formulate a generalized Klimontovich formalism and derive the corresponding BBGKY hierarchy and the kinetic equation for the grains [6], [7]. Assuming that the charge of the plasma particles is small compared with the charge of the dust particles and that also their velocity changes in collisions are small we arrive at a Fokker-Planck type kinetic equation. We also take the limit $m_s/m_g \to 0$, where m_s is the mass of electrons ($s = e$) or ions ($s = i$) and m_g is the mass of a grain. Assuming the distribution functions f_s to be Maxwellian we arrive at a solution for f_g. For typical values of plasma parameters with weak plasma coupling guaranteeing small charge fluctuations, $e_e^2/(aT_e) \ll 1$, a being the grain radius, this solution can be written as

$$f_g(\mathbf{v},q) = \frac{n_{0g}}{\sqrt{2\pi a T_{\text{eff}}^*}} e^{-[(q-q_0)^2/2aT_{\text{eff}}^*]} \left(\frac{m_g}{2\pi T_{\text{eff}}}\right)^{3/2} e^{-m_g v^2/2T_{\text{eff}}}. \qquad (1)$$

where

$$T_{\text{eff}}^* = \frac{2T_e}{1+Z_i} \frac{1+t+z}{t+z} \qquad (2)$$

$$T_{\text{eff}} \simeq 2T_i \frac{t+z}{t-z} \qquad (3)$$

Here n_{0g} is the averaged number density of grains, $z = e^2 Z_g/(aT_e)$ (Z_g is the charge number of a grain and T_e the electron temperature), $t = T_i/(Z_i T_e)$, $S_e^2 = T_e/m_e$, $Z_g = q_0/e_e$, $Z_i = |e_i/e_e|$ and q_0 is the equilibrium grain charge satisfying

$$I(q_0, 0) = 2\sqrt{2\pi} a^2 e_i^2 n_i S_i \left[1 + \frac{z}{t} - \left(\frac{m_i}{m_e}\right)^{1/2} \left(\frac{T_e}{T_i}\right)^{1/2} \frac{n_e}{Z_i n_i} e^{-z}\right] = 0. \qquad (4)$$

Distribution (1) is a Maxwellian velocity and a Gibbs grain charge distribution with temperatures T_{eff} and T_{eff}^* respectively. For $t+z \gg 1$ and $Z_i = 1$ we have $T_{\text{eff}}^* = T_e$. In general for $Z_i = 1$ T_{eff}^* exceeds the electron temperature.

The behaviour of the temperature T_{eff} is even more surprising. In the case of neutral grains ($z=0$) it is equal to $2T_i$. The factor 2 is due to the inelastic nature of the charging collisions: a part of the kinetic energy of the ions is transformed into additional kinetic energy of the grains. This is different from the situation of conventional Brownian motion where the temperatures of the Brownian and the light bombarding particles are equal.

Equation (3) implies the possibility of very high effective temperatures when z approaches t or even negative temperatures for $z > t$. Physically these phenomena are connected with the decrease of the friction coefficient with increasing grain charge. The reason for this is the fact that the charge dependent part of the ionic charging cross section is larger for ions moving with smaller relative velocities (motion parallel to the grain motion), so that the difference between the anti-parallel and parallel ion fluxes decreases with increasing charge. At $z = t$ the friction coefficient is zero.

It is clear that in this situation where the effective temperature approaches infinity or becomes negative, the approximation of dominant charging collisions is no longer valid, i.e. Coulomb collisions and collisions with neutrals should be taken into account.

ELASTIC COLLISIONS

We disregard the grain-grain and the electron-grain Coulomb collisions because of the assumed smallness of the grain density and the negligible momentum transfer respectively. The ion-grain Coulomb collisions are described by a Landau collision integral with a modified Coulomb logarithm. The modification is necessary because of the competition between the radius a of a grain and the usual minimum impact parameter $r_{Li} = \pi |Z_i e q|/(8T_i)$. For $r_{Li} \leq a$ the Coulomb logarithm would include the contribution of collisions with particles reaching the grain surface, i.e. charging collisions. On the other hand it seems natural not to use the Debye length, $r_D = [T_i/(4\pi Z_i^2 e^2 n_i)]^{1/2}$, as the maximum impact parameter, but instead $r_D + a$, as in [7], because the screened potential of a finite size grain is the DLVO potential [9]. In [10] a modification has been suggested to take into account the ion scattering at big angles in dusty plasmas. At the same time both the approximation in [7] and the one in [10] cannot provide a value of the logarithm, which is positive for all values of the dusty plasmas parameters. Our new form has been obtained in [11] and can be written as

$$\ln \Lambda_i = \ln\left[(r_D + a + r_{Li})/(a + r_{Li})\right]. \tag{5}$$

In the limit $r_D \gg r_{Li} \gg a$ this reduces to the usual Coulomb logarithm, in the limit $r_{Li} + a \gg r_D$, however, the Coulomb logarithm is a small number. In that case a more exact calculation of the interaction (for example in the Debye potential) is needed. The collisions between neutral plasma particles and grains are described by a Fokker-Planck type collision integral which can be derived from the Boltzmann equation.

The ion-grain and neutral-grain elastic collisions result in the following modification of eq.(2):

$$T_{\text{eff}} = 2T_i \frac{\left[1 + \frac{z}{t} + \frac{z^2}{t^2}\ln\Lambda_i + \frac{n_n}{n_i}\left(\frac{m_n}{m_i}\right)^{1/2}\left(\frac{T_n}{T_i}\right)^{3/2}\right]}{\left[1 - \frac{z}{t} + 2\frac{z^2}{t^2}\ln\Lambda_i + 2\frac{n_n}{n_i}\left(\frac{m_n}{m_i}\right)^{1/2}\left(\frac{T_n}{T_i}\right)^{1/2}\right]}, \quad (6)$$

where the subscript n refers to the neutral plasma particles.

If there are no neutral plasma particles ($n_n = 0$), the Coulomb collisions alone can produce a saturation of the grain temperature, but in the case of dominant charging collisions T_{eff} can still be anomalously large. This fact provides a qualitative explanation of the experimental observation of grain temperatures much higher than the ion temperature [1], [2].

Eq. (6) shows that the effective temperature increases with decreasing neutral density. The influence of the neutral density is especially important at $1 - z/t + 2(z/t)^2 \ln\Lambda_i \simeq 0$. Then a decrease of the neutral gas pressure can produce an anomalous growth of the effective temperature. That is in qualitative agreement with the experimental observation of the melting of dusty crystals by reduction of the gas pressure [1], [2].

CHANGING GRAIN MASS

In order to get insight in the influence of changing mass of dust particles we now consider a system consisting only of neutral gas atoms (or molecules) and heavy neutral grains. We assume that all gas atoms hitting a grain are absorbed by the grain. In the statistical theory of the grain system mass now plays the role of a new dynamic variable. It should be noted that the assumption of complete absorption seems justified under the conditions of the experiments aimed at plasma synthesis of fine grains [3], [4].

The appropriate kinetic equations for the grains and for the neutral gas were given in [12], [13]. In the latter equation not only neutral-grain collisions but also neutral-neutral collisions and re-supply from external sources should be taken into account. Conservation laws can be derived, the total energy, however, is no longer a conserved quantity. The reason is obvious: a part of the kinetic energy of a colliding atom is added to the kinetic energy of a grain, while the remainder is spent for heating of the grain.

The collision term in the kinetic equation for the grains is greatly simplified by expansion in powers of the small mass ratio m/M (atom mass divided by grain mass). The solution of the kinetic equation for the atoms is approximately a Maxwellian, if we assume a dominating role of interatomic collisions. Under these conditions the distribution function of the grains may be assumed to be isotropic and the kinetic equation for the grains becomes:

$$\frac{df_g(P,M,t)}{dt} = n_n \left(\frac{8T_n m}{\pi}\right)^{1/2} \left\{\frac{g(P,M)}{M} + \frac{P}{3M}\frac{\partial g(P,M)}{\partial P} \right.$$
$$\left. + \frac{2}{3}\frac{T_n}{(P^2)}\frac{\partial}{\partial P}\left[P^2\frac{\partial g(P,M)}{\partial P}\right] - \frac{\partial g(P,M)}{\partial M}\right\}, \quad (7)$$

where $g(P,M) = \sigma(P,M)f_g(P,M)$, $\sigma(P,M)$ is the cross section for atom-grain collisions, and n_n and T_n are the number density and the temperature of the atoms respectively.

If the last term of eq. (7) is omitted, i.e. if the mass growth is neglected, then an exact stationary solution is a Maxwellian with temperature $T_g = 2T_n$ in agreement with Section 2.

It is possible to investigate the temperature evolution on basis of the general solution of eq. (7). For simplicity we neglect, however, the mass dispersion and write:

$$f_g(P,M,t) = F(P,t)\delta[M - \mu(t)]. \tag{8}$$

Substitution of eq. (8) into eq. (7) leads to

$$\frac{d\mu(t)}{dt} = n_n \left(\frac{8T_n m}{\pi}\right)^{1/2} \sigma[\mu(t)] \tag{9}$$

and to the equation for $F(P,t)$ which is solved by the Maxwellian

$$F(P,t) = n_g\{2\pi\Delta(t)\}^{-3/2}\exp[-P^2/\Delta(t)] \tag{10}$$

with the time dependent temperature $\Delta(t) = T_{\text{eff}}(t)\mu(t)$. It was shown for a particular dependence $\mu(P,M)$ on M in [12] and proved for arbitrary dependence of $\mu(P,M)$ on M in [13] that

$$T_{\text{eff}}(t) = \frac{4}{5}T_n + C\mu(t)^{-5/3}, \tag{11}$$

where C is a constant of integration.

Asymptotically $\mu(t)$ approaches infinity and T_{eff} approaches $(4/5)T_n$. The mass growth results in cooling of the dust component below the gas temperature, while without mass growth eventually T_{eff} is twice T_n.

INNER ENERGY OF GRAINS

Again we consider a system consisting of neutral gas atoms and neutral grains. We add the inner energy as a new dynamic variable. We do not consider the angular momentum, cf. [8], [14], because it is considered in some detail in a paper at this conference [8].

The change of the inner energy for absorption at collisions is non-negative definite.

The Fokker-Planck equation for the grain distribution function $f_g(\mathbf{P},M,\varepsilon,t)$, where ε represents the inner energy, takes the form [15]:

$$\frac{df_g}{dt} = \sigma(M)\frac{\partial}{\partial P_\alpha}\left\{-\beta_\alpha f_g + \lambda_{\alpha\beta}P_\beta f_g + \frac{\partial}{\partial P_\beta}\left(\kappa_{\alpha\beta}f_g\right)\right\} - \frac{\partial(J\sigma f_g)}{\partial M} - \frac{\partial}{\partial \varepsilon}\left(\frac{3J\sigma\xi f_g}{m}\right), \tag{12}$$

where $\sigma(M)$ is the cross section for absorption and the coefficients J, β_α, $\lambda_{\alpha\beta}$, $\kappa_{\alpha\beta}$ and ξ are determined as different moments of the momentum calculated through integrals with the distribution function of the light particles [14].

We assume that the distribution function of the light particles is Maxwellian with temperature T_n. Then as a result we find for the evolution of the average inner energy per particle of the grains:

$$\langle \varepsilon(t) \rangle = \frac{\mu(0)}{\mu(t)} \langle \varepsilon(0) \rangle + 2T_n \{1 - \frac{\mu(0)}{\mu(t)}\}. \quad (13)$$

Clearly the grains are heated by the absorbing collisions with light particles.

CONCLUSIONS

Situations may exist where the electric charge of the dust particles changes, but not the mass and the inner energy (at the same time the inner grain temperature, in general, can be different from the kinetic temperatures of the other plasmas components). This is the case when the absorption of ions is combined with recombination and the emission of neutral atoms. The results show that then the stationary distributions of the grain velocities and charges are described by effective temperatures other than those of the plasma subsystem. These effective temperatures are determined by the competition between charging collisions and elastic collisions. Grain-neutral and Coulomb collisions tend to equalize the grain temperature to the temperature of neutrals or ions respectively, while charging collisions can produce anomalous temperature growth. That might be the main mechanism of the experimentally observed grain heating.

When the mass is not kept constant by recombination and emission of atoms, it should also be introduced as a new dynamic variable. The mass of the grains will grow indefinitely under such circumstances. The consequences are interesting. In the present paper the case of neutral atoms and neutral grains was considered. The grains are eventually cooled to 80% of the temperature of the ambient gas, while in the absence of mass growth the effective temperature is about twice the temperature of the atoms.

It is relatively simple to generalize the obtained Fokker-Planck equations to include also terms representing the angular momentum and the inner energy of the grains as new dynamic variables. The evolution of the average inner energy in our simple model appears to be directly coupled to the evolution of the mass. Asymptotically the inner energy expressed as a temperature becomes twice the temperature of the ambient gas.

REFERENCES

1. Melzer, A., Homan, A., and Piel,A., *Phys. Rev. E*, **53**, 2757 (1996).
2. Morfill,G.E., Thomas, H.M., Konopka, U., and Zuzic, M., *Phys. Plasmas*, **6**, 1764 (1999).
3. Stoffels, E., Stoffels, W.W., Kroesen, G.M.W., and de Hoog, F.J., *J. Vac. Sci. Technology*, **A14**, 556 (1996).
4. Vivet, F., Bouchoule, A., and Boufendi, L., *J. Appl. Phys.*, **83**, 7474 (1998).
5. Quinn, R.A., Goeres, A., and Sedlmayer, E., *Astron. Astrophys.*, **317**, 216 (1996).
6. Zagorodny, A.G., Schram, P.P.J.M., and Trigger, S.A., *Phys. Rev. Lett.*, **84**, 3594 (2000).
7. Schram, P.P.J.M., Sitenko, A.G., Trigger, S.A., and Zagorodny, A.G., *Phys. Rev. E*, **63**, 016403 (2000).
8. Ignatov, A.M., Trigger, S.A., Maiorov, S.A., and Ebeling, W., report on this Conference

9. Verwey, E.J., and Overbeek, J.Th.G., *Theory of the Stability of Lyophobic Colloids*, Publ. Elsevier, Amsterdam, 1948.
10. Khrapak, S.A., Ivlev, A.V., Morfill, G.E., and Thomas, H.M., private communication, to be published.
11. Trigger, S.A., Kroesen, G.M.W., Stoffels, E., Stoffels, W.W., and Schram, P.P.J.M., to be published.
12. Trigger, S.A., *Contrib. to Pl. Phys.*, **41**, 331 (2001).
13. Ignatov, A.M., Trigger, S.A., Ebeling, W., and Schram, P.P.J.M., *Physics Letters A*, **293**, 141 (2002).
14. Ignatov, A.M. , Trigger, S.A., Maiorov, S.A., Ebeling, W., and Schram, P.P.J.M., Vth European Workshop on Dusty and Colloidal Plasmas, 23–25 August 2001, Potsdam (Germany), Book of Abstracts.
15. Trigger, S.A., and Ebeling, W., *Vth European Workshop on Dusty and Colloidal Plasmas*, 23–25 August 2001, Potsdam (Germany), Book of Abstracts.

Dust-Acoustic and Shear Waves in Strongly Coupled Dusty Plasmas

G. Prasad*, J. Pramanik*, B.M. Veeresha*, A. Sen* and P.K. Kaw*

*Institute for Plasma Research, Bhat, Gandhinagar 382428, India

Abstract. We report on experimental observations of large amplitude dust acoustic oscillations and transverse shear waves in a three dimensional configuration of dusty plasma that is in the strongly coupled fluid regime. The onset and nature of these spontaneous oscillations is strongly influenced by the ambient neutral pressure. In an extended dust cloud, as the neutral pressure is progressively diminished, spontaneous longitudinal oscillations appear first followed by transverse shear waves. For more compact equilibrium configurations the shear waves do not appear but a further decrease in pressure causes the dust acoustic waves to develop higher harmonic oscillations. A simple theoretical model based on the generalized hydrodynamic (GH) equations is developed to explain this phenomena.

INTRODUCTION

The study of collective motion in strongly coupled dusty plasmas is a subject of much current theoretical and experimental interest[1, 2, 3, 4]. When the Coulomb parameter $\Gamma = e^{-a/\lambda_p}(Z_d e)^2/aT_d$ is of order 1 or larger ($Z_d e$ is the charge on the dust grain, a is the intergrain distance, T_d is the dust temperature, λ_p is the plasma Debye length) strong coupling effects set in and introduce interesting modifications in the collective properties of the dusty plasma. In past studies, linear dispersion relations obtained using a viscoelastic model have predicted interesting novel effects - such as additional dispersive corrections in the dust acoustic wave including a turnover of the dispersion curve beyond a certain wave number and a new transverse branch of oscillation - the so called dust shear wave. Recently we have experimentally confirmed the existence of the shear wave in the strongly coupled fluid state [5]. In subsequent experiments we have also obtained large amplitude coherent DAWs. We present a brief description of these experimental observations and also discuss a simple mode coupling theory based on a nonlinear version of the GH model to explain the nonlinear features of the DAW.

EXPERIMENTAL RESULTS

The experiments were carried out in a stainless steel cylindrical chamber (95-cm length, 20 cm diameter) evacuated to a pressure of 10^{-3} mbar and then injected with Argon gas [5]. A stratified glow discharge was produced by applying 500 volts of electric potential between an anode structure and the vessel wall (used as a cathode) at 1 mbar.

A levitated dust cloud of laboratory grade kaolin particles (with a size distribution in the range of 0.5 to a few microns) was created with the help of the electric field of the double layers of the glow discharge and illuminated by a He-Ne laser. The video images were digitized with a frame grabber card. Plasma parameters were measured with a Langmuir probe. A stable three dimensional dust cloud was observed above 0.25 mbar at a discharge voltage of 337 volts and a discharge current of 100 mA. In the first set of experiments a dust cloud of large axial extent ($\sim 50cms$) was formed by using a sharp tipped anode structure. Then as the neutral pressure was reduced to about 0.13 mbar the dust cloud showed spontaneous longitudinal oscillations of the acoustic type propagating vertically downwards (in the direction of the gravity). From an analysis of consecutive video frames the typical wave number and phase velocity of this oscillation was found to be $\lambda_y \sim 0.2$cm, $V_{phy} \sim 4cm/sec$ so that $\omega_y \sim 20$Hz. The wavelength and the frequency of this longitudinal wave was found to remain nearly the same when the discharge voltage was varied from 300 to 350 volts. The phase velocity of this wave was found to agree quite well with the theoretical estimate of the phase velocity of the dust acoustic waves (DAW), which for $k\lambda_p \ll 1$ is given as $C_{da} \sim \lambda_p \omega_{pd}$. The typical parameters for this experiment were $n_i \sim 2 \times 10^{16} m^{-3}$, $T_d = T_i \sim 0.03ev$, $n_d \sim 5 \times 10^{11} m^{-3}$, $M_d \sim 5.4 \times 10^{-14} kg$, $Z_d \sim 10^4$ giving the theoretical value of the phase velocity of these waves to be $C_{da} \sim 4.5 cm/sec$, in close agreement with the experimentally measured value. When the pressure in the chamber was further reduced to 0.1 mbar an additional wave motion appeared in the dust cloud over and above the dust acoustic mode. This mode had a significantly lower frequency, propagated along the axial direction of the chamber and had a vertical polarization. It thus constituted a transverse shear wave. A typical value of the measured wave number and phase velocity of this mode was $\lambda_z \sim 3.7mm$, $V_{phz} \sim 4.2mm/sec$, leading to $\omega_z \sim 1$ Hz. For $\Gamma \gg 1$, a simple theoretical estimate of the frequency of the transverse shear mode has been given in [1] as $\omega \sim (kd)\omega_{pd}/\sqrt{3}$. For $\lambda_z \sim 3.7mm$ this gives $\omega \sim 1$Hz which is close to the experimental value. Any further decrease in the pressure did not produce any significant changes in the wave behaviour. In another set of experiments, carried out under similar operating conditions, but with a differently shaped anode (hollow conical structure) the equilibrium dust cloud had a more compact size (axial length of $\sim 10cms$). When the neutral pressure was reduced in this case spontaneous oscillations of DAWs once again appeared in the dust cloud. The oscillations were found to have a sharp power spectrum centered around $k \sim 7.5 cm^{-1}$. However on further reduction of the pressure no transverse oscillations were seen possibly because of the axial size restriction. Instead, the original coherent dust acoustic oscillations continued to exist and in addition gave rise to higher harmonic oscillations. Video images of these nonlinear oscillations and the evolution of their power spectra are shown in Fig. 1. As is seen from the data the fundamental mode (which corresponds to the most linearly unstable DAW) acquires a steady state saturated level by pumping energy into damped higher harmonic modes. A simple theoretical model based on this physical picture and using the nonlinear GH model is presented in the next section.

THEORETICAL ANALYSIS

The GH model provides a simple but physically intuitive description of a strongly coupled fluid. However its application has so far been restricted to only linear wave propagation problems. We begin by presenting a simple nonlinear extension of the model for describing the dust dynamics. The starting point is the generalized momentum equation which (in one dimension) is of the form,

$$\left(\frac{\partial}{\partial t} + u_d \frac{\partial u_d}{\partial x}\right) = -\frac{1}{M_d n_{0d}} \frac{\partial P}{\partial x} + \frac{Z_d eE}{M_d} + \int_{-\infty}^{t} dt' \int dx' \eta_d(x-x', t-t') u_d(x', t') \quad (1)$$

where u_d is the dust velocity, P and E are the pressure and electric field and M_d is the dust particle mass. The quantity η_d may be identified as a nonlocal viscoelastic operator which accounts for memory effects for increasing values of Γ. If η_d is modeled as a product of a simple memory function which is exponential in time (with τ_m as the relaxation time) and a spatial function which has the usual features typical of a viscosity operator, then eq.(1) can be exactly converted to the form,

$$(1+\tau_m\frac{\partial}{\partial t})\left[M_d n_{0d}(\frac{\partial}{\partial t} + u_d\frac{\partial}{\partial x})u_d + \frac{\partial P}{\partial x} - Z_d e n_{0d} E\right] = \eta^* \frac{\partial^2 u_d}{\partial x^2} \quad (2)$$

where $\eta^* = (4\eta/3+\zeta)$. In the absence of temperature fluctuations, eqn.(2) along with the nonlinear continuity equation can be used to study nonlinear DAW propagation. The electron and ion dynamics represented by the usual Boltzman relations and Poisson's equation (or its quasineutral limit) would be the other relations to complete the full set of model equations. We have used such a model set of equations to examine various nonlinear limits of the DAW propagation. To understand the present experimental results we have included an equilibrium ion drift in the model as a source of linear instability for the DAW [6]. For $\omega\tau_m \ll 1$ and $k\lambda_p \ll 1$ (the limit corresponding to the experimental observations of DAW) we have carried out a simple multi-mode multiple scale expansion of the nonlinear equations. In suitable normalized coordinates we have expanded all quantities in the form, $q(x,t) = q_0 + \sum_{m=1}^{\infty} \varepsilon^m \sum_{l=1}^{N} q_l^{(m)}(\tau) exp(ilkx - il\omega t) + c.c.$ where $q(x,t) = (n, u_d, \phi)$, $q_0 = (1,0,0)$ and $\tau = \varepsilon t$. Restricting ourselves to just two modes ($N=2$) and using standard multiple scale analysis to order ε^3 we arrive at the following coupled evolution equations.

$$\frac{\partial \phi_1^{(1)}}{\partial \tau} = \gamma_1 \phi_1^{(1)} - \iota \phi_1^{(1)} \phi_2^{(1)*} \quad (3)$$

$$\frac{\partial \phi_2^{(1)}}{\partial \tau} = \gamma_2 \phi_2^{(1)} + \iota \phi_1^{(1)} \phi_1^{(1)} \quad (4)$$

where $\gamma_l = l - \alpha l^2$, $\iota = \sqrt{-1}$ and $*$ denotes complex conjugation. α is the ratio of the linear growth rate to the viscous damping rate. When $0.5 < \alpha < 1$, $\gamma_1 > 0$ indicating instability of the fundamental DAW mode and $\gamma_2 < 0$ implying damping of the first harmonic mode. In such a case, the above model describes the transfer of energy

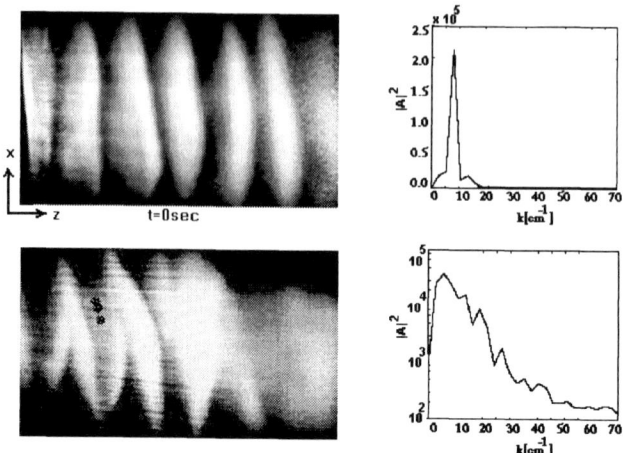

FIGURE 1. Nonlinear Dust Acoustic Waves

from the linearly unstable mode ϕ_1 to its damped first harmonic through the quadratic nonlinear coupling terms. Saturation is thus possible if there is energy balance and the saturation amplitudes can be obtained by setting the time derivatives to zero in the above equations. This gives us, $|\phi_1^{(1)}/\phi_2^{(1)}| = \sqrt{|\gamma_2/\gamma_1|} = \sqrt{|2(1-2\alpha)/(1-\alpha)|}$. For the initial two mode spectrum shown in the top half of Fig.1, $\alpha \sim 0.9$ gives a good agreement. In principal the analysis can be extended to a larger number of modes to fit the nature of the saturated spectrum shown in the bottom half of Fig. 1.

To conclude, we have experimentally observed the spontaneous excitation of large amplitude DAWs and the transverse shear wave in a strongly coupled dusty plasma that is in the fluid state. The characteristics of the shear wave seem to be well explained by the linear viscoelastic theory of the GH model. To understand the nonlinear features of the DAW we have developed a simple nonlinear extension of the GH equations and used it to explain the saturation of the DAW through harmonic wave generation. The nonlinear GH equations can also serve as a paradigmatic model for exploring other nonlinear regimes particularly when $\omega\tau_m \gg 1$. Such investigations are presently underway and will be reported elsewhere.

REFERENCES

1. Kaw, P. K., and Sen, A., *Phys. Plasmas*, **5**, 3552–3559 (1998).
2. Kalman, G., Rosenberg, M., and Dewitt, H. E., *Phys. Rev. Lett.*, **84**, 6030 (2000).
3. Mishra, A., Kaw, P. K., and Sen, A., *Phys. Plasmas*, **7**, 3188 (2000).
4. Pieper, J. B., and J.Goree, *Phys. Rev. Lett.*, **77**, 3137–3140 (1996).
5. Pramanik, J., Prasad, G., Sen, A., and Kaw, P. K., *Phys. Rev. Lett.*, **88**, 175001 (2002).
6. Fortov, V. E., Khrapak, A. G., Khrapak, S. A., Moltokov, V. I., Nefedov, A. P., Petrov, O. F., and Torchinsky, V. M., *Phys. Plasmas*, **7**, 1374–1380 (2000).

Experimental Dust Levitation in a Plasma Sheath near a Surface

Amanda A. Sickafoose[a], Joshua E. Colwell[a], Mihály Horányi[b], and Scott Robertson[b]

[a] *Laboratory for Atmospheric and Space Physics, University of Colorado, Boulder, CO 80309, USA*

[b] *Department of Physics, University of Colorado, Boulder, CO 80309, USA*

Abstract. We report the results of experiments on the levitation of dust particles in an argon plasma sheath above a flat, conducting surface. Types of particles tested include hollow glass microballoons, polystyrene DVB (divinylbenzene) microbeads, and JSC-1, a lunar regolith simulant. Plasma characteristics are determined using a Langmuir probe, while the sheath potential profiles are measured by an emissive probe. Dust particles levitating above the surface of the plate are illuminated by an Ar laser and observed by a video camera. Our experimental results suggest the following: (1) various types and sizes of particles can levitate in a plasma sheath above a surface; (2) particle levitation height and corresponding charge are comparable to the values calculated from orbital motion limited (OML) theory; (3) exposure to a UV source slightly alters the particle levitation height; (4) a mechanism to inject particles into the sheath is not necessary given a large enough surface bias; (5) complex particle dynamics occur in the plasma, and horizontal transport across the surface has been measured.

INTRODUCTION

There are many examples of active dust transport near surfaces in the solar system: dust grains suspended above the lunar surface, spokes observed in Saturn's rings by Voyager, and recent images of infilled craters from the NEAR spacecraft at Eros [e.g. 1]. Electrostatic dust levitation and transport may also occur on Mercury, asteroids, and comets [2]. Dusty regoliths are produced by the interplanetary micrometeoroid flux on nearly all airless bodies in the solar system. Therefore, understanding dust charging, levitation, and dynamics above surfaces is important for interpreting remote sensing data and analyzing the evolution of most planetary surfaces.

Objects in a plasma, such as planetary bodies in the solar wind, charge to a floating potential determined by the balance between charging currents in the local plasma environment. The primary charging currents are due to collection of electrons and ions from the plasma, photoemission, and secondary electron emission. In cases where secondary electron emission and photoemission are weak, objects will become negatively charged due to electron collection and will be surrounded by a plasma sheath. Negatively charged dust grains from these surfaces can thus be levitated in a plasma sheath above the surface at a height where the gravitational force is balanced by the electric force. The interaction between charged dust particles and a plasma sheath above a surface is one proposed mechanism for dust levitation and transport on bodies throughout the solar system.

We have conducted experiments on the levitation and transport of dust grains in an argon plasma sheath above a conducting surface. The purpose of these experiments is not to simulate conditions found in the solar system, but to obtain an understanding of dust dynamics. We determine that the experimental data for dust potentials and levitation heights in a plasma sheath are consistent with applicable model equations. We also investigate the horizontal motion of dust and show that dust particles are readily transported across the experimental surface.

EXPERIMENTAL SETUP

The experiments are conducted in a cylindrical stainless steel vacuum chamber 51 cm in diameter and 28 cm deep. The schematic diagram for the chamber is the same as *Sickafoose et al.* [3]. The chamber is evacuated and then filled with argon gas to a pressure of 1.6×10^{-4} torr. Dust grains rest on a horizontal, conducting surface ~ 30 cm in diameter in the middle of the chamber. The electric field above the surface is controlled by connecting the surface to an external power source. An insulated hammer underneath the surface can be manually activated to agitate and inject dust into the sheath, however this is not always necessary. Grains have been observed to separate from the surface and achieve levitation when it is biased to ≤ -30 V without agitation. There is a tungsten filament located underneath the surface, which creates the primary electrons to ionize gas in the chamber. The resulting plasma is collisionless and is assumed to be Maxwellian. The filament position below the surface prevents charging of dust grains on the surface by primary electrons. A Langmuir probe and an emissive probe are employed to measure plasma characteristics.

Dust grains on and above the surface are illuminated by an air-cooled argon laser. The laser beam passes through a cylindrical lens, producing a vertical sheet. A viewing window perpendicular to the incoming laser sheet allows observation of the dust by a video camera. A narrowband filter (488 ± 2 nm) rests in front of the video camera, allowing the laser light reflected from dust grains to be observed while incident light from the filament and outside of the chamber is nearly eliminated.

The Langmuir probe measurements indicate a background electron density of $n_0 = 2 (\pm 1) \times 10^7$ cm^{-3}, an electron temperature $T_e = 3.6 (\pm 0.2)$ eV, and a plasma potential $\varphi_p \cong 1.7$ V. These parameters give an electron Debye length of $\lambda_d \cong 0.3$ cm. In order to determine the sheath potential profile above the surface, data from the emissive probe are taken from approximately 0.6 to 7.0 cm from the surface at 0.05 cm increments. Under typical plasma conditions, the surface has a measured floating potential of −10 to −20 V. When a sheath is driven to more negative voltage, in this case having a surface bias, V_b, much less than the floating potential, the sheath height can be much larger than the Debye length. For $V_b = -40$ V the sheath height is approximately $z = 2.7$ cm. Beyond the sheath, the presheath extends to a distance of roughly 7 cm from the surface, at which point the plasma potential is typically reached.

DUST LEVITATION EXPERIMENTS

Three different types of dust grains are used in the levitation experiments: hollow glass microballoons (<45 microns in diameter), polystyrene DVB microbeads (10 microns in diameter), and JSC-1, a lunar regolith simulant (<25 microns in diameter). The glass microballoons have a large charge-to-mass ratio and are thus levitated at a smaller plate bias. The polystyrene microbeads have a narrow size distribution that allows calibration of the experiment through comparison with a theoretical model. The JSC-1 is used because its chemical composition and mineralogy fall within the ranges of lunar mare soil samples [4].

The polystyrene microbeads are used to compare the particle potential as a function of surface bias to the potential predicted from balancing charging currents to a spherical particle using the OML model (similar to experiments performed by [5]). As the surface voltage is made more negative in the experiments, the levitation height of a grain increases. This is expected, since the electric field in the sheath increases for decreasing surface biases and the potential well extends further into the plasma. For a range of plate biases −30 to −80 V, the measured dust potentials and charges correspond well to those calculated using OML theory [3]. Additionally, the observed dust levitation heights match the heights calculated using the OML theory combined with balancing the gravitational force and the force in the electric field [3]. Dust levitation experiments also show that under UV exposure, dust particles consistently float at a slightly lower height and thus a slightly less negative potential [3].

In order for dust to levitate, sufficient force to overcome surface adhesion is required. Although the experiment has an agitator, which can hit the surface and inject dust particles into the plasma sheath, this mechanism is not necessary. All types of dust particles are launched into the sheath without agitation when the surface bias is less than approximately −30 V. The larger the surface bias, the larger the number of particles observed lifting upward from the surface. These results suggest that levitation and transport of dust particles above a surface with a plasma sheath does not require extreme surface potentials or external disturbances such as impacts.

HORIZONTAL DUST TRANSPORT EXPERIMENTS

In the plasma sheath above the surface, particles are seen singly, in groups, in stable configurations, and exhibiting a range of complex vertical and horizontal motion. The next step in understanding the observation of dust motion near surfaces in space is to investigate the horizontal transport of grains. We place a graphite surface in the chamber and coat it with JSC-1 dust. A clean graphite collector, a rectangle (5.0 cm x 3.8 cm) is set on top of the dust. After 45 minutes of total exposure to the plasma sheath with the surface bias set to −100 V (in order to allow dust to overcome adhesion), the collector is visibly dusty. Data of the clean collector and the dusty collector with the right side wiped clean for comparison are shown in Figure 1. A reverse experiment, in which a dusty collector is placed on a clean surface, produced similar results. Current experiments include measurements of dust transport from a confined distribution in order to quantify horizontal motion.

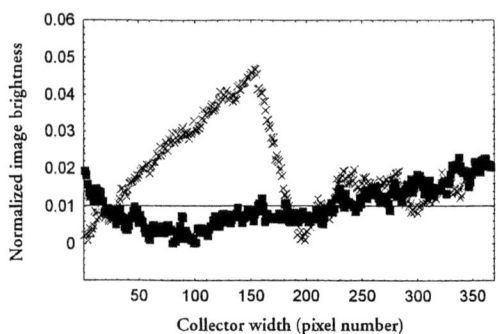

FIGURE 1. Normalized image brightness from digital photos of the clean collector (squares) and the collector after the experiment with the right half wiped clean (crosses). The solid line represents the average of the clean collector data. Clearly, there is dust on the left hand side of the collector that was transported during the experiment.

CONCLUSIONS

These experiments demonstrate that levitating dust grains of a standard size float at heights and potentials corresponding to those predicted by OML theory. They also add support to the model of electrostatic processes being the primary cause of dust levitation and transport near surfaces in space. Future experiments include a more detailed investigation of JSC-1 levitation and motion above a surface having a plasma sheath. In particular, we plan to study the horizontal transport of dust particles above surfaces of different compositions and having varied topography. These experiments can be compared to observations of dust dynamics above the Moon and those that might have occurred on the surface of Eros [1,2].

ACKNOWLEDGMENTS

The authors acknowledge support from the NASA Microgravity and Fluid Physics and Graduate Student Researcher programs. We also thank Zoltán Sternovsky.

REFERENCES

1. Rennilson, J.J. and D.R. Criswell, *The Moon* **10**, 121-142 (1974); Zook, H.A. and J.E. McCoy, *Geophys. Res. Lett.* **18**, 2117-2120 (1991); Robinson, M.S., P.C. Thomas, J. Veverka, S. Murchie, and B. Carcich, *Nature* **413**, 396-400 (2000).
2. Ip, W.H., *Geophys Res. Lett.* **13**, 1133-1136 (1986); Lee, P., *Icarus* **124**, 181-194 (1996); Mendis, D.A., J.R. Hill, H.L.F. Houpis, E.C. Whipple Jr., *Astrophys. J.* **249**, 787-797 (1981).
3. Sickafoose, A. A., Colwell, J.E., Horányi, M., and Robertson, S., *J. Geophys. Res.* to appear, (2002).
4. McKay, D.S., J.L. Carter, W.W. Boles, C.C. Allen, and J.H. Alton, "JSC-1: A new lunar soil simulant" in *Engineering, Construction, and Operations in Space IV*, Am. Soc. of Civ. Eng., New York, 1994, pp.857-866.
5. Arnas, C., M. Mikikian, G. Bachet, and F. Doveil, *Phys. of Plasmas* **7**(11), 4418-4422 (2000); Arnas, C., M. Mikikian, and F. Doveil, *Phys. Scripta* **T89**, 163-167 (2001).

Grain Oscillations Induced By Electrode Voltage Modulation

G. Sorasio*, D. P. Resendes* and P. K. Shukla[†]

*Centro de Fisica de Plasmas, Instituto Superior Tecnico, 1096 Lisboa Codex, Portugal
[†]Theoretische Physik IV, Fakultät für Physik & Astronomie Ruhr-Universität Bochum, D-44780 Bochum, Germany

Abstract. We have analyzed the dynamics of a single grain immersed in a plasma sheath, under low pressure conditions. The influence on the grain dynamics of an electrode potential modulation have been analytically and numerically explored. For the first time the effects of this excitation technique on the plasma sheath environment have been theoretically analyzed, and the grain motion resulting uom the sheath modification induced by the external driver have been studied. Comparison with experimental data inferred from published investigations is also presented.

INTRODUCTION

A negatively charged dust particle immersed in a plasma sheath levitates above the negatively biased electrode due to the electrostatic force acting vertically upwards, ion drag and gravity acting vertically downwards [1]. It has been experimentally observed that, under high pressures, the profile of the dust grain trapping potential can be considered parabolic [2], in contrast with the low pressure case, when the situation dramatically changes and the trapping potential becomes highly nonlinear [3, 4, 5]. In recent experiments dust grain oscillations have been driven by different excitation techniques and the resulting grain dynamics have been analyzed. In the present manuscript, instead of studying the consequences of the driving techniques on the grain motion as performed in previous analysis [6], we explore the influence of the externally applied driver on the sheath environment and we explain the externally induced nonlinear grain oscillations. In this study the grain oscillations are driven by varying sinusoidally the electrode voltage around a fixed value ϕ_{wj}. The potential well in which the dust grain is trapped oscillates and changes its shape with the same frequency of the electrode potential giving rise to nonlinear grain oscillations. Both primary and sub-harmonic nonlinear resonance regimes are investigated. The results show that a small amplitude modulation of the electrode potential, at the natural frequency of the well, may produce large nonlinear grain oscillations. The results show also that the amplitude of the potential modulation, near the sub-harmonic frequency, has to be above a threshold in order to induce large grain oscillations (parametric resonance). These have only very recently been identified [6]. The sheath model used in the manuscript follows closely the one presented by [7] where the sheath edge is at $= D$ (where z is the vertical axis), while the electrode is at $z = 0$. As boundary conditions, we take the electric potential at the sheath edge to be zero, where the ion acceleration is also zero. The sheath thickness D is determined by

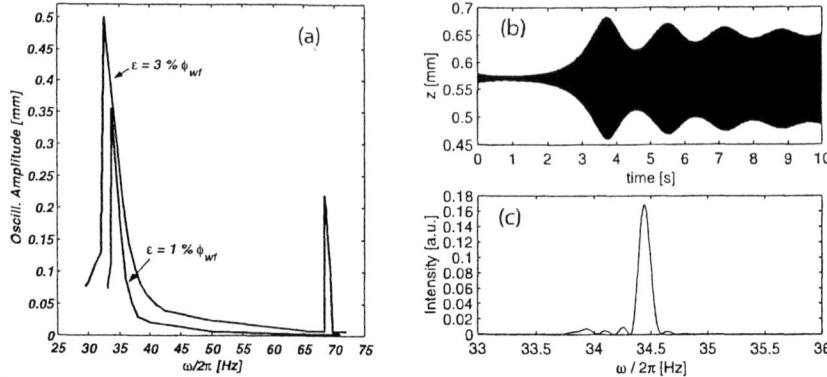

FIGURE 1. (a) Saturation oscillation amplitude for two different electrode potential voltage: $\varepsilon = 1\%\phi_{wf}$ and $\varepsilon = 3\%\phi_{wf}$. Dust grain (b) position and (c) oscillation spectrum under the influence of an electrode modulation amplitude $\varepsilon = 3\%\phi_{wf}$ and forcing frequency $\omega = 68.9\,\text{Hz}$. The parameters are: $p = 3\,\text{mtorr}$, $T_e = 0.5\,\text{eV}$, $T_i = T_n = 0.05\,\text{eV}$, $P_J = 1.2\,\text{gcm}^{-3}$, $R = 2.5\,\mu\text{m}$, $\rho_{wf} = 3\,\text{V}$, $n_0 = 2 \times 10^9\,\text{cm}^{-3}$

the requirement that the potential take on a given value at the electrode $\phi_s(z=0) = \phi_w$.

The main forces on a dust grain are the electrostatic force F_e acting vertically upwards, gravity (F_g) and ion drag (F_{id}) acting vertically downwards (positive z-direction). As for the sheath model, the total force acting on the dust grain is calculated following closely the model presented by [7]. In order to focus the attention on the results, the presentation of the model is left to the reader who can find a detailed description in [7, 8]. The dust grain equations of motion is $dv/dt = \tau_0 F_i/m_p v_{thi}$ where the displacement is normalized by the ion Debye radius λ_{Di} in the plasma, the time by $\tau_0 = (\lambda_{Di}/v_{thi})$ the velocity by V_{thi} and the total force is $F_t = F_e + F_g + F_{nd} + F_{id}$. It can be clearly seen that the only damping term is due to the neutral drag $F_{n_d}(v)$ [9,10].

Electrode potential modulation

The electrode potential modulation as a way of exciting dust grain oscillation has already been experimentally explored by several groups under high [11] and low pressure [6] conditions but, so far, the mechanism which drives the grain oscillation has not been theoretically identified. In the present analysis we investigate theoretically the nonlinear behavior of a dust grain under the influence of a low frequency sinusoidal modulation of the voltage applied on the negatively biased electrode under low pressures.

The voltage ϕ_w applied on the electrode is shielded by the plasma which self-consistently creates the plasma sheath. In the present model the sheath is source free since, under low pressures, the ion mean free path is much larger than the sheath thickness. This assumption implies that the level of the electrode voltage does not influence the plasma while, under much higher pressures, larger is the voltage larger is the ionization induced by the electrons accelerated through the sheath. When the electrode voltage increases, the plasma sheath is larger since the plasma needs a larger space to shield a higher potential. On the other hand, when the electrode voltage decreases the sheath thickness decreases. When the voltage on the electrode is modulated, the potential which traps the dust grain oscillates with the same frequency of the electrode: the grain finds itself trapped in a time varying well and starts to oscillate. In order to explore the grain dynamics we consider, at first, the steady state situation: the trapping potential can be expanded in a Taylor series around the equilibrium point z_{eq}. When the electrode potential oscillates as $\phi_w = \phi_{wf}(1 + \varepsilon \sin(\omega t))$, where ε is the amplitude of the electrode modulation, the trapping potential oscillate in the same way and its stable equilibrium point z_{eq}, i.e. the dust grain equilibrium position, will oscillate with the same frequency ω of the electrode voltage $z_{eq} = z_{eq0}(1 + h\sin(\omega t))$. If only the first three terms of the expansion are retained, the dust grain equation of motion can be written as:

$$\frac{d^2[z - z_{eq}(t)]}{dt^2} - \gamma\frac{d[z - z_{eq}(t)]}{dt} - \gamma = -[z - z_{eq}(t)]^2 \alpha - [z - z_{eq}(t)]^3 \beta - [z - z_{eq}(t)]\omega_0^2 \tag{1}$$

where z_{eq} is the position of the time varying bottom of the well respect to the electrode. As observed above, under low pressures, the sheath can be considered source free and thus the electrode voltage influences mainly the sheath thickness and the coefficients of the trapping potential can be considered constant in time. It has to be noticed that, in previous analysis [6], the dust grain equilibrium position has been considered constant in time and the electrode modulation was though to affect the electrostatic force acting on the grain without modifying its stable equilibrium position. In the present analysis the electrode voltage modulation affects self consistently the external field and the charge on the grain leading to an oscillating trapping potential. If z_{eq0} is the stable equilibrium point calculated for ϕ_{wj} and the variables are renamed as $z = (z - z_{eq0})$ and $z_{eq}(t) = [z_{eq}(t) - z_{eq0}]$ equation (1) can be rewritten as:

$$\frac{d^2}{dt^2} - \gamma\frac{dz}{dt} + \omega_0^2 z \left[1 - \frac{2z_{eq}(t)\alpha}{\omega_0^2} + \frac{3z_{eq}^2(t)\beta}{\omega_0^2}\right] = \\ -z^2[\alpha - 3z_{eq}(t)\beta] - z^3\beta + z_{eq}(t)\left[\omega_0^2 - \omega^2 - \gamma\omega\cot(\omega t)\right] - z_{eq}^2(t)\alpha + z_{eq}^3(t)\beta \tag{2}$$

which neglecting the terms of second order in z_{eq} becomes:

$$\frac{d^2}{dt^2} - \gamma\frac{dz}{dt} + \omega_0^2 z\left[1 - \frac{2z_{eq0}h\alpha}{\omega_0^2}\sin(\omega t)\right] = +f(t) - z^2\Gamma - z^3 \tag{3}$$

where $f(t) = -\omega_0^2 z_{eq0} h \sin(\omega t + \delta)$ with $\tan(\delta) = \gamma\omega/(\omega^2 - \omega_0^2)$ and $\Gamma = [\alpha - 3z_{eq0}\beta h\sin(\omega t)]$. Equation (3) is the well known Mathieu equation with nonlinear and forcing terms on the right hand side. The first consideration is that the existence of a parametric resonance depend on the nonlinearity of the trapping potential: if the potential well is linear the grain dynamics reduces to forced oscillation under friction, where the forcing term is $f(t)$. It has to be noticed that, as far, was not known why, when the trapping potential is parabolic, i.e. constant grain charge and linear electric field, no parametric features are observed [11], while a clear parametric resonance is detected when the trapping potential is nonlinear [6]. The second consideration is that, in the case of nonlinear trapping potential well, the dust grain dynamics will show features of both nonlinear and parametric regimes. In figure 1(a) it can be seen that, at the natural well frequency ω_0 the grain dynamics shows nonlinear resonance. The figure is obtained by solving the complete system of equations describing the grain dynamics without using any expansion of the trapping potential. At the natural well frequency ω_0 the main driving term is the forcing $f(t)$. On the other hand, at twice the natural frequency ω_0 the main resonance is due to parametric excitation, since the amplitude ε of the electrode voltage modulation has to be above a threshold for the resonance to be observed and the grain oscillation frequency is nearly ω_0. Since the forcing term is proportional to roo and the time varying parameter is proportional to a the largest effect would be given by the forcing term.

In conclusion we have successfully explained why, under low pressures, the electrode potential modulation driving technique can excite both parametric and nonlinear resonance regimes while, when the trapping potential is parabolic, or when the nonlinearity are weak, i.e. under higher pressures, no parametric resonance can be excited. The numerical simulation presented show good agreement with the experimental data [6].

ACKNOWLEDGMENTS

G. Sorasio would like to thank the ICPDP organization for the kind invitation and financial support. This research was supported by the Research Training Networks Programme of the European Union through contract HPRN-2000-00140 entitled "Complex Plasmas: The Science of Laboratory Colloidal Plasmas and Mesospheric Charged Aerosols".

REFERENCES

1. Shukla P.K. and Mamun A.A., *Introduction to Dusty Plasma Physic.*, Institute of Physics Publishing Ltd., Bristol, 2002.
2. Tomme E.B., Law D. A., Annaratone B. M., and Allen, I. E., *Phys. Rev. Lett.*, **85**, 2518 (2000).
3. Ivlev A.V., Sutterlin R., Steinberg V., Zuzic M., and Morfill G., *Phys. Rev. Lett.*, **85**, 4060 (2000).
4. Zafiu C., Melzer A., and Piel, A., *Phys. Rev. E*, **63**, 66403 (2001).
5. Sorasio G., Resendes D.P., and Shukla P. K., *Phys. Lett. A*, **293**, 67 (2002).
6. Ohno N., Sawai M., Misawa T., Asano K. and Takamura S., *Physica Scripta*, **T92**, 81–86 (2002).
7. Sorasio, G., Fonseca, R.A. Resendes D.P. and Shukla, P.K., "Dust grain oscillation in plasma sheaths under low pressures", in *Dust Plasma Interaction in Space*, edited by P. K. Shukla, Nova Publishers, New York, 2002.
8. Resendes, D.P. Sorasio ., and Shukla, P.K., *Phys. Plasmas*, **9**, in press (2002).
9. Sorasio G., Mendis D.A., and Rosenberg M., Planet., *Space Sci.*, **49**, 1257 (2001).
10. Baines M.I., Williams I.P., and Asebiomo A.S., *Mon. Not. R. Astron. Soc.*, **130**, 63–74 (1965).
11. Schollmeyer H., Meltzer, A., Homann, A., and Piel, A., *Phys. Plasmas*, **6**, 2693 (1999).

Boundary Phenomena in RF and DC Glow Discharge Dusty Plasmas

Edward Thomas, Jr.[*], William E. Amatucci[†], and Gregor E. Morfill[¶]

[*]*Physics Department, Auburn University, Auburn, AL 36849 USA*
[†]*Plasma Physics Division, Naval Research Laboratory, Washington, DC 20375*
[¶]*Max-Planck-Institut für Extraterrestrische Physik, D-85740 Garching, Germany*

Abstract. In experimental investigations of dusty or complex plasmas, studies have been performed primarily using dc or rf glow discharge plasmas. In spite of the similarity of the experimental parameters produced by each of these plasma generation techniques, the addition of charged microparticles to these plasmas often leads to different phenomena. This paper discusses the similarities and differences in particle transport observed at the boundary between the microparticle clouds and the surrounding plasma. Results will highlight experimental studies performed using two different dc glow discharge dusty plasma experiments - the Auburn Dusty Plasma Experiment (DPX) and the Naval Research Laboratory DUPLEX experiment – and an rf glow discharge dusty plasma experiment – the Plasma-Kristall Experiment (PKE-Nefedov) at Max-Planck-Institut für Extraterrestrische Physik. Sheath-like structures are observed at the particle cloud – plasma interface in all three cases.

INTRODUCTION

The majority of dusty or complex plasma experiments are performed using one of two types of experimental configurations – the dc glow discharge [1-3] or the rf glow discharge [4, 5]. In spite of the considerable number of experiments that have been conducted using these two configurations, there remains questions about the differences that arise when charged microparticles are introduced to these plasma systems.

For example, most of the strongly coupled and plasma crystal phenomena associated with complex plasmas are based upon studies using rf glow discharge plasmas. By comparison, much of the work on dust acoustic modes and other collective phenomena have been performed using dc glow discharge plasmas. This does not claim that the results obtained in one type of plasma system can not be reproduced in the other. Rather, the published record of experimental results suggests that the microparticles have a "preferred" configuration for each of the plasma generation techniques. Consequently, investigations have tended to focus on those experimental configurations that are most readily attainable in each device.

The experiments discussed in this paper represent one of the first direct comparisons between studies performed on both dc- and rf-generated dusty plasmas. Specifically, an analysis is performed on particle transport near the plasma - microparticle cloud boundary in a dusty plasma. It is shown that sheath-like structures are found to exist near the boundaries of both types of dusty plasma environments.

EXPERIMENTS

The dusty plasma experiments described in this paper are performed using three different devices. The Plasmakristall Experiment (PKE-Nefedov) at the Max Planck Institute is an rf-powered device. It uses a grounded upper electrode and a powered lower electrode (at 13.56 MHz) to generated argon plasmas in a region that has a spatial extent of ~ 3 cm [6]. The Auburn Dusty Plasma Experiment (AU-DPX) is a dc glow discharge experiment that uses both a powered anode ($V_a \sim 200$ V) and a powered cathode ($V_c \sim -130$ V) to generate argon plasmas. The chamber is made from 6-way, ISO 100 stainless steel crosses with an inner diameter of 10 cm [2]. Finally, the Naval Research Laboratory Dusty Plasma Experiment (NRL-DUPLEX) is also a dc glow discharge experiment. It uses a powered anode ($V_a \sim 800$ V) and a grounded cathode to generate argon plasmas. However, NRL-DUPLEX is a large experiment; it is a cylindrical polycarbonate vacuum vessel with a 80 cm inner diameter and approximately 80 cm in height [7].

For all three experiments, microparticles of comparable size are used. In PKE-Nefedov studies, uniform 3.4 µm diameter melamine-formaldehyde particles are used. In the AU-DPX and NRL-DUPLEX experiments, polydisperse silica microparticles with an average diameter of 2.9 µm are used. A comparison of the features of these three experiments is summarized below in Table 1.

One of the key parameters to note in Table 1 is the ratio of the chamber diameter (a) to the microparticle cloud diameter (d). This gives an indication of the role of the walls of the chambers on the overall confining potential of the microparticles in the plasma. It is postulated that the smaller the value of this ratio, the greater the influence of the vacuum vessel walls on the particle confinement.

Table 1. Comparison of Plasma Experiments

Experiment	Plasma Source	Cloud diameter (d)	Chamber diameter (a)	Ratio (a/d)
PKE - Nefedov	rf glow	1.5 – 2 cm	3 cm	1.3 - 1.5
AU – DPX	dc glow	3 – 4 cm	10 cm	2.5 - 3.3
NRL - DUPLEX	dc glow	10 – 15 cm	80 cm	5.3 - 8.0

In each of these experiments, a measurement is performed on the trajectories of microparticles as they approach the particle cloud – plasma boundary. This is accomplished through the use of either Particle Image Velocimetry (PIV) techniques [2,6,8] or a laser flashing technique. In PIV, a pair of laser pulses, synchronized to the frame-grabbing rate of a video camera is used to generate two-dimensional velocity profiles. The laser flashing technique uses a sequence of several short (~ 1 ms) laser flashes to make multiple illuminations of the particles on a single frame. If the separation time between the pulses is known, it is possible to reconstruct the velocity in the plane of illumination [9]. An example of a microparticle trajectory obtained using the laser flashing technique is shown in Fig. 1. In either case, the spatial evolution of the velocity profile can be used to reconstruct the local potential profile in the vicinity of the boundary [10].

ANALYSIS OF PARTICLE MOTION

In order to characterize the plasma – particle cloud boundary, an analysis is performed of microparticle trajectories as they approach the boundary. Here, the particle velocity is measured as a function of the distance, $<r>$, from the cloud boundary. When the measurements of the particle motion are analyzed for the three experiments, a number of similarities arise. An example of one of these measurements is shown in Fig. 2. Plotted is the particle velocity, in mm/s, as a function of distance from the boundary, in mm, for (a) NRL-DUPLEX, (b) AU-DPX, and (c) PKE-Nefedov.

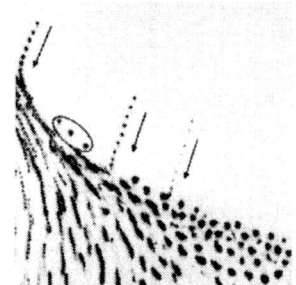

Figure 1: A composite image of 4 video frames of the microparticles – showing the trajectory of particles approaching the lower boundary of a void in the PKE-Nefedov chamber. The motion of the individual particles is indicated by the arrows. The particles in the lower right portion of the image and those in the oval are stationary particles in the plane of illumination.

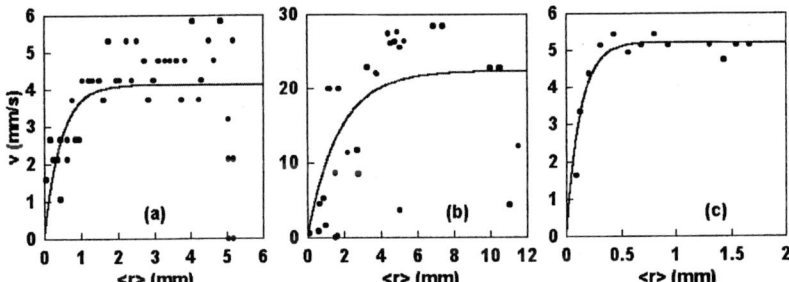

Figure 2: Comparison of velocity measurements from (a) NRL-DUPLEX, (b) AU-DPX, and (c) PKE-Nefedov experiments. The solid line in each figure is a curve fit of the form: $v(r) = a(1 - e^{-br})$. The three datasets show that as particles approach the plasma – cloud boundary, there is region of roughly constant velocity followed by a rapid deceleration of the particles. It is postulated that this corresponds to the presence of a sheath-like structure at the boundary. It is noted that in the case of the two dc glow discharge experiments [(a) and (b)], this sheath region is ~ 2 to 4 cm; for the rf experiment, this region is ~ 0.5 cm.

First, measurements from all of the experiments show that particles experience a constant velocity, in some cases, a slight acceleration, at distances of several millimeters from the cloud boundary. As the particles approach the cloud boundary, there is a rapid deceleration of the particles. It is shown in Ref. [9] that if the charge of the particle can be estimated, it is then possible to use the trajectory of the particle to estimate both the magnitude and spatial distribution of the potential difference ($\Delta\varphi$) between the particle cloud and the surrounding plasma. From the analysis of the trajectories, a sheath-like region, with a potential $\Delta\varphi \leq 0.1\ T_e$, is shown to be present at the cloud - plasma boundary for a cloud consisting of ~3 μm diameter particles. This assumes a particle charge $Z \sim 5000$ electrons as computed using orbit-motion-limited theories.

However, one of the outstanding issues in this study is the difference in the spatial structures of this sheath region between the dc- and rf-glow discharge experiments. Both of the dc experiments indicate a sheath region $x_s \sim 2 - 4$ mm. However, in the rf experiment, this region is $x_s \leq 0.5$ mm. The origin of the difference remains unclear. One possibility is that the small size of PKE-Nefedov ($a/d \sim 1$) allows a greater influence of the vacuum vessel walls on the potential structure. Nonetheless, these initial results are promising because a clear parallel between phenomena in both the dc and rf glow discharge dusty plasma systems has been identified. It is hoped that future investigations – notably with more precise measurements of the particle charge – will provide additional insight into the common features of both types of dusty plasma systems.

ACKNOWLEDGMENTS

The authors would like acknowledge the support of the following agencies: the U.S. National Science Foundation and NASA (E.T.), the Naval Research Laboratory (W.E.A.), das Bundesministerium für Bildung und Forschung durch das Zentrum für Luft- und Raumfarht e.V. (G.E.M.).

REFERENCES

1. C. Thompson, A. Barkan, N. D'Angelo, and R. L. Merlino, *Phys. Plasmas*, **4**, 2331 (1997).
2. E. Thomas, Jr. and M. Watson, Phys. Plasmas, **6**, 4111 (1999).
3. R. Gandy, S. Willis, and H. Shimoyama, *Phys. Plasmas*, **8**, 1746 (2001).
4. H. Thomas, G. E. Morfill, V. Demmel, and J. Goree, *Phys. Rev. Lett*, **73**, 652 (1994).
5. G. S. Selwyn, J. Singh, and R. S. Bennett, *J. Vac. Sci. Technol. A*, **7**, 2758 (1989).
6. G. E. Morfill, H. M. Thomas, U. Konopka, et. al., *Phys. Rev. Lett.*, **83**, 1598 (1999).
7. E. Thomas, Jr., W. E. Amatucci, C. Compton, and B. Christy, "Observations of structured and long-range transport in a large volume dusty (complex) plasma experiment", accepted, *Phys. Plasmas*, (2002).
8. E. Thomas, Jr., Phys. Plasmas, **6**, 2672 (1999); E. Thomas, Jr., *Phys. Scripta*, **T89**, 20 (2001).
9. E. Thomas, Jr., B. M. Annaratone, G. E. Morfill, and H. Rothermel, "Measurements of forces acting on suspended microparticles in the void region of a complex plasma", submitted to *Phys. Rev. E.* (2002).
10. E. Thomas, Jr., *Phys. Plasmas*, **8**, 329 (2001).

Transport of Macroparticles in Weakly Ionized Dusty Plasma of Gas Discharges

O. S. Vaulina, O. F. Petrov, V. E. Fortov

Institute of High Temperatures, RAS, Moscow, Russia

Abstract. Transport of macroparticles in dust fluids have been numerically studied for radial pair potentials of different types. Estimations of effective dust charges have been performed for dust particles in dc- and rf – discharges under microgravity and ground- based conditions.

In this work, the results of numerical simulations of particle dynamics in dust fluids (including the ordering of macroparticles and self-diffusion processes) are presented, and an application of these results for the dust diagnostics in laboratory plasma of rf- and dc- discharges is considered. The particle dynamics in dust fluids has been studied by three-dimensional Brownian dynamics method under periodic boundary conditions [1]. The transport characteristics were calculated for different dust temperatures T, and charges eZ, different types of pair potentials $\phi(r)$ of interparticle interaction, and ratio ξ of characteristic dust frequency ω_p to the friction coefficient v_{fr}

$$\xi = \omega_p / v_{fr} \equiv \sqrt{\frac{eZ}{2\pi m_p} \phi''(r_o)} / v_{fr}, \qquad (1)$$

in the limits of values typical for laboratory experiments. Here m_p is the dust mass, ϕ'' is the second derivative of pair potential $\phi(r)$ at the mean intergrain distance $r_o = n_p^{-1/3}$, where n_p is the dust concentration. We used in our calculations: the screening Yukawa potentials ($\phi = \phi_c \exp(-\kappa\, r/r_o)$, $\kappa = r_o/\lambda$, $\phi_c = -eZ/r$), the power functions ($\phi \propto \phi_c\, r_o^j/r^{j+1}$, $5 > j > 0.5$); the combination of Yukawa potentials with different screening lengthens λ; and the Yukawa potential combined with the power functions. Last two potentials were used for simulation of the reducing of screening with the distance from particle predicted in Refs. [2,3]. All modal potentials were isotropic (radial), repulsive and long-interacting: $2\pi |\phi'(r_o)| > r_o |\phi''(r_o)|$.

Pair correlation functions $g(r)$ and structure factors $S(k)$ were studied for analysis of ordering of macroparticles. The calculation shown that the order of macroparticles is independent on a friction (v_{fr}) and fully determined by the ratio of ϕ''/T from the gas state of the system to the crystallization point where the *bcc*- lattices were formed for all considered cases. Thus, for analysis of phase state of dust systems we can use an effective coupling parameter $\Gamma^* = (Z^*e)^2/Tr_o$ with the effective particle charge

$$Z^*e = \{Ze\, \phi''/2n_p\}^{1/2} \qquad (2)$$

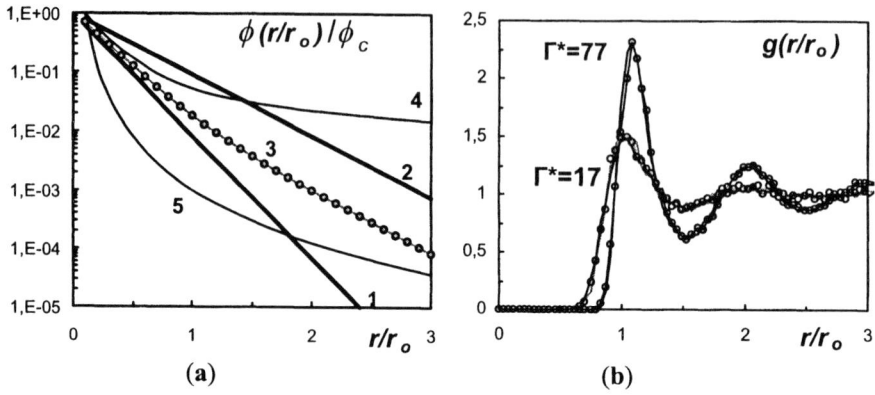

FIGURE 1. Ratio of ϕ to ϕ_c (a) and the pair correlation function g versus r/r_o for different $\phi(r)/\phi_c$:
1 - $\exp(-4.8r/r_o)$; 2 - $\exp(-2.4r/r_o)$; 3 - $0.1\exp(-2.4\,r/r_o) + \exp(-4.8\,r/r_o)$,
4 - $\exp(-4.8\,r/r_o) + 0.05 r_o/r$; 5 - $0.05(r_o/r)^3$.

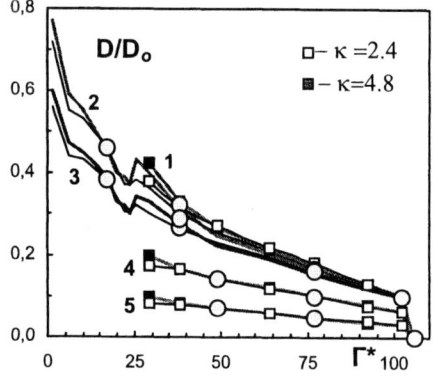 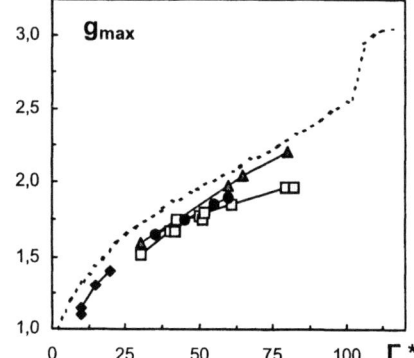

FIGURE 2. Ratio of D to the Brownian diffusion constant $D_o = T/m_p \nu_{fr}$ versus Γ^* for different ξ: 1 – 0.045; 2 – 0.14; 3 – 0.41; 4 – 1.22; 5 – 3.65. Circles are the points of calculations with different types of model potentials.

FIGURE 3. Values of first maximums g_{max} for correlation function versus the retrieved Γ^* for experiments in dc- ($\Diamond$),($\triangle$) and rf - ($\square$), (o) discharges. Dashed line is the result of numerical simulations.

The illustration of pair correlation functions for different values of Γ^* are shown in Fig. 1 for different model potentials. It should be noted that for all cases, the melting point of lattices was observed for effective Γ^* about 106, where the first maximum of correlations functions was abrupt changed and the self diffusion coefficients D were fast reduced (see Fig. 2). Self-diffusion coefficients D were obtained for different pair potentials. Basic calculations were performed for Yukawa interactions. The calculations shown that two basic parameters are responsible for the particle dynamics in dust fluids: the effective coupling parameter Γ^* and the scaling parameter ξ (Fig. 2).

The dust temperature T, the pair correlation function $g(r)$ and the self-diffusion coefficients D can be measured for the dust clouds in plasma without additional external actions on the systems. Nevertheless, the measurements of these characteristics allow a determination of effective dust parameters (Γ^*, Z^*) only. Determination of pair potential $\phi(r)$ requires additional assumptions about its form. We will consider the determination of effective parameters Γ^*, and charges Z^* on base of dust diffusion measurements in weakly ionized plasma of gas discharges.

First experiment is the measurements in striations of dc- discharge ($P = 0.4$-1 Tor, the current $I = 3$mA, the electron temperature $T_e \approx 2$-5 eV, the ion density $n_i \sim 5\ 10^9$ cm^{-3}) in neon (Ne) under ground-based conditions. The measurements were performed for small iron particles with the radius $a = 1$-3 µm. Basic purpose of these experiments was to study the dust structures the most close to the gas state. The effective coupling parameters Γ^* retrieved from diffusion measurements were varied from 10 to 20. Experiments in dc-discharge (Ne, $P = 1$ Tor, $I = 0.1$-1 mA, $T_e \approx 3$-7 eV, $n_i \sim 5\ 10^9$ cm^{-3}) under microgravity conditions carried out on the Mir space station for big bronze particles. The mean dust size $<a> = 65$ µm was more than the mean free path length l_{in} of ion-neutrals collisions and above the ion Debay radius λ_i [4].

We studied dynamics of small ($a = 1$-2.5 µm) Al_2O_3 particles in a single dust lay, which was formed above the ground electrode of rf- discharge ($P = 10$-40 Pa, the power $W = 2$-7 W, $T_e \approx 1$-3 eV, $n_i \sim 10^9$ cm^{-3}) in argon. The effective coupling parameters were estimated. Nevertheless an application of our three-dimensional simulation for analysis of effective characteristic can be not valid in this case. Last measurements were carried out in rf- discharge (Ar, $P = 36$ - 98 Pa, $W = 0.14$-1 W, $T_e \approx 1$-3 eV, $n_i \sim 10^9$ cm^{-3}, $a = 1.7$ µm) under microgravity conditions in the scope of scientific international program (PKE-3) [5].

The first maximums g_{max} for measured correlation functions are presented in Fig. 3 versus effective Γ^* retrieved from diffusion measurements for all four experiments. The obtained Γ^* values are in a god agreement with the measurements of correlation functions $g(r)$. We can see also that the both the strongly correlated dust fluids and the weakly non-ideal dust structures can be formed under conditions of gas discharge plasma ($\Gamma^* \sim 10$-85). Retrieved values of effective particle charges Z^* are presented in Figs. 4a-d. Under assumption of Yukawa interparticle interactions $Z^* = Z\ \{(1+\kappa+\kappa^2/2)\ \exp(-\kappa)\}^{1/2}$ and $(eZ/a) \approx 2$-4 T_e, we can find that the dust screening lengths λ are about λ_i ion Debay radius for the small particles ($a < \lambda_i$) in bulk three-dimensional dust clouds ($\lambda < 40$ µm for ground-based dc- discharge, and $\lambda < 90$ µm for rf-discharge in microgravity). The value of λ for big bronze particles ($<a> > \lambda_i$) is about electron Debay radius ($\lambda > 700$ µm). These results are in agreement with the numerical simulations presented in Ref. [2]. In the case of one-lay dust structures (ground-based rf- discharge), the retrieved value of $\lambda > 400$ μm is close to the electron Debay radius. But the determination of effective dust parameters on base of our calculations may be incorrect in this case.

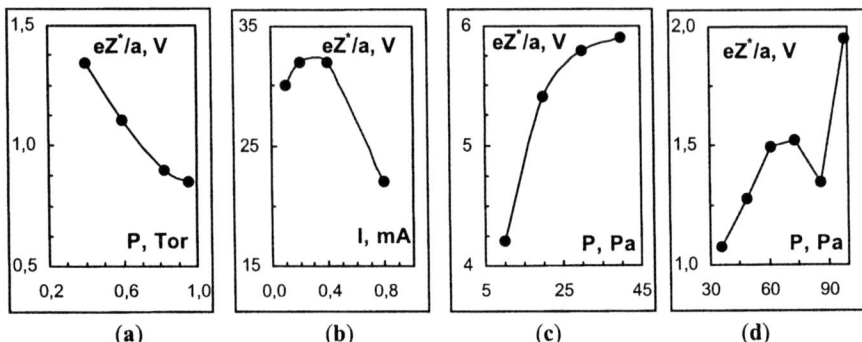

FIGURE 4. The effective dust charges Z^* for different experiments in dc- (a),(b) and rf - (c), (d) discharges under (a),(c) ground-based and (b),(d) microgravity conditions.

To conclude, the transport of macroparticles in dust fluids have been numerically studied for radial pair potentials of different types. The parameters responsible for particle dynamics in dust fluids have been considered including an effective coupling parameter and a scaling parameter. Estimations of effective dust charges have been performed for macroparticles in dc- and rf – discharges under microgravity and ground- based conditions.

ACKNOWLEDGMENTS

This work was supported by the Russian Foundation for Basic Research, Grant No.01-02-16658, No.00-02-32281 from the Russian Foundation for Basic Research, by INTAS Grant No. 2000-0522.

REFERENCES

1. Vaulina, O.S., Vladimirov, S.V., *Plasma Phys.*, **9**, 835 (2002)
2. Daugherty, J.E., Porteos, R.K., Kilgore, M.D., *et al*, *J. Appl. Phys.*, **72**, 3934 (1992).
3. Allen, J.E., *Phys. Scripta*, **45**, 497 (1992).
4. Nefedov, A.P., Vaulina, O S., Petrov, O.F., *et al*, (to be published in JTEP)
5. Stuffler, T., Schmitt, G., Pfeuffer, H., *et al*., Proceed. of 52nd International Astronautical Congress 1-5 Oct 2001/Toulouse, France, IAF-01-J.6.02.

Stationary Equilibria of Self-Gravitating Dusty Plasmas

Frank Verheest* and Vladimir M. Čadež[†]

*Sterrenkundig Observatorium, Universiteit Gent, Krijgslaan 281, B-9000 Gent, Belgium
[†]Belgian Institute voor Space Aeronomy, Ringlaan 3, B-1180 Brussels, Belgium

Abstract. Plasma and dust are key ingredients of our universe, prompting a revision of stationary equilibria of self-gravitating dusty plasmas. We review plane-parallel and spherically symmetric stationary configurations without flows. Indications about possible perturbations connect the inhomogeneity and Jeans scale lengths.

The presence of dust around and between stars has been known for a long time [1,2], and dust grains immersed in ambient plasmas and radiative environments become electrically charged by a variety of processes. For the many fascinating properties of dusty plasmas we can refer to recent books [3,4]. Extended mass systems have originally been studied as neutral matter. Gravitation being always attractive, collapse of extended regions with distributed masses is inevitable, unless counteracted by thermal agitation or other forces of repulsion. The (in)stability of a neutral cloud has been studied in astrophysics for a century, since Jeans obtained in 1902 his criterion based on the assumption that the unperturbed gaseous cloud is initially uniform [5]. The determination of self-consistent but nonhomogeneous equilibrium configurations remains a challenge [6,7], and the link between gravitation and dusty plasmas comes about because for certain grain sizes gravitational effects have to be taken into account. In what follows, we discuss some typical stationary configurations and outline the inherent difficulties.

Many stationary states are possible, and we first consider all charged particles as one fluid. Mass flow in one direction would lead to a depletion/source at one end and an accumulation/sink at the other. Similarly, in spherically symmetric states flows induce a total mass pile-up or depletion at the center, and therefor flows will be omitted. Extending our preliminary results [8], the ideal magnetohydrodynamic equation of motion reduces to $\nabla(p_0 + B_0^2/2\mu_0) + \rho_0 \nabla \psi_0 = 0$, where notations are standard and subscripts 0 refer to the stationary state. To relate this to Poisson's gravitational law, we take a polytropic pressure $p_0 = C\rho_0^\gamma$, and assume that the ratio β of the plasma pressure to the magnetic pressure $B_0^2/2\mu_0$ remains constant. In this model the Alfvén speed V_A is constant, and the magnetic field and plasma pressure are stronger in denser regions, as in highly conductive plasmas with frozen-in magnetic fields [8]. The equilibrium balance equation, together with Poisson's equation yields a single equation for ρ_0, written for one-dimensional systems as

$$\rho_0 \frac{d^2\rho_0}{dz^2} + (\gamma - 2)\left(\frac{d\rho_0}{dz}\right)^2 + \frac{4\pi G\beta}{C\gamma(1+\beta)}\rho_0^{4-\gamma} = 0 \quad (1)$$

It is simplest to assume isothermal pressures ($\gamma = 1$ and $C = v_{td}^2$), leading to $\rho_0(z) = \rho_{00}\mathrm{sech}^2(z/\lambda_{ms})$, with an extremum at $z = 0$ where the gravitational force vanishes. The scale length λ_{ms} is given by $\lambda_{ms}^2 = (V_A^2 + 2v_{td}^2)/\omega_{Jd}^2$, where ρ_{00} is the density at the center of the cloud. The Jeans frequency ω_{JD} is given through $\omega_{JD}^2 = 4\pi G\rho_{00}$. We note that the inhomogeneity scale length λ_{ms} is closely related to the corresponding Jeans length, obtained from linear stability analysis of magnetosonic modes in a homogeneous plasma [9]. The magnetic field is $B_0(z) = B_{00}\mathrm{sech}(z/\lambda_{ms})$, where the central field strength B_{00} is given through $B_{00}^2 = \mu_0 V_A^2 \rho_{00}$, as V_A is constant. For general values of γ the implicit solution of (1) is

$$\int_1^{\rho_0/\rho_{00}} \frac{r^{\gamma-2}dr}{\sqrt{1-r^\gamma}} = \frac{2z}{\gamma\lambda_{ms}} \quad (2)$$

Various possibilities have been numerically investigated and are illustrated in figure 1. Such models turn out to be well behaved when $\gamma > 1$, but are of finite extent. This might be an indication that fragmentation could occur, but this interesting suggestion warrants further investigation. Moreover, the variation of the density is then so abrupt that local perturbations would have to be on small scales that are not compatible with the usual Jeans approach. To the contrary, models with $\gamma < 1$ have an unphysical pressure behaviour that increases outwardly from the center without bounds.

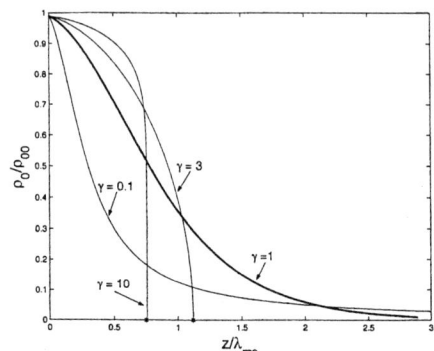

FIGURE 1. Density distributions vs. distance from the central location for various values of γ

For the study of self-gravitation in dusty plasmas, our eventual aim, a multifluid approach is needed, based on continuity and momentum equations per species α, on the gravitational Poisson equation and on Maxwell's equations [3]. For stationary states without flows the magnetic field decouples from the other physical quantities and the electric field is derivable from an electrostatic potential φ. Taking polytropic pressures $p_{\alpha 0} = C_\alpha n_{\alpha 0}^{\gamma_\alpha}$ gives for generic values of $\gamma_\alpha \neq 1$ Bernoulli type results $C_\alpha \gamma_\alpha n_{\alpha 0}^{\gamma_\alpha}/(\gamma_\alpha - 1) + q_\alpha \varphi_0 + m_\alpha \psi_0 = D_\alpha$ whereas for isothermal constituents ($\gamma_\alpha = 1$ and $C_\alpha = \kappa T_\alpha$) the densities are $n_{\alpha 0} = n_{\alpha 00} \exp\left[-(q_\alpha \varphi_0 + m_\alpha \psi_0)/\kappa T_\alpha\right]$. Densities n_{00} have been specified at the center of the cloud, and these will be needed to compute the respective plasma and Jeans frequencies, because one cannot assume equilibrium densities to be constant everywhere. Two coupled Poisson equations remain to be solved for the potentials, but analytic solutions are only possible when simplifying assumptions reduce the mathematical complexity.

In order to see the implications for dusty plasmas we turn to the standard model for low-frequency dusty plasma phenomena, and follow the lines of recent treatments [10]. The simplifications on the plasma side are that electron and ion inertia can be neglected, and densities reduce to Boltzmann expressions. It is tempting to assume almost perfect charge neutrality at all locations [10], the physical interpretation being that electrons and ions are tied to the charged dust by electrostatic forces. As a consequence, φ_0 becomes expressible as a algebraic function of ψ_0, and the gravitational Poisson equation reduces to an equation in ψ_0 alone. The charged dust is the only constituent of the plasma mixture to feel any gravitation, and the electrons and ions follow suit through their electrostatic coupling.

To see how acceptable this often invoked hypothesis really is, we simplify the dusty plasma model to one that can be discussed analytically. We assume that all electrons have been accreted onto the dust grains ($n_{e00} \simeq 0$), the ions have no inertia and the dust is so heavy that its temperature effects can be neglected. Ion momentum balance leads to a Boltzmann expression for the density, and dust momentum balance relates the two potentials, $\psi_0 = (C_{da}^2 n_{i00}/Z_d n_{d00}) \tilde{\varphi}_0$. In this expression the typical dust-acoustic velocity $C_{da} = \lambda_D \omega_{pd}$ occurs, λ_D is the plasma Debye length (here the ion Debye length), $\tilde{\varphi}_0 = e\varphi_0/\kappa T_i$ the normalized electrostatic potential and Z_d the absolute dust charge number. The two Poisson equations reduce to

$$\lambda_D^2 \nabla^2 \tilde{\varphi}_0 = Z_d n_{d0}/n_{i00} - \exp(-\tilde{\varphi}_0), \qquad \lambda_{da}^2 \nabla^2 \tilde{\varphi}_0 = Z_d n_{d0}/n_{i00} \qquad (3)$$

with a modified Jeans length $\lambda_{da} = c_{da}/\omega_{Jd}$, expressed through c_{da} rather than v_{td}. This is physically plausible: only the electrostatic coupling between the warm ions and the cold dust prevents total collapse of the dust, so that the ion pressure counts. Elimination of the dust density n_{d0} between the two Poisson equations (3) gives a single equation $(\lambda_{da}^2 - \lambda_D^2)\nabla^2 \tilde{\varphi}_0 = \exp(-\tilde{\varphi}_0)$, with solution for the one-dimensional case yielding

$$n_{i0} = n_{i00} \mathrm{sech}^2 \frac{\tilde{z}}{\sqrt{2}}, \qquad n_{d0} = \frac{\lambda_{da}^2}{\lambda_{da}^2 - \lambda_D^2} \cdot \frac{n_{io}}{Z_D} \qquad (4)$$

Here $\tilde{z}$ is a dimensionless coordinate, defined through $z^2 = \tilde{z}^2(\lambda_{da}^2 - \lambda_D^2)$, and it has been assumed that $\lambda_{da} > \lambda_D$, which is the case for almost all dusty plasmas under study. It thus turns out that the deviation from strict charge neutrality is $Z_d n_{d0} - n_{i0} = \lambda_D^2 n_{i0}/(\lambda_{da}^2 - \lambda_D^2)$, a general result that does not depend on the detailed behaviour of n_{i0}. This indicates that at the central location there is always more negative dust than positive ions, owing to the way the self-gravitation tends to concentrate the dust, which is opposed by the electrostatic coupling to the ions. Note that if the complete plasma were initially charge neutral, our results indicate that not all electrons can accrete onto the dust grains, as supposed in the mathematical description. This shows the pitfalls of simple minded approaches to self-gravitation. Nevertheless, the total charge density is small if $\lambda_D \ll \lambda_{da}$, and remarkably enough, vanishes at infinity rather than at the central location, as illustrated in figure 2. Except for some factors, density profiles like (4) follow the same behaviour as what was obtained in the single

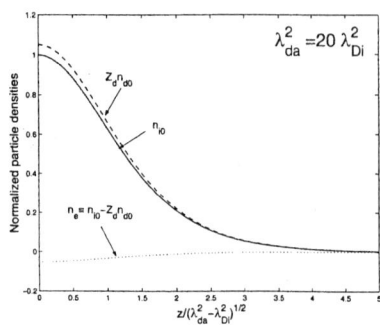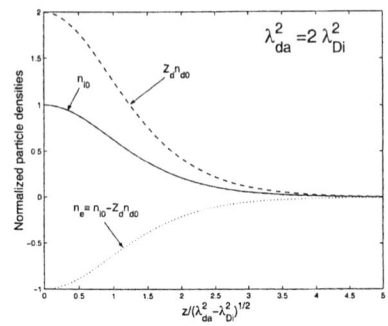

FIGURE 2. Densities and charge distributions vs. distance from the central location, for the solutions (4), and using in this model λ_{Di} as representative of the plasma Debye length λ_D.

fluid approximation, but on a different inhomogeneity scale. The opposite cases yield discontinuities in density, showing that very heavy charged dust ($w\omega_{pd} < \omega_{Jd}$) cannot be stabilized alone by electrostatic forces, imparted by massless ions, as assumed here. Again we have here a possibility of fragmentation that merits to be treated in more detail.

To conclude, it turns out that for the cases where explicit solutions can be obtained, the inhomogeneity lengths are of the order of the Jeans lengths calculated in the usual way. Maybe we can turn the argument around for more complicated configurations and use the Jeans swindle or local perturbation approach to find the scales over which the system itself is no longer homogeneous.

ACKNOWLEDGMENTS

This research was supported by the Fonds voor Wetenschappelijk Onderzoek (Vlaanderen) (FV) and the Belgian Institute for Space Aeronomy and the Belgian Federal Office for Scientific, Technical and Cultural Affairs (VMC).

REFERENCES

1. Whittet, D.C.B., *Dust in the Galactic Environment*, Bristol: IOP, 1992.
2. Evans, A., *The Dusty Universe*, Chichester: Wiley, 1994.
3. Verheest, F., *Waves in Dusty Space Plasmas*, Dordrecht: Kluwer, 2000.
4. Shukla, P.K. and Mamun, A.A., *Introduction to Dusty Plasma Physics*, London: IOP, 2002.
5. Jeans, J.H., *Astronomy and Cosmogony*, Cambridge: University Press, 1929.
6. Chandrasekhar, S., *Hydrodynamic and Hydromagnetic Stability*, Oxford: Clarendon Press, 1961.
7. Binney, J. and Tremaine, S., *Galactic Dynamics*, Princeton: University Press, 1988.
8. Verheest, F., Cadez, V.M. and Jacobs, G., in: *Waves in Dusty, Solar and Space Plasmas*, Eds. F. Verheest *et al.*, Woodbury: AIP, CP 537, 2000, p. 91.
9. Verheest, F., Hellberg, M.A. and Mace, R.L., *Phys. Plasmas*, **6**, 279 (1999).
10. Tsintsadze, N.L., Mendonca, J.T., Shukla, P.K., Stenflo, L. and Mahmoodi, J., *Physica Scripta*, **62**, 70 (2000).

Stability of Particle Arrangements in a Complex Plasma

S. V. Vladimirov, A. A. Samarian, J. Albrecht,
B. W. James, S.A. Maiorov, and N .F. Cramer

School of Physics, University of Sydney, New South Wales 2006, Australia

Abstract. It is shown that the stability of the vertical and horizontal confinement of colloidal "dust" particles levitating in a complex plasma appears as a non-trivial interplay of the external confining forces as well as the interparticle interactions and collective processes such as the plasma wake.

In the laboratory experiments, the micrometer sized highly charged dust grains levitate in the sheath region under the balance between the gravitational and electrostatic forces acting in the vertical direction as well as the externally imposed confining potential applied in the horizontal plane [1]. The vertical confinement involving the gravity force and the electrostatic force acting on the dust particles with variable charges is a complex process exhibiting oscillations, disruptions and instabilities [2-6]. A characteristic feature of the particle confinement is also the strong influence of plasma collective processes such as the plasma wake [7, 8]. Consider two colloidal particles of mass M and charges Q, separated by the distance X_d horizontally (i.e., aligned along the x-axis), see Fig. 1a or Z_d vertically (aligned along the z-axis), see Fig. 1b. In the simplest approximation, the

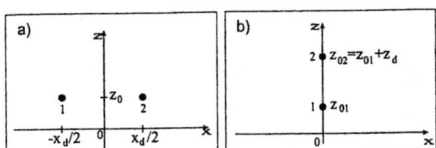

FIGURE 1. Sketch of the particle configurations.

particles interact via the screened Coulomb (Debye) potential $\phi_D = Q^2 \exp(-|r|\lambda_D)/|r|$ where λ_D is the plasma Debye length. For particles levitating in the plasma sheath, the interaction potential in the vertical direction is asymmetric because of the ions flowing towards the negatively charged electrode. However, it is also instructive to consider the case with Debye only interaction even in the vertical direction. We assume that the external confining force acting in the x-direction can be written as $F^{\text{ext}} = -\gamma_x(x - x_0)$, where $\gamma_x \sim QdE_x^{\text{ext}}/dx$ is a constant and obtain the balance of the external confining and Debye repulsion forces

$$\frac{2Q^2}{x_d^2}\left(1 + \frac{x_d}{\lambda_D}\right)\exp\left(-\frac{x_d}{\lambda_D}\right) = \gamma_x x_d. \tag{1}$$

The balance of forces in the vertical direction includes also the gravitational force $F_g = Mg$ and the sheath electrostatic force $F_{el} = QE_z^{ext}(z)$. In equilibrium, we assume that the sheath electric field near the equlibrium position can be linearly approximated $F_{el} - Mg = -\gamma_z(z - z_0)$, where $\gamma_z \sim QdE_z^{ext}/dz$ is a constant. For the vertically aligned particles (Fig. 1b) the lower and upper equilibrium positions are z_{01} and $z_{02} = z_{01} + z_d$, respectively. In this case, the equilibrium balance of the forces in the vertical direction can be written as $F_{el,1(2)}(z_{0l(2)}) - M_{1(2)}g + F_{1,2}^{D,W}(z_{02} - z_{01}) = 0$, where $F_{1,2}^{D,W}$ are the forces of the interaction between the particles due to their interaction Debye and/or asymmetric (wake) potentials Φ_D and/or Φ_d, respectively: $F_1^D(z_{02} - z_{0l}) = Qd\Phi_D(|z|)/d|z|\big|_{|z|=z_d}$, and $F_2^{D,W}(z_{02} - z_{0l}) = Qd\Phi_D(|z|)/d|z|\big|_{|z|=z_d}$. In the case of Debye only interaction between the particles, we obtain equation similar to (1), with the obvious change of x to z. In the case of the asymmetric wake potential, the equilibrium condition for the levitation of two identical particles gives us

$$\frac{Q^2}{z_d^2}\left(1 + \frac{x_d}{\lambda_D}\right)\exp\left(-\frac{z_d}{\lambda_D}\right) - \gamma_z^W(z_d - z_W) = \gamma_z z_d. \tag{2}$$

where z_W is the distance between the minimum of the asymmetric attracting potential characterized by γ_z and the upper particle. Now, consider horizontal perturbations of two horizontally aligned particles, Fig. 1a. By including the phenomenological damping β and linearly expanding the interaction forces, we obtain two oscillation modes with the frequency

$$\omega_{xx,1} = -\frac{i\beta}{2} + \left(\frac{\beta^2}{4} + \frac{\gamma_x}{M}\right)^{1/2}, \tag{3}$$

for the two particles oscillating with equal amplitudes ($A_1 = -A_2$), and

$$\omega_{xx,2} = -\frac{i\beta}{2} + \left[\frac{\beta^2}{4} + \frac{\gamma_x}{M}\left(3 + \frac{x_d^2/\lambda_D^2}{1 + x_x/\lambda_D}\right)\right]^{1/2} \tag{4}$$

for the particles oscillating counter phase ($A_1 = -A_2$). Both modes are always stable.

The next case involves vertical oscillations of two horizontally aligned particles, Fig. 1a. We obtain that the two oscillation modes have the frequency similar to (3), with the change of x to z, for the two particle oscillating in phase ($A_1 = A_2$), and

$$\omega_{xz,2} = -\frac{i\beta}{2} + \left(\frac{\beta^2}{4} + \frac{\gamma_x}{M} - \frac{x\gamma_x}{M}\right)^{1/2} \tag{5}$$

for the two particles oscillating counter phase ($A_1 = -A_2$). While the first mode is always stable, the counter phase mode can now be unstable, depending on the ratio γ_x/γ_z. This instability arises because of the confining potential in the direction *perpendicular* to the direction of particle oscillations.

By introducing small vertical perturbations δz_i of the vertically aligned particles, we obtain for the case of Debye only interactions equations analogous to the first case of horizontal vibrations of horizontally aligned particles. There are two oscillations modes;

the first one has the frequency similar to (3) for the two particle oscillating in phase with equal amplitudes $A_{1,2}$, and the second mode's frequency is similar to (4), both with the obvious change of x to z. Taking into account the asymmetry of the interaction potential, we obtain that the first oscillation mode, for the particles moving in phase with equal amplitudes $A_1 = A_2$, is unchanged while the second frequency is now given by

$$\omega_{zz,2}^W = -\frac{i\beta}{2} + \left[\frac{\beta^2}{4} + \left(\frac{\gamma_z}{M} + \frac{\gamma_z^W}{M}\left(1 - \frac{z_W}{z_D}\right)\right)\left(3 + \frac{z_d^2/\lambda_D^2}{1+z_d/\lambda_D}\right) + \frac{\gamma_z^W}{M}\frac{z^W}{z_d}\right]^{1/2} \quad (6)$$

for the counter phase oscillations; their amplitudes are not equal in magnitude:

$$A_1 = -\left(2 + \frac{z_d^2/\lambda_D^2}{1+z_d/\lambda_D}\right)\left(1 - \frac{z_W}{z_d} + \frac{\gamma_z}{\gamma_z^W}\right)A_2.$$

Both modes are always stable. Now, consider horizontal oscillations of two vertically aligned particles. When the particle interaction is symmetric of Debye type, we obtain two modes of oscillations, the first one corresponds to to the particles oscillate in phase (with equal amplitudes), and its frequency is equal to (3). The second one is similar to (5), with the frequency

$$\omega_{zx,2} = -\frac{i\beta}{2} + \left(\frac{\beta^2}{4} + \frac{\gamma_x}{M} - \frac{\gamma_z}{M}\right)^{1/2}, \quad (7)$$

and $A_1 = -A_2$. We see that the counter phase mode can be unstable, the condition for this instability is somewhat opposite to the condition of the instability of the mode of vertical vibrations of two horizontally arranged particles (5). If to take into account the plasma wake, the equation of horizontal motion of the upper particle in this case is the same as for the symmetric Debye only interaction, while the lower particle is oscillating in the wake potential characterized by γ_x^W which is its horizontal strength in the parabolic approximation. For our purposes here it is sufficient to assume that γ_x^Wr is a positive constant of order (or slightly more) than γ_x^W, see, e.g., numerical simulations [8]. The frequency of the first mode coincides with (3) while the frequency of the second mode is given by

$$\omega_{zx,2} = -\frac{i\beta}{2} + \left[\frac{\beta^2}{4} + \frac{\gamma_x}{M} + \frac{\gamma_z}{M} - \frac{\gamma_z}{M} - \frac{\gamma_z^W}{M}\left(1 - \frac{z_W}{z_d}\right)\right]^{1/2}, \quad (8)$$

Now, we see that the wake potential can stabilize possible horizontal instability of two vertically algned particles; note that for the supersonic wake potential this stabilization occures only within the Mach cone. The amplitudes of the second mode of oscillations are related by $A_1 = \frac{\gamma_x^W A_2}{\gamma_z + \gamma_z^W(1-z_W/z_d)}$. Thus for the asymmetric interaction potential, the second mode of oscillations does not correspond to the counter phase motions: the vibrations of particles are *in* phase now, with unequal amplitudes.

The proposed mechanism can be related to experimentally observed phenomena, for example, for the two-particle system in planar rf-discharge, involving horizontal

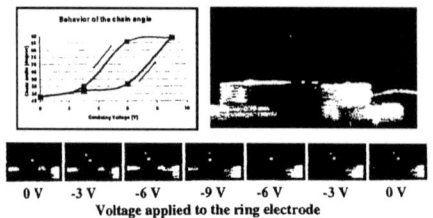

FIGURE 2. Experiment: levitation of two particles.

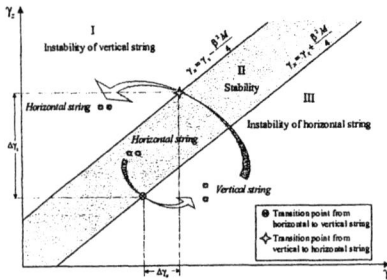

FIGURE 3. Stability diagram of the particle arrangements.

oscillations of two particles aligned in the vertical string and hysteretic phenomena in the disruptions of the horizontal and vertical arrangements, see Fig. 2. The stability diagram for the two-particle system, Fig. 3, reveals two extreme regions: one is the region (I) where $\gamma_z > \gamma_x + M\beta^2/4$, corresponding to the vertical string unstable with respect to the horizontal motions of the particles, another is the region (III) where $\gamma_x > \gamma_z + M\beta^2/4$ corresponding to the horizontal string unstable with respect to the vertical motions of the particles, as well as the central region (II) where both structures can be stable.

ACKNOWLEDGMENTS

This work was supported by the Australian Research Council.

REFERENCES

1. Thomas, H., and Morfill, G., *Nature (London)*, **379**, 806 (1996).
2. Vladimirov, S.V., et al, *Phys. Rev. E*, **56**, R74 (1997); *ibid*, **60, 7369 (1999)**; *ibid*, **62**, 2754 (2000).
3. Nunomura, S., et al, *Phys. Rev. Lett.*, **83**, 1970 (1999).
4. Melzer, A.., et al, *Phys. Rev. Lett.*, **83**, 3194 (1999).
5. Steinberg, V., et al, *Phys. Rev. Lett.*, **86**, 4540 (2001).
6. Samarian, A.A., et al, *Phys. Rev. E*, **64**, 025402(R) (2001).
7. Vladimirov, S. V., et al, *Phys. Rev. E*, **52**, 2172 (1995); *Phys. Plasmas*, **3**, 444 (1996).
8. Maiorov, S.A., et al, *Phys. Rev. E*, **63**, 017401 (2001).

SECTION 4: POSTER PRESENTATIONS

Experimental Study of Coulomb crystal formation in Hollow Cathode discharge

A. K. Agarwal* and G. Prasad*

Institute for Plasma Research, Bhat, Gandhinagar 382428, India

Abstract. The formation of macroscopic ordered structures in hollow cathode discharge in Ar plasma has been observed. A Coulomb quasicrystal is formed by alumina grains with diameter of 70 to 140 μm. A simple phenomenological model where in the intergrain spacing results from an attractive electric-field-induced dipole-dipole force balanced by a repulsive monopole Coulomb force is used to explain the observed features of the Coulomb crystal.

Dusty (complex) plasma is a normal electron-ion plasma, with an additional charged component of micron sized dust grains. Since electrons are more mobile than ions, a dust grain in plasma will acquire a negative charge (Z_D) with respect to the unperturbed plasma potential and alters the collective behavior of the ambient plasma. Dusty plasma basic physics has been the subject of intensive study in the context of astrophysics and plasma-aided manufacturing. Laboratory studies have focused on the crystal formation and phase transition[1, 2] and study of low frequency oscillations[3]. Ikezi [4] first suggested the possibility of formation of Coulomb solids (negatively charged dust grains) in colloidal plasmas. The coulomb coupling parameter Γ given by $\Gamma = e^{-a/\lambda_p}(Z_d e)^2/aT_d$, is of order 1 or larger. ($Z_d e$ is the charge on the dust grain, a is the intergrain distance, T_d is the dust temperature, λ_p is the plasma Debye length). Coulomb solids occur when $\Gamma > 170$ for a one component plasma.

In Coulomb crystal experiments, the grains are confined in the lower sheath of the electrode configuration. In the vertical direction the dust grains are levitated by the balance of gravitational force and electrostatic force on the dust grains associated with the electric field in the sheath. The mutual monopole Coulomb repulsion between the individual dust grains leads to their separation to the order of intergrain distance. In the horizontal direction the grains are also confined by the force associated with the horizontal componenet of the electric filed (E) in the sheath. The sheath electric filed can also induce polarization on the dust grains with dipole moment $P = Ea^3$, where a is the radius of the dust grain. It is clear that the dipole moment increases with increase of the grain diameter. The dipole moments lead to an additional attractive force between the aligned grains in the crystal.

In this work, we report the observation of dipole-dipole force influencing the crystal formation in a DC hollow cathode discharge. The dust grains, levitated in the present experiment, are much higher in mass and size than conventionally used, in dusty plasma. Results of the experiment are compared with a phenomenological model used by Mohideen [5].

The schematic of experimental setup is shown in Fig1. Stainless steel cylinder 2204

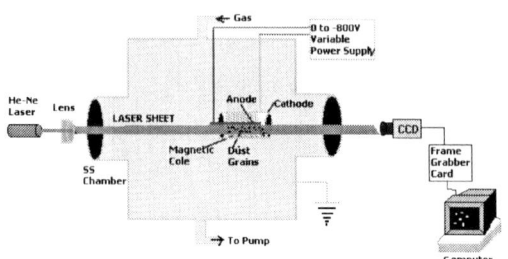

FIGURE 1. Schematic of experimental setup

mm in length and 220 mm in diameter is used as vacuum chamber. It has five ports, which are used for diagnostics and pumping. A cylindrical cathode of 40 mm inner diameter and 60 mm in length is placed inside the vacuum chamber. Anode is a stainless steel wire of 2 mm in diameter placed co-axially with the cathode.

Few grams of spherical alumina dust grains (density =3.97 gm/cm^3 and size = 70 to 140μm) were sprinkled inside the cylindrical cathode. Rectified power supply of 0 to −800volts was used to produce plasma. Plasma temperature and density were measured using single Langmuir probe (tungsten wire 2 mm in diameter and 2 mm in length). Plasma parameters were measured in dust free plasma. Helium-neon laser is used to visualize the dust grains, while the forward scattered light was acquired using a CCD camera. The analog output of the camera is digitized via a framed grabber card and then stored on the computer.

The vacuum chamber was evacuated to 0.02 mbar and purged with Argon gas up to bar pressure and again evacuated. This process is repeated several times in the beginning of the experiment. Glow discharge is struck by applying 300 volts discharge voltage at 0.5 mbar pressure. The discharge voltage is gradually raised to 900 volts and the neutral gas pressure increased to 1 mbar. The neutral pressure is then reduced to 0.7 mbar and the discharge voltage was reduced to 400 volts.

It is observed that initially when the discharge is struck, the number of levitated dust grains were less and nucleation of the crystal begins with very few grains. The inter grain distance between the grains is also large. As the time passes, new dust grains flow from all the sides to the nucleation region and the dust cloud increases in size and thus the dust grain density. It is observed that when these new dust grains approache towards the dust cloud, their velocity was higher but at the cloud boundary their velocity was found to decrease. These new dust grains entering in to the dust cloud are found to disturb the existing dust grains in the dust cloud. The oscillations of new dust grains so entered, is damped gradually and finally they come to rest to an equilibrium position. It was observed that inter grain distance reaches a saturation value in around an hour. It is important to note that during this process, very few grains are observed to escape from the dust cloud.

As the discharge power increases from 0.5 W to 0.8 W intergrain spacing changes from 1.8 mm to 0.4 mm, and afterwards, the intergrain spacing does not vary with power. In figure 2(a) intergrain spacing as a function of discharge power at different pressures is shown. It appears that the intergrain spacing is insensitive to pressure variation but decreases rapidly with increase in power, and neutral pressure does not significantly reduce the thermal energy of the dust particles. Fig. 2b gives the variation of electron

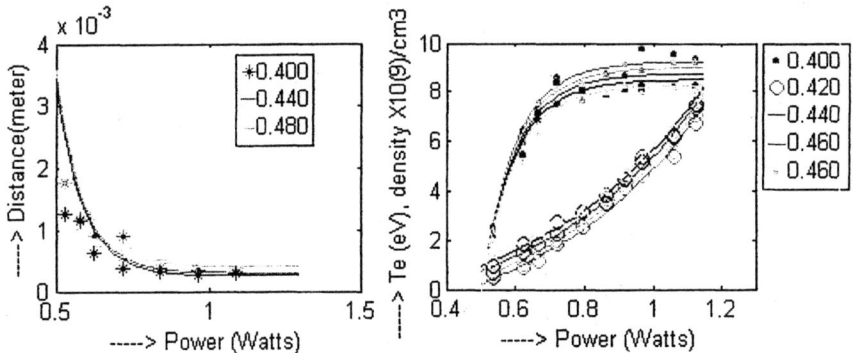

FIGURE 2. Varitation of (a) interparticle distance and (b) plasma parameters with discharge power

temperature and ion density with discharge power for various gas pressures. Electron temperature increase from 2 to 8eV as the discharge power is increased from 0.5W to 0.8W and thereafter saturates as also the intergrain spacing at the same discharge power. Plasma density however, slowly increases as the discharge power is increased. The debye length $\left[\lambda_D = (T_e/4\pi n e^2)^{1/2}\right]$ for the same range of power decreases from 0.45 mm to 0.25mm. It suggest that decrease in interparticle distance (figure 4) is, consistant with the assumption that the screened monopole Coulomb interaction plays a dominant role in the intergrain spacing.

When a dust grain is levitated in the sheath, electric static force due to the sheath electric field balances against gravity given by,

$$m_d g = qE \qquad (1)$$

The sheath electric field may induce polarization leading to dipole moment $P = Ea^3$. The attractive dipole-dipole force between aligned dipoles separated by a distance d, taking into account screening by the background plasma, is given by [5]

$$F_{dd} = \frac{6P^2}{d^4}\left(1+x+\frac{1}{2}x^2+\frac{1}{6}x^3\right)\exp(-x), \qquad (2)$$

where, $x = d/\lambda_D$. Taking into account the screening effect, monopole repulsion force on negatively charged dust grains is given as,

$$F_{mm} = \frac{Z_D^2 e^2}{d^2}(1+x)\exp(-x). \qquad (3)$$

At a given plasma condition represented by plasma temperature Te and plasma density, a rough equilibrium intergrain lattice spacing d can be obtained by balancing the above two forces i.e. by $F_{dd} = F_{mm}$. Neglecting the screening effect $(x = 0)$ and solving, the condition becomes,

$$d \propto \frac{1}{T_e^{2\alpha}} \Rightarrow d \propto (\text{Discharge power})^{-2\alpha}; \quad \text{as} \quad T_e \propto (\text{Discharge power}); \quad (4)$$

This equilibrium spacing d derived from the model is plotted as a function of the discharge power in Fig 4a. The values of T_e from the best-fit curve are used. There is a good agreement between the dipole - dipole interaction model and the observed dependence of the intergrain spacing with discharge power. It is to be noted that the dipole force begins to increase as the diameter of the dust grain are increased. The dipole moment can be induced due to the plasma flow past the dust. The ion streaming, ion-focusing etc. may lead to the change in the charge grain concentration surrounding the dust and leading to the polarization.

In conculsion, the formation of ordered structure of charged macroscopic grains of large size and mass in hollow cathode discharge is observed. The confinement of a complex plasma is achieved through a high charge on the grain. We measure the lattice structure and intergrain spacing of Coulomb crystals formed in hollow cathode discharge as a function of plasma temperature and density. The main feature of our experiment is the observed constant intergrain separation, which is determined by a balance between an attractive electric-field-induced dipole-dipole and a repulsive monopole interaction between the charged dust grains. It has been shown that the change of the discharge conditions (pressure or power) induces the melting transition of the dust grains form ordered structure to fluid or gas like states in the sheath boundary of the lower electrode.

REFERENCES

1. G. E. Morfill *et al.* J. Vac. Sci. Tech. A **14**, 490 (1996) and reference therein.
2. A. Melzer *et al.* Phys. Rev. E **53**, 2757(1996) and reference therein.
3. V. E. Fortov *et al.* Phys. Plasmas **7**, 1374 (2000).
4. H. Ikezi, Phys. Plasmas **29**, 1974 (1986).
5. U. Mohideen et al, Phys. Rev. Let, **81**, 349 (1998).

Observation of Rotating Dust Particles

A. K. Agarwal and G. Prasad

Institute for Plasma Research, Bhat, Gandhinagar- 382428, India

Abstract. Experimental observation of rotation in a strongly coupled dusty plasma in the absence of external magnetic field is reported. It appears that self consistent field of the rotating particles is responsible for the rotation. The rotating structure resembles a convective cell pattern in a vertical plane, which indicates that gravity plays a crucial role.

INTRODUCTION

Dust grains get negatively charged because of high mobility of electrons as compared to that of ions in the laboratory plasma. These dust grains can be levitated against gravity by the sheath electric field E. At sufficiently high density, the dust grains influence the properties of usual electron ion plasma and give rise to variety of new phenomena. Dust particles due to their heavy mass need very high magnetic fields for magnetization. It is observed experimentally that magnetic field B of around 100 Gauss is sufficient to observe rotation [1, 2, 3, 4] and that under the applied field the ions undergo $E \times B$ drift. The ions drag the dust particles and consequently they exhibit rotation. In the experiments of Law it et al. [5], the dust particle circulation was observed to be due to the positively biased probe placed over the plasma crystal. The circulating particles exhibit convective pattern within the crystal and the circulation was attributed to the non-uniform electric field in the crystal induced by the probe. In this letter, we report novel observations of the rotation of strongly coupled dust component in an unmagnetized plasma. The observations are made in a hollow cathode DC glow discharge plasma. The dust grains are found to rotate in a vertical plane. The particle motion clearly indicates that the force of gravity is playing an important role for the observed structures. It is observed that there is a threshold on the number of particles for observing the rotation.

EXPERIMENTAL RESULTS

The experimental setup is described as follows. A stainless steel cylindrical vacuum chamber as shown in Fig. 1 with length of 22 cm and a diameter of 22 cm is used for the experiment. The chamber has five ports which are used for pumping to create vacuum as well as for diagnostic purposes. A hollow stainless steel cylindrical cathode of 4 cm diameter and a length of 6 cm was placed inside the vacuum chamber. It was covered with a PVC tube so that the outer surface of the cathode is shielded. The anode is a stainless steel wire of 2 mm diameter and 4 cm in length is placed coaxially with the cathode. It

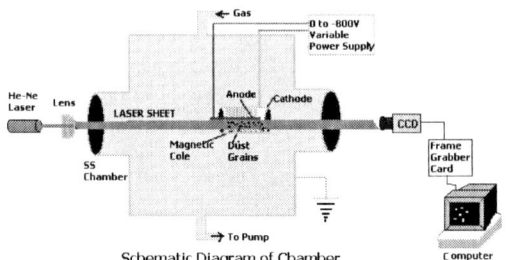

FIGURE 1. The experimental set up

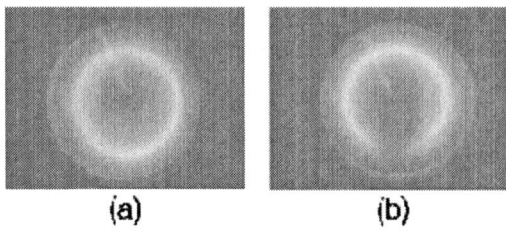

FIGURE 2. Plasma glow discharge (a) without and (b) with dust particles

should be noted that the anode is recessed up to 2 cm from the surface of the cathode to modify the electric fields. Few grams of alumina grains (density $\rho = 3.97\,\text{g/cm}^3$ and size 30 to 60 μm) were sprinkled inside the cylindrical cathode. The dust grains were observed visually with the help of a Helium-Neon laser. A CCD camera together with framer grabber card was used to acquire image on to a computer. Plasma temperature and density were measured using single Langmuir probe. The vacuum chamber was evacuated to 0.02 mbar and purged with Argon gas at a pressure of 1 mbar. The chamber was repeatedly evacuated and filled with Argon gas a few times in the beginning of the experiment. Glow discharge was struck by applying a discharge voltage of 300 V at 0.5 mbar pressure. The discharge voltage was gradually raised to 800 V and the neutral pressure was raised to 1 mbar. The cylindrical ring structure in Fig. 2 shows the plasma column which is formed due to the discharge. It is clear from Fig. 2 that the dust, spread on the cathode quenches the plasma. The break in the plasma column near the cathode is the sheath region and arises because of the dust grains shielding the cathode. As we reduce the discharge voltage we observe an accumulation of dust particles in the sheath region, which forms a triangular lattice structure. The formation of lattice structure clearly indicates that the dust particles are in a strongly correlated regime. This triangular lattice structure expands with decrease in discharge voltage such that the dust layers can be ultimately evaporated from it. The evaporated dust particles the sheath region into the plasma column and form a structure shown in region I and region II of Fig. 3. We followed the dynamics of dust structure in region I with increased magnification, which is basically of a similar nature as that in region II. This dust structure exhibits interesting dynamic behaviour as the discharge voltage is increased. When the discharge voltage is small (~ 390 V) the structure seems to remain static but individually few particles wander around randomly. However, when we increase the discharge voltage after a certain threshold (~ 410 V),

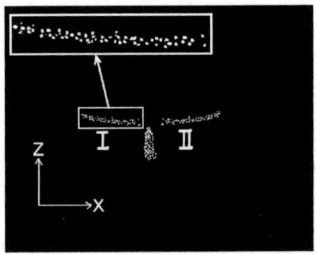

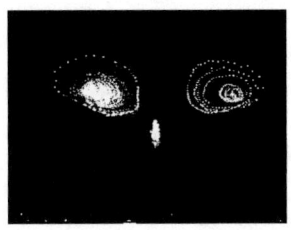

FIGURE 3. Equilibrium dust configuration is shown in region I and region II.

FIGURE 4. The clockwise and anti-clockwise rotating structures of region I and region II. ($V_d = 460$ V)

the collection of dust particles start rotating in a vertical plane. It is observed that the particles rotate in anti-clockwise direction in region I and in clockwise direction in region II as shown in Fig. 4. The radial dimension of the rotating cloud structure and the rotation speed are found to increase with increasing discharge voltage as is apparent from Fig. 5 and Fig. 6. When we repeat the experiment by ramping up the voltage as soon as a few particles accumulate in the plasma column we do not observe any rotation. Figure 6 shows that the rotation frequency decreases with decreasing number of dust particles. These observations clearly suggests that the self consistent fields of the dust species is crucial to achieve the rotating state. The orbit of a single dust grain is traced by taking the individual frames of the movie and superimposing them. The background laser light and the glow from the background plasma is substracted from each individual frame before superimposing them. We observe that the rotation speed of the dust grains increases

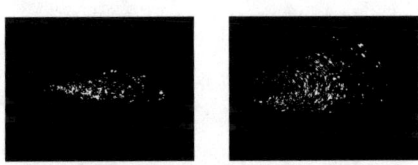

FIGURE 5. Rotating dust clouds at discharge voltage $V_d = 410$ V (left) and $V_d = 450$ V (right)

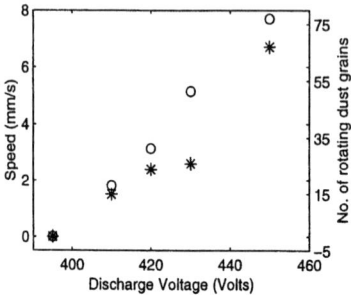

FIGURE 6. Speed and number of rotating grains at different discharge voltages.

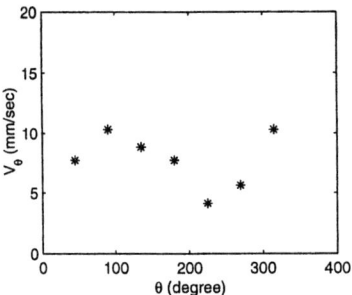

FIGURE 7. Variation of rotation speed with (θ).

with radii. However, the speed also exhibits angular (θ) dependence as shown in Fig. 7. Although the predominant motion of the dust grains is that of rigid rotation, they speed up gravity and decelerate as they rise up against gravity. In conclusion we present the results of observation of rotation of dust particles in an unmagnetized plasma under the influence of self consistent field of particles. We have followed the evolution of the rotation and convective pattern so set up right from the stable equilibrium. It appears that some fluid instabilty (like Rayleigh-Taylor) may be playing a role. Theoretical investigation is under way and will be reported elsewhere.

REFERENCES

. Maemura, I. Y., Yang S.C., and Fujiyama, H., *Surface and Coatings Tech.*, **98**, 1351 (1998).
. Sato, N., *et al.*, *Phys. Plasmas*, **8**, 1786 (2001).
. Konopka, U., *et al.*, *Phys. Rev. E*, **61**, 1890 (2000).
. Kaw, P. K., Nishikawa, K., Sato, N., *Phys. Plasmas*, **9**, 387 (2002).
. Law *et al.*, *Phys. Rev. Lett.*, **80**, 4189 (1998).

Numerical Investigation of Ponderomotive Force Effect Based Contamination Control in Dusty Plasmas

M. Amroun*, M. Djebli[†] and R. Annou[†]

*Quantum Electronics Lab., Physics Department, U.S.T.H.B, Algiers, Algeria
[†]Theoretical Physics Lab., Physics Department, U.S.T.H.B, Algiers, Algeria

Abstract. An impurity control scheme in plasmas [M. Amroun and R. Annou, Phys. Plasmas **8**, 5331(2001)] where electrons ions and dust grains get separated when experiencing a cyclotron wave generated ponderomotive force, is numerically investigated. The electron and ion ponderomotive forces are included and the system of differential equations is solved. It is shown, that plasma-impurity separation is effective beyond a critical distance from the source.

INTRODUCTION

When plasma contains small solid micro to submicron sized dust grains with high charge ($Q \sim 10^4 e$) and mass ($m \sim 10^{12} m_i$), its properties get strongly affected. The effect of dust grains through their dynamics and charge fluctuations, has been greatly studied [1]. In particular, the presence of dust in semi-conductor manufacturing devices, leads to thinfilms defects that is a cause of material yield loss. There is a need to get rid of the dust component. It has been proposed to this end, by Amroun and Annou [2] to separate dust grains from plasma particles by imposing a magnetic field and launching a cyclotron wave, where its frequency matches the grain cyclotron frequency. Hence, the grain gyroradius grows resonantly whereas the electron and ion ones stay almost constant. Far from the source, an analytical solution has been derived. In this note, a numerical resolution is conducted taking into account electron and ion ponderomotive forces.

BASIC EQUATIONS

We consider a plasma consisting of three components namely, ions with mass m_i and charge e, electrons with mass m_e and charge $-e$ and negatively charged dust grains with mass m_d and charge $-Ze$.

The dusty plasma is immersed in an homogeneous magnetic field $B_0 \hat{z}$ which propagates a left hand circularly polarized wave $\vec{E} = E(z)(\hat{x} + i\hat{y})e^{i\omega t} + cc$. By virtue of

Faraday's and Ampere's laws, and putting $E(z) = |E(z)|e^{i\phi(z)}$, one finds, [1]

$$|E|\left(\frac{\partial\phi}{\partial z}\right)^2 + \frac{\partial^2|E|}{\partial z^2} + \frac{\omega^2}{c^2}\left[1 - \sum_\sigma \frac{\omega_{p\sigma}^2}{\omega(\omega - \Omega_\sigma)}\right]|E| = 0 \qquad (1)$$

$$\frac{\partial^2\phi}{\partial z^2} + 2\left(\frac{1}{|E|}\frac{\partial|E|}{\partial z}\right)\frac{\partial\phi}{\partial z} = 0 \qquad (2)$$

$$n_\sigma q_\sigma \frac{\partial\phi}{\partial z} - n_\sigma F_\sigma + kT_\sigma \frac{\partial n_\sigma}{\partial z} = 0 \qquad (3)$$

where $\omega_{p\sigma} = \left(\frac{n_\sigma q_\sigma^2}{m_\sigma \varepsilon_0}\right)^{1/2}$ is the plasma frequency, $\Omega_\sigma = q_\sigma B_o/m_\sigma$ is the cyclotron frequency, ϕ the electrostatic potential and $\vec{F}_\sigma$ being the ponderomotive force.
From Eqs. (1) – (3) and the expression of the ponderomotive force given by,

$$F_\sigma = -\frac{q_\sigma^2}{2m_\sigma\omega}\frac{1}{(\omega - \Omega_\sigma)}\frac{dE^2}{dz} \qquad (4)$$

we find,

$$\frac{\partial^2\tilde{\phi}}{\partial\tilde{z}^2} + 2\left(\frac{1}{\varepsilon}\frac{\partial\varepsilon}{\partial\tilde{z}}\right)\frac{\partial\tilde{\phi}}{\partial\tilde{z}} = 0 \qquad (5)$$

$$\frac{\partial^2\varepsilon}{\partial\tilde{z}^2} - \varepsilon\left(\frac{\partial\tilde{\phi}}{\partial\tilde{z}}\right)^2 + (1+\alpha)^2\left[1 + \frac{2\tilde{n}_i}{Z(\alpha+1)} - \frac{\tilde{n}}{\alpha}\right]\varepsilon = 0 \qquad (6)$$

$$\frac{\delta_i}{\tilde{n}_i}\frac{\partial\tilde{n}_i}{\partial\tilde{z}} + \frac{\delta_d}{\tilde{n}}\frac{\partial\tilde{n}}{\partial\tilde{z}} + \frac{(1-\alpha)}{(1+\alpha)\alpha Z}\frac{\partial\varepsilon^2}{\partial\tilde{z}} = 0 \qquad (7)$$

$$(1+\delta_i)\frac{\partial\tilde{n}_i}{\partial\tilde{z}} + (\delta_d - Z)\frac{\partial\tilde{n}}{\partial\tilde{z}} - \left(\frac{2}{(1+\alpha)Z}\tilde{n}_i - \frac{1}{\alpha}\tilde{n}\right)\frac{\partial\varepsilon^2}{\partial\tilde{z}} = 0 \qquad (8)$$

The quantities used in Eqs. (5)–(8) are normalized as follows: $\tilde{n} = n_d/n^*$; $\tilde{n}_i = n_i/n^*$; $n^* = \varepsilon_0 B_0^2/m_d$; $\varepsilon^2 = |E|^2/\lambda^2$; $\lambda^2 = 2T_e Z B_0^2/m$; $Z = |q|/e$; $\tilde{z} = z/L$ and $L = c/\Omega_d$. δ_i and δ_d stand respectively for T_i/T_e and T_d/T_e. We have chosen the frequency of the wave close to the dust gyrofrequency, to drive resonantly the dust grains only.

NUMERICAL SOLUTION AND CONCLUSION

We solve the set of nonlinear differential equations (5)–(8) numerically in the case where $T_i = T_e$ and $T_d = T_e/10$ and plot $X = (\tilde{n} - \tilde{n}_i)/\tilde{n}$ versus normalized space coordinate $\tilde{z}$.
In Figure.1 we have considered the case $Z = 100$. It is clearly shown that the ion density decreases very strongly beyond $\tilde{z} = z/L = 3.6$ to reach a region where $X \to 1$, i.e., $\tilde{n} \gg \tilde{n}_i$ beyond $\tilde{z}_c > 4.35$. The effect of the electric field is to accelerate the separation, the limit is reached for $\tilde{z}_c = 4.35$ when $\varepsilon_o = 10^{-4}$ and $\tilde{z}_c = 2.5$ for $\varepsilon_o = 10^{-3}$.

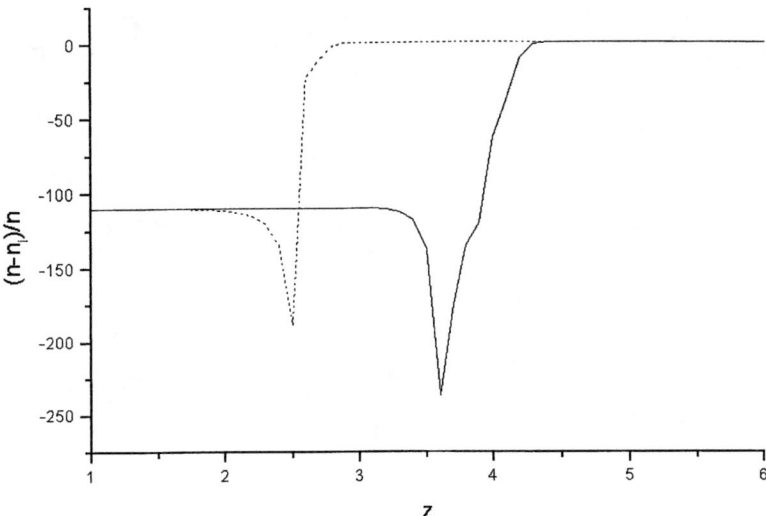

FIGURE 1. $\alpha = 0.01$, $\delta_i = 1$, $\delta_d = 0.1$, $n_e = n_i/10$, $Z = 100$. The solid line corresponds to $\varepsilon_0 = 10^{-4}$ and the dashed one to $\varepsilon_0 = 10^{-3}$.

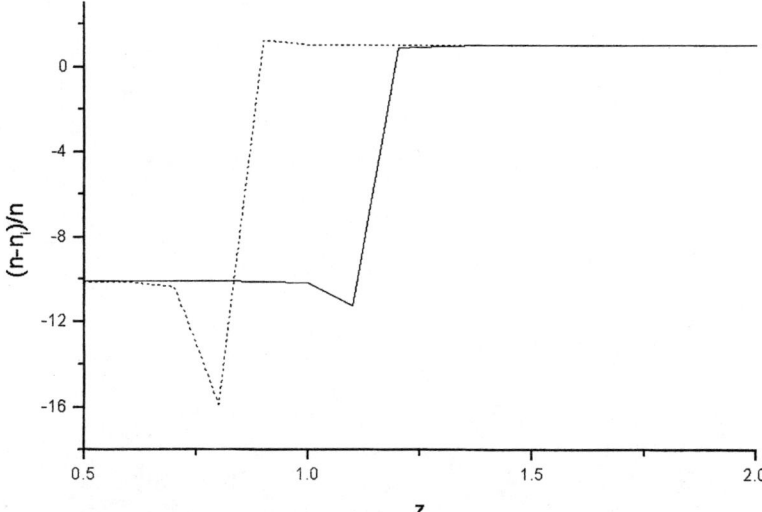

FIGURE 2. $\alpha = 0.01$, $\delta_i = 1$, $\delta_d = 0.1$, $n_e = n_i/10$, $Z = 10$. The solid line corresponds to $\varepsilon_0 = 10^{-4}$ and the dashed one to $\varepsilon_0 = 10^{-3}$.

separation, the limit is reached for $\tilde{z}_c = 4.35$ when $\varepsilon_o = 10^{-4}$ and $\tilde{z}_c = 2.5$ for $\varepsilon_o = 10^{-3}$.
In Fig.2 we have kept the same values as in Fig.1 and have considered a smaller dust charge $(Z = 10)$. The critical point $\tilde{z}_c$ is found to be close to the origin ($\tilde{z}_c = 1.2$ for $\varepsilon = 10^{-4}$ and $\tilde{z}_c = 0.8$ for $\varepsilon_0 = 10^{-3}$).

To conclude, the use of a cyclotron wave generated pondermotive force in plasma particles-dust grains separation is clearly shown. The set of non-linear differential equations in the presence of the the pondermotive force is established and solved numerically. There is a critical distance $z_c = \tilde{z}_c L$ beyond which $\tilde{n} \gg \tilde{n}_i$ that is a dust free space. This result is in accordance with the asymptotic analytical solution. The effect of the electric field is to accelerate the separation.

REFERENCES

1. Selwyn, G. S., *In The Physic of Dusty Plasmas*, **177** (1996).
2. Amroun, M., and Annou, R., *Phys.Plasmas*, **8** (2001).

Dusty RF Discharges With Secondary Electron Emission

Yu. I. Chutov*, W.J. Goedheer[†], O.Yu. Kravchenko* and O.A. Lavrov*

Taras Shevchenko Kiev University, Kiev, Ukraine, yuch@univ.kiev.ua
[†]*FOM-Institute for Plasma Physics "Rijnhuizen", Nieuwegein, The Netherlands,
goedheer@rijnh.nl*

Abstract. Dusty RF discharges are simulated by using the PIC/MCC method. Secondary electron emissions are taken into account in the framework of the effective yield obtained earlier. It is shown that the secondary emission essentially influence the electron and ion densities in the discharges due to more intensive ionization caused by secondary electrons. However the dust particle charge is conserved at the change of the secondary emission yield due to a specific charging of dust particles.

There are two main regimes of RF discharges [1,2] called the α - regime and γ - regime. They differ by the role of a secondary emission from electrodes or walls. RF discharges are supported mainly by a volume ionization in the α - regime unlike the γ - regime where the secondary emission is important for the discharge support. The α - regime is realized at a relatively high pressure, the γ - regime at a low pressure. Although RF discharges with the secondary emission were investigated earlier, the role of the secondary emission in the discharges is not yet clear. The main difficulty arises from the necessary to take into account (self-consistently with discharge parameters) all kinds of secondary electrons emitted from the electrodes (walls) by ion, electron, fast atom, or metastable impact as well as by ultraviolet radiation from the discharge.

Recently, models of the effective secondary-emission yield γ per ion were developed for the breakdown [3] and DC glow discharges [4] in argon. The models take into account all kinds of secondary electrons and give the dependence of the effective yield γ on the cathode electric field reduced by the argon atom density. These models can be useful for RF discharges.

Dust particles can appear in RF discharges as the product of the plasma-wall interaction with their subsequent transport into an interelectrode space or can be created due to coagulation of various components in chemically active plasmas. It is known [5,6] that the dust particles can essentially influence the parameters of the RF discharges due to a continuous selective collection of background electrons and ions that can essentially influence their energy distribution functions. Dusty RF discharges with secondary electrons were not investigated earlier, although it is obviously that the secondary emission has to influence the properties of dusty discharges especially at low pressures when the role of secondary electrons is growing. The computer simulation of dusty RF discharges with secondary electrons is the aim of the work.

A one-dimensional RF discharge is considered between two plane electrodes separated by a gap of $d = 2.0$ cm which is filled with Ar at various pressures. Immobile dust particles of the given radius R_d are distributed uniformly in the interelectrode gap with a given density N_d. The dust particles collect and scatter electrons and ions distributed in the discharge with density n_e and n_i, respectively. A harmonic external voltage $V_e(t) = V_o sin(\omega t)$ at a frequency $f = 13,56\ MHz$ and various amplitudes V_o sustains the RF discharge.

The PIC/MCC method described in detail earlier for discharges without dust particles is developed for computer simulations of the RF discharge with dust particles. The Monte Carlo technique [10] is used to describe electron and ion collisions. The collisions include elastic collisions of electrons and ions with atoms, an ionization and excitation of atoms by electrons, the charge exchange between ions and atoms, Coulomb collisions of electrons and ions with dust particles, as well as the electron and ion collection, and scattering by dust particles. In addition to a usual PIC/MCC scheme, the weighting procedure is used also for the determination of a superparticle charge part, which is interacting with a dust particle.

The electron-argon collision cross-sections used in the model are the same as those used in [7]. The Coulomb cross-section for electron and ion scattering by immobile dust particles is taken from [8]. The secondary emission is taken into account in the framework of the models of [3,4] or at given various constant yields γ of the effective secondary emission.

The simulation starts at an initial uniform distribution of electrons and ions with given

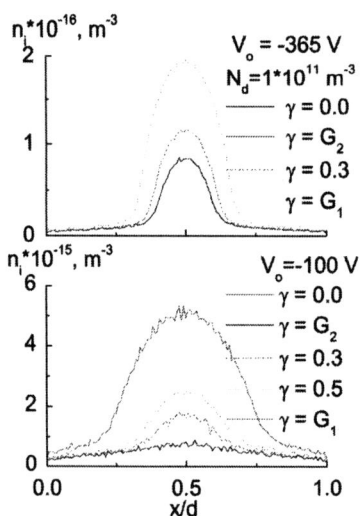

Source: Fig.1r.EPS

FIGURE 1. Spatial distributions of the ion n_i density for various γ and V_o

densities and is prolonged by iterations up to a moment when a change of discharge parameters is less a given limit. Simulation show that 400-1000 cycles are enough to obtain the periodically steady state of RF discharges

Spatial distributions of the electron n_e and ion n_i densities across the interelectrode gap obtained for various phases φ of the sustaining external voltage, show that the distributions of the ion n_i density are the same for various phases φ unlike the electron density n_e distributions which are changing close to electrodes. Note, the electron density n_e distributions are the same for various phases φ in the central part of the interelectrode gap however there is a difference between electron and ion densities due to the space charge of dust particles like to [5,6]. The distributions show the existence of the central quasi-neutral region (with taking into account the total dust particles charge) and non-neutral RF sheaths close to both electrodes like to the case of the RF discharge without the secondary emission [5,6].

The influence of the secondary emission on the dusty RF discharge can be seen in Fig.1 where spatial distributions of the ion n_i density are plotted for various combinations of the effective secondary-emission yields γ and the amplitude V_o of the harmonic external voltage. The distributions are obtained at the dust particle density $N_d = 3 \ast 10^{11}$ m^{-3} and the dust particle radius $R_d = 1\mu m$. As can be seen in Fig. 1, the increase of the constant yields γ causes an essential increase of the ion n_i density in the central part of

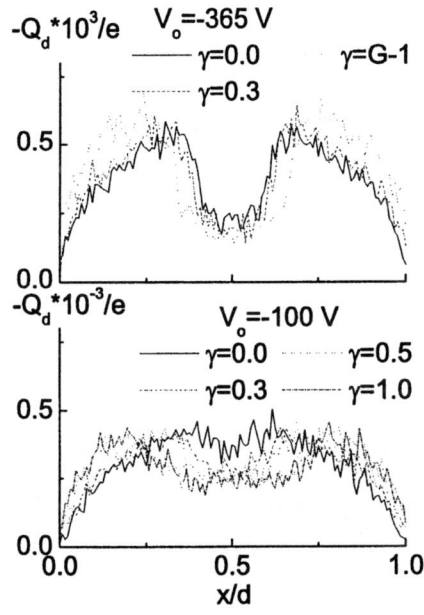

Source: Fig.2r.EPS

FIGURE 2. Spatial distributions of the dust particle charge q_d

the discharge at $V_o = const$ whereas the ion density is practically unchanged in sheaths. The density increase is caused by an additional ionization in the central discharge part by secondary electrons whereas the ion shielding of the given applied voltage causes the invariability of the ion density distributions in sheaths. As can be seen in Fig.1, the remarkable influence of the secondary emission on the discharge parameters takes place only at $\gamma > 0.2$. Note, the discharge parameters obtained in the framework of the model [4] ($\gamma = G2$) does not differ from the parameters in the discharge without the secondary emission because the effective secondary-emission yields γ in the model [4] is less than $\gamma = 0.2$. The model [3] provides the very strong increase of the ion density in the central discharge part because the model gives an effective secondary-emission yield γ that can amount to high values.

Spatial distributions of the dust particle charge q_d are shown in Fig. 2 for the conditions of Fig. 1. As can be seen in Fig. 2, the charge q_d depends very weakly on the yield γ at $V_o = 365\ V$ unlike the ion n_i density shown in Fig. 1. It is a typical result for intensive RF discharges and is caused by the dust particle charging in low-pressure RF discharges considered earlier in [5,6]. It was shown here that non-monotonic profiles of the dust particle charge in RF discharges are caused by the change of the ion current into a dust particle in non-uniform quasi-neutral plasma with common fast electrons due to their fast mixing in regions wit low electric fields. The simulations of RF discharges with secondary emission show that the ratio of the electron and ion currents into a dust particle is approximately conserved while changing the effective secondary-emission yields γ in intensive RF discharges. This causes the invariability of dust charge distributions.

Typical examples of spatial distributions of the dust particle charge q_d for low-power RF discharges are shown in Fig. 2 in the case of $V_o = 100\ V$. As can be seen in this case, the secondary emission changes the negative charge of a dust particle in the quasi-neutral central part of the discharge compared to the high-power RF discharge. However, like in the high-power RF discharges, the change of the yield γ does not result in a change of the spatial distributions

This work was partially supported by INTAS and by a grant from the Ukrainian Ministry of Education and Science.

REFERENCES

1. Lieberman M.A., Lichtenberg A.J. *Principles of Plasma Discharges and Material Processing, Wiley, New York, 1994, pp. 1-19*.
2. Levitskij S.M. *Zh. Tekh. Fizs*,**27**, 1001-1109 (1957).(in Russian)
3. Phelps A.V., Petrovic Z .Lj. *Plasma Sources Sci. Technol.*, **8**, R21-R44 (1999).
4. Donko Z. *Phys. Rev. E* ,**64**, 7420-7425 (2001).
5. Chutov Yu.I.., Goedheer W.J., Kravchenko O.Yu., Zuz V.M., Yan M. *Materials Science Forum: Plasma Processing and Dusty Particles*, **382**, 69-79 (2001).
6. Chutov Yu., Goedheer W., Kravchenko O., Zuz V., Yan M. *J. Plasma and Fusion Research. SERIES*,4, 340-344 (2001).
7. Surendra M., Graves D.B., Jellum G.M. *Phys. Rev., A* , **41**, 1012-1014 (1990).
8. Trubnikov B.A., *"Particle collisions in full ionized plasmas" in Problems of plasma theory,*, Gosatomizdat, Moscow, 1963, PP.98-182.

Dusty Sheaths in Plasmas with Two-Temperature Electrons

K. Asano*, Yu. I. Chutov[†], O.Yu. Kravchenko[†], N. Ohno*, A.F. Pshenychnyj[†], R.D. Smirnov[†] and S. Takamura*

*Nagoya University, Nagoya, Japan, takamura@ees.nagoya-u.ac.jp
[†]Taras Shevchenko Kiev University, Kiev, Ukraine, yuch@univ.kiev.ua

Abstract. Self-consistent dusty sheaths are investigated by using computer simulations of a temporal evolution of one-dimensional slab plasma with two-temperature electrons and dust particles. The evolution is caused by a collection of electrons and ions by both an electrode (wall) and dust particles, which are initially immersed into plasma. The peculiarity of the sheaths is non-monotonic spatial distributions of an electric potential that causes a protection of the electrode (wall) from fast ions. The protection is depending on rations of both electron temperatures and densities.

Plasmas with two-temperature electrons can be realized in fusion devices consisting of energetic (hot) electrons due to plasma heating by strong RF fields [1,2]. Sheaths in the plasmas are investigated usually under an assumption about two Maxwellian distributions. In the framework of the assumption, it was shown earlier [2-4] that sheath properties in the plasmas with two species of electrons are equivalent to one in plasmas with one-temperature electrons whose temperature is equal to the harmonic average of the two temperatures. However the electron equilibrium can be essentially disturbed in collisionaless sheaths [5].

The sheaths can consist of dust particles appeared as the product of the plasma-wall interaction in various technological devices including controlled fusion devices. Dust particles can essentially influence sheath parameters due to the continuous selective collection of background electrons and ions that can cause an essential change of both electron and ion energy distribution functions as well as an ion flux in sheaths [6]. However the dusty sheaths in plasmas with two-temperature electrons were not investigated earlier. Therefore the aim of this work is to investigate dusty sheaths in plasmas with two-temperature electrons by using the kinetic computer simulation.

In this work like to [7], a temporal evolution is simulated of one-dimensional slab plasma due to a collection of electrons and ions by plane electrode (wall) immersed into plasma initially. In this case, it is possible to get a self-consistent asymptotic distribution of plasma parameters including a sheath region without special boundary conditions for sheaths.

It is assumed that the collisionless slab plasma is uniform initially and consist of two groups of electrons with densities n_c and n_h and temperatures T_c and T_h, respectively, as well as hydrogen ions with a density n_o and a temperature T_{io}. The plasma is quasi-neutral so that $n_c + n_h = n_o$. Besides, motionless spherical neutral dust particles of given radius R_d are distributed close to the electrode (wall) according to a given initial distribution which is typical for sputtering of dust particles from electrodes (walls). The plasma starts to evolve after a start of a collection of background electrons and ions by the electrode (wall) and dust particles, which are charged continuously due to the collection. Scattering of electrons and ions by dust particles takes place due to a large size of dust particles.

The PIC/MCC method (1D3V model) is developed for computer simulations of the plasma evolution with dust particles. The electrode (wall) is chosen as the left boundary of the simulation region whereas the right boundary is located in a presheath where the plasma is quasi-neutral during the plasma evolution. The continuous exchange by superparticles takes place on the right boundary of the simulation region that has to be described at the computer simulation. In this work, the original model of the exchange was developed which takes into account the self-consistent change of the electric potential as well as electron and ion energy distribution functions on the right boundary.

The Monte Carlo technique [8] is used to describe interactions of electrons and ions with dust particles. The interactions include Coulomb's collisions of electrons and ions with dust particles, as well as the electron and ion collection by dust particles. In addition to a usual PIC/MCC scheme, the weighting procedure is used also for the determination of a superparticle charge, which is interacting with a dust particle. The cross-sections of an electron and ion collection by dust particles are taken according to the Orbit Motion Limited (OML) theory [9]. The Coulomb cross-section for electron and ion scattering by immobile dust particles is taken from [10].

Simulations results were obtained for various ratios $\alpha = n_h / n_o$ of the density n_h of hot electrons to the total density $n_o = n_c + n_h$ of all electrons and various ratios of initial temperatures $\beta = T_h / T_c$ of hot T_h and cold T_c electrons. Typical results of the computer simulations are shown in Fig. 1 - 2. Note that the spatial coordinate x and linear sizes are divided by the initial Debay length $L_d = (kT_c / 4\pi n_o e^2)^{1/2}$ of could electrons with the total density n_o. The time t is multiplied by the initial ion plasma

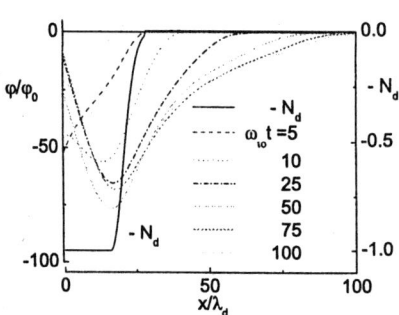

Source: Fig.1s.EPS

FIGURE 1. Spatial distributions of the electric potential φ. Curve N_d is the spatial distribution of the number of dust particles in a Debye cube.

frequency $\omega_i = (4\pi n_{oe}^2/M)^{1/2}$. The figures consists of the spatial distribution of the dust particle number $N_d(d)$ in a Debay cube in order to mark out the dusty region and to make clearer understudying spatial distributions of other parameters. The results are shown at a condition that $R_d = 0.03$ and the number $N_d = N_{do} = 1$ is constant at $x < x_o = 16$ and decreases according to $N_d = N_{do} exp(-(x-x_o)^2/w^2)$ at $x_o < x < x_1$. Dust particles absent at $x > x_1 = 28$.

Obtained results show that a disturbance of the ion densities is formed close to the electrode (wall) initially due to the collection of electrons and ions. The disturbance penetrates continuously in the undisturbed plasma converting at last stages of the evolution into a rarefaction wave which propagetes about with the ion sound speed for the harmonic averaged electron temperatures.

The propagation of the rarefaction wave in the undisturbed plasma without dust particles is accompanied by a relaxation of plasma parameters behind the wave and forming a sheath in front of an electrode (wall). That forming can be seen in Fig. 1 where spatial distributions of the self-consistent electric potential φ are shown for various times t after the start of the collection. As can be seen in Fig. 1, the spatial distributions evolve from monotonic distributions at initial evolution times to non-monotonic distributions during a penetration of a rarefaction wave into undisturbed plasma. Simulations show that a characteristic time of a sheath formation is about equal to the transit time of an ion sound wave through the sheath.

Of course, non-monotonic spatial distributions of the self-consistent electric potential φ are caused by the total space electric charge Q_t. A typical spatial distribution of the charge Q_t is shown in Fig. 2 by the corresponding curve for the late evolution stage when the sheath formation is finished. As can be seen in the figure, a distinctive peculiarity of the distribution is a region of a negative charge, which causes the non-monotony of the electric potential. The total charge Q_t consist of electric charges of electrons Q_e,

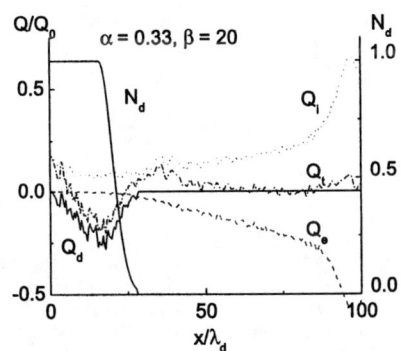

Source: Fig.2s.EPS

FIGURE 2. Spatial distributions of the total space electric charge Q_t as well as electric charges of electrons Q_e, ions Q_i, and dust particles Qd

ions Q_i, and dust particles Q_d whose spatial distributions are shown also in Fig. 2 by corresponding curves. Note, the electron space charge Q_e is very small compared to the ion space charge Q_i in the region of dust particles. It is caused by the usual strong decrease of the electron density in sheaths due to an action of decelerating electric fields.

Therefore the negative total space charge is caused by the total negative charge Q_d of dust particles shown by the corresponding curve in Fig. 2. The charge Q_d is a result of balance between electron and ion currents into a dust particle, as usually. Our computer simulations show that the ion current density is about constant across the sheath due to rare collisions of ions with dust particles like to [6]. Therefore it is possible to propose that the ion flux into a dust particle be also about constant because the ion drift velocity exceeds essentially the ion random velocity. The electron current into a dust particle decreases continuously including the region of dust particles where the decrease is caused by the continuous collection of electrons even in the region of the reverse change of the self-consistent electric potential. As a result, the total charge Q_d of dust particles shown in Fig. 2 changes non-monotonic across the dusty region. The charge Q_d increases at the boundary of dust particles due to the increase of the dust particle density and than decreases due to the balance of electron and ion currents into a dust particle.

The non-monotonic self-consistent electric potential in the front of the electrode (wall) with a potential minimum in dusty plasma causes a protection of the electrode (wall) from the intensive ion bombardment because the ions reach the electrode (wall) with a lower energy. Obtained results show that the negative minimum potential φ_{min} decreases about linearly with increasing the temperature ratio $\beta = T_h/T_c$ of hot (T_h) and cold (T_c) electrons like to the probe floating potential. The potential depends also on the density ratio $\alpha = n_h/n_o$ of the of hot electrons (n_h) to the total density $n_o = n_h + n_c$ of all electrons. The analysis show that the dependence of the potential φ_{min} on α and β can not be described by a dependence of the potential on the effective temperature because a primary collection of fast electrons by electrodes and dust particles causes a quick deviation of the electron energy distribution function from an equilibrium.

This work was partially supported by INTAS and by a grant from the Ukrainian Ministry of Education and Science. One of the authors (Yu.I.Ch.) thanks JSPS for a support of his stay at the Nagoya University where the paper was begun.

REFERENCES

1. Shiraishi K., Ohno N., and Takamura S. *J. Plasma Fusion Res*, **69**, 1371-1377 (1993).
2. Takamura S. *Phys. Letters*, **133**, 312-314 (1988).
3. Boyd R. L. F., and Thompson J. B. *Proc. R. Soc. A*, **252**, 102-117 (1959).
4. Boswell R. W., Lichtenberg A. J., and Vender D. *IEEE Trans. Plasma Sci.*, **20**, 62-71 (1992).
5. Song S. B., Chang C. S., and Choi D.- *Phys. Rev.*, **55**, 1213-1216 (1997).
6. Chutov Yu., Kravchenko A., Schram P.P.J.M. *Physica B*, **128**, 11-20, (1996).
7. Cipolla J.W. (Jr.), Silevitch M.B. *J. Plasma Phys.*, **25**, 373-389 (1981).
8. Birdsall C.K. *IEEE Trans. Plasma Sci.*, **19**, 65-85 (1991).
9. Allen J. *Phys. Scr.*, **45**, 497-503 (1992).
10. Trubnikov B.A., "Particle collisions in full ionized plasmas" in *Problems of plasma theory*, Gosatomizdat, Moscow, 1963, PP.98-182.

Waves in Magnetized Plasmas with a Spectrum of Dust Sizes

N. F. Cramer*, F. Verheest[†], S. V. Vladimirov* and M. Wardle*

*Theoretical Physics Department and Research Centre for Theoretical Astrophysics, School of Physics, The University of Sydney, N.S.W. 2006, Australia
[†]Sterrenkundig Observatorium, Universiteit Gent, Krijgslaan 281, B-9000 Gent, Belgium

Abstract. The resonance absorption of low-frequency electromagnetic waves in a dusty magnetized plasma is considered. The dust grains carry a proportion of the negative charge of the plasma, and there is assumed a distribution of dust sizes, typical of that deduced to be present in interstellar molecular clouds.

Even though in astrophysical plasmas such as interstellar clouds $n_e \approx n_i$, i.e. only a small proportion of the negative charge is carried by the dust grains, the dispersion properties of Alfvén and magnetoacoustic waves are strongly modified by the charged dust. With very small charge on the dust grains, the waves have the usual shear and compressional Alfvén wave properties, while for a moderate negative charge on the grains the left-hand circularly polarized wave is a whistler or helicon wave extending to low frequencies, while the right-hand circularly polarized wave has a cutoff frequency. The modification of the Alfvén resonance absorption mechanism due to negative charge residing on the grains in a dusty plasma was investigated [1,2], and it was shown that the wave energy propagating at oblique angles to the magnetic field in an increasing density gradient can be very efficiently absorbed at the Alfvén resonance in a dusty plasma. In that work, it was assumed that there was a single dust grain mass and charge, defining a single dust-cyclotron frequency.

Here we extend the earlier studies [1] of linear wave propagation in homogeneous cold magnetized dusty plasmas by allowing for a distribution of dust sizes, and investigate the effects on the Alfvén resonance absorption process. In natural environments such as interstellar molecular clouds, dust grains have a power law "MRN" distribution of sizes [3,4], rather than a single size (typically $\simeq 0.1\,\mu$m). The propagation of low frequency waves in magnetized plasmas with power law dust size distributions has been considered [5-7], for wave propagation parallel to the magnetic field. Here we extend this work by considering propagation of the waves obliquely to the magnetic field. Obliquely propagating waves can encounter the Alfvén resonance, and it is shown here how the presence of a dust size distribution affects the resonance process.

A fluid model of the cold dusty plasma is used. The uniform background magnetic field $\mathbf{B}_0$ is in the z-direction. The grains of a given size form a cold fluid, and there is a continuous distribution of grain sizes. The notation of Ref. [5] is used. Thus a probability distribution function $f(r)$ is introduced, such that $n_{d0}f(r)dr$ is the number density of grains $\delta n_d(r)$ with radius between r and dr, i.e. $\delta n_d(r) = n_{d)}f(r)dr$ with n_{d0}

the total number density of grains of all sizes. A dimensionless grain radius is introduced, $r_d = r/r_0$ where r_0 is the grain radius corresponding to the maximum value of $f(r)$. A new distribution function $f_d(r_d)$ is then defined, such that $\delta n_d(r_d) = n_{d0} f_d(r_d) dr_d$. A power law distribution is used here, as in [5]:

$$f(r) = c'_p r^{-p}, \qquad f_d(r_d) = c_p r_d^{-p}, \qquad (1)$$

with $r_{\min} < r < r_{\max}$, so $r_0 = r_{\min}$. We define $r_m = r_{\max}/r_{\min}$. The dust cyclotron frequency is written as $\Omega_d(r) = \Omega_d(r_0) r_d^{-2}$, where $\Omega_d(r_0) = \Omega_{d0} = q_d(r_0) B_0 / m_d(r_0)$. The differential of the square of the dust plasma frequency for grains with radius in the range r_d to $r_d + dr_d$ can be written

$$\delta(\omega_d^2) = \frac{\omega_d^2(r_0)}{r_d} \delta n_d = \frac{\omega_d^2(r_0)}{r_d} f_d(r_d) dr_d, \qquad (2)$$

where the dust plasma frequency $\omega_d(r_0)$ for the size r_0 is given by

$$\omega_d^2(r_0) = n_{d0} q_d^2(r_0) / \varepsilon_0 m_d(r_0).$$

The cold plasma dielectric tensor components may be written

$$u_1 = \frac{\omega^2}{c^2} \sum_i \frac{\omega_{pi}^2}{\Omega_i^2 - \omega^2}, \qquad u_2 = \frac{\omega^2}{c^2} \sum_i \frac{\omega_{pi}^2 \omega}{\Omega_i (\Omega_i^2 - \omega^2)}, \qquad (3)$$

where the sums are over all the charged species except the electrons, but including the dust grains. Two coupled differential equations in x can be derived for the wave field components E_y and B_z [8]:

$$\frac{dE_y}{dx} - \frac{k_y u_2}{u_1 - k_z^2} E_y = i\omega \frac{u_1 - k_z^2 - k_y^2}{u_1 - k_z^2} B_z, \qquad (4)$$

$$\frac{dB_z}{dx} - \frac{k_y u_2}{u_1 - k_z^2} B_z = \frac{i}{\omega} \frac{(u_1 - k_z^2)^2 - u_2^2}{u_1 - k_z^2} E_y, \qquad (5)$$

We define the Alfvén speed using the plasma ion mass density, $v_A = B_0/\sqrt{\mu_0 \rho_{i0}}$, where $\rho_{i0} = m_i n_{i0}$, and the Alfvén speed corresponding to the total dust mass density ρ_{d0}, $v_{A0} = B_0/\sqrt{\mu_0 \rho_{d0}}$. Replacing the sums over dust grains by integrals over the continuous distribution, we obtain

$$u_1 = \frac{\omega^2 \Omega_{i0}^2}{v_A^2 (\Omega_{i0}^2 - \omega^2)} + \frac{\omega^2 \Omega_{d0}^2}{s v_{A0}^2} \int \frac{f_d(r_d)}{r_d (\Omega_{d0}^2/r_d^4 - \omega^2)} dr_d, \qquad (6)$$

$$u_2 = \frac{\omega^3 \Omega_{i0}}{v_A^2 (\Omega_{i0}^2 - \omega^2)} + \frac{\omega^3 \Omega_{d0}}{s v_{A0}^2} \int \frac{r_d f_d(r_d)}{(\Omega_{d0}^2/r_d^4 - \omega^2)} dr_d, \qquad (7)$$

We define the dimensionless frequency $f = \omega/\Omega_{d0}$, the ratio of charged dust mass density to plasma ion mass density, $b = \rho_{d0}/m_i n_{i0} = v_A^2/v_{A0}^2$, and the ratio of characteristic

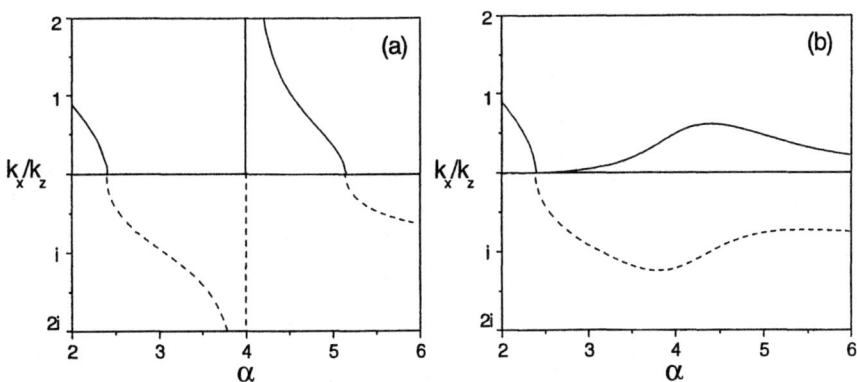

FIGURE 1. The ratio of perpendicular to parallel wavenumber k_x/k_z against the normalized parallel wavenumber $\alpha = v_{A0}k_z/\Omega_{d_0}$. The real part of k_x/k_z (solid curves) is plotted above the α axis, and the imaginary part of k_x/k_z (dashed curves) is plotted below. The vertical line indicates the Alfvén resonance. (a) the dust grains have a size spectrum with $p = 4$, and $f = 0.826$, just below the lowest dust cyclotron frequency, (b) as for (a) except $f = 0.827$, just above the lowest dust cyclotron frequency.

dust grain cyclotron frequency to plasma ion cyclotron frequency, $g = \Omega_{d0}/\Omega_{i0} \ll 1$. We also define the dimensionless wavenumber $\alpha = v_{A0}k_z/\Omega_{d0}$. Then we have from (6) and (7),

$$\frac{u_1}{k_z^2} = \frac{f^2}{\alpha^2}\left[\frac{1/b}{1-f^2g^2} - \frac{1}{sf^2}I_1\right], \quad \frac{u_2}{k_z^2} = \frac{f^2}{\alpha^2}\left[\frac{fg/b}{1-f^2g^2} - \frac{1}{sf}I_4\right] \quad (8)$$

where, if we use the power law distribution (1),

$$I_1 = c_p\int_1^{r_m}\frac{r_d^{3-p}}{(r_d^4-u^4)}dr_d, \quad I_4 = c_p\int_1^{r_m}\frac{r_d^{5-p}}{(r_d^4-u^4)}dr_d, \quad (9)$$

where $u^2 = \Omega_{d0}/\omega = 1/f$ and I_1 is one of three related integrals (I_1, I_2 and I_3) employed in Ref. [5]. Integrals I_1 and I_4 have an imaginary part due to the singularity at $u = r_d$, i.e. at $\omega = \Omega_d(r_{dr})$, the dust cyclotron resonance for those grains of radius r_{dr}. This leads to dust cyclotron damping of the waves.

Consider propagation in a uniform plasma, with wavenumber component k_x. The dispersion relation resulting from Eqs. (4) and (5) is shown in Figure 1, as a plot of k_x/k_z against the normalized parallel wavenumber α. The dust grains have a size spectrum with $p = 4$. In Figure 1(a), $f = 0.826$, corresponding to ω just below the lowest dust cyclotron frequency. The true Alfvén resonance in k_x still exists, because there is no dust cyclotron damping of the wave. Figure 1(b) is as for Figure 1(a) except $f = 0.827$, corresponding to ω just above the lowest dust cyclotron frequency, with resultant dust cyclotron damping. For α below the resonance value α_r, the imaginary part of k_x is almost unchanged, although a small real part of k_x arises just below the resonance. Below the lower cutoff value of α, the wave propagates practically undamped. Similarly, above the upper cutoff value of α, the wave has almost the same imaginary part, although acquiring a small real part of k_x. However, it is seen that the dust cyclotron damping leads to a drastic modification of the wave's behaviour in the vicinity of the former resonance. The resonance is smoothed out, with an imaginary part of k_x comparable in size to the real part. A wave propagating into this region suffers heavy damping, with the energy dissipation occurring via cyclotron heating of the dust grains over a relatively broad region of α.

When collisions of the ions and dust grains with background neutral atoms and molecules are included, the dielectric tensor components must be modified to include the collision frequencies of the ions and dust grains [4,9]. For a spectrum of grain sizes, the collision frequency of the dust grains v_{dn} is a function of grain radius. The relative importance of collisional and dust-cyclotron damping must then be determined for a range of parameters relevant to waves in magnetized interstellar molecular clouds. The results should be important for questions of star formation and angular momentum loss in such clouds. Charged dust may also play a role in instabilities of magnetized plasmas, such as the ring-beam instability driving Alfvén and magnetoacoustic waves in cometary environments [10]. The spectrum of dust sizes will modify the dispersion relations and introduce a dust-cyclotron damping threshold for the instability.

ACKNOWLEDGMENTS

Support for this work has been provided by the Australian Research Council.

REFERENCES

1. Cramer, N.F., and Vladimirov, S.V., *Phys. Scripta*, **53**, 586 (1996).
2. Cramer, N.F.,, Yeung, L.K., and Vladimirov, S.V., *Phys. Plasmas*, **5**, 3126 (1998).
3. J.S. Mathis, Rumpl, W., and Nordsiek, K.H., *Astrophys. J.*, **217**, 425 (1977).
4. Wardle, M., and Ng, C., *Man. Not. R. Astran. Soc.*, **303**, 239 (1999).
5. Tripathi, K.D., and Sharma, S.K., *Phys. Plasmas*, **3**, 4380 (1996).
6. Bliokh, P., Sinitsin, V., and Yaroshenko, V., *"Dusty and Self-Gravitational Plasmas in Space"*, Astrophysics and Space Science Library, Kluwer Academic Publishers, Dordrecht (1995).
7. Verheest, F., *"Waves in Dusty Space Plasmas"*, Astrophysics and Space Science Library, Kluwer Academic Publishers, Dardrecht (2000).
8. Cramer, N.F., *"The Physics of Alfvén Waves"*, Wiley-VCH, Berlin (2001).
9. Cramer, N.F., and Vladimirov, S.V., *Publ. Astron. Soc. Australia*, **14**, 170 (1997).
10. Cramer, N.F.,, Verheest, F. and Vladimirov, S.V., *Phys. Plasmas*, **6**, 36 (1999).

Ballooning Instability in the Jovian Magnetosphere

Nilakshi Das* and K.S. Goswami[†]

*Department of Physics, Tezpur University, Napaam- 784 028 Assam, India.
[†]Centre of Plasma Physics, Dispur- 781 006, Assam, India.

Abstract. The ballooning instability arising due to the curvature of the magnetic field in the Jovian magnetosphere and co-rotational motion of plasma particles with the planet, is investigated in this problem. The MHD equation for the system of electrons, ions and dust grains appropriate to the situation are considered. The stability condition for the ballooning mode is derived using the energy principle. The effect of dust particles is studied in destabilizing the mode.

INTRODUCTION

Ballooning modes are internal pressure driven instabilities that occur in multidimensional configurations. From the examination of Voyager 2 photographs in forward scattered light, Showalter[1-2] et al reported the discovery of a very tenuous ring of Jupiter, largely composed of micron-sized grains, extending outward from the brighter thin inner rinlg to a radial distance of about 210,000 km. Showalter et al believed that plasma drag is the dominant evolutionary process which transports the Jovial ring material. The Jovian magnetodisc contains heavy ions which have apparently originlated from the volcanic activity of the satellite Io. Magnetohydrodynamic instabilities such as interchange and ballooning instabilities under the action of the centrifugal force have been suggested as cause of the rapid outward transport of the Iogenic ions. In this paper, we have investigated the ballooning instability arising due to the curvature of the magnetic field in the Jovian magnetosphere and co-rotational motion of plasma particles with the planet. The MHD equation for the system of electrons, ions and dust grains appropriate to the situation are considered. We use a rotating cylindrical plasma and magnetic field equilibrium in which the magnetic field is purely toroidal, $\vec{B} = B(r)\hat{\theta}$ in which all equilibrium quantities depend only on the cylindrical radius r.

THEORETICAL FORMULATION

In magnetospheric situation the plasma co-rotates with the planet at a constant frequency ω about say, the z-axis. The ballooning modes are well described in a co-ordinate frame co-rotating with the plasma. Then the equation of motion of dust grains, ions and electrons in the rotating frame is [4-5]

$$\rho \frac{d\vec{v}_d}{dt} = \vec{v}_d \times \left[-\frac{z_d e n_d}{c} \vec{B}(\vec{r}) + 2\rho_d \vec{\Omega} \right] - \nabla p_d - \rho_d \vec{\Omega} \times \vec{\Omega} \times \vec{r} - \rho_d \nabla \phi \quad (1)$$

$$0 = \frac{n_i e}{c} \vec{v}_i \times \vec{B}(\vec{r}) - \nabla p_i \quad (2)$$

$$0 = -\frac{n_e e}{c} \vec{v}_e \times \vec{B}(\vec{r}) - \nabla p_e \quad (3)$$

where ρ is the dust mass density, v_d, v_i and v_e are velocities of the dust grains, ions, electrons. $\phi(r)$ is the cylindrically symmetric gravitational potential, r is the radial coordinate in polar variables Ω is the angular velocity of co-rotation. The second and fourth terms of (1) are the Coriolis force and the centrifugal force, respectively. Combining equations (1) – (3)

$$\rho \frac{d\vec{v}_i}{dt} = \frac{1}{4\pi} \left(\nabla \times \vec{B} \right) \times \vec{B}(\vec{r}) + 2\rho \vec{v}_d \times \Omega \hat{z} - \nabla P + \rho \Omega^2 r \hat{r} \quad (4)$$

where use has been made of the equation

$$\nabla \times \vec{B} = \frac{4\pi}{c} e(-Z_d n_d \vec{v}_d + n_i \vec{v}_i - n_e \vec{v}_e)$$

The frozen-in law

$$\vec{E} + \frac{1}{c} \times \vec{B} = 0$$

gives

$$\frac{\partial \vec{B}}{\partial t} = \nabla \times \left(\vec{v} \times \vec{B} \right) \quad (5)$$

Other standard MHD equations are:

$$\frac{\partial \vec{B}}{\partial t} + \nabla \cdot (\rho \vec{v}) = 0 \quad (6)$$

$$\frac{d}{dt} \left(\frac{p}{\rho^\gamma} \right) = 0 \quad (7)$$

The linearized form of equation (4) is expressed as,

$$\rho \frac{d\vec{v}_i}{dt} = \frac{1}{4\pi} \left(\nabla \times \vec{B}_\circ \right) \times \vec{B}_1 \frac{1}{4\pi} \left(\nabla \times \vec{B}_1 \right) \times \vec{B}_\circ + 2\rho \vec{v}_1 \times \Omega \hat{z} - \nabla p_1 + \rho_1 \Omega^2 r \hat{r} - \rho \nabla \phi_1 \quad (8)$$

It is convenient to express all perturbed quantities in terms of the Lagrangian displacement vector $\vec{\xi}(x,t)$ which expresses the deviation of the fluid element residing at the point x at time t from the position it would have had at time t if carried by the equilibrium flow from its position at $t = 0$. Then from equations (5)–(7), perturbed quantities can be expressed as

$$\vec{v}_1 = \frac{d\vec{\xi}}{dt}; \quad \rho_1 = \nabla(\rho_\circ \vec{\xi}); \quad p_1 = -\gamma p_\circ \nabla \vec{\xi}; \quad \vec{B}_1 = \nabla \times (\vec{\xi} \times \vec{B}_\circ)$$

We assume that no macroscopic flow exists in the equilibrium condition and we consider incompressible perturbations from it, as they are subject to the least favourable stability conditions. Hence,

$$\nabla \cdot \vec{\xi} = 0 \qquad (9)$$

Ballooning modes are characterized by plasma displacements that are mostly out of the magnetic surface. Here inhomogeneity is considered along the r-direction. Then finally, equation (8) takes the form

$$\vec{F}(\vec{\xi}) = \rho \frac{d^2\vec{\xi}}{dt^2} = \frac{1}{4\pi}\left[\left(\frac{B_\circ}{r}+\frac{\partial B_\circ}{\partial r}\right)\frac{\xi_r}{r}\frac{\partial B_\circ}{\partial \theta}\hat{\theta}\right]$$
$$+\frac{1}{4\pi}\left(B_\circ \frac{\partial B_{1r}}{\partial \theta}\hat{r}\right) - 2\rho\Omega\hat{z}\times\frac{\vec{\xi}}{dt}+\rho\Omega^2 r\hat{r}-\rho\frac{\partial\phi}{\partial r}\hat{r} \qquad (10)$$

The change in potential energy of the system due to the perturbation is

$$\nabla W = -\frac{1}{2}\int dv\vec{\xi}\cdot\vec{F}(\xi)$$
$$= -\frac{1}{2}\int dv(\xi_r\hat{r}+\xi_z\hat{z})\cdot\vec{F}(\xi)$$
$$= -\frac{1}{2}\int dv\xi_r \left(\frac{B_\circ}{4\pi}\frac{\xi_r}{r}\frac{\partial^2 B_\circ}{\partial \theta^2}-\rho\frac{\partial\phi}{\partial z}+\Omega^2 r\rho\right) \qquad (11)$$

and according to the energy principle, an equilibrium is stable if and only if

$$\delta W \geq 0$$

that is $\quad \dfrac{\partial \phi}{\partial r} \geq \Omega^2 r + \dfrac{B_\circ^2}{4\pi\rho}\dfrac{\xi}{r}\dfrac{\partial^2 B_\circ}{\partial \theta^2}$

Considering that all the perturbed quantities are proportional to $\exp[-i(\omega t - m\theta - k_z z)]$, the condition can be written as,

$$\left|\frac{\Omega^2 r^2}{\xi_e m^2}\right|^2 \leq \left|\frac{B_\circ^2}{4\pi\rho}\right|^2 + \left|\frac{k_z}{m^2 \xi_r}\phi r\right|^2 \qquad (12)$$

This condition shows that an increase in the dust mass density may have a destabilizing effect. The ballooning mode grows in the environment of Jovian magnetosphere when the dust mass density is so high that condition (12) is not satisfied. For Jupiter's magnetosphere [6-7] at $r = 50R_J$, ($R_J = 71,398$ km), $B_\circ = 1$ nT, the dust mass density $\rho = 1$ gm/cm^3, the angular velocity of rotation of the planet $\Omega = 1.75 \times 10^{-4}$s^{-1}, and the gravitational acceleration $g = 71.2$ cm/sec^2. Using these data, it is found that the centrifugal term $\Omega^2 r (\approx 10^4)$ is predominant over the gravitational term (≈ 10) and hence, the stability condition is not satisfied. The dust mass density ρ may have a significant effect at small r in destabilizing the magnetspheric plasma.

SUMMARY

It is seen from the discussion that the centrifugal and gravitational terms play key roles in determining whether the instability will grow or not. The system may be stable if the gravitational term is predominant over the centrifugal term. It is to be noted that the stability is also disturbed by the presence of dust grains. An increase in dust mass density ρ may cause the system to be unstable against the ballooning mode.

REFERENCES

1. Northrop, T. G., Mendis, D. A. and Schaffer, L., *Icarus*, **79**, 101–115 (1989).
2. McNutt, R. L. .Jr., Coppi, P. S., Selesnick, R. S. and Coppi, B., *J. Geophys. Res.*, **92**, 4377 (1987).
3. Hameiri, E., Laurence, P. and Mond, M., *.J. Geophys. Res.*, **Vol. 96**, No. 2, p 1513–1526 (1991).
4. Abe, T. and Nishida, A., *J. Geophys. Res.*, **Vol. 91**, No. A9, p 10,003 –10,011 (1986).
5. Northrop, T. G. and Birmingham, T. J., *J. Geophys. Res.*, **Vol. 67**, No. A2, p 661–669 (1982).
6. Mihaly, H., *Phys. Plasmas*, **Vol. 7**, No. 10, p 3847–3850 (2000).
7. Northrop, T. G. and Hill, J. R., *J. Geophys. Res.*, **Vol. 88**, No. 1, p 1–11 (1983).

Dust-Acoustic Wave Instability at the Diffuse Edge of RF Inductive Low-Pressure Gas Discharge Plasma

A.V. Zobnin, A.D. Usachev, and V.E. Fortov.

Institute for High Energy Densities Russian Academy of Sciences
Izhorskaya 13/19, 125412, Moscow, Russia

Abstract. A spontaneous excitation of a wave of grain density in a dusty cloud suspended at the diffuse edge of RF inductive gas discharge is revealed. The main physical parameters of this wave and background plasma are measured. The theoretical model of the observable phenomenon basing on the theory of dust acoustic waves in a collisional dusty plasma well correlates with the experimental data in a broad band of experimental conditions. The analytical research of influence of a variable charge of dust grains on the development observable dusty plasma instability is conducted. It is shown, that a necessary condition for development of dusty plasma instability is an availability of the permanent electrical field ($E_0 \geq 3$ V/cm) at the area of dusty cloud.

INTRODUCTION

The instability of dust component in complex plasma exhibited in a spontaneous excitation of random or organized motion of an ensemble of dusty grains is the same general and immanent property of dusty plasma as well as "classical" plasma instabilities in plasma without particles. This work is devoted to an experimental research and analytical simulation of DAW instability at the diffuse edge of RF inductive gas discharge plasma.

SETUP AND DIAGNOSTICS

A scheme of the experiment for research of a DAW instability in RF inductive discharge plasma and on visual diagnostics of dust wave parameters is presented on Fig. 1. The RF glow inductive discharge (100 MHz, ~1W) was excited in a vertically oriented cylindrical glass tube of diameter 3 cm and length 65 cm with a help of two-rings inductor in neon. The measurements were performed at 5 pressures of a neon (Table 1). Monodisperse melamine formaldehyde particles of 1.87±0.04 μm diameter were used. The wave parameters (phase velocity υ_{DAWI}, frequency ω_{DAWI}, distribution of grain concentration $n_d(H)$ along vertical axis H) were measured with a help of the high speed CCD camera Redlake500. The distributions of the background plasma parameters (electron density $n_e(H)$, electron temperature $T_e(H)$ and space potential $U_s(H)$) were measured by a single mobile Langmuir probe. The dust grain charge has been calculated numerically [1] for conditions of collisional plasma on the basis of measured plasma parameters. Note that all the data were measured in the region where a wave amplitude

CP649, *Dusty Plasmas in the New Millennium: Third International Conference on the Physics of Dusty Plasmas*, edited by R. Bharuthram et al.
© 2002 American Institute of Physics 0-7354-0106-3/02/$19.00

was low to compare them with the DAW analytical model [2-4].

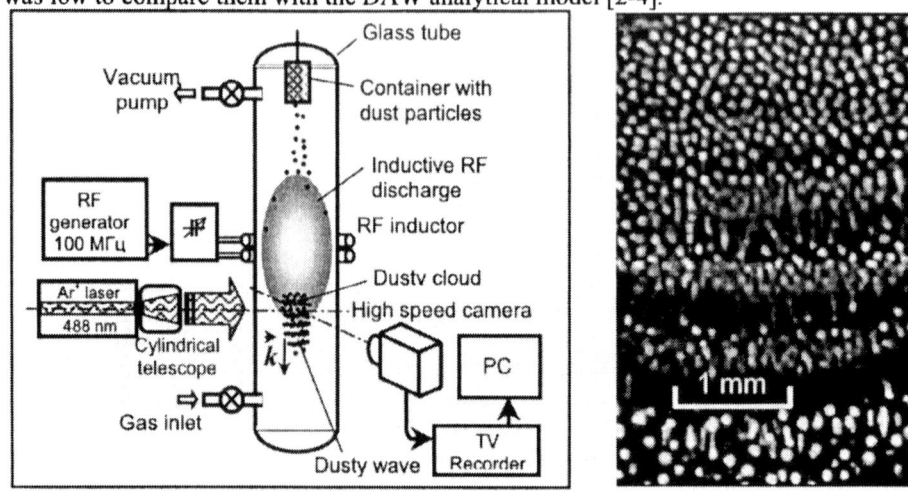

FIGURE 1. The scheme of experiment for investigation of the dust acoustic wave instability at the diffuse edge of RF inductive discharge.

FIGURE 2. Single frame video image of DAW instability at 50 Pa of neon pressure.

ANALYTICAL MODEL.

As the observable phenomenon, DAW instability, represents a kind of traveling waves of a grain density, it theoretical interpretation was conducted within the framework of the DAW theory in a collisional plasma [2-4]. To find a dispersion relation $k = k(\omega)$ the Poisson equation was linearized with the $\delta\varphi \sim \exp(ikx - i\omega t)$ dependence for electrostatic potential,

$$k^2 \delta\varphi = 4\pi e(-\delta n_e + \delta n_i - Z_d \delta n_d - n_d \delta Z_d). \tag{1}$$

The electron density perturbation δn_e was found in Boltzmann approximation. The ion density δn_i and dust density δn_d perturbations were found using the standard fluid approach [2-4]. In this case the dimensionless dispersion equation is

$$1 + \frac{\chi P}{1+P} + \tilde{k}^2 + i\tilde{E}\tilde{k} = \frac{\tilde{k}^2 + i\tilde{E}\tilde{k}(1+\chi)}{\tilde{\omega}(\tilde{\omega} + i\tilde{\eta})}, \tag{2}$$

where $\tilde{k} = k r_{Di}$, $\tilde{E} = eE_0 r_{Di}/T_i$, $\tilde{\omega} = \omega/\omega_{pd}$, $\tilde{\eta} = \eta/\omega_{pd}$, $P = Z_d n_d / n_e$, r_{Di} is the ion plasma Debye length, $E_0 \sim 4$ V/cm is the permanent electric field strength, $\omega_{pd} = 2eZ_{d0}(\pi n_d/m_d)^{-1/2}$ is the dusty plasma frequency, η is the viscosity coefficient for dust particles, $\chi(n_{i0}/n_{e0})$ is the logarithmic derivative of Z_d on the base of n_e/n_i and $\delta Z_d = \chi Z_d$ ($\chi = 0 \div 0.33$). The numerical solutions of the Eq.2 with respect to $\tilde{k}_{Re} = \tilde{k}_{Re}(\tilde{\omega})$ and $\gamma =$

$\gamma(\omega) = -k_{Im}(\omega)$ are presented on Fig. 3 and Fig.4 respectively for limiting values of χ.

TABLE 1. The measured and calculated plasma and DAWI parameters.
Designations: p – pressure of Ne; η – viscosity coefficient of dust grains in neutral neon; n_e and T_e – electron density and temperature at the region of dusty wave; n_i and $\tilde{T}_i$ – ion concentration and effective temperature; r_D – plasma Debye length; n_d and Z_d – dust grain concentration at unperturbed region and calculated dust grain charge [1]; υ_{DAWI}, λ_{DAWI}, ω_{DAWI} and k_{DAWI} – measured phase velocity, wave length, frequency and wave number of the wave, ω_{max} – calculated value which result in $\gamma_{DAWI}(\omega)$ = max; γ_{DAWI} – measured wave growing rate; γ_{max} – calculated growing rate at ω_{max}.

Experiment	#1(*)	#2	#3(*)	#4	#5	#6	#7	Error
p, Pa	10	15	20	20	30	50	50	±2
η, c^{-1}	37	56	74	74	110	185	185	±10%
n_e, cm^{-3}	2×10^8	2×10^8	3×10^8	3×10^8	3×10^8	4×10^8	4×10^8	±40%
T_e, eV	4.2	4.1	4.0	4.0	3.7	3.5	3.5	±0.5eV
$n_i=n_e+Z_dn_d$	2,8×10^8	5×10^8	3.3×10^8	5×10^8	4.7×10^8	6.6×10^8	5.7×10^8	±50%
$\tilde{T}_i$, K	1030	680	515	515	340	300	300	±25%
r_D, μm	132	81	86	86	59	47	50	±30%
n_d×10^{-4}, cm^{-3}	2,4	10	1.5	7	7	12	7	±30%
Z_d [1]	3400	3000	2900	2900	2360	2160	2160	-
ω_{pd}, sec^{-1}	397	715	219	578	470	564	431	±30%
υ_{DAWI}, cm/s	8,3±1	5,8±2.3	4,8±0.5	4,8±1.6	4,2±1.4	2,9±0.3	2,3±0.3	←
λ_{DAWI}, mm	5.2±0.7	1.26±0.5	3.0±0.3	1.05±0.4	0.95±0.3	0.65±0.1	0.67±0.1	←
$\omega_{DAWI}=2\pi\nu$, s^{-1}	100±16	290±30	100±30	290±40	280±40	285±10	220±40	←
γ_{DAWI}, cm^{-1}	-	>10	>10	>10	>10	6±4	5±4	←

(*) – small dust grain concentration.

COMPARISON TO THE EXPERIMENT.

An examination of the data presented on Figs. 3,4 and Table 1 follows us to the next conclusions:

- the dispersion curves $\tilde{k}_{Re} = \tilde{k}_{Re}(\tilde{\omega})$ calculated by Eq. 2 fit the experimental data better then the "collisionless" approach [2], $\tilde{k}_{Re}(\tilde{\omega}) = \tilde{\omega}\sqrt{1-\tilde{\omega}^2}$ (see Fig.3);
- the frequencies ω_{max}, corresponding to maxims of calculated curves $\gamma_{max}= \gamma(\omega_{max})$, well correlate with the frequencies of instabilities measured at the experiment;
- the computational values $\gamma_{max}=\gamma(\omega_{max})$ well correlate with the DAWI growing rates estimated at the experiment (compare the data from Tab.1 with the data on Fig.4);
- the computational value of γ_{max} decreases as neutral gas pressure decreases; at a pressure is higher of 80 Pa the computational value γ_{max} becomes negative, as it is observed at experiment - DAWI have not observed at $p> 60$ Pa;
- with other things being equal, the computational value of ω_{max} is proportional to n_d (compare the curves 3 and 4 on Fig. 4), that have been observed in experiment;

- the availability of a constant electrical field E_0 is a necessary condition for a self-excitation of DAWI, the forced movement of a cloud (by a passage of gas) up on 5 mm to the area with lower values E_0 (~ 1-2 V/cm) resulted in damping of DAWI.

Thereby it is possible to conclude, that the considered DAW instability is the dust acoustic wave, amplified in a permanent electrical field of plasma ambipolar diffusion.

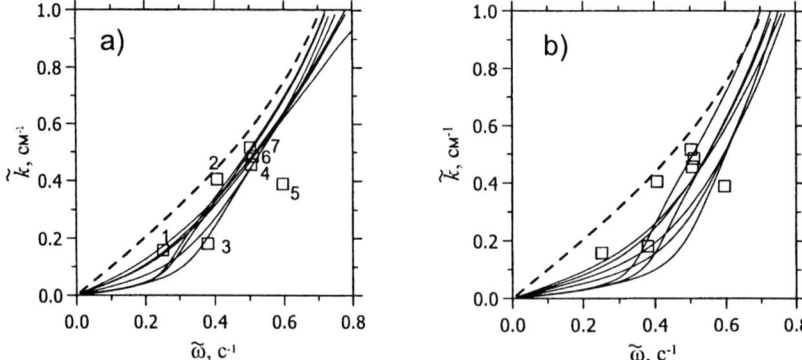

FIGURE 3. Theoretical (lines) and measured (squares) dispersion dependences. The digits near the squares mean the number of experiment according to Table 1. The dashed line is the calculated dispersion dependence for collisionless plasma. 7 solid lines are results of calculations by Eq. 1 for collisional dusty plasma for experimental data presented on Table 1 and $\chi = 0$ (a), $\chi = 0.33$ (b).

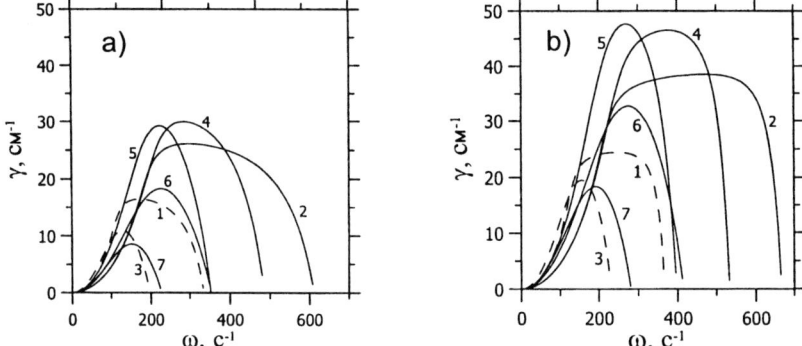

FIGURE 4. Numerical solution of Eq. 1 relative the grow rate $\gamma(\omega)$ for the experimental conditions from Table 1 and $\chi = 0$ (a), $\chi = 0.33$ (b). The digits near the curves mean the number of experiment according to Table 1.

REFERENCES.

1. Zobnin, A.V., Nefedov, A.P., Sinel'shchikov, V.A., and Fortov, V.E., *JETP*, **91**, 554 (2000).
2. Rao, N.N., Shukla P.K., and Yu, M.Y., *Planet. Space Sci.*, **38**, 543 (1990).
3. Ivlev, A.V., Samsonov, D., Goree, J., and Morfill, G., *Phys. Plasmas*, **6**, 741 (1999).
4. Fortov, V.E., Khrapak, A.G., Khrapak S.A., et al., *Phys. Plasmas*, **7**, 1374 (2000).

Spatial Separation of Dust Particles by their Sizes at the Diffuse Edge of RF Inductive Discharge Plasma

A.V. Zobnin, A.D. Usachev, and V.E. Fortov.

Institute for High Energy Densities Russian Academy of Sciences
Izhorskaya 13/19, 125412, Moscow, Russia

Abstract. The technique for measuring *in situ* 2D-spatial distribution of the polymeric (melamine formaldehyde) spherical particles by the size in complex plasma is developed. The experimental research of the phenomenon of spatial separation of powder particles by their sizes (0.4÷2 μm) at the diffuse edge of RF inductive low-pressure discharge plasma is conducted. It is revealed, that during experiment the size of particles decreases, and the speed of decreasing $\dot{R}$ divided by the plasma electron density n_e is a constant value for different n_e, $\dot{R}/n_e = 3 \cdot 10^{-9}$ Å·s^{-1}·cm^{-3}.

INTRODUCTION

The phenomenon of a spatial separation of powder particles in complex plasma by their sizes is a rather general property of such dusty plasma systems and is a result of various functional dependences of the forces acting on a particle in plasma from it size. These forces are: electric force acting on a charged particle in a constant electric field (~r), ion drag and thermophoretic forces (~r^2), force of gravity (~r^3) and some others. As a rule, the boundary line between the regions of dusty cloud containing particles with the different particle sizes is sharp. From a point of view of possible industrial applications, the considered phenomenon can be applied for microparticles separation on their sizes in the submicron size region inaccessible for existing methods. In this doing the development of existing methods for particle size measurements is necessary.

SETUP AND DIAGNOSTICS

The scheme of experiment on creation of a powder cloud in RF inductive discharge plasma and on diagnostics of spatial particle distribution on their sizes *in situ* is presented on Fig. 1. The RF glow inductive discharge (100 MHz, ~1W) was excited in a vertically oriented cylindrical glass tube of diameter 3 and length 65 cm with a help of two-rings inductor in neon. The measurements were conducted at two pressures of a neon - 80 and 150 Pa. A dusty cloud was formed in the discharge plasma by shaking of the container filled by monodisperse spherical transparent melamine formaldehyde (MF) particles with the diameters 0.99±0.04 and 1.87±0.04 μm (n = 1.68, ρ = 1.51 g/cm^3). The particles are spilled through lattice bottom of the container, dropped downwards and suspended in an electrostatic trap at the bottom of discharge area. This electrostatic trap is formed due to a combination of an electric fields created by charged tube surface and ambipolar diffusion [1]. The distributions of plasma parameters, these are electron density $n_e(H)$, electron temperature $T_e(H)$ and space potential $U_s(H)$, were measured by a

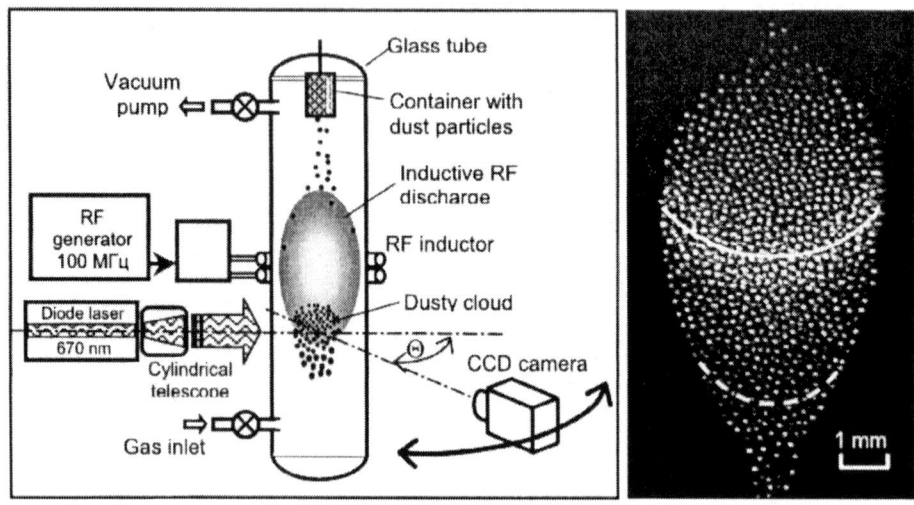

FIGURE 1. The scheme of dusty cloud formation at the diffuse edge of RF inductive discharge and 2D diagnostics of particle size distribution in dusty cloud.

FIGURE 2. The view of MF-1.87 dusty cloud at $\theta = 74°$ with respect to the detecting laser beam. Solid (dashed) white line shows the area with maximum (minimum) intensity of scattered laser light.

single mobile Langmuir probe (H is the distance down from the RF inductor). The determination of particle sizes was based on measurements of the angular distribution of scattered light intensity (Fig. 1) and further interpretations the measured data using Mie theory. For simultaneous measurements of scattered light intensity from different parts of powder cloud it was illuminated by a diode laser sheet ($\lambda=670$ nm) with the cross-section size 15×0.4 mm^2, and the intensity of scattered light was measured by means of CCD video camera from each particle separately. An analytical research have shown, that for obtaining the information about the size of particles it is necessary and enough to measure the intensity of scattered light under three fixed angles θ with respect to the direction of the laser light propagation - 74°, 82° and 92°. The typical single frame video image of dusty cloud at 74° is shown in Fig. 2.

EXPERIMENTAL RESULTS

It was found, that at the injection of the monodisperse particles of 0.99 and 1.87 µm diameter to the discharge by small portions (~ 10-100 particles) a rather small dusty cloud with a common number of particles about 500 was formed at the bottom part of the discharge [1,2]. The addition of the new particles did not result in increase of common number of the particles in a cloud. In this doing the measured sizes of the particles in such cloud corresponded to their passport meanings of the sizes, these are 0.99 and 1.87 microns accordingly. However, at an injection of the particles by large.portions (~10^3 particles) and by many times to the discharge the dusty cloud began to increase in the sizes up to 5×12 mm^2 with total amount of trapped particles up to 10^4. The measurements of the particle sizes in this case have shown, that a cloud consists of particles of different sizes, and the particle size varies monotonously from 0.4 (1.6) microns at the cloud top to 0.9 (1.8) microns at the cloud bottom when use the particles with passport size 0.99 (1.87) microns (Fig. 3). In this doing the scattering indicatrix for particles 0.4÷0.99 microns (1.4÷1.8 microns) had the brightly expressed maxims (Fig. 2), that corresponds to a spherical form of the trapped particles, while the fraction located below of particles of the basic size gave the monotonous scattering indicatrix, that corresponds to particle conglomerates. We suppose that the conglomerates are coupled MF particles.

It was found that the particles are decreased in size during an exposition in RF discharge plasma. During this experiment (~10 min) the shape of dusty cloud was invariable and the position of each particle seems likely the same. Hence the size diagram evolution can be attributed namely to size diminishing but not to particle diffusion within the cloud area. The rate of particle size etching was found depends on particle position along the vertical axis H of dusty cloud (Fig. 4). The comparison of the rate dependence $\acute{R}(H)$ with the electron density dependence $n_e(H)$ have shown that the ratio $\acute{R}(H)/n_e(H)$ is constant for all possible H. That means the sublimation of dust material occurs due to bombardment of the MF particles by charged atomic plasma species – ions and electrons.

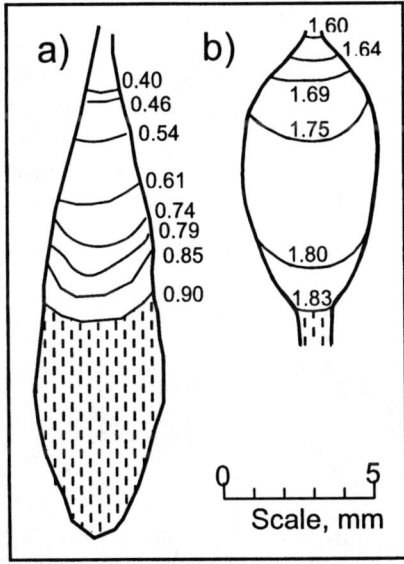

FIGURE 3. The measured spatial distribution of particles on their sizes at the suspended dusty cloud at the diffuse edge of a RF inductive discharge plasma at 150 Pa of neon: **a)** after injection of the mono-disperse 0.99 µm-particles; **b)** after injection of the mono-disperse 1.87 µm-particles. The digits mean the measured particle size (µm) along corresponding solid line. Hatched areas mean the cloud region with the coupled particles.

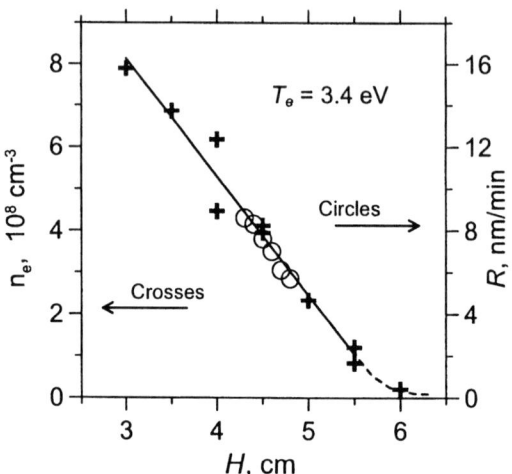

FIGURE 4. The measured rate $\dot{R}$ of particle size diminishing (circles) and electron density n_e (crosses) vs the distance H from RF inductor. The solid line is the linear fit of $n_e(H)$. Gas pressure – 80Pa.

CONCLUSION

The diffuse edge of RF inductive discharge is a good tool for experiments with complex plasma. This plasma region is characterized by smoother plasma gradients of $n_e(x)$ and $T_e(x)$ in comparison with the plasma sheath region of a planar RF-E discharge and the strata regions of DC glow discharge. That permits to create 3D powder clouds and observe various interesting phenomena – dust acoustic wave instability, spatial separation of particles by size, etc. This experiment has shown that the monodisperse MF particles widely used in dusty plasma experiments are subjected to plasma etching and enough quickly diminishing their size. The rate of etching normalized by plasma electron density was measured.

REFERENCES

1. Zobnin, A.V., Nefedov, A.P., Sinel'shchikov, V.A., Sinkevich, O.A., Usachev, A.D., Filinov V.S., and Fortov, V.E., *Plasma Physics Reports,* **26**, 415 (2000).
2. Fortov, V.E., Nefedov, A.P., Sinel'shchikov, V.A., Usachev, A.D., Zobnin, A.V., *Physics Letters,* **A267**, 179 (2000).

In Situ Study of Surface Modification of Nano-particles in Reactive Plasmas by Mie-Ellipsometry

G. Gebauer, T. Galka, J. Winter

Experimental Physics II, Ruhr-University Bochum, 44780 Bochum, Germany

Abstract. Nanoparticles play an important role in modern materials technology, in catalysis and, for instance, in astrophysics. Mie-Ellipsometry makes it possible to gain insight in the growth process of nanoparticles in reactive plasmas and consequently allows their process control. In this work improvements of the technique will be presented and discussed at the examples of the in situ surface modification of methylmelamine-formaldehyde particles in oxygen plasmas.

INTRODUCTION

Particle diagnostics by Mie-Ellipsometry is done by a measurement of the polarisation of scattered light. The particles are laterally irradiated with laser light. The polarisation state of the light, which is scattered from the particles is measured under a defined angle and is analysed based on the Mie-Theory [1].The Mie-Theory describes the correlation between the polarised scattered light, the particle form and size and the dielectric properties of the particles. This measurement technique was suggest for the first time in 1994 by Hayashi and Tachibana [2][3][4]. This field of activities was continued e.g. by Swinkels [5] in 1999. The Mie-Theory assumes the scattering of a plane electromagnetic waves from a homogeneous isotropic spherical particle. Therefore the measurements described in this work have been done on isotropic spherical norm particles [6]. The particle size distribution and multi-layered particles have been taken into account for the theoretical description of the results.

The polarisation state of scattered light is determined by its modulation. Argon und Oxygen are used as discharge gases with a pressure between 50 and 200Pa. The energy density was 0.025 to 0.5W/cm^3 at a frequency of 13.56MHz in a capacitive coupled parallel plate reactor. With this experimental arrangement the polarisation state of the scattered light is measured. Equation 1 defines the ellipsometric angles Ψ and Δ, which result from the ratio of the scattered electric field strengths parallel and perpendicular to the plane of incidence.

$$tan(\Psi)exp(i\Delta) = \frac{E_{parallel}}{E_{perpendicular}} \tag{1}$$

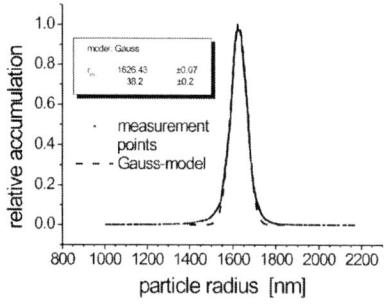 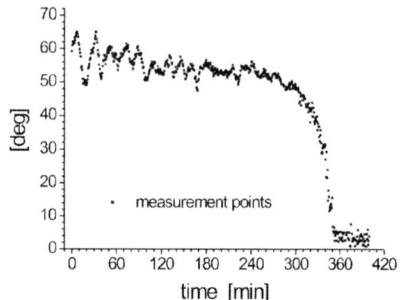

FIGURE 1. Particle size distribution determined by the producer.

FIGURE 2. Ellipsometric angle Ψ as a function of the residence time of the particles in the discharge.

MODIFICATON OF PARTICLES IN REACTIVE PLASMAS

The surface of MF-particles is altered in an reactive oxygen plasma predominantly by chemical etching. This chemical etching causes a decrease of the particle radius and possibly also a change in energy flux due to the reaction enthalpy. The reactive species in a oxygen plasma are both neutral atomic hydrogen atoms and atomic and molecular oxygen ions. Their reaction with hydrocarbon contained in the particle leads predominatly to CO which is volatile and released into the chamber. The chemical composition of MF near the surface is gradually changed. Due to the charging of large particles these particles gain another possibility to increase their energy by collision with ions. Ions gain energy in the sheath around the particle which is in the order of a few kT_e, i.e. of the order of 12eV in our case. It is well established, that ion bombardment also enhances the reactivity of atomic oxygen with a yield close to unity for an amorphised surface with many free valences. Since the thermal conductivity of dielectric particles is small, the energy flux to the surface leads to its temperatur increase and only to a slow equilibration over the volume of the particle. Swinkels [5] has measured an equilibrium temperature of the particles of up to 200^oC under similar conditions. This is close to the thermal stability of MF (250^oC). This may lead to enhanced material loss by faster chemical reactions and/or sublimation. The rate of erosion is thus determined by the differnt kinds of energy absorption and by a change of the material composition at the surface.

The particles were introduced into the plasma from the top. The size distribution of the particles taken into the plasma was determined by the producer [6] and the results are displayed in figure 1. Figure 1 shows, that the measured particle size distribution can be well approximated by a Gauss-distribution. For the following numerical calculation we have assumed that the form of the distribution is unchanged during the erosion process and therefore only the average radius and the standard derivation σ change. The particle parameters are determined by the measurement of the ellipsometric angle Ψ. The in situ measured results are present in figure 2.

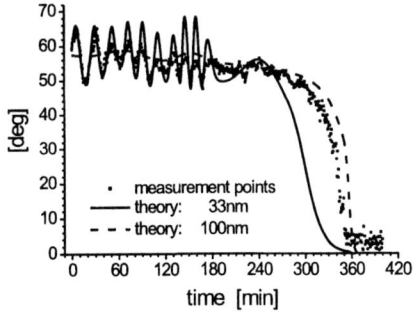
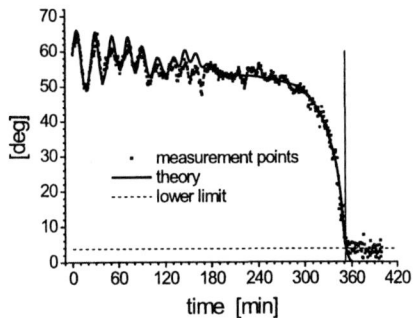

FIGURE 3. Comparison between measured values and theory with regard to several widths of the Gaussian distribution.

FIGURE 4. Comparison between measured values and theory.

In figure 3 the measured values are compared with the calculated results, where it was assumed for both calculations that the reduction rates are equal and that the widths of the distributions (standard deviation) are time independent. Figure 3 shows, that at the beginning of this process the measured values can be described by the calculated results with a standard deviation of 33nm very well, wheras at the end of the measurement a standard deviation of 100nm yields a better agreement. Obviously the standard deviations and their changes must be considered as time dependent for the numerical model. Figure 4 present the calculated results, which contain the variation of both the average radius and of the standard deviation. The variation of the average radius and of the standard deviation are shown in figures 5 and 6. The first point of the calculated average radius and the standard deviation in figures 5 and 6 reproduce the measured values in figure 1. The accuracy in determining the particle distribution is about 10%.

The lower limit of the measured average radius was determined to be 20nm. A significant change of the reduction rate (figure 5) appears after a measurement time of 187 minutes when the average radius is of the order of 200nm. In the first phase the average reduction rate is R_1=6.8nm/min in the second it is R_2=1.4nm/min. The calculation of the average radius and of the incidental standard deviation are independent of one another. Therefore figures 5 and 6 show that after a measurement time of 187 minutes a change of the plasma-particle-interaction occurs. Assuming that the external conditions around the particle are essentially constant, we would expect in the case of a surface process that the reduction rate is constant, too [7]. The decrease of the reduction rate thus indicates that the property of the particle surface is modified. This may be due to a simple change in the chemical composition due to the preferential removal of C-H groups from MF by chemical erosion. This hypothesis would be in agreement with an observed discoloration of the discharge close to the trapped particles. The discoloration however indicates clearly that the plasma close to the particles is changed in its chemical composition and probably also in its parameters. This in turn may be responsible for the change in the effective erosion rate.

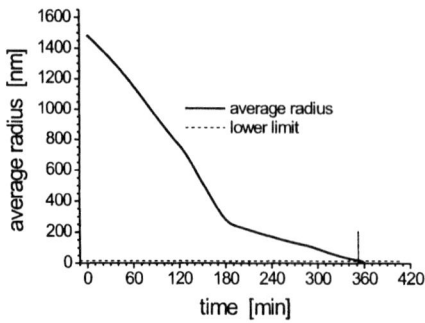
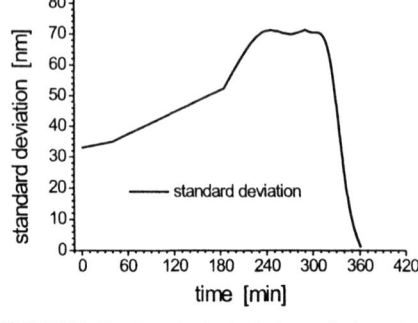

FIGURE 5. Average radius as a function of the residence time of the particles in the discharge.

FIGURE 6. Standard deviation of the size distribution as a function of the residence time.

The abruptness of the change indicates a threshold process. This may be a change in incident ion energy to a value below a displacement threshold or a breakdown of the OML assumtions at this particel diameter leading to ion fluxes no longer proportional to the particle surface area. In order to further identify the governing mechanism additional analysis of a local plasma (spectroscopy) and gas composition (residual gas analysis) is necessary.

CONCLUSION

It was shown that in situ Mie-Ellipsometry allows an accurate and sensitive determination of the size distribution (diameter, standard deviation of a Gaussian) within about 10%. Due to this new quality of the method we could determine the erosion of MF-particles in a oxygen plasma and have shown that a sudden change in plasma-particle interaction occurs at a given radius of the particle which corresponds to a welldefined residence time of the particle in the reactiv plasma. The change is due to a surface modification of the particle, possibly in conjunction with changes of the local plasma. The exact mechanism, however, is still under investigation.

REFERENCES

1. Mie, G., *Ann. Phys.*, **25**, 377 (1908).
2. Hayashi, Y., and Tachibana, K., *Jpn. J. App. Phys.*, **33**, L476 (1994).
3. Hayashi, Y., and Tachibana, K., *Jpn. J. App. Phys.*, **33**, 4208 (1994).
4. Hayashi, Y., and Tachibana, K., *Jpn. J. App. Phys.*, **33**, L804 (1994).
5. Swinkels, G., *Proefschrift, Technische Universität Eindhoven* (1999).
6. Microparticles GmbH, Berlin (2001).
7. Gebauer, G., *Dissertation, Ruhr-Universität Bochum* (2001).

On the Magnetosonic Wave and Instability in a Dusty Plasma

M.A. Hellberg*, A.P. Matthews* and F. Verheest[†]

*School of Pure & Applied Physics, University of Natal, Durban, South Africa
[†]Sterrenkundig Observatorium, Universiteit Gent, Ghent, Belgium

Abstract. We present some results of a numerical study of the dust-modified magnetosonic wave, driven unstable by perpendicular dust drift.

In a dusty plasma the magnetosonic wave is modified, with a frequency shift related to the dust charge density. Dust grains drifting across the magnetic field may drive this low frequency electromagnetic wave unstable. We describe a numerical study of such a situation, based on a multi-fluid model.

We consider a model consisting of electrons, ions, and negatively-charged, drifting dust grains ($\mathbf{U_d} = U\mathbf{e}_x$) in a magnetic field $\mathbf{B_0} = B_0\mathbf{e}_z$, and study electromagnetic waves propagating along the x-ax is. It has been suggested that this configuration may have application in the study of planetary rings, such as those of Jupiter and Saturn [1], where electrons and ions co-rotate with the magnetosphere, while dust grain orbits are near-Keplerian, yie lding a relative drift.

After results based on a kinetic model were initially reported [1], the problem was re-examined using a more transparent magnetofluid model, and different results obtained [2]. We extend that analytical work by evaluating numerically not only the full dispersion relation, but also the reduced form, valid for $\omega < \Omega_i$ [2].

Working in the plasma frame, and neglecting dust charging at these frequencies, it has been shown that the fluid equations lead to a linear dispersion relation for the X-mode, $D_{xx}D_{yy} = D_{xy}^2$, where the dispersion tensor elements are [2]

$$\begin{aligned}
D_{xx} &= 1 - \sum_\alpha \frac{\omega_{p\alpha}^2}{(\omega - kU_\alpha)^2 - k^2 v_{T\alpha}^2 - \Omega_\alpha^2}, \\
D_{xy} &= \sum_\alpha \frac{\omega_{p\alpha}^2 \Omega_\alpha}{(\omega - kU_\alpha)^2 - k^2 v_{T\alpha}^2 - \Omega_\alpha^2}, \\
D_{yy} &= \omega^2 - c^2 k^2 - \sum_\alpha \frac{\omega_{p\alpha}^2[(\omega - kU_\alpha)^2 - k^2 v_{T\alpha}^2]}{(\omega - kU_\alpha)^2 - k^2 v_{T\alpha}^2 - \Omega_\alpha^2}.
\end{aligned} \quad (1)$$

Here the gyrofrequencies Ω_α include charge-signs, and $\omega_{p\alpha}$ are plasma frequencies. The dust drift $U_d = U$, while $U_e = U_i = 0$.

For $|\Omega_d| \ll |\omega| \ll \Omega_i$; $v_{Td} \simeq 0$; and $c \gg V_A$, the plasma Alfvén speed, a reduced dispersion law is obtained [2]

$$[(\omega - kU)^2 - \omega_{dlh}^2][\omega^2 - k^2 V_{ms}^2 - \omega_{dlh}^2] = \delta^2 \Omega_i^2 (\omega - kU)^2. \tag{2}$$

Here V_{ms} is the plasma magnetosonic velocity, the normalized dust charge density $\delta = N_d q_d / N_i q_i$ measures free electron depletion due to absorption on dust, and the dust-lower-hybrid frequency ω_{dlh}, given by $\omega_{dlh}^2 = \frac{V_A^2}{c^2} \omega_{pd}^2 = \delta^2 \Omega_i^2 \frac{N_i m_i}{N_d m_d} = \delta \Omega_i \Omega_d$, is usually small if most of the mass is in the dust.

For stationary dust ($U = 0$), one finds stable modes

$$\omega^2 = \frac{1}{2}(k^2 V_{ms}^2 + \delta^2 \Omega_i^2 + 2\omega_{dlh}^2) \pm \frac{1}{2}\sqrt{(k^2 V_{ms}^2 + \delta^2 \Omega_i^2)^2 + 4\omega_{dlh}^2 \delta^2 \Omega_i^2}. \tag{3}$$

Hence one obtains a dust-modified magnetosonic mode, with frequency-shift due to dust terms through δ and ω_{dlh} [2],

$$\omega^2 \simeq k^2 V_{ms}^2 + \delta^2 \Omega_i^2 + \omega_{dlh}^2 \frac{k^2 V_{ms}^2 + 2\delta^2 \Omega_i^2}{k^2 V_{ms}^2 + \delta^2 \Omega_i^2}, \tag{4}$$

together with a low-frequency dust lower hybrid mode,

$$\omega^2 \simeq \omega_{dlh}^2 \frac{k^2 V_{ms}^2}{k^2 V_{ms}^2 + \delta^2 \Omega_i^2}. \tag{5}$$

Perpendicular dust drift may lead to a beam instability [2],

$$\omega \simeq kU + i\omega_{dlh}\left(\frac{k^2 U^2 - k^2 V_{ms}^2}{k^2 V_{ms}^2 + \delta^2 \Omega_i^2 - k^2 U^2}\right)^{1/2}, \tag{6}$$

which holds provided U is super-magnetosonic, but below a k-related upper limit,

$$0 < U^2 - V_{ms}^2 < \frac{\delta^2 \Omega_i^2}{k^2}. \tag{7}$$

We note that for given dust parameter values and $U > V_{ms}$, the upper limit in (7) imposes an upper limit on the wavenumber k for instability. The condition (7) implies further that, in the planetary rings, this instability cannot occur close to synchronous orbit, where the co-rotation speed is approximately equal to the Keplerian speed, and drift velocities are small [2].

Alternatively, seeking an approximate solution ω_0 of the form $\omega_0^2 = k^2 V_{ms}^2 + \delta^2 \Omega_i^2$, the dispersion relation yields [2]

$$\omega = \omega_0 - \frac{1}{2}\left(\omega_{dlh}^2 \frac{\delta^2 \Omega_i^2}{2\omega_0}\right)^{1/3} + i\frac{\sqrt{3}}{2}\left(\omega_{dlh}^2 \frac{\delta^2 \Omega_i^2}{2\omega_0}\right)^{1/3}. \tag{8}$$

We solve the full dispersion relation numerically, using the expressions in (1), and compare with numerical solutions of the reduced dispersion relation (2), as well as with

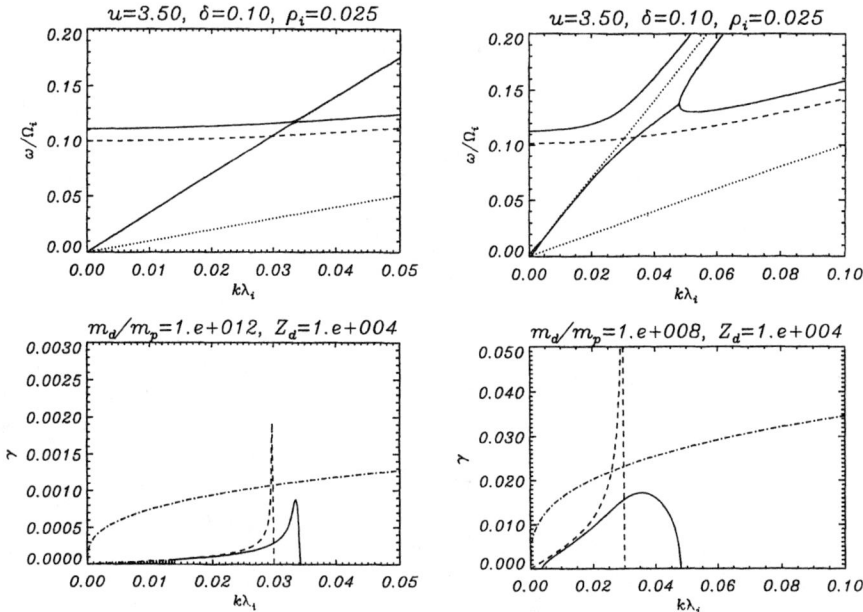

FIGURE 1. Typical dispersion and growth-rate curves for $U = 3.5V_{ms}$ and (a) $m_d = 10^{12}m_p$ and (b) $m_d = 10^8 m_p$. The solid lines and dashed lines are the solutions of equations (1) and (2), respectively, and the dot-dashed lines are from the literature [1]. Dotted lines represent the drift and magnetosonic speeds.

the results of Li and Havnes [1]. We use standard parameter values that demonstrate the mode behaviour: singly-ionized oxygen; cold dust with charge $Z_d = 10^4$ and mass $m_d = 10^8 m_p$ or $10^{12} m_p$; magnetic field $B_0 = 1\mu T$; ion density $n_i = 10^8$ m^{-3}; and dust density $n_d = 10^3$ m^{-3}, yielding a dust charge fraction $\delta = Z_d n_d / n_i = 0.1$. Lengths are normalized by the ion inertial length $\lambda_i = c/\omega_{pi}$.

The stationary case ($U = 0$), yields good agreement between the full and the reduced dispersion relations for the stable dust-modified magnetosonic mode (4), although a further upward shift in frequency is found numerically. In addition one observes a low frequency mode with $\omega \simeq \omega_{dlh}$.

In Figure 1(a) we see clear evidence of coupling between the zero-frequency beam mode ($\omega = kU$) and the magnetosonic mode (4), leading to instability. There is very localized modification of the dispersion curves of the two modes and the growth-rate curve is sharply peaked in a narrow region at the crossover point. Instability is, however, spread over a wider range in frequency and wavenumber. For low to intermediate values of k, equation (2) provides excellent agreement in growth-rate, and overall there is similarity in shape between the approximate and full dispersion relation curves. However, the latter yields an unstable range in k that is somewhat wider than predicted by (2). Equation (7) implies a very sharp cutoff at $k_c^2 = \delta^2 \Omega_i^2 / (U^2 - V_{ms}^2)$. As k approaches k_c, the approximations underpinning the reduced dispersion relation break down, yielding a shift in the most unstable wavenumber, k_{max}.

The effect of higher charge-to-mass ratio of the dust grains can be seen in Figure 1(b). From the low frequency regime we see that associated with the beam there are stable modes at $\omega = kU \pm \omega_{dlh}$. The main differences from Figure 1(a) are that dispersion is modified over a wide range in k, the growth-rate curve is much less sharply peaked, and the unstable region is wider in both ω and k, with the upper cutoff now being significantly greater than k_c.

Over the range $1 < U/V_{ms} < 10$, we find that for fixed dust mass the peak growth-rate (γ_{max}) is effectively independent of U (cf. (7) and (8)). Similarly, the ratios to k_c of both the upper cutoff wavenumber and k_{max} are independent of U.

From (7) one would expect a sharp cutoff in instability around $U \simeq V_{ms}$. Surprisingly, however, numerical solution of (1) yields instability for $U < V_{ms}$. Values of k and ω in a similar range to those that are unstable for $U > V_{ms}$, are weakly unstable. However, this instability, although arising from the beam, is not related to the magnetosonic mode. In particular, unstable wavenumbers and frequencies cover a wider range than is the case for $U > V_{ms}$, with the values of γ_{max}, k_{max}, and the most unstable frequency, all increasing dramatically as U drops below V_{ms}, underlining the fact that there is a change in the cause of instability. Indeed, as U is reduced further, the most unstable frequencies rise well above Ω_i and the unstable region shifts to much shorter wavelengths, e.g. for $U = 0.2V_{ms}$, we find $\gamma_{max} \simeq 6\Omega_i$ at $\omega \simeq 180\Omega_i$ and $k \simeq 850\lambda_i$.

It is noted that, even for $U > V_{ms}$, a similar rapid change in characteristics, and a shift of peak growth-rate to $\omega \gg \Omega_i$ is observed as the dust charge density δ is increased, particularly for $\delta > 0.4$. Significant reduction in the free electron density clearly inhibits support of the magnetosonic mode.

This model may be applied to planetary rings, as indicated earlier [1]. However, we note that for $U < V_{ms}$ the observed instability is not related to the dust-modified magnetosonic mode. Thus it appears that, for instance, for Saturn's A ring, with near-synchronous orbits and relative drift speeds well below the magnetosonic speed, any instability found must arise from some other drift-related mode.

ACKNOWLEDGMENTS

This work was supported by the Flemish Government and the South African NRF in the framework of the Flemish–South African Bilateral Scientific and Technological Cooperation, and the Universiteit Gent.

REFERENCES

1. Li, F., and Havnes, O., *Planet Space. Sc.*, **48**, 117–125 (2000).
2. Verheest, F., and Hellberg, M., *Trans. I.E.E.E. Plasma Science*, **29**, 283–287 (2001).

Study of Particle Formation and its Applications in Ar/CH$_4$ and Ar/C$_2$H$_2$ Mixtures

S. Hong, J. Berndt, J. Winter

Experimental Physics II, Ruhr-University Bochum, 44780 Bochum, Germany

Abstract. In this paper, we have investigated the spatial and temporal evolution of dust formation in Ar/CH$_4$ and Ar/C$_2$H$_2$ discharges. In the case of Ar/CH$_4$ discharges, transient high power is needed at the ignition time of the discharge to initiate the particle growth. In the Ar/C$_2$H$_2$ discharge however the particles are formed spontaneously at constant low power. Due to this different initiation process the temporal evolution of the dust formation is significantly different for both kind of discharges. The dust particles are detected by means of laser light scattering and by measuring the extinction of the laser after passing the discharge. The response of the plasma to the formation of dust has been analyzed by emission spectroscopy and mass spectroscopy. Also, we have demonstrated a possibility for using those particles to mofify properties of DLC films.

INTRODUCTION

Since Langmuir and his co-workers have discovered the existence of dust particles in glow discharges [1], the formation of dust can be observed in several plasma processes such as etching, sputtering or thin film deposition. In many cases and especially in the fabrication of electronic circuits dust particles are considered as a main source of contamination [2]. Sometimes however the incorporation of nanoparticles in growing films can be used for a controlled modification of film properties [3]. Besides these technical aspects dusty plasmas are of great interest in various branches of physics such as astrophysics or thermonuclear fusion research. The aim of this paper is to study the temporal and spatial evolution of the dust formation in a capacitively coupled standard GEC cell in Ar/C$_2$H$_2$ and Ar/CH$_4$ mixtures.

EXPERIMENTAL SETUP

Figure 1 shows a sketch of the experimental setup. A capacitively coupled standard GEC cell is used as reaction chamber. During plasma operation the chamber is pumped by rotary vane vacuum pump resulting in a pressure of about 10^{-1} mbar for typical flow rates of 8 sscm Ar and 2 sccm C$_2$H$_2$ and CH$_4$ respectively. Additionally, after each operation the chamber is cleaned using an O$_2$ discharge and is then pumped by turbomolecular pump for several hours down to a pressure of 10^{-7} mbar. The gases are introduced through the showerhead upper electrode. A 13.56 MHz RF generator is capacitively coupled to the upper electrode through an impedance matching network. The lower electrode is grounded. A mass spectrometer is employed to analyze residual gas

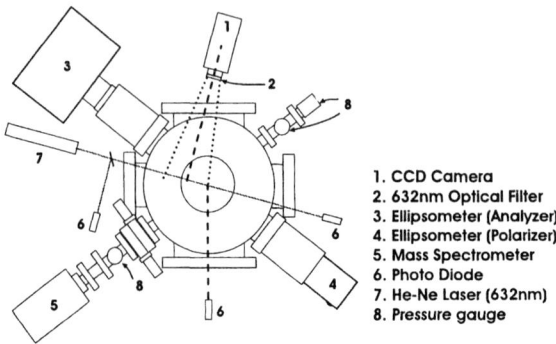

FIGURE 1. Experimental Setup

components during the process. The dust particles are detected by laser light scattering using a He-Ne laser. A ICCD camera, and two photo diodes are placed at the adequate position to measure the intensity of the scattered laser light and the extinction of the laser after passing the discharge.

EXPERIMENTAL RESULTS

In hydrocarbon plasmas, the formation and the presence of two kinds of particles have been reported [4]: white powder type and brown flake type. The white powder type is formed in the discharge whereas the brown flake type is formed from sputtering of the coating that is growing onto the rf electrode by energetic ion bombardment [4]. To initiate the white powder type particle formation in the Ar/CH_4 discharge, it is necessary - in our experiment - to apply transiently high power (ignition power>60W, 15-30 sec; maintenance power 20W) to the discharge. Without such transient high power input, no indication for the growth of dust particles was observed. In the Ar/C_2H_2 discharge however, the dust particles are formed spontaneously at a constant low power of 20W. Due to this different initiation process the temporal evolution of the dust formation is significantly different for both kinds of discharges. The CCD camera measurements show that 1.5 minutes after the ignition of the Ar/CH_4 discharge the space between the electrodes is nearly completely filled with dust particles. After a time of approximately 3.5 minutes the development of a dust free domain in the center region between the electrodes can be observed. In the case of the Ar/CH_4 discharge this domain is increasing in time. The situation for the Ar/C_2H_2 discharge is quite different. The volume of the dust free domain, which also develops in the center region between the electrodes is not continuously growing but is oscillating in time. An overview of the temporal development of the dust formation is given in Fig 2. Here the intensity of the laser after passing trough the discharge is plotted versus time. In the Ar/CH_4 plasma, we

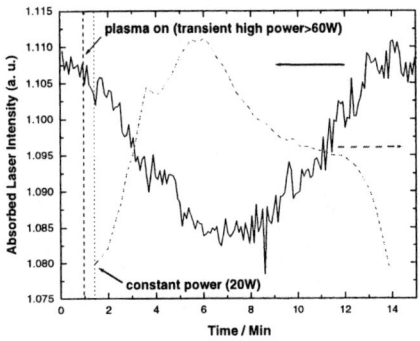
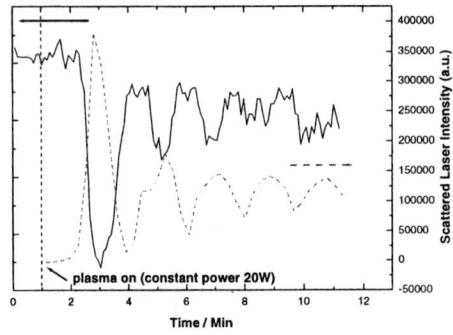

FIGURE 2. Ar/CH$_4$ discharge(left), Ar/C$_2$H$_2$(right) : intensity of laser after passing the discharge(solid), integrated spatial profile of scattered laser(dashed)

observed the maximum of the extinction approximately 5 minutes after the initiation of the plasma. After that maximum the extinction of the laser is continuously decreasing i.e. the intensity of the laser after passing the discharge is increasing until it reaches a constant value. In the case of the Ar/C$_2$H$_2$ discharge the extinction of the laser shows a quasi periodical behavior with a period of about 2 minutes.

It is well known that negatively charged particles in a plasma are submitted to various forces, such as gravitational force, ion drag force, neutral drag force, thermophoretic force and electric force [5]. The particles are confined in the bulk plasma as long as the electric force balances the sum of the others. As all these forces scale with different powers of the particle diameter the confinement strongly depends on the particle size. In the first phase of particle growth the particulates are small enough to be confined even in the center region between the electrodes. During the further growth process the other forces become more and more important. Consequently the particles are continuously pushed out of the discharge and a dust free domain develops. To re-initiate the particle formation in Ar/CH$_4$ plasma, we have to apply transiently high power again to the plasma. Consequently a region that is once dust free remains dust free as long as the power is kept constant at a level of about 20 Watt. In the case of the Ar/C$_2$H$_2$ discharge the particles are formed spontaneously without such a procedure. Therefore the process of nucleation and the further growth of particles can start again in the dust free regions. The result is a periodic behavior in the temporal evolution of the dust formation. It is well known that the formation of powder in a plasma leads to significant changes in the plasma parameters such as electron temperature, electron density and plasma composition [6]. Due to the different behavior of the dust formation in the Ar/CH$_4$ and the Ar/C$_2$H$_2$ discharge the response of the plasma to the powder formation exhibits also significant differences for both kind of discharges. With the onset of dust formation the electron temperature starts to increase [6]. Consequently the light emission from the bulk plasma is also increasing (Fig 3). Corresponding to the development of the dust

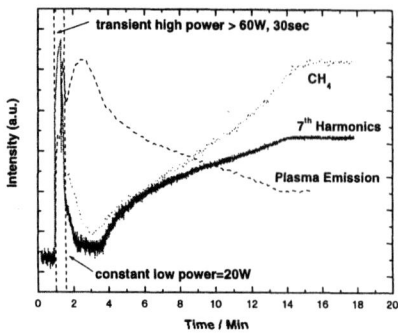 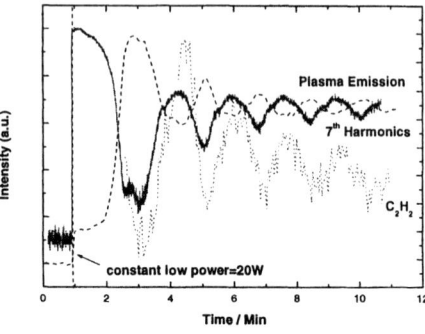

FIGURE 3. Ar/CH$_4$ discharge(left), Ar/C$_2$H$_2$(right) : 7th harmonic (solid), plasma emission (dashed), mass spectrum (dotted)

formation we observe one single intensity maximum in the case of the Ar/CH$_4$ discharge and an oscillating behavior of the emitted light in the Ar/C$_2$H$_2$ discharge. The same can observed for the mass spectrometer signal for mass 16 (CH$_4$) and mass 26(C$_2$H$_2$) respectively. The increase of the electron temperature leads to an increased dissociation and subsequent depletion of the hydrocarbon compounds due to consumption by particle growth. Hence the formation of dust is accompanied by a decrease of CH$_4$ and C$_2$H$_2$ respectively.

It is known that presence of dust particles in the discharge has a dramatic effect on the discharge impedance [7]. We have measured 3rd, the 4th, the 5th, and the 7th harmonics of voltage signal during the whole process. Especially, the 7th harmonic is very sensitve to the presence and growth process of dust particles. Using these facts, size of dust particles could be monitored and deposited onto silicon wafer successfully.

ACKNOWLEDGMENTS

This study was partially supported by MSWF NRW.

REFERENCES

1. Langmuir, I., Found, G., and Dittmer, A.F., *Science NewYork*, **60**, 392 (1924).
2. Smadi, M.M., Kong, G.Y., Carlile, R.N., and Beck, S.E, *J. Vac. Sci. Tech. B*, **10**, 30 (1992).
3. Tanenbaum, D.M., Laracuente, A.L., and Gallagher, A., *Appl. Phys. Lett.*, **68**, 1705 (1996).
4. Grenier, I.G., Guilbaud, V.M., and Plain, A., *Eur. Phys. J. AP.*, **8**, 53 (1999).
5. Andre Bouchoule, *Dusty Plasmas : Physics, Chemistry and Technological Impacts in Plasma Processing*, John Wiley & Sons, 1999.
6. Hollenstein, Ch., *Plasma Phys. Control. Fusion*, **42**, R93 (2000).
7. Bouchoule, A., and Boufendi, L., *Plasma Sources Sci. Technol.*, **2**, 204 (1993).

Electrostatic Discharging of Dust near the Surface of Mars

C.E. Krauss*, M. Horányi* and S. Robertson[†]

Laboratory for Atmospheric and Space Physics, University of Colorado, USA
[†]*Department of Physics, University of Colorado, USA*

Abstract. Due to the prevalence of Martian dust devils and dust storms, an understanding of the physics of electrical discharges in Martian dust is critical to future Mars exploratory missions. Measurements done in our lab on the charging of single dust grains show that particles of JSC-Mars-1, a Martian regolith simulant, can have large electrical potentials due to triboelectric charging. We have conducted experiments to determine the range of pressures and wind speeds over which discharges can be observed in a low-pressure CO_2 atmosphere. We have also examined the affects of particle-size distribution on the observed discharge rates.

INTRODUCTION

When dust particles come into contact, charge can be transferred between the grains. This effect is referred to as triboelectric charging. Wind driven dust studies show that in the case of particles with identical compositions, the particle with a larger radius in a collision preferentially becomes positively charged [1]. The stratification of particle sizes generated by upwinds within a dust cloud causes an electric field to form. When the electric potential within the cloud exceeds the breakdown voltage of the surrounding atmosphere, the charge is released in a discharge similar to lightning.

Triboelectric charging of dust particles and the resultant electrical discharges have been observed in several terrestrial phenomenon, including volcanic plumes and dust devils. Due to Mars' low atmospheric presssure (4.5 - 6 Torr [2]), arid climate, and frequently windy environment, it has been suggested that dust near the surface of Mars is even more susceptible to triboelectric charging and subsequent electrical discharges than dust on Earth.

The expected susceptibility of Martian dust to triboelectric charging is of particular interest in light of images taken by the Mars Global Surveyor's orbital camera (MGS MOC) which show dust devil tracks along the surface of Mars [3]. Martian dust devils can be nearly 6 km in height and 100's of meters in width [4]. While these features are orders of magnitude larger than their terrestrial counterparts, they are still much smaller than the global dust storms which can cover large portions of the planet and last for several months. The large comparative size of these phenomenon suggests that electrical discharges due to triboelectric dust charging on Mars could be numerous and observable.

The two experiments described here are designed to study the conditions which enhance and inhibit electrical discharges on Mars. The experimental results have both scientific and engineering applications since the discharges may affect optical and electrical systems of equipment, interfer with radio communications, and affect the safety of future human explorers on the Martian surface.

EXPERIMENT 1

In order to simulate the Martian environment, the experiments use JSC-Mars-1, a Martian regolith simulant. Removed from the southern flank of Mauna Kea, JSC-Mars-1 is composed of weathered volcanic ash particles < 1 mm in diameter which contain 43.5% SiO_2. It is believed to approximate the grain size, density, porosity, reflectance spectrum, mineralogy, chemical composition, and magnetic properties of the soil of Mars [5]. Work done in our lab has shown that JSC-Mars-1 particles can have extremely large charging potentials, up to ± 15 V [6].

Given these large charging potentials and a CO_2 excitation energy of only 10.0 eV (with ionization at 14.4 eV) [7], it follows that electrical discharges are expected to occur in JSC-Mars-1 dust. To examine this phenomenon, a 4.7 L polycarbonate vacuum jar with a radius of 8.5 cm is evacuated to ≈ 0.2 Torr and then filled with CO_2 to attain the desired pressure. Approximately 700 mL of JSC-Mars-1 is placed in the bottom of the chamber to form a layer several centimeters in depth. To simulate the windy conditions inside a dust devil or dust storm, a motor-driven nonconductive stirring rod is used. Both the pressure and the stirring rate can be varied to simulate differing climatic conditions. Figure 1 shows a schematic diagram of the experimental setup.

When taking data, the entire device is enclosed in a dark container to prevent contamination from outside light. Inside the container there is a photomultiplier tube connected to a computer which determines the number of discharges observed over a given time period. A wire probe is placed in the chamber and connected directly to an oscilloscope. Signals are seen from the probe coincident with signals from the photomultiplier tube indicating that the discharges are associated with rapidly changing electrical potentials.

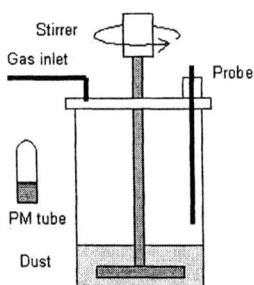
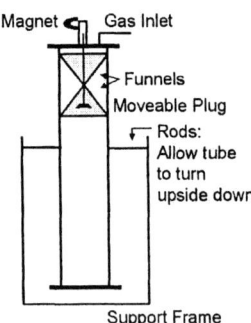

FIGURE 1. (left) Schematic of experiment 1. The entire apparatus is enclosed in a dark chamber while measurements are taken. (right) Schematic of experiment 2. The glass tube flips upside down to reset the experiment.

The frequency and intensity of the discharges have been examined over a range of pressures from 1 - 8 Torr and a range of simulated wind speeds between 1.5 - 5.2 m/sec.

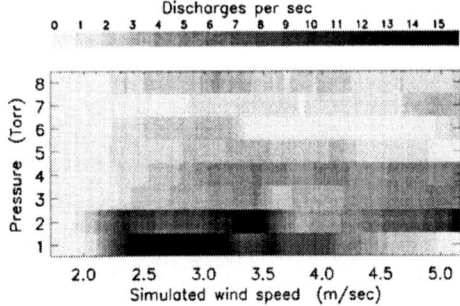
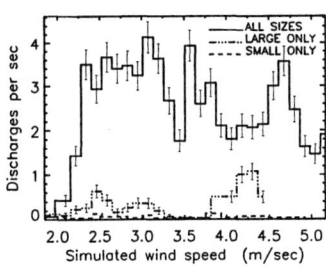

FIGURE 2. (left) Discharge rates are plotted in grey-scale as a function of pressure and simulated wind speed. The key at the top denotes which colors correspond to which discharge rates. (right) Discharge rate vs. wind speed at a pressure of 8 Torr. The results for three particle size distributions are shown. A large particle size distribution is required for significant discharges to occur.

Under extremely dark viewing conditions, discharges are visible to an observer with dark-adapted eyes.

The actual discharge rates are dependent on the amount of regolith stirred, but the trends of local maxima and minima within the discharge rates are independent of the dust loading. The discharge rate is negligible until the simulated wind speed reaches ≈ 2.1 m/sec when rates begin to increase rapidly (Fig 2). The discharge rate is greater near the lower thresholds of Martian atmospheric pressure and decreases as the pressure approaches ≈ 6 Torr.

Discharge rates were determined for three particle-size distributions. Distribution 1 is that which occurs naturally in JSC-Mars-1: 51% 250-1000 μm, 24% 150-249 μm, and 25% < 150 μm. Distribution 2 consists of particles > 355 μm, and distribution 3 consists of particles < 120 μm. Fig. 2 shows that a mixture of small and large particles is required in order to produce a significant number of discharges.

EXPERIMENT 2

The second experiment is designed to show that electrical discharges can be created due to vertical charge separation under Mars-like environmental conditions. A 1.2 meter glass tube with a radius of 17 cm is evacuated to ≈ 20 mTorr and then filled with CO_2 to attain the desired pressure. Two funnels have been placed in the tube to create an hourglass-like set-up. A magnet at the top of the experiment controls whether the movable plug is open or closed. A schematic diagram of the experimental setup is shown in Figure 1.

A mixture of 100 μm JSC-Mars-1 particles and 53 μm glass microballs is placed in the upper funnel with the plugged closed. Due to their placement on the triboelectric series, the glass microballs tend to charge positively compared to the JSC-Mars-1. When the magnet is removed, the plug opens and the dust drops down the tube. Because the glass microballs have smaller cross-sections, they are able to stay aloft a short while longer than the JSC-Mars-1, aiding in the creation of an electric potential which can breakdown the local atmosphere. The entire device is suspended vertically on a frame which allows it to be turned upside down. This recharges the experiment by moving the dust back into the upper funnel.

A high-voltage probe is attached to an oscilloscope which records the voltages observed during each dust drop.

This experiment is still in the development stage, but preliminary results show that discharges can be observed both visually and electronically when the dust is dropped over vertical distances, allowing for charge separation.

DISCUSSION AND FUTURE WORK

The experiments demonstrate that JSC-Mars-1 grains can be sufficiently charged via triboelectric charging to breakdown the local CO_2 atmosphere. The resultant discharges, detected both visually and electronically, have been observed when the simulant is stirred in low-pressure CO_2 at speeds comparable to those expected of Martian winds. Additionally, these discharges require a large range of particle sizes to occur and can be produced due to charge separtion over vertical distances. This supports the high possibility of electrical discharges occurring on Mars due to dust agitation.

Future work must be done to determine the effects of factors such as atmospheric composition, UV radiation, humidity, and temperature on the observed discharge rates. This work will aid in the development and design of a prototype remote-sensing instrument which can detect discharges on or near the surface of Mars.

ACKNOWLEDGMENTS

The authors acknowledge support from the NASA Space Science Graduate Student Research Program (NGT5-50345) and thank Matt Triplett and Zoltan Sternovsky for their assistance in building the apparatuses.

REFERENCES

1. Stow, C. D., *Weather*, **24**, 134–140 (1969).
2. Kieffer, H. H., Jakosky, B. M., Snyder, C. W., and Matthews, M. S., editors, *Mars*, Tucson: Univ. of AZ Press, 1992.
3. Edgett, K. S., and Malin, M. C., *31st Lunar and Planetary Science Conference* (2000).
4. Thomas, P., and Gierasch, P. J., *Science*, **230**, 175–177 (1985).
5. Allen, C. C., Jager, K. M., Morris, R. V., Lindstrom, D. J., Lindstrom, M. M., and Lockwood, J. P., *EOS Trans. AGU*, **79**, 405 (1998).
6. Sickafoose, A. A., Colwell, J. E., Horányi, M., and Robertson, S., *J. Geophys. Res.*, **106**, 8343–8356 (2001).
7. McDaniel, E. W., *Collision Phenomena in Ionized Gases*, New York: John Wiley & Sons Inc., 1964.

Validity of Epicyclic Description of Saturnian Dust Grain Orbits

C. J. Mitchell*, M. Horányi* and J. E. Howard*

LASP, University of Colorado, Boulder, CO 80309, USA

Abstract. We investigate the validity of the epicyclic approximation to the motion of charged dust grains about Saturn. Non-ideal forces, such as planetary oblateness, solar radiation pressure, plasma drag, and the effects of time-dependent charging are neglected. A new and simplified derivation of the first-order epicyclic expansion is presented. We calculate the mean motion and the aspect ratio of the epicyclic ellipse for a wide range of particle sizes of either charge sign. Grains are started both inside and outside of synchronous radius on both prograde and retrograde orbits with initial Kepler velocity. For gravitationally dominated grains the mean motion is given by the classical expression but for magnetically dominated grains the mean motion differs markedly from that of the equilibrium orbit if the orbit is looped.

INTRODUCTION

We investigate the metamorphosis of equatorial charged dust grain orbits about Saturn with varying parameters such as grain radius and initial orbit, extending earlier work [1, 2, 3, 4, 5, 6]. Our primary goal is to determine conditions for the validity of an epicyclic model. Both prograde and retrograde orbits, of either charge sign, inside and outside of the synchronous radius are considered. While the lowest order epicyclic description is generally valid for nearly circular orbits, for highly magnetized grains the mean motion can deviate significantly from the frequency of the circular equilibrium orbit. In such cases, the orbits are highly looped and the relevant mean motion is the drift velocity familiar from guiding center theory [3].

HAMILTONIAN MODEL

The equations of motion of a dust particle of charge q and unit mass in a planetary magnetosphere may be derived from the Hamiltonian and effective potential:

$$H = \frac{1}{2}p_r^2 + U^e(r), \quad U^e(r) = \frac{1}{2r^2}(p_\phi - \frac{\omega_{C0}}{r})^2 + \frac{\Omega\omega_{C0} - \omega_k^2}{r}, \quad (1)$$

where $p_r = \dot{r}$, $\omega_{C0} = qB_0/mc$ is the cyclotron frequency, with B_0 the magnetic field on the planetary equator, $\omega_k = \sqrt{\mu/R^3}$ is the Kepler frequency, μ is the planetary mass times the gravitational constant, R is the planetary radius, Ω is the planetary rotation rate, and $p_\phi = r^2\dot\phi + \omega_{C0}/r$ is the conserved azimuthal canonical momentum. Here distances are measured in units of R, frequencies in units of ω_k, and mass in units of m. Thus, the radial motion is equivalent to a one-dimensional nonlinear oscillator described by U^e. Figure 1 shows typical trajectories in the inertial frame for a gravitationally dominated grain and one which is magnetically dominated and highly looped.

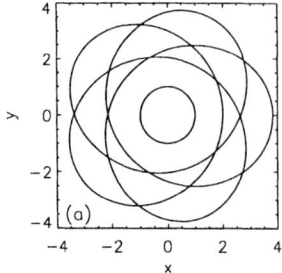 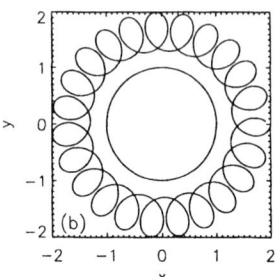

FIGURE 1. (a) Gravitationally dominated ($r_g = 169$ nm, retrograde, $q > 0$) and (b) magnetically dominated ($r_g = 30$ nm, retrograde, $q < 0$) particles observed in an inertial reference frame.

EPICYCLIC EXPANSION

We suppose the existence of a stable circular orbit for a given value of p_ϕ at radius r_0. The radial equilibrium condition ($U_r^e = 0$), together with the definition of p_ϕ then yields a quadratic for the orbital frequency $\omega_0 = \omega(r_0)$,

$$r_0^3 \omega_0^2 - \omega_{C0} \omega_0 - (\omega_k^2 - \omega_{C0} \Omega) = 0 \ . \tag{2}$$

A circular reference orbit (the "deferent") is Lyapunov stable if U_{rr}^e and U_{zz}^e are both positive, which yield the radial and transverse epicyclic (libration) frequencies. Here we shall assume transverse stability and restrict attention to in-plane motions. The linearized radial motion is harmonic at the epicyclic frequency κ, given approximately by

$$\kappa_0^2 = U_{rr}^e = \omega_0^2 - 4\omega_c \omega_0 + \omega_c^2 \ , \tag{3}$$

where $\omega_c = \omega_{C0}/r_0^3$ is the local cyclotron frequency. It follows that $r \approx r_0 + a\cos(\kappa t + \beta)$, where the amplitude a and phase β are determined by the initial conditions. The azimuthal motion is then given by the conservation of p_ϕ:

$$\phi = \int \left(\frac{p_\phi}{r^2} - \frac{\omega_{C0}}{r^3} \right) dt \approx \phi_0 + nt - \frac{a}{\kappa_0 r_0^3} \left(2p_\phi - \frac{3\omega_{C0}}{r_0} \right) \sin(\kappa t + \beta) \ . \tag{4}$$

Therefore, as long as a is small, the first order motion about the deferent in the frame rotating with the mean motion (n) of the grain is an ellipse, with axis ratio

$$b/a = (\omega_c - 2\omega_0)/\kappa_0 \ , \tag{5}$$

where b is measured along the axis perpendicular to a. When a and b are not small, higher order expansions in powers of a become necessary. In these calculations the frequency κ must also be so expanded. The mean motion is then

$$n = <\dot{\phi}> = \frac{1}{T}\int_0^T \dot{\phi}\,dt = \frac{1}{T}\int_0^T \left(\frac{p_\phi}{r^2} - \frac{\omega_{c0}}{r^3}\right)dt\,,\ T = \sqrt{2}\int_{r_{min}}^{r_{max}} \frac{dr}{\sqrt{E-U^e}}\,. \quad (6)$$

The analytical evaluation of these integrals is difficult and not very illuminating. We therefore perform numerical integrations to compare with the results of the simulation.

NUMERICAL RESULTS

In each of the eight cases examined, as the grain radius is decreased from large to small, the orbits in the mean motion frame initially expand (from a 2:1 ellipse which was distorted as the orbit grew), either reaching a critical size or crashing into or escaping from the planet, and then shrinking back down, eventually becoming a small circle. Larger particles orbit the planet close to the Kepler frequency, while smaller ones do so at approximately the planet's rotational frequency.

For gravitationally dominated particles, first order epicyclic theory is able to accurately predict the mean motion, regardless of the initial conditions and charge sign. Small deviations occur for particles with large values of a, but are never greater than a few percent. However, if a is large, then the epicycle is curved around the planet and far from elliptical. The validity for highly magnetized grains depends on the shape of the orbit. Those grains which have a monotonically increasing ϕ are accurately predicted by the epicyclic model. If, on the other hand, the particle's orbit is looped in the inertial reference frame so that $\dot{\phi}$ changes sign, then the theory fails for both the mean motion and the axis ratio. Examining the effective potentials for these grains reveals large asymmetries which invalidate the lowest order motion.

Our results may be summarized as follows:

Positive Prograde: Particles outside of synchronous radius, r_s were repelled form the planet and often had large radial excursions, while those inside r_s were attracted to the planet. The magnetically dominated grains inside showed a transition from looped to not-looped orbits as the grain size decreased.

Negative Prograde: These orbits always agreed well with theory. They are attracted to r_s and are the only case to have a smooth transition from gravitationally to magnetically dominated grains without collisions with the planet or ejections.

Positive Retrograde: These orbits are always repelled from the planet, so gravitationally dominated particles can have large radial excursions. The magnetically dominated particles are always looped.

Negative Retrograde: These orbits are always attracted to the planet, so only slight deviations from ellipticity are observed. Again, the magnetically dominated particles are always looped.

Figure 2 shows a set of epicycles and corresponding effective potentials for magnetically dominated, negatively charged particles started on retrograde orbits at $2R$. The orbits of these grains are highly looped in the inertial frame and, as can be seen from Fig. 3, epicyclic predictions for them fail.

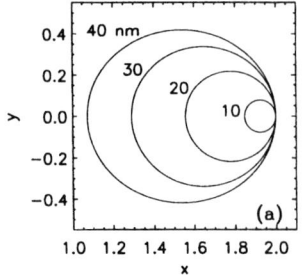 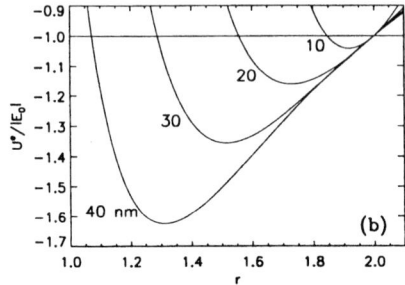

FIGURE 2. (a) Epicycles and (b) effective potentials for negatively charged particles started on retrograde orbits with the Kepler velocity at $2R$ for several small grains. These orbits are highly looped in the inertial frame and have very asymmetric potentials.

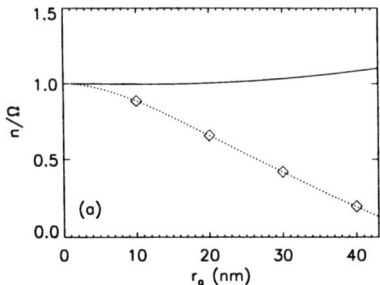 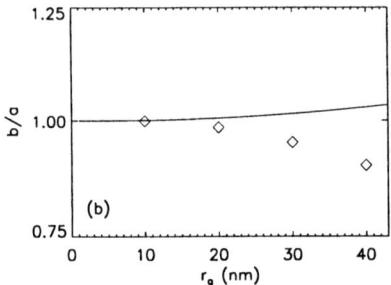

FIGURE 3. (a) Mean motion and (b) aspect ratio for the same grains as in Fig. 2. The solid line is the prediction of the lowest order epicyclic expansion, while the dotted line is from numerical integration of eqns. (7,8).

To summarize, lowest order epicyclic theory is sufficient to predict the mean motions of all gravitationally dominated particles around Saturn, even those with large radial excursions, although the epicycles are not ellipses for these orbits. The theory also works for magnetically dominated particles for which $\dot{\phi}$ is monotonic, but fails for particles executing looped orbits.

REFERENCES

1. Mendis, D. A., Hill, J. R., and Houpis, H. L. F., *J. Geophys. Res.*, **88**, A929–A942 (1983).
2. Schaffer, L., and Burns, J. A., *J. Geophys. Res.*, **99**, 17211–17223 (1994).
3. Northrop, T. G., and Hill, J. R., *J. Geophys. Res.*, **88**, 6102–6108 (1983).
4. Xu, R. L., and Houpis, H. L. F., *J. Geophys. Res.*, **90**, 1375–1384 (1985).
5. Howard, J. E., Horányi, M., and Stewart, G. R., *Phys. Rev. Lett.*, **83**, 3993–3996 (1999).
6. Horányi, M., *Annu. Rev. Astron. Astrophys.*, **34**, 383–418 (1996).

Brownian motion of absorbing dust grains

A. M. Ignatov*, S. A. Trigger[†], S.A. Maiorov* and W. Ebeling[†]

*General Physics Institute, Moscow, Russia
[†]Humboldt University, Berlin, Germany

Abstract. We study the rotational and translational kinetics of massive particulates (dust grains) absorbing the ambient gas. Equations for microscopic phase densities are deduced resulting in the Fokker-Planck equation for the dust component. It is shown that although there is no stationary distribution, the translational and rotational temperatures of dust tend to certain values, which differ from the temperature of the ambient gas.

INTRODUCTION

The average kinetic energy of the dust grains may be considerably higher than the temperature of the ambient plasma. Recent studies demonstrated that such anomalous behaviour of Brownian particles is due to electrostatic plasma oscillations in the sheath area [1, 2]. It was also shown that the dynamical charging of grains can result in inequality of the dust temperature and the temperatures of the light components even for the case of a steady plasma [3, 4]. The latter result is in definite contradiction with traditional approach to the Brownian motion, which is based on the assumption of equal temperatures.

In the present work we consider a simplified model of Brownian particles absorbing the ambient gas [5]. The grain charging as long as electrostatic interactions are ignored. On the other hand, we take into account the mass growth of the grains and their rotational motion. The developed analytical theory and computer simulations demonstrate that inelastic atom-grain collisions result in inequality of the translational and rotational dust temperatures and the temperature of the ambient gas.

ANALYTIC THEORY

Elementary collision. It is assumed that the process of collision elapses in two stages. First, each atom hitting the grain is attached to its surface transferring, therefore, some energy and momentum to the grain. A part of the atom's energy is spent for heating the grain surface. Second, some inner forces redistribute the grain mass in such a way that it shapes into a sphere. The state of a grain is described by the ten-dimensional vector, $\Gamma = (\mathbf{R}, \mathbf{P}, \mathbf{G}, M)$, where $\mathbf{P}$ is the grain's linear momentum, $\mathbf{G}$ is the angular momentum relative its center of inertia, $\mathbf{R}$, and M is the grain mass. Analysis of the conservation laws shows that the state vector after the collision is

$$\Gamma' = (\frac{m\mathbf{r}+M\mathbf{R}}{M+m}, \mathbf{P}+\mathbf{p}, \mathbf{G}+\frac{(\mathbf{r}-\mathbf{R}) \times (M\mathbf{p}-m\mathbf{P})}{M+m}, M+m), \qquad (1)$$

where $\mathbf{r}$ and $\mathbf{p}$ are the coordinate and the momentum of the projectile atom at the instant of collision, and m is the atom mass. It assumed that the grain radius, $a(M)$, depends on its mass, and $|\mathbf{r}-\mathbf{R}| = a(M)$ in Eq. (1). The mapping (1) is non-conservative, $\|\partial\Gamma'/\partial\Gamma\| < 1$. It should be pointed out that the leap of the center of inertia in Eq. (1) is a necessary consequence of the momentum conservation.

Fokker-Planck equation. Using the elementary collision (1) we derive the equations for the microscopic phase densities generalizing the Klimontovich approach [6]. Then, performing a lot of hard work that includes averaging and expanding in powers of the small mass ratio, $m/M \ll 1$, we finally arrive at the Fokker-Planck equation for the dust distribution function, $N_d(\Gamma)$:

$$\frac{dN_d(\Gamma)}{dt} = \frac{\partial}{\partial P_i}\left\{-s_i N_d(\Gamma) + \kappa_{ij}\frac{\partial N_d(\Gamma)}{\partial P_j}\right\} + \frac{\partial}{\partial G_i}\left(\eta_{ij}\frac{\partial N_d(\Gamma)}{\partial G_j}\right) + \frac{\partial \gamma_i N_d(\Gamma)}{\partial R_i} - \frac{\partial J N_d(\Gamma)}{\partial M}. \qquad (2)$$

The kinetic coefficients introduced here are: $J(\Gamma) = \pi a^2(M)m\langle v\rangle$, $s_i(\Gamma) = \pi a^2(M)\langle vp_i\rangle$, $\kappa_{ij}(\Gamma) = \frac{1}{2}\pi a^2(M)\langle vp_i p_j\rangle$, $\gamma_i(\Gamma) = \frac{2m}{3M}\pi a^3(M)\langle v_i\rangle$, and $\eta_{ij}(\Gamma) = \frac{m^2}{8}\pi a^4(M)\langle v(\delta_{ij}v^2 - v_i v_j)\rangle$, where $\mathbf{v} = \frac{\mathbf{p}}{m} - \frac{\mathbf{P}}{M}$ and the short-hand notation for the averaging over the ambient gas distribution function, $N_n(\mathbf{p},\mathbf{r})$, is used, $\langle K\rangle \equiv \int d\mathbf{p}\, K N_n(\mathbf{p},\mathbf{R})$.

The physical meaning of the most of the kinetic coefficients and corresponding terms in Eq. (2) is fairly obvious. J is the mass flow at the grain surface, the last term in Eq. (2) provides the mass growth of the dust component. The coefficient s_i is the drag force acting upon a grain, the quantities κ_{ij} and η_{ij} characterizes the diffusion in the momentum space. Since the angular momentum transferred to the grain is independent of its angular velocity, there is no friction torque within the accepted model. Of interest is the term proportional to γ_i in Eq. (2). It originates from the change in the center of inertia in Eq. (1) and describes grain migration due to asymmetric bombardment by gas atoms.

Equilibrium temperature. The Fokker-Planck equation (2) may be explicitly solved. The general solution asymptotically tends to the Maxwellian distribution over $\mathbf{P}$ and $\mathbf{G}$ with growing mass and with certain values of translational and rotational temperatures. Assuming the gas distribution function is isotropic and there is no dispersion over the grain mass, both temperatures are expressed in terms of normalized energy flux at the grain surface, $\alpha = \frac{a^2(M)}{6mJ}\langle p^3\rangle$; for the case of the Maxwellian distribution $\alpha = \frac{2}{3}T_0$, where T_0 is the gas temperature. The translational temperature tends to $T_t \to \frac{6}{5}\alpha = \frac{4}{5}T_0$. The rotational temperature is sensitive to the dependence of the grain moment of inertia on its mass. In particular, in the case of an an arbitrary spherically symmetric mass distribution, e.g. a solid grain, $T_r \to \frac{3}{2}\alpha = T_0$. If the grain radius is independent of its mass ("spongy grain"), then $T_r \to \frac{5}{2}\alpha = \frac{5}{3}T_0$. Of interest is that ignoring the mass growth results in $T_t = 2T_0$ [3, 4].

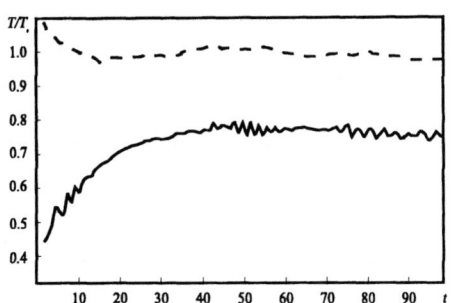
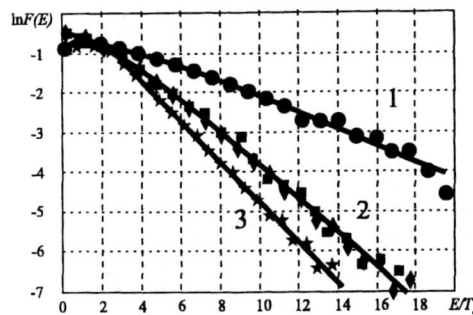

FIGURE 1. Left: time evolution of translational (solid line) and rotational (dashed line) temperatures for the absorbing grain with growing mass. Right: energy distributions for various kinds of collisions. The distributions over kinetic energy are plotted for the complete momentum accomodation in neglecting the mass growth (●), for the specular reflections (■) and for the absorbing grain with growing mass (★). The distribution over rotational energy is shown for the absorbing grain (♦). The solid curves correspond to Maxwellian distributions with $T/T_0 = 2$ (1), $T/T_0 = 1$ (2) and $T/T_0 = 4/5$ (3).

NUMERIC SIMULATIONS

The simulations of the Brownian kinetics of a single grain were performed in a following way. The computational area was a three-dimensional cube of unit length on edge in contact with the unbounded equilibrium gas, which was simulated by randomly injected point atoms with semi-Maxwellian distribution. Since there were no forces acting upon atoms, their trajectories were straight lines. The average number of atoms depended on the thermal velocity; in most runs it fluctuated around 10000. We also checked that our results were insensitive to the actual value of the gas temperature.

The grain was represented by a movable sphere of a radius small compared to the cube edge; typically, its initial size was $a_0 = 0.01$ and its initial mass was $M_0 = 1000$. If an atom hit the grain surface then it transferred a part of its momenta to the grain. Various laws of atom-grain interaction were used: (i) specular reflections, (ii) an absorbing grain with growing mass, $a(M) \propto M^{1/3}$, (iii) complete energy and momentum accomodation without mass growth [3, 4]. In the case of the absorbing grain, no new atom was injected into the cube after the collision that resulted in some reduction of density. Since the grain was initially small, a very little part of atoms might have experienced the collisions. The equations of motion for both translational and rotational degrees of freedom of the grain were solved; the time step was small compared to the average time between the collisions.

To minimize the influence of the walls the grain was confined near the center of the cube with the help of the auxiliary spherically-symmetric parabolic potential well. Evidently, such confinement resulted in multiplication of the grain distribution by a Boltzmannian factor and could not alter the distribution over the kinetic energy.

The main goal of our simulations was to accumulate enough data in order to reconstruct the grain distribution function over its kinetic and rotational energy. The translational or rotational energy axis, say, $0 < E < 20 T_0$, was split in a number of sub-bands (usually, 50). Two methods of averaging were used. First, we could trace the energy variation of the single grain and evaluate time it spent in each energy sub-band. Then, these time intervals were summed up resulting in time-averaged distribution function. This method results in meaningful data only if the grain mass is fixed. Otherwise, no statistically significant result can be obtained with the time-averaging method: the grain motion slows down as its mass grows, and it takes more and more time for the grain to migrate from one energy sub-band to another.

Another method of averaging was the simulation of the canonical Gibbs ensemble. Initially, the grain was situated at the center of the cube, its rotational and kinetic energies were chosen randomly using the random-number generators of Maxwellian distributions with corresponding initial kinetic, T_{t0}, and rotational, T_{r0}, temperatures. The evolution of the grain energies was recorded for the time period $0 < t < t_{max}$. The obtained dependence represented a single sample from the Gibbs ensemble. The whole procedure was repeated many ($\approx 10^4$) times with varied random initial energies but the same initial temperatures. By counting down a number of samples in each energy sub-band for a given time instant we were able to reconstruct the time evolution of the energy distribution function. The different results obtained by different methods of averaging clearly indicate that the system is nonergodic.

The results of our simulations are summarized in Fig. 1. The energy distributions obtained for various laws of collisions are in fairly good correspondence with analytical theory.

ACKNOWLEDGMENTS

This work was supported by NWO (project 02-02-16439) and RFBR (project 047-008-013).

REFERENCES

1. R. A. Quinn and J. Goree, Phys. Plasmas **7**, 3904 (2000); Phys. Rev. E **61**, 3033 (2000).
2. G. Joyce, M. Lampe, and G. Ganguli, Phys. Rev. Lett., **88**, 095006 (2002)
3. A. G. Zagorodny, P. P. J. M. Schram, and S. A. Trigger, Phys. Rev. Lett. **84**, 3594 (2000).
4. P. P. J. M. Schram, A. G. Sitenko, S. A. Trigger, and A. G. Zagorodny, Phys. Rev. E **63**, 016403 (2001).
5. A. M. Ignatov, S. A. Trigger, S. A. Maiorov, and W. Ebeling, Phys. Rev. E **65**, 046413 (2002)
6. Yu. L. Klimontovich, *Statistical physics* (Harwood Academic Publishers, New York, 1986).

Influence of dust-ion collisions on waves in self-gravitating dusty plasmas

Gerald Jacobs[*], Victoria V. Yaroshenko[†] and Frank Verheest[*]

[*]Sterrenkundig Observatorium, Universiteit Gent, Krijgslaan 281, B-9000, Gent, Belgium
[†]Max-Planck Institute for Extraterrestrial Physics, Postfach 1312, D-85741 Garching, Germany

Abstract. In self-gravitating dusty plasmas, the dust particles can interact through electric, self-gravitational and collisional forces. If neutral dust is absent, the main collisional forces will be those between the ions and the dust grains. When only the dust-ion collisions are retained in the description, rootlocus theory can be used to easily determine the stability of self-gravitating plasmas. Moreover, rootlocus theory enables a visualization of the evolution of the real frequencies and damping decrements of the dust-acoustic and ion-acoustic modes. The development of these important frequencies are mapped for increasing dust-ion frequencies.

GENERAL FORMALISM

We consider a plasma consisting of electrons, ions and charged dust grains. As the presence of neutral particles is excluded, the major collisional effects occur between ions and charged dust grains. The wave period is assumed considerably different from the dust charge fluctuation time and as a result the dust charges can be treated as constant. The electrons are assumed to be Boltzmann distributed $n_e = n_{e0} \exp(e\psi_E/k_B T_e)$, and our basic equations further include the continuity equations

$$\frac{\partial n_i}{\partial t} + \frac{\partial}{\partial x}(n_i v_i) = 0, \qquad \frac{\partial n_d}{\partial t} + \frac{\partial}{\partial x}(n_d v_d) = 0, \qquad (1)$$

and the equations of motion for the ions and dust particles,

$$\frac{\partial v_i}{\partial t} + v_i \frac{\partial v_i}{\partial x} + \frac{q_i}{m_i} \frac{\partial \psi_E}{\partial x} + \frac{\partial \psi_G}{\partial x} + \frac{v_{Ti}^2}{n_i} \frac{\partial n_i}{\partial x} + v_{id}(v_i - v_d) = 0, \qquad (2)$$

$$\frac{\partial v_d}{\partial t} + v_d \frac{\partial v_d}{\partial x} + \frac{q_d}{m_d} \frac{\partial \psi_E}{\partial x} + \frac{\partial \psi_G}{\partial x} + \frac{v_{Td}^2}{n_d} \frac{\partial n_d}{\partial x} + v_{di}(v_d - v_i) = 0, \qquad (3)$$

where $v_{di} = m_i n_{i0} v_{id}/m_d n_{d0}$, with v_{id} and v_{di} the collision frequencies and the other notations being standard. The description is closed using the Poisson equations

$$\frac{\partial^2 \psi_E}{\partial x^2} = \frac{1}{\varepsilon_0}(n_e e - n_i q_i - n_d q_d), \qquad \frac{\partial^2 \psi_G}{\partial x^2} = 4\pi G(m_i n_i + m_d n_d), \qquad (4)$$

from which the electrostatic (ψ_E) and gravitational (ψ_G) potentials can be found.

CP649, *Dusty Plasmas in the New Millennium: Third International Conference on the Physics of Dusty Plasmas*, edited by R. Bharuthram et al.
© 2002 American Institute of Physics 0-7354-0106-3/02/$19.00

DISPERSION RELATION

Assuming the perturbations to be proportional to $\exp[-i\omega t + ikx]$, we linearize all the previous equations, simplify the full dispersion relation and obtain

$$\left[\omega^2 - k^2 v_{Td}^2 + \omega_{Jd}^2 - A\omega_{pd}^2\right]\left[\omega(\omega + iv_{id}) - k^2 v_{Ti}^2 - A\omega_{pi}^2\right] = A^2 \omega_{pi}^2 \omega_{pd}^2, \quad (5)$$

where $\omega_{Ji} \ll \omega_{Jd}$ has been used, $\omega_{J\alpha}^2 = 4\pi G n_\alpha m_\alpha$ represents the Jeans frequency per species and the abbreviated notation $A = \left(1 + 1/k^2 \lambda_{De}^2\right)^{-1}$ has been introduced. In the absence of dust-ion collisions and for non-streaming cold dust, equation (5) equals equation (27) of Meuris *et al.* [1]. On the other hand, for dusty plasmas devoid of self-gravitational effects, with $k^2 \lambda_{De}^2 \ll 1$ and $Z_i = 1$, $q_d = -eZ_d$, equation (5) matches equation (21) of D'Angelo [2], if there $\tau_L = \infty$ and $\alpha = 0$ is set. If we substitute $\omega = i\Omega$ in the dispersion relation (5), we obtain a quartic equation with real coefficients and further on all equations are written in the new variable Ω. The use of this new frequency will prove to be advantageous for the stability analysis.

The collisionless dispersion relation has real roots r_{ia} and r_{da} in Ω^2 and these can be approximated as

$$r_{ia} \simeq -A\omega_{pi}^2 - k^2 v_{Ti}^2, \qquad r_{da} \simeq \omega_{Jd}^2 - k^2 c_{da}^2, \quad (6)$$

making clear that r_{ia} and r_{da} represent the ion-acoustic respectively the dust-acoustic branch of the dispersion relation. Here the dust-acoustic speed was introduced as $c_{da}^2 = \lambda_D^2 \omega_{pd}^2 = (\lambda_{De}^{-2} + \lambda_{Di}^{-2})^{-1} \omega_{pd}^2$ [3]. Clearly, the collisionless dispersion relation has a critical wavenumber $k_{cr} = \omega_{Jd}/c_{da}$, below which the plasma becomes gravitationally unstable [4].

Reverting to equation (5), we note that this equation can be rewritten in the form

$$(\Omega^2 - r_{ia})(\Omega^2 - r_{da}) + v_{id}\Omega(\Omega^2 - \Lambda) = 0, \quad (7)$$

where Λ can be calculated as

$$\Lambda = \omega_{Jd}^2 - A\omega_{pd}^2 \left(1 + \frac{n_{i0}Z_i}{n_{d0}Z_d}\right)^2. \quad (8)$$

For a dusty plasma with cold dust particles, which additionally obeys the inequality $\left(1 + n_{i0}Z_i/n_{d0}Z_d\right)\omega_{pd} > \omega_{Jd}$, a wavenumber k_Λ is introduced as

$$k_\Lambda^2 = \frac{\omega_{Jd}^2}{\left[\left(1 + \frac{n_{i0}Z_i}{n_{d0}Z_d}\right)^2 \omega_{pd}^2 - \omega_{Jd}^2\right] \lambda_{De}^2}. \quad (9)$$

As Λ changes sign at $k = k_{cr}$, it will become clear that wavenumber regions separated by k_Λ behave differently towards the presence of dust-ion collisions.

TABLE 1. Classification for different wavenumber regions

k	...	k_Λ	...	k_{cr}	...
Λ	+	0	−	−	−
r_{da}	+	+	+	0	−
case	C		B		A

ROOTLOCUS METHOD

Since equation (7) can be written as $D(\Omega) + v_{id}N(\Omega) = 0$, where $N(\Omega)$ and $D(\Omega)$ are polynomials in the complex variable Ω and both have real coefficients, one can always create a so called rootlocus plot [5]. Such a rootlocus plot shows how the roots of the dispersion relation (7) move in the complex plane as v_{id} increases from zero towards infinity.

Using only elementary principles of rootlocus theory the stability of the system can easily be determined, without even the need of a rootlocus plot [6]. It can be proven that for the considered dispersion relation, a plot does not cross the imaginary axis except in roots corresponding with $v_{id} = 0$ or $v_{id} = \infty$, furthermore these roots for zero or infinite collision frequencies are either real or purely imaginary. Simple calculations then provide how the curves arrive ($v_{id} = \infty$) and start ($v_{id} = 0$) as well as which parts of the real axis are part of the rootlocus plot. The thus acquired information proves to be sufficient for the determination of the stability of the plasma.

For dusty plasmas where the electrostatic interactions are dominant *i.e.* $\left(1 + n_{i0}Z_i/n_{d0}Z_d\right)\omega_{pd} > \omega_{Jd}$, there turn out to be three distinct regions in wavenumber space, separated by k_Λ and k_{cr} [6]. These regions correspond with the signs of r_{da} and Λ and are summarized in Table 1. On the other hand, for dusty plasmas where self-gravitational interactions dominate *i.e.* $\left(1 + n_{i0}Z_i/n_{d0}Z_d\right)\omega_{pd} < \omega_{Jd}$, the analysis is independent of the wavenumber and always corresponds with case C.

Dusty plasmas with negligible self-gravitational interactions *i.e.* with $\omega_{Jd} \ll \omega_{pd}$ belong to category A and, using rootlocus theory, it can be easily proven that all dusty plasmas in this category are stable for every value of v_{id}. Conversely, plasmas belonging to categories B or C are always unstable, independent of the value of v_{id}.

Rootlocus plots of all cases are given in the figures, for case A the given plot is only valid for a situation $|r_{da}| < |\Lambda| < 9|r_{da}|$. In the plots, crosses correspond with the roots of the collisionless dispersion relation and the curves represent the evolution of the roots towards a situation $v_{id} = \infty$, corresponding with the circles. We emphasize that the collision frequency here is treated as a purely mathematical parameter, because the mathematical solutions for very large collision frequencies may not be physically meaningful. Basically, the solutions for all values of the collision frequency are calculated, afterwards the solutions for very large v_{id} can be discarded if need be.

Since only case A represents stable plasmas, we can conclude that the criterion for gravitational instability is the same as for collisionless plasmas, namely there is instability for $k < k_{cr} = \omega_{Jd}/c_{da}$. In other words, collisions between ions and charged dust grains have no influence on the gravitational stability, but merely influence the frequencies and damping rates of the dust-acoustic and ion-acoustic modes.

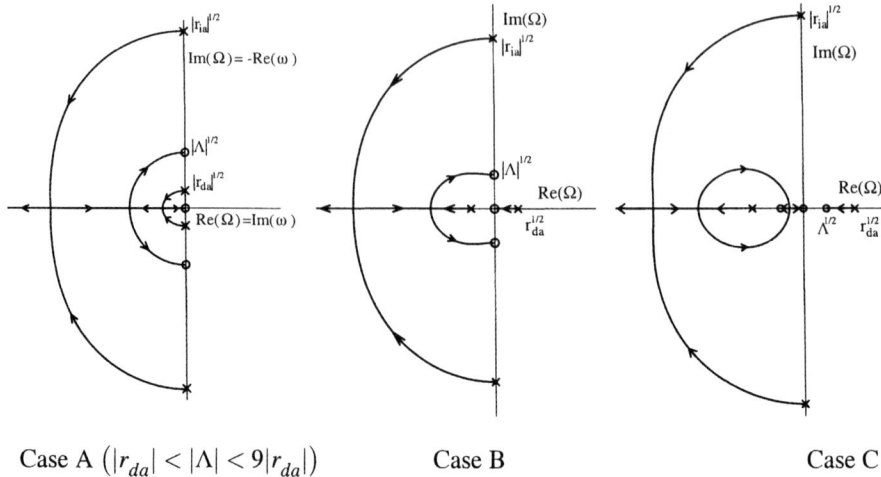

Case A $\left(|r_{da}| < |\Lambda| < 9|r_{da}|\right)$ Case B   Case C

CONCLUSIONS

The stability of longitudinal disturbances in self-gravitating dusty plasmas has been investigated previously within a collisionless model [7]. We studied modifications due to the possible presence of dust-ion collisions and illustrated qualitatively how the real frequency and damping decrement of the ion-acoustic and Jeans modes change over the spectrum of possible collision frequencies, using the semi-analytical rootlocus method.

It shows that dust-ion collisions reduce the growth rate of the unstable Jeans dust mode but can never overturn the gravitational instability. The results of Ivlev *et al.* [8] are partly reproduced for the ion-acoustic branch, as the inclusion of self-gravitational effects barely modifies these ion-acoustic modes.

ACKNOWLEDGMENTS

GJ and FV acknowledge a research grant from the Bijzonder Onderzoeksfonds of the Universiteit Gent. VVY thanks the "Onderzoeksfonds K.U. Leuven" for supporting her through a research fellowship (grant F/00/070).

REFERENCES

1. P. Meuris, F. Verheest and G.S. Lakhina, *Planet. Space Sci.* **45**, 449-453 (1997).
2. N. D'Angelo, *Phys. Plasmas* **5**, 3155-3160 (1998).
3. N.N. Rao, P.K. Shukla and M.Y. Yu, *Planet. Space Sci.* **38**, 543–546 (1990).
4. F. Verheest, G. Jacobs and V. Yaroshenko, *Phys. Plasmas* **7**, 3004-3008 (2000).
5. J. Willems, *Stability Theory of Dynamical Systems*, Nelson and Sons, London (1970).
6. G.Jacobs, V.V. Yaroshenko, F. Verheest, Low-frequency electrostatic waves in self-gravitating dusty plasmas with dust-ion collisions, *submitted*.
7. P.K. Shukla and F. Verheest, *Astrophys. Space Sci.* **262**, 157-162 (1999).
8. A.V. Ivlev, D. Samsonov, J. Goree and G. Morfill, *Phys. Plasmas* **6**, 741-750 (1999).

Dust grains as a diagnostic tool for rf-discharge plasma

B.W. James, A.A. Samarian and F.M.H. Cheung

School of Physics, University of Sydney, NSW 2006, Australia

Abstract. The location of the sheath edge in a planar rf-discharge has been determined using test dust grains. The diagnostic technique is based on measurement of the equilibrium position of fine dust grains levitated above the powered electrode in an rf-discharge. Estimates show that for grains with radii less than 500 nm the grain equilibrium position and sheath edge location differ by less than 5 percent, and this difference continues to decrease with grain radius. We use this technique to diagnose the sheath in an argon plasma which was generated at pressures in the range 20 – 100 mTorr. The transient motion technique was used for the radial potential profile measurements. The results obtained was compared to the profile obtained using the rf-compensated single Langmuir probe. The two results show good agreement.

The possibility of using the dust grains as a diagnostic tool based on the fact that the grains in a discharge plasma acquire electric charge by collecting electrons and ions from the surrounding plasma. The charge on a grain can be extremely high (say 10^3 - 10^4 electron charges for a micron-sized particles) and depends on the ion and electron fluxes to the particle surface. The equilibrium position of such test grains is determined mainly by gravity and electrostatic forces in the sheath region. This means that the equilibrium position and motion of the grains is strongly dependent on plasma and sheath conditions. Analysis of grain behaviour can provide information on the spatial profile of electric field, ion density and velocity, and the sheath edge location.

The location of the sheath edge in a planar rf-discharge has been determined using test dust grains. The diagnostic technique is based on measurement of the equilibrium position of fine dust grains levitated above the powered electrode in an rf-discharge. Estimates show that for grains with radii less than 500 nm the grain equilibrium position and sheath edge location differ by less than 5 percent, and this difference continues to decrease with grain radius. We have used this technique to diagnose the sheath in an argon plasma which was generated at pressures in the range 20 – 100 mTorr by applying a 15 MHz signal to the power electrode [1,2]. The experimental arrangement is shown in Figure 1.

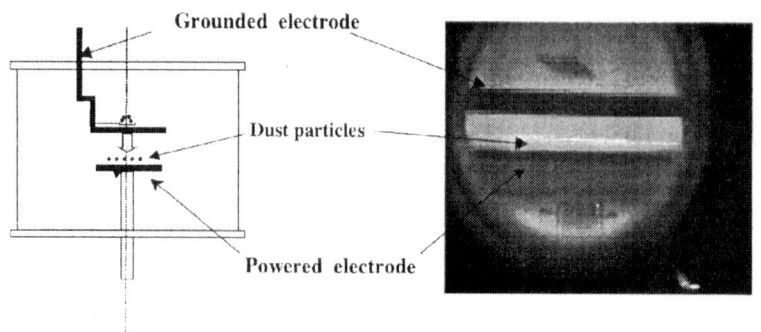

FIGURE 1. Experimental chamber and image of test gains levitated above the electrode.

The test grains are generated in the discharge by electrode sputtering under high power (up to 200W) and high pressure (up to 1 Torr) conditions. The dust grains are illuminated using a Helium-Neon laser, which enters the discharge chamber through a 40-mm diameter window. Windows mounted on a side port and on the top of the chamber provide a view of the light scattered at different angles by the suspended dust particles. The size of the growing grains was estimated from analysis of the scattered light using techniques proposed in [3]. In our experiments grains grown to 300 - 500 nm. To obtain a vertical cross-section of the dust grain layer and provide a sheath dimension measurement, the laser beam was expanded in the vertical directions into a sheet of light by a system of cylindrical lenses. Images of the illuminated dust layer are obtained using a charged-coupled device (CCD) camera with a micro lens. The video signals are stored on a videotape recorder or are transferred to a computer. The resulting images allow direct measurement of test grain equilibrium position and therefore a determination of the sheath edge location. The variation of sheath thickness with pressure for different rf-input powers is shown in Figure 2.

The radial potential profile was measured using a transient motion technique (TMT) in which a dust particle levitated in the sheath was displaced from its equilibrium position by applying a negative voltage to a two pin electrode. When the

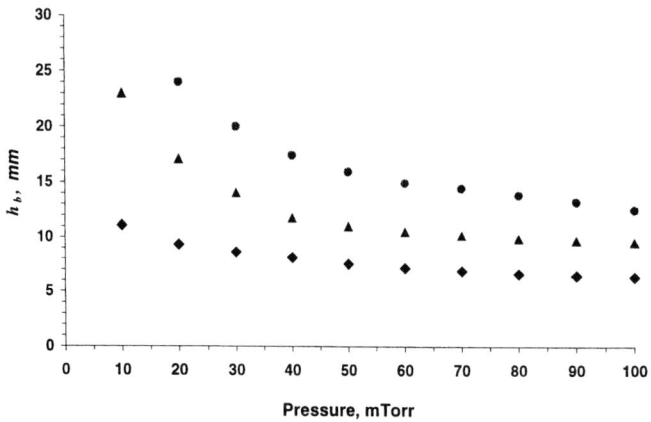

FIGURE 2. Position of sheath edge, h_b, as a function of pressure for different rf-input powers: squares – 100W, triangles – 60W, circles – 35W.

voltage is switched off the particle returns to its original position due to the action of the radial electrostatic force $F_{el}=Z_d E_r$. By tracking the particle motion it is possible measure instantaneous velocity and acceleration. Then, using the equation of motion the potential difference between two points can be found as follows:

$$\Delta\varphi = \frac{ma\Delta s + F(v_1)\Delta s_1 + F(v_2)\Delta s_2}{Z_d}$$

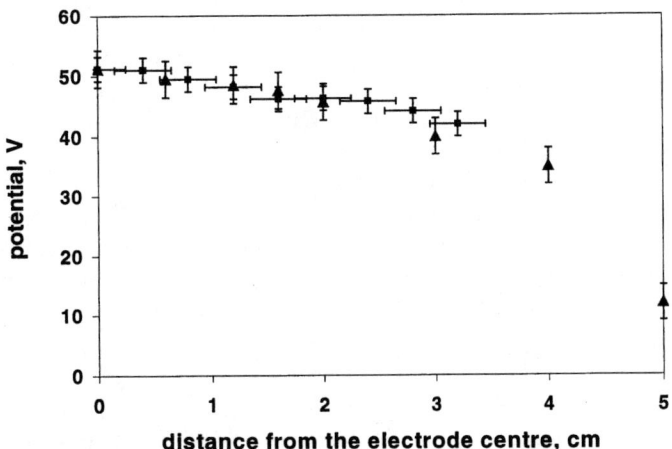

FIGURE 3. Radial potential profile: ■ - Langmuir probe technique, ▲-transient motion technique

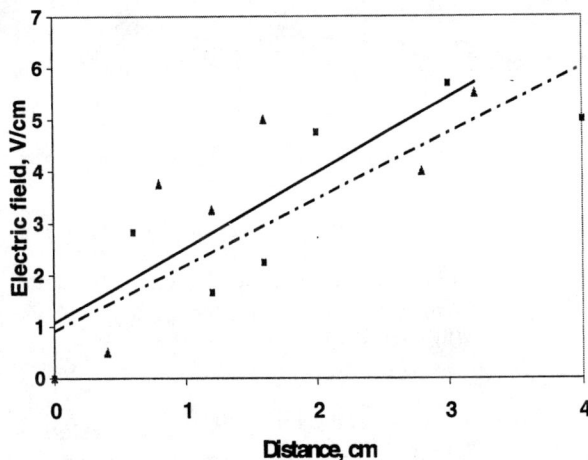

FIGURE 4. Electric field dependence versus distance from the electrode centre. ■ - Langmuir probe technique, ▲-transient motion technique. The solid line is the trend for TMT data; the dashed line is mean square fit for probe data.

where $F = 4/3\delta m n_n v_{T_n} \pi a^2 v$ is the neutral drag force. The potential variation obtained in this way was normalised to the plasma potential at $r = 0$, as measured by the Langmuir probe.

The value of Z_d, the particle charge was obtained by the vertical resonance method [4]. As the particle motion is in the horizontal plane we can assume that the charge is constant. The resulting profile is shown on the Figure 3. The profile obtained using a rf-compensated single Langmuir probe is also shown. The two results show good agreement.

The radial electric field, calculated from the potential data, is shown in Figure 4. The high degree of scatter among the points is due to the use of simple numerical differentiation. The result, along with linear fits to the data, is shown, however, to indicate the possibilities of the technique. By fitting functions to the potential variation it would be possible to obtain more reliable estimates of radial electric field. The TMT itself could also be improved by using a laser, instead of a pin electrode, to displace the particle, causing minimal disturbance to the plasma.

Even smaller particles than those used above find no equilibrium position in the sheath. Instead, they fill the plasma volume leaving, the sheath as a dust void. The shape of the potential well above the confining electrode in a radio-frequency (rf) discharge with a printed-circuit board electrode system [5] was visualised (see Figure 5) using fine dust grains that were generated in the discharge. The well shape was found to depend strongly on the confining potential.

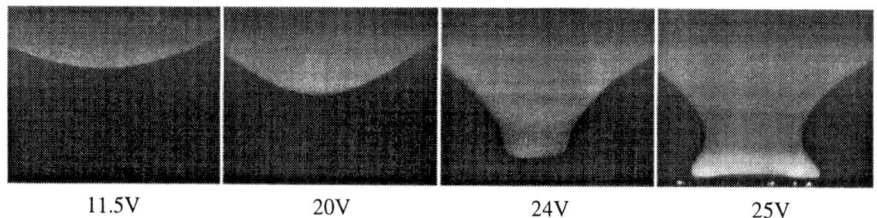

11.5V   20V   24V   25V

FIGURE 5. The dependence of the dust cloud shape on the confining electrode voltage

ACKNOWLEDGMENTS

This work was supported by Australian Research Council and the Science Foundation for Physics within the University of Sydney. AAS was supported by a University of Sydney U2000 Fellowship.

REFERENCES

1. Samarian, A., James, B., Vladimirov, S., and Cramer, N., *Phys. Rev.* **64**, 025402(R) (2001).
2. Samarian, A.A. and James, B.W., *Phys. Lett.* **A287**, 125 (2001).
3. Nefedov A.P, Vaulina O.S., Petrov O.F., *Appl. Opt.* **36**, 1357 (1997).
4. Melzer A., T. Trottenberg T., A. Piel A., *Phys. Lett.* **A191**, 301 (1994).
5. Cheung F, A. Samarian, B. James, *Physica Scripta*, **T98**, 143 (2002).

Low Frequency Acoustic Waves in Nebulae with Gravitational Field Induced by Dust

Václav Kaizr, Petr Kulhánek, David Břeň and Jan Pašek

*Czech Technical University, Faculty of Electrical Engineering, Department of Physics,
Technická 2, 166 27 Prague 6, Czech Republic*

Abstract. In presence of dust grains there are several sufficient differences. First of all the self–gravitational field of the dust cannot be neglected and equation for gravitational potential has to be included into the model. Furthermore both negatively and positively charged dust particles must be considered. That is why one–liquid model is not acceptable in the case of dusty plasma. Multi–liquid MHD set of equation supplemented with gravitational potential equation and polytrophic equation of state has to be used.

INTRODUCTION

In plasma we can find a great number of waves. In magnetic field presence, there are also several low frequency modes which have considerable influence on plasma fiber stability and behavior. The most important ones are called Alfven wave, fast magnetoacoustic wave and slow magnetoacoustic wave. The influence of dust grains and gravitational field is discussed in this paper. The zero order solution was found a numerically solved propagation of the low frequency and ultra low frequency modes through the plasma [1].

We have solved a MHD set of equations of the cylindrical plasma pinch with magnetic and gravitational self–interaction and found the behavior of magnetoacoustic waves in this structure. The first step was the acquirement of the zero order solution – stationary

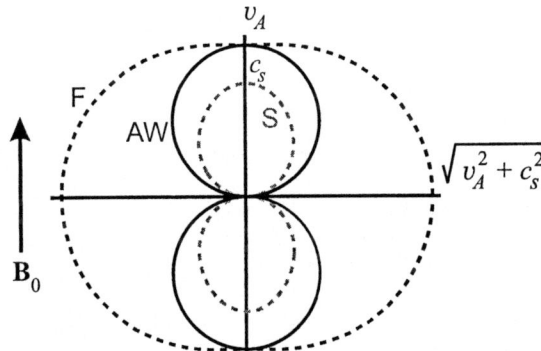

FIGURE 1. Alfvén, Slow and Fast magnetoacoustic waves.

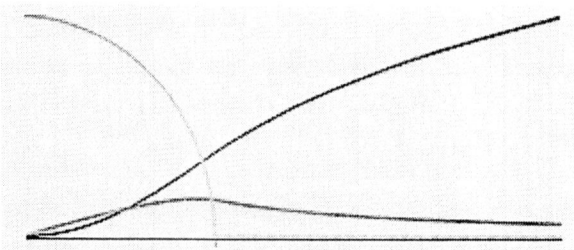

FIGURE 2. "Zero order" solution.

generalized Bennett solution for density, radial and axial velocity, azimuthal magnetic field, pressure and gravitational field potential.

$$\Delta \phi = 4\pi G \rho \qquad (1)$$

SOLVING OF EQUATIONS

The set of equations :

$$\frac{\partial \rho}{\partial t} + \nabla \cdot (\rho \mathbf{v}) = 0 \qquad (2)$$

$$p = C\rho^{\gamma} \qquad (3)$$

$$\frac{\partial \mathbf{B}}{\partial t} = \frac{1}{\mu_0 \gamma} \Delta \mathbf{B} + \nabla \times (\mathbf{v} \times \mathbf{B}) \qquad (4)$$

$$\rho \frac{\partial \mathbf{v}}{\partial t} + \rho (\mathbf{v} \cdot \nabla) \mathbf{v} = -\nabla p - \nabla pm - \rho \nabla \phi \qquad (5)$$

$$\Delta \phi = 4\pi G \rho \qquad (6)$$

I found "zero order" solution for cylindrical coordinates (Fig. 2). You can see the dependence of density on relative position of radial coordinate (green line), the dependence of gravitational potential (black line) and the dependence of magnetic field (red line) on relative position of radial coordinate [4]

There is also a data file with values of the variables mentioned above. These values enable extrapolation to smooth curves, for example the density (Table 1). Profile of the density is very similar to a Benett's profile of density but close to the surface is the deviation very sufficient [3].

WAVES IN UNLIMITED SPACE

Nowadays a try to find a solution of perturbated set of equations in unlimited space. We use three kinds of participles. The set of equations are:

$$-i\omega\delta\rho^{(\alpha)} + \rho_0^{(\alpha)} i\mathbf{k} \cdot \delta\mathbf{v}^{(\alpha)} = 0$$

$$-i\omega\delta\mathbf{B} + \frac{k^2}{\mu_0 \sigma}\delta\mathbf{B} - [\mathbf{B_0} \cdot i\mathbf{k}] \left(\frac{\sum_\alpha m_\alpha \delta\mathbf{v}^{(\alpha)}}{\sum_\alpha m_\alpha}\right) + \left[i\mathbf{k} \cdot \left(\frac{\sum_\alpha m_\alpha \delta\mathbf{v}^{(\alpha)}}{\sum_\alpha m_\alpha}\right)\right] \mathbf{B_0} = 0 \quad (7)$$

$$-i\omega\rho_0^{(\alpha)}\delta\mathbf{v}^{(\alpha)} + i\mathbf{k}c_\alpha^2\delta\rho^{(\alpha)} + \frac{i\mathbf{k}}{\mu_0}(\mathbf{B_0} \cdot \delta\mathbf{B}) - \frac{1}{\mu_0}(\mathbf{B_0} \cdot i\mathbf{k})\delta\mathbf{B} + \rho_0^{(\alpha)} i\mathbf{k}\frac{4\pi G \sum_\alpha \delta\rho^{(\alpha)}}{k^2} = 0$$

RESULTS

We have solved a MHD set of equations of the cylindrical plasma pinch with magnetic and gravitational self–interaction and found the behavior of magnetoacoustic waves in this case. The first step was the acquirement of the zero order solution – stationary generalized Bennett solution for density, radial and axial velocity, azimuthal magnetic field, pressure and gravitational potential. Having the zero order solution it is possible to initialize some small low frequency perturbations and numerically observe the formation and propagation of the slow, fast and Alfvén magnetoacoustic waves in the perturbed cylindrical geometry.

All the endeavor points to understanding of small laboratory and huge nebular pinch structures as well as the propagation of low frequency plasma waves which can form the shape of these filaments. In presence of dust grains ultra low frequency modes can propagate through the plasma. Numerical simulation of these modes will be the topic of future research.

TABLE 1. Numerical solution of program.

Step	ϕ	B	ρ
1.0000000E–06	0.0000000E+00	0.0000000E+00	6.000000
1.0000000E–06	0.0000000E+00	2.5145308E–11	6.000000
5.0009997E–03	1.2572654E–13	2.5150333E–11	5.999988
1.0001000E–02	2.5147820E–13	3.7721680E–11	5.999964
1.0001000E–02	2.5147820E–13	3.7721680E–11	5.999964
1.5001000E–02	4.4008661E–13	5.0293783E–11	5.999928
2.0001000E–02	6.9155554E–13	6.2865969E–11	5.999880
2.5001001E–02	1.0058854E–12	7.5438079E–11	5.999820

ACKNOWLEDGMENTS

This research has been supported by the research program No J04/98:212300017 *"Research of Energy Consumption Effectiveness and Quality"* of the Czech Technical University in Prague, by the research program INGO No LA 055 *"Research in Frame of the International Center on Dense Magnetized Plasmas"* and *"Research Center of Laser Plasma"* LN00A100 of the Ministry of Education, Youth and Sport of the Czech Republic.

REFERENCES

1. Kulhánek, P.: *Particle and Field Solvers in PM Models*. Czechoslovak Journal of Physics, **50**, Suppl. S3, 2000.
2. Kubíček, M.: *Numerické metody pro řešení diferenciálních rovnic*. SNTL, 1984.
3. Vitásek, E.: *Základy numerických metod pro řešení diferenciálních rovnic*. ACADEMICA, 1994.
4. Chen, F. F.: *Úvod do fyziky plazmatu*, ACADEMICA, 1977.
5. Kaizr, V.: *Solving of nonlinear equations*, ICDP Kudowa Zdroj, 2001.

On the Interaction of a Complex Plasma with an External Ion Beam

H. G. Thieme[*], R. Wiese[*], D. Gorbov[*], H. Kersten[*], and R. Hippler[†]

[*]E.-M.-Ardt-University, Institute for Physics, 17487 Greifswald, Germany
[†]St.Petersburg State University, Department of Physics, St.Petersburg, Russia

Abstract. The interaction of micro-sized (SiO_2) grains confined in an rf-plasma with an external ion beam has been qualitatively investigated. The effect of the ion beam on shape and position of the particle cloud is described and discussed in respect to void formation.
PACS 5225Zb, 5240-w, 5240Mj

INTRODUCTION

If dust particles are injected into a plasma, they become negatively charged by the currents towards the particles and can be confined in the discharge. The spatial distribution and movement of such grains in a low-temperature plasma (complex plasma) is a consequence of several forces acting on the particles. The charged particles interact mainly with the electric field in front of the electrode. The electrostatic force has to be balanced by various other forces in order to confine the particles. These forces, which have been discussed extensively by several authors [1,2], are the gravitation, the neutral and ion drag, thermophoresis, and photophoresis. But only some of the forces will play a role in laboratory complex plasmas. Commonly, the electrostatic field and the gravitational force are important. Superposition of both effects result in a parabolic potential trap [3]. In our experiments, the influence of an external ion beam supplied by an ECR ion source ECI A 125 [4] has been investigated. The effect of the ion beam is threefold :

- change of the sheath structure and the electric field,
- recharging of the dust particles, and
- variation of the ion drag force.

Observations of the behaviour of the complex plasma interacting with the ion beam have been compared with measurements of the energy flux density in the beam, too. The axial and radial distribution of the energy flux as well as the dependence on the source parameters have been obtained by a simple thermal probe [5,6] Studies on the effect of the additional ion beam might also help to clarify the questions coming up with void formation in a plasma crystal which could be caused by ion drag forces [7].

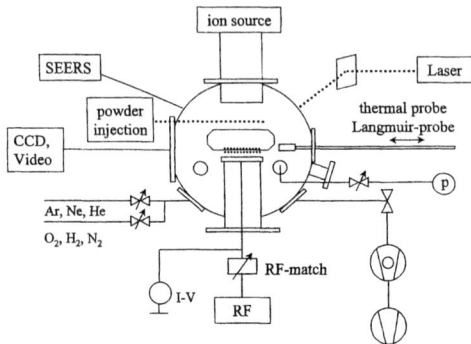

FIGURE 1. Scheme of the experimental set-up PULV Al.

EXPERIMENTAL

The experiments have been performed in a reactor PUL V Al which is schematically drawn in Fig. 1. The plasma sources (rf-plasma, ion beam gun) as well as different diagnostics (video, thermal probe) are mounted in a spherically shaped vessel. The effects of the different forces can be distinguished by a variation of the plasma density which is influenced by the gas pressure (10^{-1} ... 10 Pa), the rf-power (0 ... 20 W), and the power of the ion source (500 ... 800W). The kinetic energy of the ions are determines has been varied by the beam voltage (0 ... 1000 V).

In order to characterize the ion beam, the energy flux density of the beam at dust position has been measured by means of a thermal probe. The procedure is based on the measurement of the temporal slope of substrate temperature during the plasma process. A substrate dummy which is thermally isolated is inserted into the plasma at dust particles position [5,6]. By this procedure, the energy flux density of the external ion beam has been characterized in dependence on various experimental parameters as, for example, on the beam voltage. As expected, the energy influx increases with increasing beam voltage due to the higher kinetic energy of the ions. The ion current density itself shows only a weak dependence on the beam voltage but a strong dependence on the power which determines the ion production in the source. Whereas the mean ion density in the confining rf-plasma is in the order of $5 \times 10^8 \,\text{cm}^{-3}$, the external ion beam supplies an additional energetic ion flux of about $300\,\mu\text{A}/\text{cm}^2$ which interacts with the particle cloud.

RESULTS AND DISCUSSION

Charging, confinement, and movement of the dust particles are essentially influenced by the presence of electric fields. In our experiments, we added an additional force by the external ion beam to simulate the influence of ions. The measured height of the dust particle cloud above the rf-electrode in dependence on the ion beam voltage is plotted in

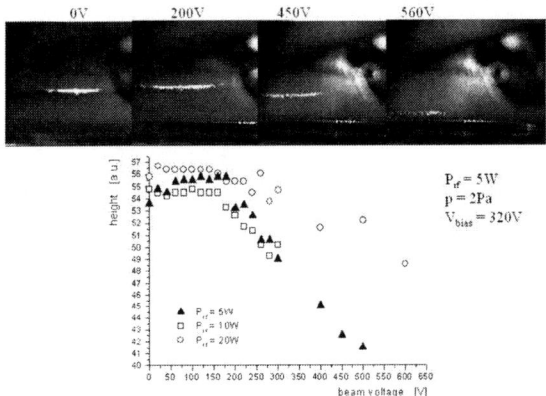

FIGURE 2. Particle cloud in front of the rf-electrode at different ion beam voltages, and the measured height of the particles

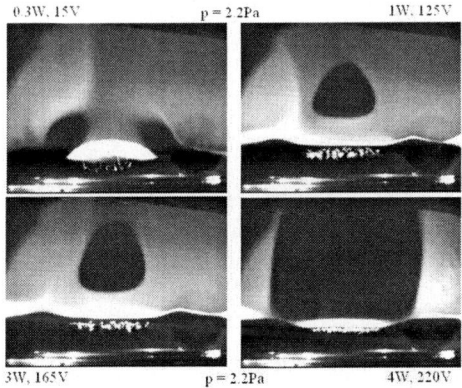

FIGURE 3. Void formation in a cloud of small particles at various rf-power and bias voltages.

Fig. 2. For small beam voltages < 200 V) there is almost no influence. Due to a variation of the sheath width, which is influenced by gas pressure and rf-power, the trapped particles follow only the variation of the rf-plasma. The beam voltage is not sufficiently high enough to extract a remarkable amount of ions into the dust region which is about 21 cm away from the grid system of the ion source. When the beam voltages increases, the height of the dust particles decreases. That means the originally planar dust cloud forms a parabolic shape and moves into the direction of the rf-electrode. For higher power of the rf-plasma and, hence, for higher bias voltage of the electrode the particle cloud is levitated in a higher position than for smaller rf-power. This observation is due to the electric field (potential) in front of the powered rf -electrode. At higher bias voltages, the field can compensate the influence of the external ion beam more efficiently (Fig. 2). Since the additional ion supply by the external ion beam interacts with the rf-plasma, the plasma density in the trapping region is changed. This effect causes a variation in the sheath (width and shape) and, thus, a variation in the particles force balance. Another effect of the additional ion supply is an enhanced charge carrier recombination at the surface of the floating dust grains. It can be expected that the net charge is decreased by the recombination. The ion bombardment leads also to an additional heating of the particles. Certainly, also the ion drag force by the ions of the beam influences the force balance of the confined grains. The importance of this effect has often been discussed in respect to the void formation in plasma crystals. Void formation for very small particles ($\sim$ 100 nm) could be observed under our experimental conditions, too (Fig. 3). The small particles have been produced either by an Ar/C_2H_2 plasma or by sputtering of the Al-electrode. Under certain conditions, the particles form a dome and, sometimes, a void structure. The size and shape of the void can be changed by a variation of the discharge power and, hence, by a variation of the bias voltage. As a consequence, the ion density and flow pattern of the ions in the plasma bulk is changed in that region where the very small particles are confined. Additional experiments in the future will clarify the phenomena in respect to the ion beam influence in an appropriate quantitative manner.

ACKNOWLEDGMENTS

This work has been supported by the Deutsche Forschungsgemeinschaft (DFG) under SFB198/A14 and by the IOM Leipzig.

REFERENCES

1. Daugherty, J.E., Porteous, R.K, Graves, D.B., *J.Appl.Phys.*, **74**, 1617 (1993).
2. Bouchoule, A. (Ed.), *"Dusty Plasmas"*, J. Wiley & Sons, (1999).
3. Homann, A., Melzer, A., Piel, A., *Phys. Bl.*, **52**, 1227 (1996).
4. Zeuner, M, Meichsner, J., Neumann, H., Scholze, F., Bigl, F., *J. Appl. Phys.*, **80**, 611 (1996).
5. Kersten, H., Deutsch, H., Steffen, H., Kroesen, G.M.W., Rippler, R., *Vacuum*, , **83**, 385 (2001).
6. Kersten, H., Rohde, D., Berndt, J., Deutsch, H., Rippler, R., *Thins Solid Films*, 377–378, 585 (2000).
7. Thomas, H. Morfill, G., Vth European Workshop on Dusty Plasmas, Potsdam, 2000.

Characteristics of Nonlinear Dust Acoustic Waves in Planetary Ring

M. Khan*, S. Sarkar*, T. K. Chaudhuri*, S. Ghosh* and M. R. Gupta*

Centre For Plasma Studies, Faculty Of Science, Jadavpur University, Calcutta - 700 032, INDIA

Abstract. Nonlinear properties of small amplitude dust acoustic waves, incorporating dust charge variation, ion inertial effect and dust drift effect have been studied. A Korteweg-de Vries (KdV) equation with positive or negative damping term depending on the wave velocity and the ring parameters governs the nonlinear dust acoustic wave. Numerical investigations reveal that in Saturn's F, G and E rings, dust acoustic solitary wave admits both negative and positive potentials. Instability arises from the available free energy of drift motion (V_0) of dust grains only for the wave with wave velocity $\lambda < V_0$.

INTRODUCTION

Bharuthram and Shukla[1], Verheest[2], Yinhua and Yu[3] and Ghosh *et. al.*[4] have studied the nonlinear dust acoustic waves (DAW) for the fixed charged dust grains and showed that nonlinear DAW can form solitons of either positive or negative electrostatic potentials. Nonlinear DAW without dust drift was studied by Ma and Liu[5], Xie *et. al.* [6] considering dust charge variation having hydrodynamical time scale is much larger than dust charging time scale and experimentally observed by Barkan *et. al.*[7]. In this paper, the nonlinear properties of small amplitude DAW including dust charge variation in Saturn's F, G and E rings have been studied considering drift velocity comparable to the ion thermal velocity and also considering hydrodynamical time scale $\left(\tau_d \approx \omega_{pd}^{-1}\right)$ is much smaller than the dust charging time scale $\left(\tau_{ch} \approx v_d^{-1}\right)$.

FORMULATION OF THE PROBLEM

We consider a collisionless, non relativistic, unmagnetized dusty plasma consisting of electrons, ions, and highly charged dust grains having charge variation, dust drift and ion inertia. We also considered the wave propagation and the dust drift velocity along X direction. The basic equations in the normalized form are

$$\frac{\partial N_d}{\partial T} + \frac{\partial (N_d V_d)}{\partial X} = 0 \tag{1}$$

$$\frac{\partial V_d}{\partial T} + V_d \frac{\partial V_d}{\partial X} = -(\Delta Q - 1)\frac{\partial \Phi}{\partial X} \tag{2}$$

$$\frac{\partial N_i}{\partial T} + \frac{\partial (N_i V_i)}{\partial X} = 0 \tag{3}$$

$$\mu_i \left(\frac{\partial V_i}{\partial T} + V_i \frac{\partial V_i}{\partial X} \right) = -\frac{1}{\sigma} \frac{\partial \Phi}{\partial X} - \frac{1}{N_i} \frac{\partial N_i}{\partial X} \tag{4}$$

$$\frac{\partial^2 \Phi}{\partial X^2} = -\frac{\lambda_d^2 e^2}{\varepsilon_0 T_e} \left[n_{i0} N_i - n_e + z_d n_{d0} (\Delta Q - 1) N_d \right] \tag{5}$$

$$n_e = n_{e0} exp(\Phi) \tag{6}$$

$(\Delta Q - 1) \left[\Delta Q = \frac{\Delta Q_d}{z_d e} \right]$ is the normalized dust charge, normalized by $z_d e$. The dust charge is given by $Q_d = -z_d e + \Delta Q_d$, $\mu_i = \frac{z_d m_i}{\sigma m_d}$ = ion inertia, $N_j = \frac{n_j}{n_{j0}}$ ($j = i, d$), n_{j0} = equilibrium number densities of the jth particle and V_j ($j = i, d$) = particle velocity of the jth particle normalized by DA speed $c_d = \sqrt{\frac{z_d k T_e}{m_d}}$. Φ = electrostatic wave potential normalized by $\frac{kT_e}{e}$, $\sigma = \frac{T_i}{T_e}$. The spatial coordinate X and time T are normalized by dust Debye length λ_d and inverse of dust plasma frequency $\left(\omega_{pd}^{-1} \right)$, respectively. To obtain the normalized dust charge variation ΔQ, we consider the orbital motion limited current balance equation, in normalized form, it reads as

$$\frac{z_d e}{\tau_{ch}} \left(\frac{\partial \Delta Q}{\partial T} + V_d \frac{\partial \Delta Q}{\partial X} \right) = \frac{\tau_d}{\tau_{ch}} (I_e + I_i) \tag{7}$$

where, τ_d is the dust hydrodynamical time scale = ω_{pd}^{-1}, τ_{ch} is the dust charging time scale = v_d^{-1}, I_e is the elctron current and I_i is the ion current.

In order to study the nonlinear propagation of small amplitude DAW, we have used standard reductive perturbation technique, the independent variables can be stretched as $\xi = \sqrt{\varepsilon} (X - \lambda T)$; $\tau = \varepsilon^{\frac{3}{2}} T$, where λ = phase velocity, ε = small parameter. To make the nonlinear perturbation consistent, we assume that the ratio $\frac{\tau_d}{\tau_{ch}} \approx v \varepsilon^{\frac{3}{2}}$, where v is a finite quantity of the order of unity, and this assumption is valid in Saturn's ring. The boundary condistions are as follows, as $|X| \to \infty$, $N_d, N_i \to 1$; $V_d \to V_0$; $\Phi, V_i, \Delta Q \to 0$

The dependent variables are expanded and introduced in the basic equations. The coefficients of terms in lowest order in ε are equated and by virtue of boundary conditions that perturbations vanish at $X = -\infty$ ($\xi = -\infty$) for all time slow or fast implying $\Delta Q^{(1)} = \Delta Q^{(1)}(\tau) = 0$. This is a consequence of the assumption that the dust hydrodynamical time τ_d is smaller by order $O\left(\varepsilon^{\frac{3}{2}}\right)$ than that of dust charging time τ_{ch}.

We obtain

$$N_d^{(1)} = -\frac{\Phi^{(1)}}{(\lambda - V_0)^2}; \quad N_i^{(1)} = \frac{\Phi^{(1)}}{\sigma (\lambda^2 \mu_i - 1)} \tag{8}$$

Finally, we obtain the following biquadratic equation in λ as

$$\frac{1}{\sigma(\lambda^2\mu_i - 1)} + \frac{(1-\delta)}{(\lambda - V_0)^2} = \delta \qquad (9)$$

For $\delta = \frac{n_{e0}}{n_{i0}} \approx 0$ i.e. for an effectively two components plasma comprising of positive ions and negatively charged dust grains with fixed charge and $T_e \approx T_i$, $\mu_i \ll 1$. The roots of equation are as follows

$$\lambda \approx V_0 \pm \sqrt{1 - \mu_i V_0^2} \qquad (10)$$

Equation (10) shows that λ is real only if $|V_0| \leq \sqrt{\frac{1}{\mu_i}} = V_{thi}$.

Therefore, for the two component dust-ion plasma, the DA solitary wave exists only if the dust drift velocity is less than the ion thermal velocity (Ghosh et. al.[4]).

Now equating the coefficients of terms in next higher order in ε and after suitable substitution, we obtain the following modified K-dV equation as

$$\Phi_\tau^{(1)} - \alpha \Phi^{(1)} \Phi_\xi^{(1)} + \beta \Phi_{\xi\xi\xi}^{(1)} + \gamma \Phi^{(1)} = 0 \qquad (11)$$

where

$$\alpha = \beta \left[\frac{3}{(\lambda - V_0)^4} + \left(\frac{\delta}{1-\delta}\right) - \frac{(3\lambda^2\mu_i - 1)}{\sigma^2(1-\delta)(\lambda^2\mu_i - 1)^3} \right]$$

and

$$\beta = \frac{1}{2}\left[\frac{2}{(\lambda-V_0)^3} + \frac{\lambda\mu_i}{\sigma(1-\delta)(\lambda^2\mu_i-1)^2}\right]^{-1}; \quad \gamma = v\frac{\beta\beta_d(\sigma\lambda^2\mu_i - 2)}{\sigma(\lambda^2\mu_i-1)(\lambda-V_0)^2}$$

Employing Karpman and Maslov's method and using boundary conditions, we arrive at the following approximate solution exhibiting slow time (i.e. τ dependent) evolution of the amplitude, the width and the velocity of the $sech^2$ solitary wave

$$\Phi^{(1)} = -\Phi^{(1)}(\tau) sech^2 \sqrt{\frac{\alpha\Phi^{(1)}(\tau)}{12\beta}} \eta \qquad (12)$$

where

$$\eta = \xi - U\tau = \sqrt{\varepsilon}[X - (\lambda + \varepsilon U)]; mbox\Phi^{(1)}(\tau) = -\Phi^{(1)}(0)e^{-\gamma\tau}; \quad U = \frac{\alpha}{6}\Phi^{(1)}(0)e^{-\gamma\tau}$$

$\Phi^{(1)}(\tau)$ and $U(\tau)$ are the τ dependent soliton amplitude and soliton velocity.

Several features of physical interest arise depending on the values of the coefficients α, β and γ of the KdV equation. These are functions of λ: the roots of the dispersion equation and also other plasma parameters.

NUMERICAL RESULTS AND DISCUSSIONS

We assume that there exists a corotating plasma with O^+ ions in Saturn's rings and the electron ion temperature ratio $\sigma = 1$. The results are applied in Saturn's F, G and E rings using the various plasma parameters. For all the three rings of Saturn $\sigma = \frac{T_i}{T_e} = 1, a = 1\mu m, n_i \approx 10^7 m^{-3}, T_e = 10^2 ev$[8], $\frac{\tau_d}{\tau_{ch}} \approx 10^{-4}$ and $\frac{a}{\lambda_d} \approx 10^{-6}$, all these datas justifies the scaling $\frac{\tau_d}{\tau_{ch}} \approx \nu\varepsilon^{\frac{3}{2}}$ and other assumptions.

For all the three rings F, G and E, $V_0 < \frac{1}{\sqrt{\mu_i}}$ under the above condition, all the four roots of the dispersion equation are real. For all the rings three roots are positive and one negative:

$$\lambda_4 < 0 < \lambda_3 < V_0 < \lambda_2 < \frac{1}{\sqrt{\mu_i}} < \lambda_1$$

In F ring, $\frac{\alpha}{\beta} > 0$ for $\lambda = \lambda_1 > \frac{1}{\sqrt{\mu_i}}$, solitary wave has negative potential, dust density condensation and ion density depletion wave, also implies that wave is almost undamped. For $\lambda = \lambda_2 < \frac{1}{\sqrt{\mu_i}}$, γ is large positive and strongly damped positive potential wave with dust density and ion density depletion.

The G ring scenario is qualititavely the same as the E ring one. For $\lambda = \lambda_1 > \frac{1}{\sqrt{\mu_i}}$, it is a very weakly, almost undamped positive potential wave with small damping. For $\lambda = \lambda_2 < \frac{1}{\sqrt{\mu_i}}$ $|\gamma| \approx 10^{-1}$. So it is a moderately damped negative potential wave with both dust and ion density depletion. For $\lambda = \lambda_3 \left(0 < \lambda_3 < V_0\right)$, we find that γ has a large negative value in F ring, whereas it is of moderate magnitude in G and E rings. Thus for F ring waves with $\lambda = \lambda_3$ is strongly growing, whereas for G and E rings waves with $\lambda = \lambda_3$ are moderately growing. The instability arises from the available free energy of drift motion of dust grains having V_0 greater than the wave velocity λ_3. The opposite phenomenon, i.e., damping occurs for waves with wave velocity $\lambda = \lambda_1, \lambda_2 > V_0$.

REFERENCES

1. Bharuthram, R., and Shukla, P. K., *Planet Space Sc.*, **40** (1992).
2. Verheest, F., *Planet. Space Sc.*, **40** (1992).
3. Yinhua, C., and Yu, M. Y., *Phys. Plasmas*, **1** (1994).
4. Ghosh, S., Sarkar, S., Khan, M., and Gupta, M. R., *Planet. Space Sc.*, **48** (2000).
5. Ma, J. X., and Liu, *Phys. Plasmas*, **4** (1997).
6. Xie, B., He, K., and Huang, Z., *Phys. Lett.*, **A247** (1998).
7. Barkan, A., Merlino, R. L., and D'Angelo, N., *Phys. Plasmas*, **2** (1995).
8. Goertz, C. K., *Rev. Geophys.*, **27** (1989).

Ion drag in complex plasmas

S. A. Khrapak, A. V. Ivlev, G. E. Morfill, and H. M. Thomas

Centre for Interdisciplinary Plasma Science, Max-Planck-Institut für Extraterrestrische Physik, D-85740 Garching, Germany

Abstract. The problem of estimating the ion drag force in complex plasmas is considered. It is shown that the standard calculation of the ion-dust elastic scattering (orbital) cross-section often fails in complex plasmas. This is because the range of ion-grain interaction typically exceeds the Debye length whilst the standard approach uses the cut-off at the impact parameter equal to the Debye length. A new simple analytical approach to estimate the orbital cross-section is proposed. Our analytical results agree well with the available numerical results. The ion drag turns out to be significantly larger than previously estimated. It exceeds the electrostatic force in the limit of weak electric field for micron-sized grains. We suggest that this is the cause of the central "void" observed in microgravity complex plasma experiments.

The ion drag force, F_I, consists of two parts often referred to as *collection* and *orbital* forces. They are associated with ion collection by the grain, and ion elastic scattering in the electric field of the grain, respectively. The calculation of F_I has been addressed recently in several works [1, 2]. Barnes *et al.* [1] modified the standard theory of pair collisions of charged particles in plasmas by taking into account the finite grain size and ion collection by the grain. A numerical calculation of the momentum-transfer cross section for elastic ion scattering was reported by Kilgore *et al.* [2] for a point-like grain, with the potential distribution derived from a self-consistent numerical solution of the Poisson-Vlasov equation. We show below that the analytical expression derived for the orbital part of the ion drag in [1] underestimates significantly the numerical results of [2] in a case of a bulk plasma (subthermal ion drift) for typical dust and plasma parameters.

In this work we propose a simple approach to improve the estimation of the ion drag force. We make the following assumptions:

- Small, $a \ll \lambda_D$, isolated, $\Delta \gg \lambda_D$, negatively charged spherical grain (a, Δ, and λ_D are the grain radius, intergrain separation, and plasma Debye length, respectively).
- Singly charged collisionless ions, $l_i \gg \lambda_D$ (l_i is the ion mean free path).
- Weak electric field $E \ll T_i/el_i$ and subthermal ion drift, $u \ll v_{T_i}$ (u and $v_{T_i} = \sqrt{T_i/m}$ are the ion drift and thermal velocities, respectively).
- Shifted Maxvellian distribution function for ions $f(\mathbf{v}) = f_0(v)(1 + \mathbf{uv}/v_{T_i}^2)$.
- Screened Coulomb potential of interaction between ion and grain during collision.
- Ormital motion limited (OML) theory for grain charging is applicable.

The general expression for the ion drag force is

$$\mathbf{F}_I = m \int \mathbf{v} v f(\mathbf{v}) [\sigma_c(v) + \sigma_s(v)] d\mathbf{v}, \tag{1}$$

where **v** is the ion velocity, m is the ion mass, and $\sigma_c(v)$ and $\sigma_s(v)$ are the (velocity dependent) momentum-transfer cross sections for the ion collection and scattering, respectively. They are given by

$$\sigma_c(v) = \int_0^{\rho_{c(v)}} 2\pi\rho d\rho, \quad \sigma_s(v) = \int_{\rho_{c(v)}}^{\rho_{max}} (1-\cos\chi) 2\pi\rho d\rho, \qquad (2)$$

where ρ is an impact parameter and $\chi(\rho)$ is the scattering angle. The impact parameter corresponding to ion collection is given by the OML theory: $\rho_c(v) = a(1 + 2e|\phi_s|/mv^2)^{1/2}$, where $|\phi_s|$ is the grain surface potential. The collection cross-section is

$$\sigma_c(v) = \pi a^2 (1 + 2e|\phi_s|/mv^2). \qquad (3)$$

The estimation of the orbital cross-section is a main problem addressed in this work. For the screened Coulomb potential and small grain, the scattering of ions is characterized by one dimensionless parameter

$$\beta(v) \simeq Ze^2/mv^2\lambda_D, \qquad (4)$$

where Z is the absolute magnitude of grain charge. Physically β is the ratio of Coulomb radius to the Debye length. For subthermal ion drift the characteristic quantity is $\beta(v_{T_i})$. The standard approach – "Coulomb potential + cut-off at $\rho_{max} = \lambda_D$" – is valid only if $\beta(v) \ll 1$. This is because the ratio of momentum transfer due to ions with $\rho < \lambda_D$ to the momentum transfer due to ions with $\rho > \lambda_D$ is proportional to $\ln[1/\beta(v)]$ in this case. Therefore, to within logarithmic accuracy, it is sufficient to consider ions with impact parameters below λ_D. For these ions the use of a bar Coulomb potential (instead of screened one) is a good approximation. For $\beta(v)$ of the order of unity or greater this approach fails: The range of ion-dust interaction can exceed the Debye length, the ions are scattered with large angles even if $\rho > \lambda_D$. This is illustrated in Fig. 1. The standard approach works well for electron-ion collisions in a usual plasma, where $\beta(v_{T_i}) \sim N_D^{-1} \ll 1$ (N_D is a number of ions inside the Debye sphere). For ion-grain collisions in complex plasmas, $\beta(v_{T_i}) \simeq z\tau a/\lambda_D$, where $z = Ze^2/aT_e$, $\tau = T_e/T_i$. For typical bulk plasma parameters and micron-sized grains $\beta(v_{T_i})$ is comparable or larger than unity as shown in Fig. 2. In this case the standard approach neglects a significant fraction of the ion momentum transfer (from ions with impact parameters above λ_D).

To improve the evaluation of the orbital cross-section it is necessary to take into account ions with impact parameters above λ_D. We propose the following simple procedure which is then justified by comparison with the available results of numerical calculations. To obtain analytical results we keep the approximation of the bar Coulomb potential, but the determination of ρ_{max} is revised. We take into account all the ions that *approach* the grain closer than λ_D. The the determination of ρ_{max} is $r_0(\rho_{max}) = \lambda_D$. The orbital momentum transfer cross-section is then

$$\sigma_s(v) = 4\pi\lambda_D^2\beta^2(v)\Gamma, \quad \Gamma(v) = \ln\left[\frac{1+\beta(v)}{a/\lambda_D+\beta(v)}\right], \qquad (5)$$

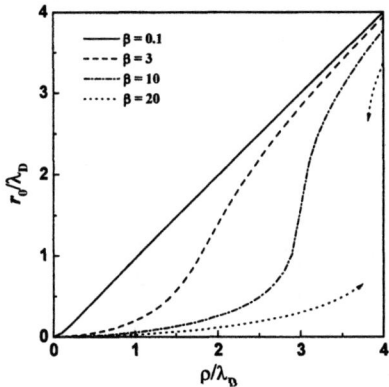

FIGURE 1. Normalized distance of the ion closest approach to the grain during collision, r_0, vs. the normalized impact parameter, ρ, for different values of β. The curves are calculated for a screened Coulomb potential. When β exceeds the critical value ($\beta \geq 13.2$) a discontinuity appears due to a potential barrier for the ions moving towards the grain. We do not consider this regime in the present work.

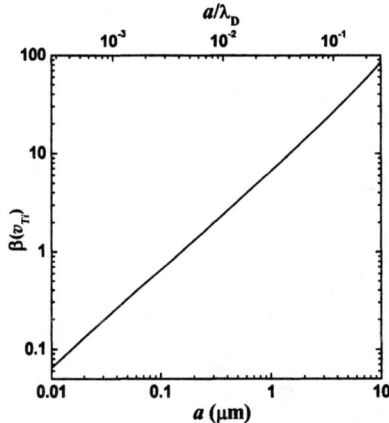

FIGURE 2. The parameter $\beta(v_{T_i})$ vs. grain radius, a. The curve is calculated for typical bulk plasma parameters: Ar gas, electron temperature $T_e = 2.5$ eV, electron to ion temperature ratio $\tau = 100$, electron concentration $n_e \simeq n_i = 10^9$ cm^{-3}

where Γ is a *modified Coulomb logarithm*. For a point-like grain $\Gamma(v) = \ln[1 + 1/\beta(v)]$, whilst within the standard approach $\Gamma(v) = \frac{1}{2}\ln[1 + 1/\beta^2(v)]$ (both forms are equivalent for $\beta(v) \ll 1$). The proposed modification of the Coulomb logarithm is in good agreement with self-consistent simulations [2] and with numerical results for an attractive Coulomb screened potential [3] up to $\beta(v) \sim 5$, as shown in Fig. 3.

Substitution of (3) and (5) into (1) gives for the ion drag force

$$F_I = \frac{8\sqrt{2\pi}}{3} a^2 n_i m v_{T_i} u \left[1 + \frac{1}{2}z\tau + \frac{1}{4}z^2\tau^2\Lambda \right], \qquad (6)$$

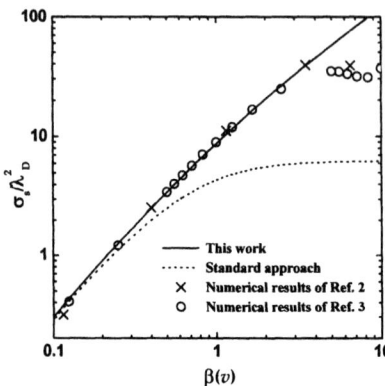

FIGURE 3. Orbital cross section of ion-grain collisions, σ_s, normalized to the squared Debye length, λ_D^2, vs. the parameter β for a point-like grain. Shown are results of the proposed approach (solid line), standard approach (dotted line), and numerical results of Refs. 2 (crosses), and Ref. 3 (circles).

where Λ is the modified Coulomb logarithm integrated over $f(\mathbf{v})$. $\Lambda \simeq 2F[\beta(v_{T_i})/2]$, where $F(x) = \int_x^\infty \exp(x-t)/t\, dt$. In a special case $\rho_c(v_{T_i}) \simeq \lambda_D$ or $a/\lambda_D \sim 1/\sqrt{2z\tau}$ the present approach gives for F_I the result $\sim \sqrt{8z\tau} \sim 40$ times higher than that of Ref. 1 for plasma parameters of Fig. 2. This large difference is due to the orbital part neglected in [1] for $\rho_c > \lambda_D$, whilst it still dominate over the collection part, according to our results.

Next we compare the magnitudes of the electrostatic, F_E and the ion drag force, F_I. Assuming $u = \mu_i E$, with $\mu_i = e l_i v_{T_i}/T_i$ we obtain for their ratio $F_I/F_E \simeq \delta l_i/\lambda_D$, where $\delta = \frac{1}{3\sqrt{2\pi}}\beta(v_{T_i})\Lambda$. δ is a slowly increasing function of $\beta(v_{T_i})$, ranging from ~ 0.3 to ~ 0.5 for $1 < \beta(v_{T_i}) < 10$. Our results were derived for the "collisionless" limit, so that $l_i \gg \rho_{max} \geq \lambda_D$. Hence, in the limit of weak electric fields the ion drag is stronger than the electrostatic force (as long as the proposed approach is applicable). This conclusion leads to a more physical insight into the mechanism of a "void" (dust-free region in the central part of a rf discharge) formation in complex plasma experiments under microgravity conditions. The electric field is weak in the center, and the ion drag (which is pointed outward) exceeds the electrostatic force (which is pointed to the center). The individual grains are pushed out of the center, leaving a void – as observed.

In conclusion, a simple procedure is proposed to improve the evaluation of the orbital part of the ion drag, which is justified by comparison with earlier (numerical) results. The ion drag force is significantly larger than previously expected. This might be important for understanding of the void formation, wave propagation, long-range interactions and other processes in complex plasmas.

REFERENCES

1. M.S. Barnes, *et al.*, Phys. Rev. Lett. **68**, 313 (1992).
2. M.D. Kilgore *et al.*, J. Appl. Phys. **73**, 7195 (1993).
3. H.-S. Hahn, E. A. Mason, and F. J. Smith, Phys. Fluids **14**, 278 (1971).

Langmuir Probe Measurements in a Complex Plasma under Microgravity Conditions

M. Klindworth*, A. Melzer*, A. Piel*, U. Konopka[†] and G. E. Morfill[†]

Institut für Experimentelle und Angewandte Physik Christian-Albrechts-Universität Kiel, 24098 Kiel, Germany
[†]*Max-Planck-Institut für extraterrestrische Physik, 85748 Garching, Germany*

Abstract. The plasma parameters of a complex plasma have been measured under microgravity conditions using a cylindrical Langmuir probe. One- and two-dimensional profiles of dusty and void regions are obtained, which show a flattened distribution of plasma potential and electron density. Characteristic changes of the plasma parameters correlate with the geometry of the particle cloud.

INTRODUCTION

In laboratory plasmas micron-sized particles are confined only in the lower plasma sheath due to the dominance of gravity [1]. Therefore Langmuir probe measurements in the bulk plasma are still rare [2]. Under microgravity conditions the particles fill a three-dimensional volume of the bulk plasma [3]. In many cases the particle distribution is inhomogeneous and a large particle free region (void) forms in the plasma center. The mechanism of this phenomenon is still not fully explained. Therefore it is highly desirable to use Langmuir probes under microgravity to obtain the plasma parameters with spatial resolution in dusty plasmas. Such Langmuir probe measurements have been performed for the first time in complex plasmas on parabolic flights in a cooperation between IEAP, University Kiel and MPE, Garching.

The main goals of this campaign are to demonstrate rapid scan techniques with automated Langmuir probes, to study the interacting of probe and dust cloud, and to obtain first sample parameters of the plasma parameters. Interesting investigations are related to the question of electron depletion in the presence of dust as a consequence of particle charging, or the appearance of the void in the center of the plasma [3, 4]. To identify the relevant forces one needs to know the plasma parameters in the presence of dust particles. These experiments are a pre-stage towards a probe diagnostic for the International Microgravity Plasma Facility.

EXPERIMENT, RESULTS AND DISCUSSION

The experiments have been performed in the PKE discharge chamber, which was designed and extensively used by MPE [3]. It has two symmetrical electrodes at a distance of 30 mm that are driven at 13.56 MHz rf power. The dust particles are spherical plastic particles of 3.4 μm diameter. The results presented below have been obtained in argon at a gas pressure of 37.5 Pa and 80 V_{pp} rf voltage. The Langmuir probe system was developed at IEAP and was specially designed for the use in the small sized rf plasma of PKE and for work under microgravity conditions during parabolic flights.

The probe can be shifted vertically and rotated about the probe shaft to scan a two-dimensional cross section trough the center of the plasma. The dust particles are illuminated by a laser fan and are observed by a ccd-camera. The experiment, field of view for the camera and an example of the dust cloud is shown in figure 1.

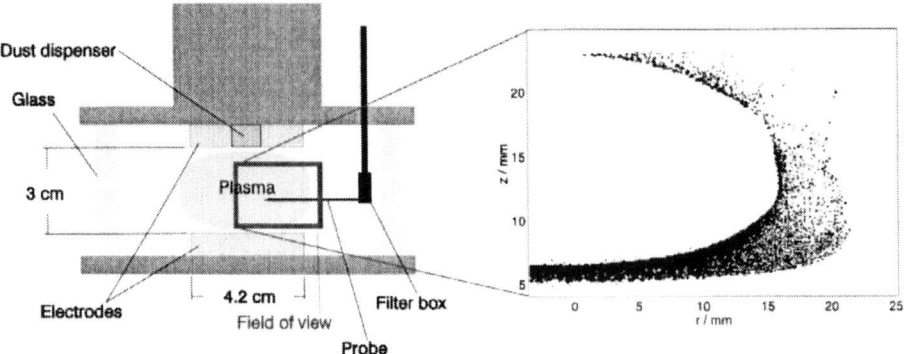

FIGURE 1. Scheme of the discharge chamber, camera field of view and dust distribution for the parameters of the presented experiment

For each parabola microgravity conditions are available for 20 s. During this time typically 10 to 36 probe characteristics can be recorded. The presented parameter profiles are composed of 2D-scans during three subsequent parabolas, then comprising 9×12 data points.

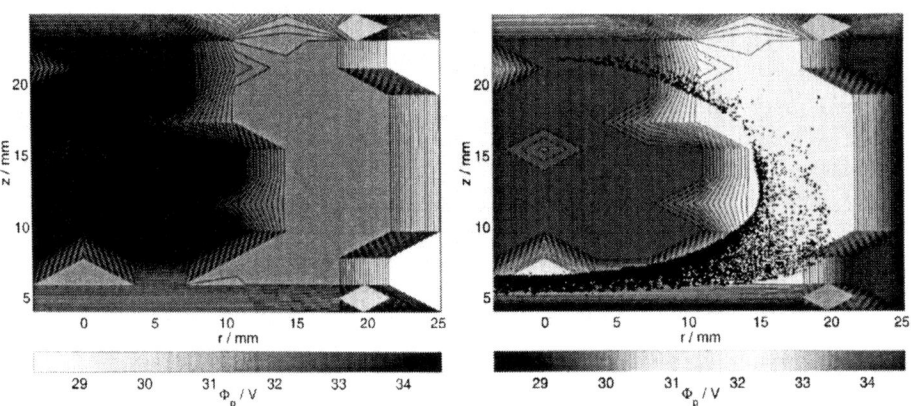

FIGURE 2. (a) plasma potential Φ_p and (b) the plasma potential profile with the superposed particle cloud. The electrodes extend to $r = 21$ mm

The applied probe voltage sweep was a simple sawtooth function. One result of this campaign was that the negatively charged particles were effectively attracted and deposited on the probe when the voltage sweep exceeded the plasma potential despite of the short duration of this phase. As can be seen in the camera observations and from later experiments in the laboratory the particles reach the probe from outermost tip and contaminate the probe during the measurement period. The probe contamination eventually results in a reduction of the electron current by a factor of 4 to 5. The data presented subsequently were obtained in the early phase of this contamination process, were the reduction in ion current is less than 40%. The presented data have not been corrected for this effect, yet. Recent laboratory experiments have shown that the contamination of the probe can be effectively suppressed by using a random probe voltage that consists of alternating voltage settings at a high frequency.

Figure 2(a) shows the plasma potential profile. As expected the plasma potential increases in vertical direction from the electrodes to the center. In dust free plasmas the radial decay of the potential is smooth (compare figure 3). However, here we find a sharp potential drop of $\Delta\Phi = 2V$ at a radial position of $r = 10$ mm. The inner, high potential side corresponds to the dust-free void region, whereas the low-potential region is correlated with the presence of the dust cloud (see fig. 2(b)). The boundary of the dust cloud and the steps in plasma potential show remarkable agreement.

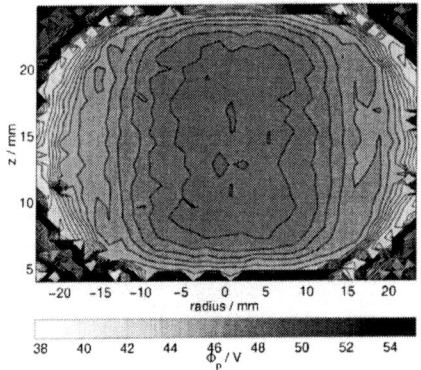

FIGURE 3. Plasma potential Φ_p for $p = 40$ Pa and $U_{rf} = 80$ V_{pp} in a dust free plasma in the laboratory.

As further finding, the plasma potential profile very well matches the electron density profile. There, a density drop of $\Delta n_e = 3 \cdot 10^7 cm^{-3}$ is found in the transition to the dust region (see fig. 4(b)).

From the probe characteristics a quite high electron temperature T_e in the plasma area between the electrodes is derived as can be seen in figure 4(a). Obviously, the electron temperature profile attains its maxima near the electrodes. A similar trend for the electron temperature was found in numerical simulations from Akdim *et al.* [5].

To judge the influence of the dust on the plasma, the charge density of the dust cloud was calculated using the average interparticle distance of 300 μm, the particle diameter, the measured electron temperature and the corrected density. The Havnes depletion parameter[6] $P = \frac{4\pi\varepsilon_0 a}{e} \frac{kT_e}{e} \frac{n_d}{n_e}$, with particle radius a, the dust density n_d and the undisturbed electron density n_e, of the order of unity is found. This means that the charge bound on the particles is comparable to the free electrons in the plasma. The presence of dust then results in a relevant depletion of free electrons and a plasma potential reduction in the particle cloud. This is in a good agreement with the observed plasma parameter profiles.

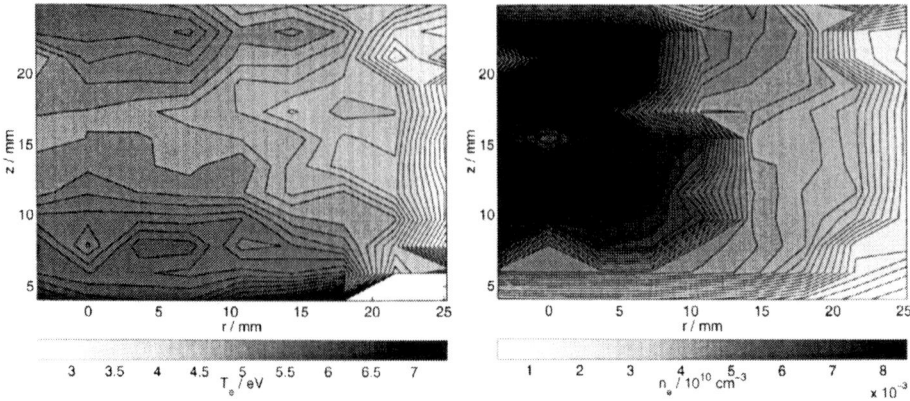

FIGURE 4. (a) electron temperature T_e and (b) electron density n_e for $p = 37.5$ Pa and $U_{rf} = 80$ V_{pp}.

This first Langmuir probe experiment under microgravity conditions demonstrated the applicability of this diagnostic to study inherent effects of complex plasmas.

ACKNOWLEDGMENTS

Helpful discussions with Dr. P. Bryant and technical assistance of K. Tarantik are gratefully acknowledged here. The project is supported by the DLR 50WM0039.

REFERENCES

1. Homann, A., Melzer, A., and Piel, A., *Phys. Bl.*, **52**, 1227–1231 (1996).
2. Barkan, A., D'Angelo, N., and Merlino, R. L., *Phys. Rev. Lett.*, **73**, 3093–3096 (1994).
3. Morfill, G. E., Thomas, H. M., Konopka, U., Rothermel, H., Zuzic, M., Ivlev, A., and Goree, J., *Phys. Rev. Lett.*, **83**, 1598–1601 (1999).
4. Goree, J., Morfill, G. E., Tsytovich, V. N., and Vladimirov, S. V., *Phys. Rev. E*, **59**, 7055–7067 (1999).
5. Akdim, M. R., and Goedheer, W. J., *Phys. Rev. E*, **65**, 015401 (2002).
6. Havnes, O., Goertz, C. K., Morfill, G. E., Grün, E., and Ip, W., *J. Geophys. Res.*, **92 A3**, 2281–2287 (1987).

Complex Plasma Experiments: the Role of Negative Ions

B.A. Klumov, A.V. Ivlev and G. Morfill

Max-Planck-Institut für Extraterrestrische Physik, D-85740 Garching, Germany

Abstract.
We investigate an influence of electronegative gases on the key plasma parameters in RF discharge with microparticles. Some contaminations, e.g., oxygen, water vapor, CO_2, etc. are often present in RF argon plasma. Under certain conditions the influence of such contaminations can be very significant. We have developed a numerical model describing the plasma composition and the spatial distribution of the plasma parameters (potential, density, etc) in the contaminated argon RF discharges with dust particles embedded. We showed that even tiny fractions of molecular oxygen can change plasma composition drastically. In turn, it can disturb significantly the spatial distribution of electrostatic potential and affect transport properties of the plasma.
We analyzed in detail the impact of a small fraction of molecular oxygen on argon plasma with microparticles. The plasma composition is changed, so that the main ions are O_2^+ and O^-. In particular, this decreases significantly the ion drag force acting on particles.
Another interesting effect is that the presence of molecular oxygen can cause heating of a neutral gas. Significant fraction of metastable oxygen $O(^1D)$ is produced in the discharge. The reaction of argon atom with metastable oxygen can increase the gas temperature by several degrees. The heating induces thermophoretic force acting on the dust particle which can be comparable in some cases with the ion drag force. Finally, we discuss the possible consequences of the negative ions presence for the complex plasma experiments under microgravity.

INTRODUCTION

Some contaminations, e.g., oxygen, water vapor, CO_2, etc. are often present in RF argon plasma. Under certain conditions the influence of such contaminations can be very significant due to high electronegativity of cited gases. In the paper we investigate effect of a small amounts of molecular oxygen added to argon discharge.

We have use one-dimensional mobility-diffusion model of a RF discharge in argon with molecular oxygen added. The spatio-temporal dependencies of number densities of ions, Ar^+, Ar_2^+, O_2^+, O^+, O_2^-, O^-, and electrons are calculated. We take into account the dissociation processes of O_2 and excitation of metastable electron states both in molecular oxygen ($O_2(a_1\Delta_g)$, $O_2(b_1\Sigma)$) and atomic oxygen ($O(^1D)$). We have solved the set of balance equations, describing density of i-constituent:

$$\frac{\partial n_i(x,t)}{\partial t} + \nabla J_i(x,t) = R_i^{prod}(x,t) - R_i^{loss}(x,t) \tag{1}$$

Here n_i and J_i denote the number density and flux of the i-species, and R_i are production and loss rates. The transport of ions and electrons is a combination of diffusion and drift in the electric field:

CP649, *Dusty Plasmas in the New Millennium: Third International Conference on the Physics of Dusty Plasmas*, edited by R. Bharuthram et al.
© 2002 American Institute of Physics 0-7354-0106-3/02/$19.00

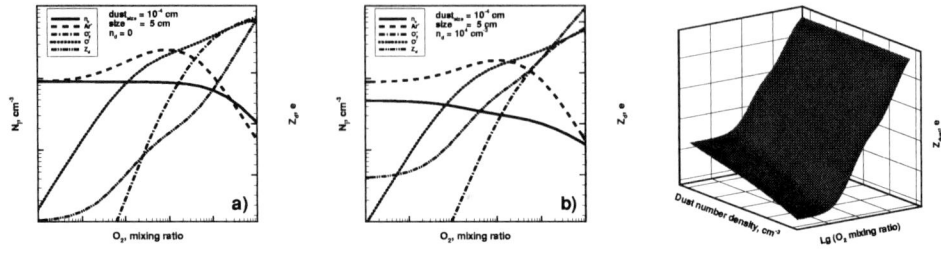

FIGURE 1. Argon plasma composition versus molecular oxygen mixing ratio, $[O_2]/[Ar]$. Figure a show the case of a small Havnes parameter, $P = Z_d n_d / n_e^0 \ll 1$, the figure b corresponds to moderate values of P. Right figure presents the dust particle charge, Z_d, versus dust number density, n_d, and molecular oxygen mixing ratio, $[O_2]/[Ar]$.

$$J_i(x,t) = \mu_i n_i(x,t) E(x,t) - D_i \nabla n_i(x,t) \qquad (2)$$

where μ_i and D_i are the mobility and diffusion coefficient of the i-constituent, $E(x,t)$ is the electric field distribution. In the bulk plasma we can use the Boltzmann approximation for the electrons:

$$D_e \nabla n_e(x,t) + \mu_e n_e E(x,t) \approx 0 \qquad (3)$$

The boundary conditions for the equations are $\partial n_i / \partial x = 0$ at the center of the discharge and $n_i = 0$ at the walls. Figure 1. shows the impact of molecular oxygen added to RF discharge on both plasma composition and microparticle charge.

IMPACT OF NEGATIVE IONS ON ION DRAG FORCE

We use Khrapak's approximation for the ion drag force at small drift velocities (PRL, submitted, 2001):

$$F_i \approx \frac{8\sqrt{2\pi}}{3} \rho_0 \lambda_i n_i m_i v_{th,i} u \qquad (4)$$

where ρ_0 is the Coulobm radius, λ_i is the ion Debye length. For high drift velocities we use the standard approach (e.g. *Barnes* et al, PRL, 1992). Figure 2. shows the RF plasma composition and associated fluxes of charge species for cases of weak and moderate electronegativity of RF plasma. Figure 3. shows the impact of the molecular oxygen added to discharge on ion drag force. We note that small amount of molecular oxygen added to discharge drastically changes the forces acting on dust particle.

HEATING OF NEUTRAL GAS BY METASTABLE SPECIES

Addition of molecular oxygen to argon plasma result in the formation of a large number of metastable molecules and atoms of oxygen. Among them are: $O_2(a_1 \Delta_g)$, $O_2(b_1 \Sigma)$, $O(^1D)$ plus

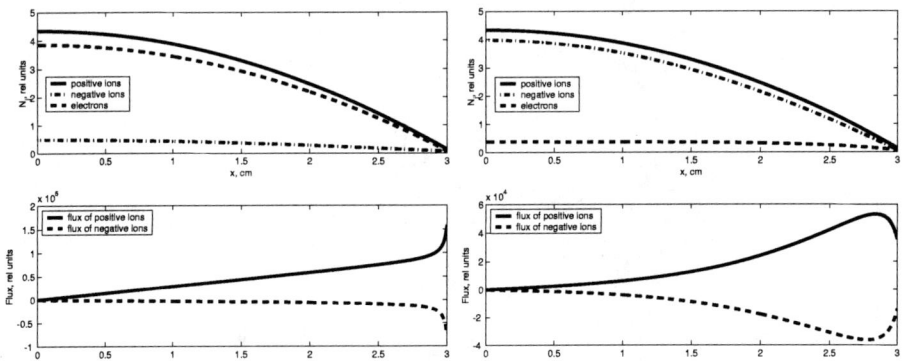

FIGURE 2. Spatial profiles of positive and negative ions and electrons (upper figures), as well as fluxes of positive and negative ions (bottom figures) in argon RF discharge with different molecular oxygen additions. Left figures correspond to low [O$_2$]/[Ar] ratio case, while right figures represent the case of rather high level of electronegativity of the plasma.

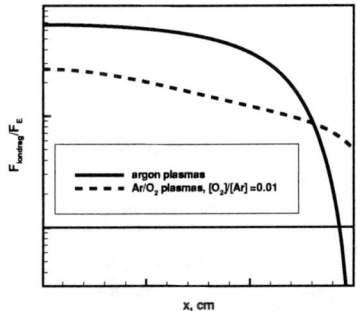

FIGURE 3. Spatial distribution of the ion drag-to-electrostatic force ratio for different values of O$_2$ mixing ratio.

vibrationally excited molecular oxygen $O_2^{(v)}$. The collisions between metastable constituents and neutral gas can cause strong heating of the neutral gas. The efficiency of this mechanism in the local approximation ($\tau_i = (k_i^{dex}[\text{Ar}])^{-1} \ll \tau_D$, where k_i^{dex} is the constant of collisional de-excitation of i-process) can be estimated as follows:

$$\Delta T_n \approx \sum_i k_e^i n_e \frac{[\text{O}_2]}{[\text{Ar}]} \tau_D \varepsilon_i, \quad \tau_D = L^2/D_n \tag{5}$$

where k_e^i is the constant of the process e + O$_2$/O $\rightarrow$ e + O$_2^*$/O*, ε_i is the energy of the proper transition, τ_D - is the diffusion time-scale.

Figure 4. shows the impact of metastable specie O(^{1}D) on neutral heating and associated thermophoretic force. Thus, addition of molecular oxygen can cause heating of a neutral gas up to a few degrees, depending on O$_2$ mixing ratio, electron temperature, and size of the discharge

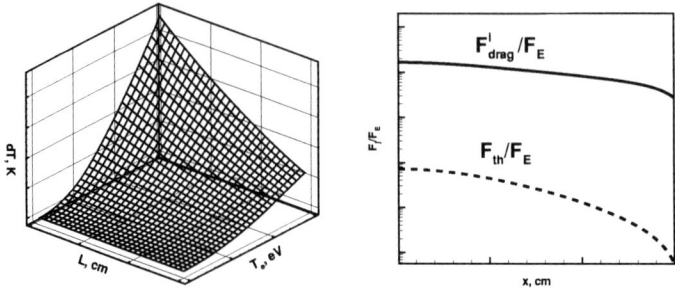

FIGURE 4. Heating of the neutrals (argon) by $O(^1D)$ versus electron temperature T_e and size L of discharge region (left figure). Right figure presents the comparison of thermophoretic, electrostatic, and ion drag forces acting on 1 μm particle in RF argon/oxygen plasma. O_2 mixing ratio is 0.01.

region. The heating-associated thermophoretic force, $f_{th} \approx -2r_d^2/v_{th}k_T \nabla T$, is much smaller than the ion drag force for micron-size particles and typical discharge parameters. But increase of dust particle size and decrease of pressure of the discharge can make the thermophoretic force to be dominant.

CONCLUSIONS

Presence of a tiny amount of molecular oxygen drastically changes the RF plasma composition. In Argon plasmas, the main ions are positive ion O_2^+ and negative ion O^-, with significantly decreased electron density. As a result, the transport properties of the plasmas are changed significantly.

Negative ions result in considerable red reduction of electric field in the bulk of plasma. In turn, this leads to significant decrease of both the ion drag and electrostatic forces acting on dust particle.

The spatial dependence of the ion drag-to-electrostatic force ratio changes significantly in the presence of negative ions. Thus, negative ions can change drastically the void formation in complex plasmas. This is in a good agreement with the experimental data.

The charge of dust particle is strongly decreased in the presence of molecular oxygen in argon plasmas.

REFERENCES

1. Khrapak S.A., Ivlev A.V. and G. Morfill, submitted to *Phys Rev. Lett.*, 2002
2. Barnes M.S., Keller J.H., Forster J.C., O'Neill J.A., Coultas D.K., *Phys. Rev. Lett.*, **68**, 313, (1992)

Particle-in-Cell Simulation of Helical Structure Onset in Plasma Fiber with Dust Grains

Petr Kulhánek, David Břeň, Václav Kaizr and Jan Pašek

Czech Technical University, Faculty of Electrical Engineering, Department of Physics, Technická 2, 166 27 Prague 6, Czech Republic

Abstract. Fully three dimensional PIC program package for the helical pinch numerical simulation was developed in our department. Both electromagnetic and gravitational interactions are incorporated into the model. Collisions are treated via Monte Carlo methods. The program package enabled to prove the conditions of onset of spiral and helical structures in the pinch.

INTRODUCTION

Three–dimensional PIC code ([1]) with periodical boundary conditions for simulation of the plasma fiber behavior was developed in our department during last two years. The aim of our project is modelling of helical plasma structures induced by the diocotron instability. These structures are observed in cometary's tails, H II regions, nebulas and AGN jets. Nowadays about 500 000 particles are included in the model (electrons, ions, dust grains and neutrals). The particle motion solver for electrons and ions contains only electric and magnetic forces. The particle motion solver for dust grains contains both electromagnetic and gravitational forces. The collective processes of both electromagnetic and gravitational origin are incorporated in the model.

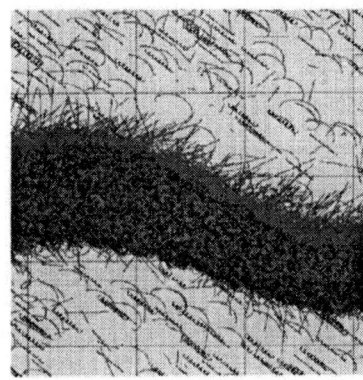 

FIGURE 1. Simulation of the plasma fiber (left). Magnetic field visualized by LIC method (right).

The collisions between the four groups of particles are computed by Monte Carlo method based on the inversion of the effective cross section formulas. Furthermore the possibility of neutrals ionization, ions recombination and dust grain charge change are included in the model. The code was written in FORTRAN 95 (Compaq Visual Fortran 6.5) and it is fully 3–D The model output concerns various diagnostic routines that provide information about average values and quadratic fluctuations of both particle and field variables. Plasma conductivity, specific heat, electric and magnetic susceptibility are the quantities to be compared with the experimental ones. The main aim of our project is the numerical simulation of helical structure onset in electromagnetically and gravitationally driven plasma fiber. Radial perturbation of electric charge distribution leads to radial electric field. This field together with axial magnetic field causes azimuth drift of all kinds of charged particles and to the onset of surface instabilities which evolve into typical helical structures observed in the space plasmas.

FIELD AND PARTICLE SOLVERS

The basic interactions are described by scalar and vector potential of the electromagnetic force ad by the gravitational potential. For low frequency modes the Laplace–Poisson equations for the potentials

$$\Delta \varphi = -\frac{\rho_e}{\varepsilon_0}; \qquad \Delta \mathbf{A} = -\mu_0 \mathbf{j}; \qquad \Delta \varphi_g = 4\pi G \rho_m \qquad (1)$$

are solved via FFT and multigrid methods [2]. The field are localized in the grid points of rectangular mesh. For high frequency modes corresponding wave equations have to be solved. The fields are visualized by LIC (Line Integral Convolution) method [3].

There are implemented four kinds of particles in the model: 1) ions, 2) electrons, 3) dust grains and 4) neutrals. The dust grains can be charged both positively and negatively. Dust grains interact via gravitational and electromagnetic forces. In this kind of plasma, there are three typical frequencies

$$\omega_{pe} = \sqrt{\frac{n_e Q_e^2}{m_e \varepsilon_0}}; \qquad \omega_{pi} = \sqrt{\frac{n_i Q_i^2}{m_i \varepsilon_0}}; \qquad \omega_{pd} = \sqrt{\frac{n_d Q_d^2}{m_d \varepsilon_0}}. \qquad (2)$$

The dust grains are responsible for ultra low frequency wave modes, ions for magnetoacoustic modes and electrons dominantly modify the propagation of electromagnetic waves through plasma.

There are implemented several particle solvers, the most important are 1) Runge–Kutta, 2) Buneman, 3) Leap–Frog and 4) Canonical. These solvers are described in detail in [4]. During motion a part of the particles acquires considerable velocities in comparison with the speed of light. That is why in 2002 were incorporated relativistic variants of these schemes [6] into the PIC package. Collisions with neutrals in the model are treated only statistically. The velocity and direction after the collision is calculated randomly according to the scattering cross–section from literature. During the collision the probability of ionization and recombination processes is calculated [6] and so the number of particles in the model is not fixed.

RADIATION

The radiation of the fibers in nebulas is very important feature because it is the most efficient channel of energy losses. A programme package solving synchrotron and bremsstrahlung radiation. The radiative fields are visualized on a two–dimensional sphere located far away from the radiation sources. The package includes visualization procedures that project the irradiated intensity on far sphere.

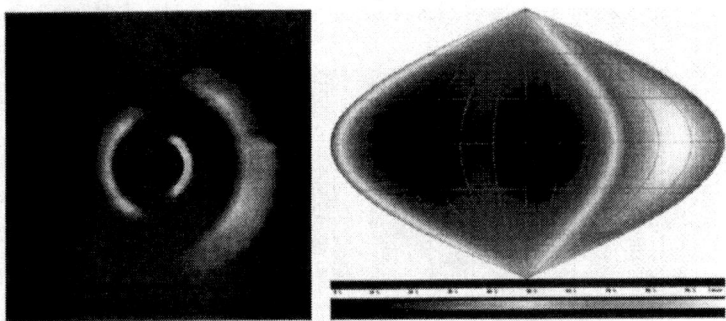

FIGURE 2. On the figure there is a typical intensity in the case of circular motion; tests of the programme package (left). Projection od radiation intensity on a far sphere; velocity of the charge is 45% of the speed of light (right).

The plasma current generates magnetic field that is important for the equilibrium. However the current increases fiber temperature via Joule heating. Increasing temperature leads to higher plasma pressure and the equilibrium would be instable. Role of the opposite channel to the Joule heating plays radiation of the fiber. Radiative processes cool the fiber and enable maintenance of the equilibrium. First calculations of radiative equilibrium configuration did Pease and Braginski. They discovered that under some conditions the cooling process can overcome the heating one and the fiber electromagnetically collapses to the axis. The most effective radiative processes are only three: bremsstrahlung, synchrotron and recombination radiation.

It is necessary to use the relativistic schemes for the calculation of the bremsstrahlung and synchrotron radiation, dominating during high velocities. From motion of the charged particle, we can calculate intensity and dependence on direction.

$$\begin{aligned} I &= \mathbf{E}\mathbf{E}^* \ ; \\ \mathbf{E} &\propto \tfrac{\mathbf{n}}{\kappa^3 R} \times \left(\left(\mathbf{n} - \tfrac{\mathbf{v}}{c}\right) \times \tfrac{\dot{\mathbf{v}}}{c} \right)\Big|_{t'} \ ; \\ \kappa &\equiv \left(1 - \tfrac{\mathbf{n}\cdot\mathbf{v}}{c}\right) \ ; \quad \mathbf{n} \equiv \tfrac{\mathbf{R}}{R} \ ; \\ \tfrac{dP}{d\Omega} &= \tfrac{e^2}{16\pi^2 \varepsilon_0 c} \tfrac{\left|\mathbf{n}\times\left(\left(\mathbf{n}-\tfrac{\mathbf{v}}{c}\right)\times\tfrac{\dot{\mathbf{v}}}{c}\right)\right|^2_{t'}}{\kappa^5} \ . \end{aligned} \quad (3)$$

PRESENT STATE AND RESULTS

Particles move in rectangular parallelepiped with periodical boundary condition. Their initial positions and velocities are similar to Bennett pinch. It was proved that radial perturbations lead to onset of surface turbulence (diocotron instability). The mechanism is as follows: Induced radial electric field together with axis component of the magnetic field cause azimuthal drift of the charged particles. These particles form vortex magnetic fields near the pinch boundary which results in spiral and helical structures. Similar structures are detected in the whole range of scales both in laboratory and space plasmas.

During 2001 relativistic particle solvers and Monte Carlo simulation of the charged particle – neutral collisions were included in the PIC program package [6]. Concurrently started work on the simulation of radiation processes and radiation losses in the pinch [5].

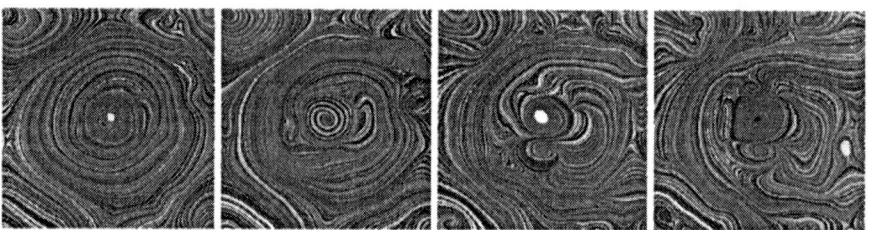

FIGURE 3. The onset of surface turbulence in a plasma filament. This turbulence results in spiral and helical structures of the fiber. The computer times are succeedingly 10.3, 11.7, 12.6, 13.4. Detailed introduction of the nondimensional variables used can be found in [4]. Similar structures are detected in the whole range of scales both in laboratory and space plasmas.

ACKNOWLEDGMENTS

This research has been supported by the research program No J04/98:212300017 *"Research of Energy Consumption Effectiveness and Quality"* of the Czech Technical University in Prague, by the research program INGO No LA 055 *"Research in Frame of the International Center on Dense Magnetized Plasmas"* and *"Research Center of Laser Plasma"* LN00A100 of the Ministry of Education, Youth and Sport of the Czech Republic.

REFERENCES

1. Birdsall, C. K.: *Particle in Cell Charged – Particle Simulations, Plus Monte Carlo Collisions With Neutral Atoms*, PIC–MCC. IEEE Trans. on Plasma Science **19** (1991), 65.
2. John, C. A.: Multigrid package MUDPACK 5.0. National Center for Atmospheric Research; http://www.scd.ucar.edu/css/software/mudpack; 1999.
3. Risquet, C. P.: *Visualizing 2D flows: Integrate and Draw.* http://www.informatic.uni-rostock.de/~carlos/FlowVisualization/IntegrateDraw.html
4. Kulhánek, P.: *Particle and Field Solvers in PM Models*, Czechoslovak Journal of Physics, Vol. **50** Suppl. S3 (2000), pp. 231–244.
5. Břeň, D.: *Numerical Modeling of Energy Losses in z–Pinches*, Czechoslovak Journal of Physics, Vol. **52** Suppl. D (2002), pp. 226–230.
6. Škandera, D.: *Additional modules for the PIC program package*, Diploma Thesis, Czech Technical University, 2002.

Collisional and Nonlinear Effects on Grain Charge and Intergrain Force

Martin Lampe*, Gurudas Ganguli*, Valeriy Gavrishchaka[†], Rajiv Goswami** and Glenn Joyce*

*Plasma Physics Division, Naval Research Lab, Washington, DC 20375-5346, USA
[†]SAIC, McLean, Virginia 22101, USA
**Physics Dept., University of Maryland, College Park, MD 20740, USA

Abstract. In recent work we have shown that trapped ions, created by occasional collisions, typically dominate the plasma response to a stationary dust grain in non-flowing plasma. A procedure is outlined for calculating the ion current to the grain, including trapped ions, and it is shown that the trapped ion current is typically larger than the OML ion current. The electrostatic interaction between two grains is also discussed. This problem is solved in context of OML theory, with no trapped ions included. The result is that the interaction is purely repulsive, and nonlinear terms due to the presence of two grains increase the repulsion as compared to the one-grain potential. The two-grain problem has not yet been solved with trapped ions, but the possibility of attractive electrostatic forces is discussed.

TRAPPED IONS AND GRAIN CHARGING

For a stationary dust grain in a stationary plasma, the orbital-motion-limited (OML) theory [1-7] is usually used to determine the shielding around a dust grain and the ion current to the grain. This theory assumes that the plasma ions can be treated as collisionless, and therefore all of the ions and electrons in the vicinity of the collector must be positive energy particles which come in from the ambient plasma ($r = \infty$) and either hit the collector or fly back out to $r = \infty$. However, negative-energy ions can be created and trapped in the potential well around the grain, as a result of occasional collisions in which ions lose energy [2,8]. A trapped ion, once created, will stay trapped for a very long time. The trapped ion density $n_t(r)$ can thus slowly build up to a large value. In fact, it turns out that the steady state density of trapped ions, in the limit of small collision frequency v, is finite and independent of v [8]. [The limit $v \to 0$ is resolved as follows. The time to reach steady state is proportional to v^{-1}, and thus steady state is never reached if v is rigorously zero. However, for conditions of interest, v is typically small in the sense that the mean free path λ_{mfp} is large compared to the Debye length λ_D, but the time to reach steady state is short compared to macroscopic times.] Monte Carlo calculations [8,9] have shown that the total number of trapped ions can be large.

In a recent paper, we developed a fully analytic method for calculating the distribution of trapped as well as untrapped ions [10]. The calculation follows the full collisional kinetics of the ions, and determines the distribution function of the trapped ions fully self-consistently, with no arbitrary parameters. Two primary assumptions are made: that

v is small in the sense that $\lambda_D/\lambda_{\text{mfp}}$ (which is usually well justified), and that v is energy-independent. The latter is not usually quantitatively accurate, but is a model assumption that makes the problem analytically tractable. We find that typically the trapped ion density $n_t(r)$ in the inner part of the shielding cloud is an order of magnitude larger than the untrapped ion density $n_u(r)$ which is used in OML theory.

The presence of trapped ions also sharply increases the ion flow to the grain, even when v is small. The reason for this is that, when an untrapped ion has a charge-exchange collision, the result is a new ion whose kinetic energy is of the order of the neutral temperature T. If the collision occurred at a point r within the negative potential well around a grain, i.e. where

$$|\phi(r)| \gg T \tag{1}$$

the new ion is almost certainly trapped in the potential well. It may fall onto the grain immediately, or it may enter an orbit encircling the grain. Subsequent charge-exchange collisions nearly always cause the grain to fall deeper into the well, and eventually (perhaps after several collisions) the ion is nearly certain to fall onto the grain. Thus any collision of an untrapped ion that occurs in the region where $|\phi(r)| \gg T$ essentially results in an ion hitting the grain. The depth of the potential well (i.e. the potential at the grain surface) is typically $\sim T_e \gg T_i$, and Eq. (1) is satisfied for a large region $r < r_0$, where a is the grain radius and r_0 is of the order of $a(T_e/T)$ or λ_D, whichever is smaller. Thus the cross-section for an ion coming in from $r = \infty$ to cross the region (1) is πr_0^2. The probability for a collision to occur while the ion is in that region is of order r_0/λ_{mfp}. Thus the effective cross-section for an incoming ion to deposit collisionally on the grain is of order $\pi r_0^3/\lambda_{\text{mfp}}$. Although this cross-section is proportional to the (small) collision frequency v, it is nonetheless usually large compared to the OML cross-section for collisionless deposition, which is of order $\pi a^2(T_e/T)$.

Once the distribution of trapped ions is known, the ion flow to the grain can be calculated [11]. The following formulation was developed by Robertson and Sternovsky [12] for a cylindrical probe and adapted to our calculation for a spherical dust grain. The ion flux to the grain can be expressed as

$$F_i = \frac{v}{4\pi a^2} \int_a^\infty dr\, 4\pi r^2 [n_t(r) + n_u(r)] p(r), \tag{2}$$

where r is the location where the ion had its last collision, and $p(r)$ is the probability that the ion's trajectory after the collision will take it onto the grain, and that it will not have another collision before reaching the grain. By using conservation of energy and angular momentum, it is possible to specify in closed form the conditions such that the ion trajectory will cross the grain, but the result [13] is lengthy and will not be reproduced here. On the other hand, it is not possible to give an exact closed-form expression for the probability $P_{\text{coll}}(r,v)$ that the ion will reach the grain before having another collision, but we can show [13] that the simple approximate expression

$$P_{\text{coll}}(r,v) = \exp(-vr/v) \tag{3}$$

is sufficiently accurate in the limit of small v. We are able to show [13] that the OML flux is included in the contribution to [2] from the region $r > \lambda_{\text{mfp}}$, and that in the limit

$v \to 0$ Eq. (2) reduces to the OML flux. Note that Eq. (2) has a factor v in front of the integral, but the integral itself diverges as v^{-1} in the limit $v \to 0$, because the mean location where the ion had its last collision is inversely proportional to v. Thus the result for F_i is non-zero in the limit $v \to 0$. Numerical evaluation of Eq. (2) indicates that Fi can be much larger than the OML ion current, even when v is small. For example, we find $F_i/F_{OML} = 6$ for the case with $a/\lambda_D = 0.015$, $T/T_e = 0.01$, and $v\lambda_D/c_s = 0.03$. The floating potential at the grain surface is determined by the requirement that the ion flux to the grain equal the electron flux. The substantial increase in ion current, due to collisions, results in a decrease in the grain potential by as much as a factor of order two. The grain charge is proportional to the grain potential, and thus is similarly reduced as compared to the usual OML result.

INTERACTION BETWEEN TWO GRAINS

The force exerted by one grain on another grain is not necessarily the same as the force derived from the potential around a single isolated grain. There can be nonlinear deformation of the shielding cloud around one grain due to the presence of the other grain, and if this results in extra positive charge collecting in the region between the two grains, it is possible that there could be a net attractive electrostatic force between two negatively charged grains [14,15]. In an earlier paper [7], we derived the OML result for the ion density $n_i(r)$ in the vicinity of two nearby grains. It is not possible to obtain an exact analytic result, because only one component of angular momentum is conserved in the two-grain system, but it is possible to write a good approximate expression in the limit where the grain separation is large compared to the grain radius a. In [7] we outlined a rather complicated analytic argument that nonlinear corrections actually *increase* the repulsion between two grains. We have now performed numerical solutions of the OML equations for two grains, and obtained the complete self-consistent result for the potential $\phi(r)$, including nonlinear interactions between the grains. The force on one grain, due to the other grain, is given by the gradient of this potential, integrated over the charge on the grain surface. The computation is complicated a bit because $\phi(r)$ near one of the grains is dominated by the self-Coulomb potential of that grain, which is very large but of course exerts no net force on the grain. When the self-potential is subtracted out, the remaining potential $\phi(r)$ is smoothly varying and easily differentiated to give the force on the grain. The result is shown in Figure 1 as a function of the grain separation r. The result is indeed that the force is purely repulsive, and that the repulsion is slightly increased by nonlinear interactions.

The two-grain problem with trapped ions has not yet been solved. The problem is complicated because trapped ions can have stochastic orbits, since there are not enough constants of the motion to specify the orbits. As a result, it is possible for a trapped ion to circle one or both grains several times, and then deposit onto a grain without undergoing a collision. We are looking into various approximate solutions for the trapped ion density self-consistent with Poisson's equation for the potential. It is possible that an attractive electrostatic force results from the overlap of the trapped ion clouds surrounding the two grains. A grain together with its trapped ion cloud is in some sense like a classical

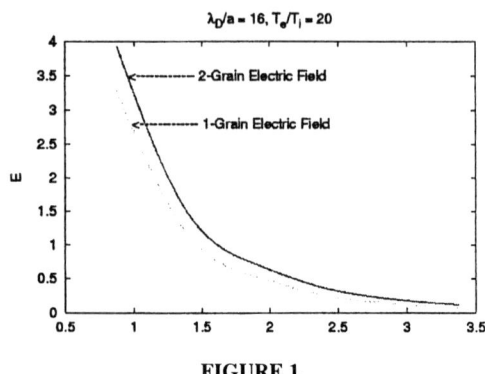

FIGURE 1.

atom. When exposed to external fields, the trapped ion cloud becomes polarized, thereby shielding the grain. When two grains are nearby, Van der Waals attractive forces should result from the mutual polarization of the two trapped ion clouds. However it is not known whether Van der Waals forces can be strong enough to overcome Coulomb repulsion and lead to net attractive forces.

ACKNOWLEDGMENTS

This work was supported by NASA and ONR.

REFERENCES

1. Mott-Smith, H.Jr., and Langmuir, I., *Phys. Rev.*, **28**, 27 (1926).
2. Bernstein, I.B., and Rabinowitz, I.N., *Phys. Fluids*, **2**, 112 (1959).
3. Laframboise, J.G., Univ. of Toronto, Inst. For Aerospace Studies, Report #100 (1966).
4. Laframboise, J.G., and Parker, L.W., *Phys. Fluids*, **16**, 629 (1973).
5. Allen, J.E., *Physica Scripta*, **45**, 497 (1992).
6. Daugherty, J.E., Porteus, R.K., Kilgore, M.D., and Graves, D.B., *J. Appl. Phys.*, **72**, 3934 (1992).
7. Lampe, M., Joyce, G., Ganguli, G., and Gavrishchaka, V., *Phys. Plasmas*, **7**, 3851 (2000).
8. Goree, J., *Phys. Rev. Lett.*, **69**, 277 (1992).
9. Zobnin, A.V., Nefedov, A.P., Sinel'shchikov, V.A., and Fortov, V.E., *JETP*, **91**, 483 (2000).
10. Larnpe, M., Gavrishchaka, V., Ganguli, G., and Joyce, G., *Phys. Rev. Lett.*, **86**, 5278-5281 (2001).
11. M. Lampe, V. Gavrishchaka, G. Joyce, and G. Ganguli, *Physica Scripta*, **T98**, 91–91 (2002).
12. Robertson, S. and Sternovsky, Z., *in these Proceedings*.
13. Lampe, M., Ganguli, G., Joyce, G., Goswami, R., Robertson, S., Stemovsky, Z., and Gavrishchaka, V., *to be published*.
14. Tsytovich, V.N., *Comments Plasma Phys. Controlled Fusion*, **15**, 349 (1994); Tsytovich, V.N., and Resendes, D., *Plasma Physics Reports*, **24**, 65 (1998).
15. Resendes, D.P., Mendonca, J.T., and Shukla, P.K., *Phys. Lett. A*, **239**, 181 (1998).

A Kinetic Study of Electron Density Fluctuation in a Dusty Plasma

Fang Li

Institute of Electronics, Academia Sinica, Beijing, P.R.China

Abstract. In the Earth's mesosphere observations have shown that the radar-scattering is strongly depend on the electron density fluctuation produced by the dust particles in the layer of NLC (the nuctilucent clouds). In this presentation the electron density fluctuation of a dusty plasma is studied by kinetic methods.

INDUCED FLUCTUATION BY THE DUST CHARGE

The induced density fluctuation due to collective interaction of a plasma is [1]

$$n_\alpha^i = -\frac{1}{q_\alpha}\chi_{l\alpha}(\omega,\vec{k})\rho(\omega,\vec{k}), \tag{1}$$

where $\chi_{l\alpha}(\omega,\vec{k})$ is the longitudinal susceptibility of the plasma component, for a dusty plasma $\alpha = e, i, d$ for electrons, ions, and dusty particles, respectively. In a dusty plasma there is an independent dusty charge fluctuation ρ_d, then

$$\rho(\omega,\vec{k}) = \rho_d - \sum_{\alpha=e,i,d}\chi_{l\alpha}(\omega,\vec{k})\rho(\omega,\vec{k}). \tag{2}$$

It can be shown from Eqs. (1) and (2) that a dust charge induced electron density fluctuation is

$$n_e^i(\omega,\vec{k}) = -\frac{1}{e}\left\{\frac{\rho_d(\omega,\vec{k})\chi_{le}(\omega,\vec{k})}{1+\chi_{le}(\omega,\vec{k})+\chi_{li}(\omega,\vec{k})+\chi_{ld}(\omega,\vec{k})}\right\}, \tag{3}$$

and the power spectrum of the fluctuation is

$$\mathcal{N}(\omega,\vec{k}) = \left\langle |n_e^i(\omega,\vec{k})|^2 \right\rangle$$
$$= \frac{1}{e^2}\left\{\frac{\langle|\rho_d(\omega,\vec{k})|^2\rangle|\chi_{le}(\omega,\vec{k})|^2}{|1+\chi_{le}(\omega,\vec{k})+\chi_{li}(\omega,\vec{k})+\chi_{ld}(\omega,\vec{k})|^2}\right\}. \tag{4}$$

DUST CHARGE FLUCTUATION

According to the fluctuation-dissipation theorem a dust charge fluctuation is given by [1]

$$\left\langle |\rho_d(\omega,\vec{k})|^2 \right\rangle = -\frac{k^2}{\omega} \text{Im} \left\{ \frac{1}{\varepsilon_{ld}(\omega,\vec{k})} \right\} \frac{k_B T_d}{2\pi}. \tag{5}$$

The susceptibility of a dust component of a dust plasma can be derived by the methods of kinetic theory. The equation of the dust particle distribution function f_d in a phase-space $\{\vec{r}, \vec{v}_d, q_d\}$ is

$$\frac{\partial f_d}{\partial t} + \vec{v}_d \cdot \frac{\partial f_d}{\partial \vec{r}} + \frac{\vec{F}}{m_d} \cdot \frac{\partial f_d}{\partial \vec{v}_d} + \frac{\partial}{\partial q_d}[I_c(q_d)f_d] = 0, \tag{6}$$

where $\vec{v}_d$ and q_d is the velocity and the charge of the dusty particle respectively. In this equation $I_c(q_d)$ is the charge current due to the collision of the plasma particles with the dust particles which is a function of q_d. The equilibrium charge q_0 can be obtained from

$$I_c(q_0) = \sum_{\beta=e,i} I_\beta(q_0) = 0.$$

On the assumption that all plasma species have Maxwellian velocity distributions, from Eq. (6) the longitudinal susceptibility of the dust component has been found to be [2].

$$\chi_{ld}(\omega,\vec{k}) = \frac{k_{Dd}^2}{k^2}[1 + \zeta_d \mathscr{Z}(\zeta_d)] + iP\frac{\Omega_C}{\omega + i\Omega_C}\frac{k_{D0}^2}{k^2}, \tag{7}$$

where $\mathscr{Z}(\zeta_d)$ is the plasma dispersion function with argument $\zeta_d = \frac{\omega}{kv_{td}}$, $k_{Dd}^2 \equiv D_d^{-2} = \frac{4\pi n_d q_0^2}{k_B T_d}$, $k_{D0}^2 = \frac{4\pi n_0 e^2}{k_B T_e}$; v_{td}, m_d, T_d is the thermal velocity, the mass, and the temperature of the dusty particle; m_e the mass of electron; and n_0 the density of the plasma without the dust; respectively. The charge frequency of a dust particle is defined

$$\Omega_c = -\frac{\partial}{\partial q_d} I_c |_{q_d = q_0}, \tag{8}$$

which is a parameter that shows the dust charge changes in the plasma. For plasma particles that are of Maxwellian distribution

$$\Omega_c = \frac{a}{\sqrt{2\pi}} \left[\frac{\omega_e^2}{v_{te}} \exp\left(-\frac{v_*^2}{2v_{te}^2}\right) + \frac{\omega_i^2}{v_{ti}} \right], \tag{9}$$

where a is the radius of the dusty particle, $v_{t\beta}^2 = \frac{2k_B T_\beta}{m_\beta}$ ($\beta = e, i$), while

$$v_*^2 = -\frac{2q_0 e}{am_e}.$$

The dust parameter

$$P = \frac{n_{d0} a k_B T_e}{n_0 e^2} \tag{10}$$

describes the importance of the dust charges in the plasma.

It can be derived from Eqs. (5), (7) that there are three parts of charge fluctuation associated with dust particles which comes from the three imaginary components of the $\frac{1}{\varepsilon_{ld}(\omega,\vec{k})}$, that is

$$\langle |\rho_d(\omega,\vec{k})|^2 \rangle = \mathscr{P}_T + \mathscr{P}_C + \mathscr{P}_W. \tag{11}$$

where $\mathscr{P}_T$ represents the fluctuation from the thermal motion of the dust particles, $\mathscr{P}_C$ from the charge change of a dusty particle, and $\mathscr{P}_W$ comes from the eigen mode oscillation of the dust particles. Each part is separately discussed in the following section.

ELECTRON DENSITY FLUCTUATION DUE TO DUST CHARGE

(I) Case of $kv_{td} \gg \Omega_c$ and $kv_{td} \gg \omega_d$

In this case one has $\langle |\rho_d(\omega,\vec{k})|^2 \rangle \simeq \mathscr{P}_T$, it is *thermal fluctuation*.

$$\langle |\rho_d(\omega,\vec{k})|^2 \rangle = \frac{n_d q_0^2}{\omega} (kD_d)^2 \zeta_d Im(\mathscr{Z}(\zeta_d)). \tag{12}$$

We have that the spectrum of the fluctuation in this condition from Eq. (4) is

$$\begin{aligned}\mathscr{N}(\vec{k}) &\equiv \int_{-\infty}^{\infty} \mathscr{N}(\omega,\vec{k}) d\omega \\ &= \frac{n_d Z^2}{[1+(kD_e)^2+n_i/n_e] \times [1+(kD_e)^2+n_i/n_e+Z^2 n_d/n_e]},\end{aligned} \tag{13}$$

where $Z = q_0/e$ is the dust equilibrium charge number, n_i the ions density, and $T = T_e = T_i = T_d$ is assumed. When the dust is negatively charged, we have

$$n_i = n_e + Z n_d.$$

(II) Case of $\Omega_c \gg kv_{td}$ and $\Omega_c \gg \omega_d$

In this case one has $\langle |\rho_d(\omega,\vec{k})|^2 \rangle \simeq \mathscr{P}_C$, it is the **charge change fluctuation**

$$\langle |\rho_d(\omega,\vec{k})|^2 \rangle = \frac{k^4 D_d^2}{k_{De}^2 + k_{Di}^2} \frac{n_d q_0^2}{P\Omega_c}, \tag{14}$$

where $k_{D\alpha}^2 = \dfrac{4\pi n_\alpha e^2}{k_B T}$ ($\alpha = e, i$). We have that the spectrum of fluctuation in this condition is

$$\mathcal{N}(\vec{k}) = P\frac{k_{D0}^2}{k^2} \frac{n_d Z^2}{[1+(kD_e)^2 + n_i/n_e] \times [1+(kD_e)^2 + n_i/n_e + Z^2 n_d/n_e + P n_0/n_e]}. \quad (15)$$

(III) Case of $\omega_d \gg k v_{td}$ and $\omega_d \gg \Omega_c$

Here we have $\left\langle |\rho_d(\omega,\vec{k})|^2 \right\rangle \simeq \mathcal{P}_W$, the eigenwave fluctuations. The charge fluctuation is

$$\left\langle |\rho_d(\omega,\vec{k})|^2 \right\rangle = P\frac{4\pi n_d q_0^2}{\omega_d} \frac{k^3 D_d^2}{\sqrt{k_{Di}^2 + k_{De}^2}} \left[\delta\left(\frac{\omega_{\vec{k}}}{\omega} - 1\right) + \delta\left(\frac{\omega_{\vec{k}}}{\omega} + 1\right) \right], \quad (16)$$

where $\omega_d^2 = \dfrac{4\pi n_d q_0^2}{m_d}$, and $\omega_{\vec{k}}$ is the solution of the equation $\varepsilon_L \equiv 1 + \chi_{le}(\omega,\vec{k}) + \chi_{li}(\omega,\vec{k}) + \chi_{ld}(\omega,\vec{k}) = 0$

$$\omega_{\vec{k}}^2 = \omega_d^2 \frac{k^2}{k_{De}^2 + k_{Di}^2 + k^2}. \quad (17)$$

which is the dispersion relation for the weelknown dust-acoustic wave. The spectrum of fluctuation is

$$\mathcal{N}(\vec{k}) = P\frac{k^3 D_d^2}{\sqrt{k_{Di}^2 + k_{De}^2}} \frac{\omega_k}{\omega_d} \frac{4\pi n_d Z^2}{\{1 + (kD_e)^2 [1-(\omega_d/\omega_{\vec{k}})^2] + (n_i/n_e)^2\}^2}. \quad (18)$$

CONCLUSIONS

In a dust plasma the electron density fluctuation produced by the dust particles can be of three different sources: from their thermal motion which is proportional to $n_d Z^2$, from their charge change which is proportional to $n_d^2 Z^3$, and from their longitudinal eigen mode oscillation which is proportional to $\sqrt{n_d}$. The results may explain observations of radar-scattering in the Earth's mesosphere[3].

REFERENCES

1. Sitenko A. G., Fluctuation and non-linear wave interactions in plasmas (Oxford: Preg Pregamon), **Chap.6** (1982)
2. Li F, Lu B W, and Havnes O., Charge density fluctuation of low frequency in a dusty plasma, *Science in China*, **A40**, 206-213 (1997)
3. Havnes O, Tosten A and Brattli A, Charged dust in the Earth's Middle atmosphere, *Physica Scripta* **89**, 133-137, (2001)

Direct Numerical Simulation of Yukawa Systems by Particle-in-cell Methods

Wolf-Christian Müller*†, Andreas Zeiler*† and Gregor E. Morfill*‡

*Centre for Interdisciplinary Science
†Max-Planck-Institut für Plasmaphysik, D-85748 Garching, Germany
‡Max-Planck-Institut für extraterrestrische Physik, D-85740, Garching, Germany

Abstract. Aiming at a fully self-consistent numerical model for the simulation of complex plasmas in rf-driven discharges, a highly efficient parallel particle-in-cell code has been developed, allowing for realizations of up to one billion interacting particles. As a first test case, we consider a Yukawa system which represents the simplest approximation of a complex plasma. The Yukawa approach where the dust particles are dressed with an isotropic Debye potential can be regarded as a low-order description of the dust-plasma interaction in the bulk a rf-driven complex plasma, away from the electrode sheaths. The simulation code is tested by examining a liquid-solid phase transition, i.e., the formation of a face-centered-cubic Yukawa crystal. This is done in a periodic-cube sub-volume, containing 13,824 dust particles, which corresponds to a total system size of $\approx$ 884,000 particles.

INTRODUCTION

During the last years there has been a tremendous growth of scientific interest in plasmas which contain solid micro-particles of μm size [1]. These complex or 'dusty' plasmas are of importance for the dynamics of various astrophysical systems [2], plasma-processing devices in semi-conductor production [3] and laboratory experiments for thermonuclear fusion [4]. After the discovery of solid-state-like ordered dust configurations in radio-frequency driven (rf) discharges [5, 6, 7, 8], complex plasmas have become a subject of interest to fundamental research, as well.

As an alternative to the various experimental possibilities of observing and manipulating complex plasmas, it is desireable to carry out self-consistent numerical simulations of such systems since they allow for a considerable flexibility in the choice of plasma parameters under 'perfectly' controlled conditions. In addition, theoretical models of dust-dust and dust-plasma interaction can directly be tested with an appropriate numerical simulation. In the following we present a first test of a fully kinetic complex plasma model based on the particle-in-cell (PIC) technique [9]. To this end, the liquid-solid phase transition of a Yukawa system is considered where the proper application of the parallel PIC code allows for a much larger number of interacting dust particles as compared to the largest Molecular Dynamics (MD) simulations of complex plasmas performed today while exhibiting a comparable precision of the short-range particle interactions.

The chosen PIC approach is able of advancing large point-particle ensembles (up to one billion particles on a Cray T3E super-computer) in time, where an arbitrary number of particle species, differing in mass and electric charge, can be treated simultaneously.

For simplicity, in the presented test case only dust particles are kinetically simulated while all other plasma constituents are approximated as inertialess Boltzmann fluids. The electrostatic interaction-potential of the ensemble is computed efficiently by a distributed multigrid algorithm.

PROPERTIES OF YUKAWA SYSTEMS

The Yukawa system simulation considers dust particles of mass m and electric charge Q which are subject to the equation of motion written here for particle i,

$$\ddot{\mathbf{r}}_i = -\frac{1}{m}\sum_{i \neq j} \mathbf{F}_{ij} \qquad (1)$$

The plasma response to the dust particles is assumed to be isotropic Debye shielding of the dust with a characteristic screening length λ_D, leading to a two-particle interaction force

$$\mathbf{F}_{ij} = -Q^2 \nabla \left(\frac{1}{r_{ij}} e^{-r_{ij}/\lambda_D} \right), \quad r_{ij} = |\mathbf{r}_i - \mathbf{r}_j|$$

Non-dimensionalizing (1) using a characteristic length ℓ_0 and the thermal velocity $v_0 = (T_0/m_0)^{1/2}$ defined with a characteristic mass m_0 and temperature T_0 yields

$$\ddot{\mathbf{r}}_i = \Gamma \sum_{i \neq j} e^{-\kappa r_{ij}} \left(\frac{1}{r_{ij}^2} + \frac{\kappa}{r_{ij}} \right) \mathbf{e}_{ij}, \quad \mathbf{e}_{ij} = \frac{\mathbf{r}_i - \mathbf{r}_j}{r_{ij}}.$$

Two inherent non-dimensional parameters characterize the system: the screening parameter $\kappa = \ell_0/\lambda_D$ and the coupling parameter $\Gamma = Q^2/(\ell_0 T_0)$. While Γ sets the extent of spatial correlation between the dust particles, κ controls the spatial arrangement of the correlated structures. Fig. 1 shows the phase diagram obtained by MD simulations (see [10] and references therein) with the triangle denoting the position of our model system after crystallization (see below).

YUKAWA CRYSTAL FORMATION

The considered Yukawa test-system consists of a periodic three-dimensional cube containing 13,824 dust particles which corresponds to a small sub-volume of the total computationally feasible system of $\approx 884,000$ particles. To reach a precision of short-range interactions comparable to the MD technique, a very fine electrostatic-field grid is used, having at least 3 mesh cells between interacting particles even for closest encounters. After an initial high-temperature/low-Γ state (see Fig. 2 [left]), where the system is behaving like a liquid as indicated by the short correlation-length in the pair-correlation of the dust particles (Fig. 2 [right]), continuous cooling by smoothly re-scaling the par-

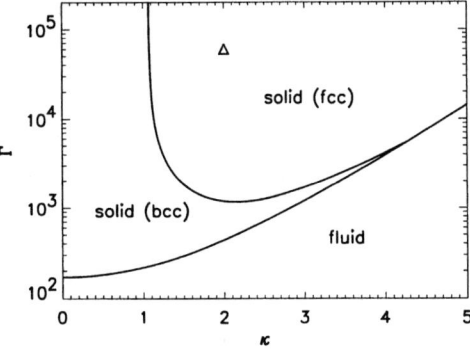

FIGURE 1. Phase diagram of Yukawa system obtained by numerical MD simulations, triangle indicates crystallized system at ($\kappa = 2$, $\Gamma = 6 \times 10^4$). fcc=face-centered cubic, bcc=body-centered cubic.

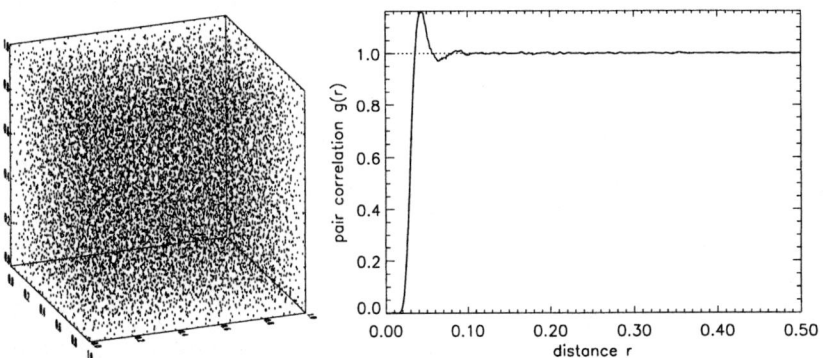

FIGURE 2. (Left) Particle positions in liquid state, (right) associated pair-correlation function.

ticle velocities leads to a spontaneous build-up of long-range correlation. The resulting Yukawa crystal is shown in Fig. 3 (left). Following the phase diagram (Fig. 1) the Yukawa-crystal structure is expected to be face-centered cubic (fcc). This is confirmed by comparing the pair-correlation of the solidified particle ensemble with the one of an ideal fcc crystal (Fig. 3 [right]).

SUMMARY

We presented first test calculations of a Yukawa system with a three-dimensional numerical complex-plasma model using the kinetic PIC approach. The simulation is able

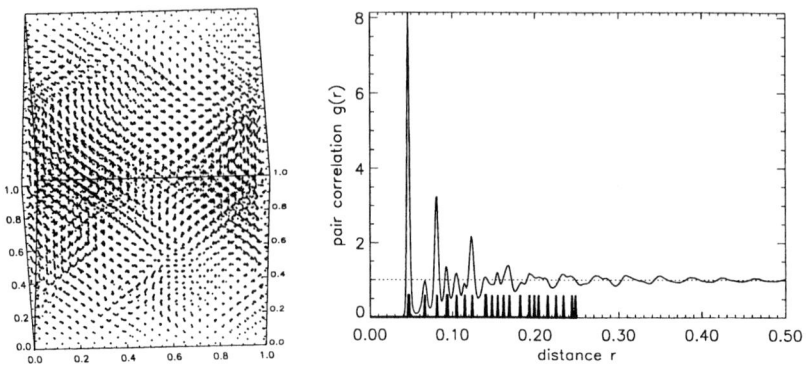

FIGURE 3. (Left) Face-centered-cubic Yukawa crystal after solidification of Yukawa liquid due to cooling. (Right) Associated pair-correlation, fat black spikes indicate corresponding peak positions for ideal fcc crystal.

to reproduce the liquid-solid phase transition where the generated crystal structure is in agreement with the one predicted by MD simulations as could be shown by calculating the pair-correlation functions. Proper simulation set-up allows for the treatment of particle ensembles that are a factor 10^2 larger than the ones of recent three-dimensional complex plasma calculations using the MD technique (see, e.g., [10]) while maintaining a comparable precision for short-range interactions.

As the next step, the numerical model will be used to treat plasma ions and dust particles kinetically to allow evaluations of numerical wake-potential models presented in [11] and [12].

REFERENCES

1. Shukla, P. K., and Mamun, A. A., *Introduction to Dusty Plasma Physics*, Institute of Physics, Bristol, 2002.
2. Goertz, C. K., *Reviews of Geophysics*, **27**, 271–292 (1989).
3. Selwyn, G. S., *Japanese Journal of Applied Physics*, **32**, 3068–3073 (1993).
4. Winter, J., *Physics of Plasmas*, **7**, 3862–3866 (2000).
5. Thomas, H., Morfill, G. E., Demmel, V., Goree, J., Feuerbacher, B., and Möhlmann, D., *Physical Review Letters*, **73**, 652–655 (1994).
6. Melzer, A., Trottenberg, T., and Piel, A., *Physics Letters A*, **191**, 301–308 (1994).
7. Chu, J. H., and I, L., *Physical Review Letters*, **72**, 4009–4012 (1994).
8. Hayashi, Y., and Tachibana, K., *Japanese Journal of Applied Physics*, **33**, L804–L806 (1994).
9. Hockney, R. W., and Eastwood, J. W., *Computer Simulation Using Particles*, Institute of Physics, Bristol, 1988.
10. Hamaguchi, S., Farouki, R. T., and Dubin, D. H. E., *Physical Review E*, **56**, 4671–4682 (1997).
11. Joyce, G., Lampe, M., and Ganguli, G., *IEEE Transactions on Plasma Science*, **29**, 238–245 (2001).
12. Hammerberg, J. E., Lemons, D. S., Murillo, M. S., and Winske, D., *IEEE Transactions on Plasma Science*, **20**, 247–255 (2001).

Interaction of Ion Beam with Dusty Plasmas

Y. Nakamura* and V. N. Tsytovich[†]

*The Institute of Space and Astronautical Science Sagamihara, Kanagawa 229-8510, Japan
[†]General Physics Institute, Russian Academy of Science Moscow, Vavilova str. 28, 117942 Moscow, Russia

Abstract. The collision cross sectional area of charged dust grains of 8.9 μm in diameter with beam ions is measured by a dusty plasma device. When the dust density is increased, the beam current ratio is reduced. The critical beam energy between scattering and absorption beam ions by dust particles is estimated.

INTRODUCTION

The presence of charged dust grains in a plasma changes the physical properties of the plasma. For example, the existence of fine particles increases the damping of the ion-acoustic wave [1]. The damping is due to collisions of ions with charged particles. From the measured damping lengths of the ion-acoustic wave, the cross sectional area of collisions was obtained [1]. Collisions of beam ions with dust particle stabilize the ion-ion instability [2]. The effective cross sectional area of particles is considered to be much larger than that of the particle since the negative electric charge of the particle is much larger than that of the ions. The purpose of the present study is to measure the collision of an ion beam with fine particles.

EXPERIMENTAL PROCEDURE

The experiment was performed in a dusty double-plasma device described in detail in a previous paper [3]. Argon gas was introduced into the chamber at a pressure $(2-4) \times 10^{-2}$ Pa under continuous pumping. Plasmas were produced in both the source and target by dc discharges between tungsten filaments of 0.1 mm diameter as a cathode and magnetic cages as an anode. Plasma parameters were: electron density $N_e = 10^8 - 10^9$ cm^{-3}, electron temperature $T_e \simeq 1$ eV and ion temperature $T_i < 0.1$ eV. The dust dispersing apparatus consists of an ultrasonic vibrator coupled with a dust reservoir (5 cm wide and 20 cm in length). The dust particles used in this experiment were glass beads of the average diameter of 8.9 μm. The ultrasonic vibrator was tuned to vibrate at a frequency of 30.7 kHz with an adjustable power by using a signal generator and a power amplifier. A semi-conductor laser was used to measure the dust density (N_d). The laser light entered and left the chamber through a pair of glass windows and was aligned to pass axially through the dust column and its reduced intensity was mea-

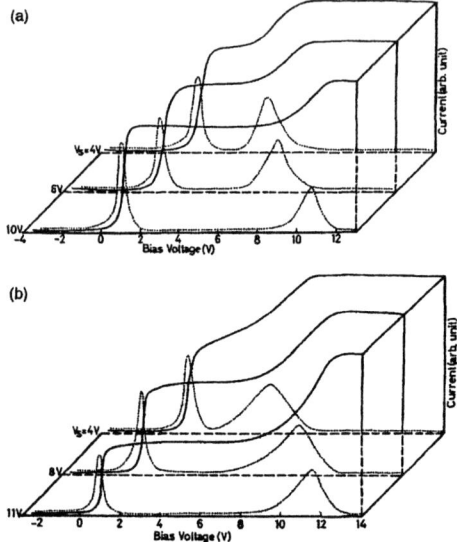

FIGURE 1. Collector current versus bias voltage measured by the directional retarding potential analyzer. Dotted curve represents the dI/dV of the respective curves. (a) Without dust when $V_s = 4$ V, 6 V and 10 V. (b) With dust density Na = 7.4×10^4 cm-3 and $V_s = 4$ V, 8 V and 11 V.

sured with a photo diode. The dust density was estimated by the relation

$$I = I_o \exp(-N_d \sigma \ell),$$

where I and I_o are the laser intensities measured with and without dust, respectively σ is the cross-sectional area of the dust grains, and ℓ is the length of the dust column (=20 cm). The dust density (N_d) was varied from zero to the maximum of the order 10^5 cm^{-3}. The density and energy of beam ions were measured by a retarding potential analyzer (RPA) and a directional RPA of 2.2 cm diameter. The amount of charge (Q) attained by the particles was estimated from the relation $Q = 4\pi\varepsilon_o r_d V_d$, where r_d and V_d are the radius and the surface potential of the grain, respectively. The amount of charge Q is found to be $(-Q/e) \approx 10^5$ when $N_d < 10^3$ cm^{-3}, and it reduces to a much smaller value of 10^2 when $N_d \approx 10^5$ cm^{-3}.

RESULTS AND DISCUSSIONS

An ion beam was injected from the source to the target plasma through a floating separation mesh grid by applying a positive dc voltage V_s to the source anode with respect to the grounded anode of the target plasma. The directional RPA was placed 23 cm away from the separation grid facing the dust region and coming beam ions. Examples of measured $V - I$ characteristics are shown in Fig. 1(a) and 1(b) together

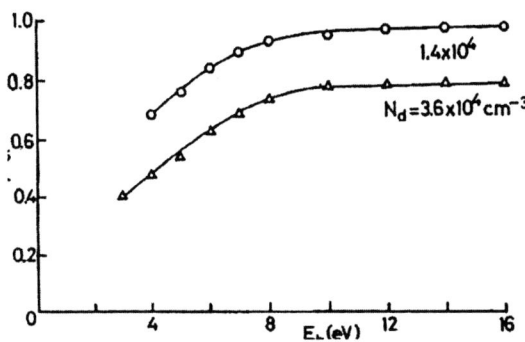

FIGURE 2. Measured beam current ratio I_i/I_{oi} as a function of the beam energy when $N_d = 1.4 \times 10^4 \text{cm}^{-3}$ and $N_d = 3.6 \times 10^4 \text{cm}^{-3}$.

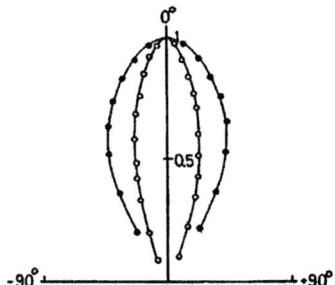

FIGURE 3. Angular profiles of the ion beam current I_i. $E_b = 6$ eV. Open circles are when $N_d = 0$. Closed circles are when $N - d = 1.4 \times 10^5 \text{cm}^{-3}$.

with their derivatives. The broadening of the energy distribution for low energy ion beams (≤ 8 eV) is clearly seen. However, high-energy ion beams (> 8 eV) are not much affected.

Measured beam-current ratios I_i/I_{io} where I_i and I_{io} are the beam current to the RPA with and without fine particles, respectively, are shown in Fig. 2 The ratio decreases when the beam energy is lower than about 8 eV, while for the higher beam energy, the ratio remains almost constant. The constant value when $E_b \geq 12$ eV is nearly equal to the measured laser light intensity ratio I/I_o. This experimental fact indicates that the high energy beam behaves like a light beam.

In order to observe the scattering of beam ions, the angular profiles of the beam current I_i has been measured by the directional RPA and are shown in Fig. 3. The 0° in Fig. 3 signifies that the analyzer faced toward the incoming beam. The plus and minus

90° indicate that the analyzer looked upward and downward, respectively. The beam current is normalized with that of 0°. It is found experimentally that the angular profile of scattering is sharper when the beam energy is larger.

The experimental results stated above indicate that scattering is dominant for low-energy beams and extinction, i.e. the beam ions are absorbed by hitting dust particles, and is larger for a higher energy beam. The boundary between two effects seems to be 8 eV. We will discuss this problem. We use the standard approach [5] by normalizng the ion density with its initial value $n = n_i/n_{io}$, the distance with λ_{Di}^2/a; $x \to xa/\lambda_{Di}^2$ where a is the size of dust particles and $\lambda_{Di} = \Delta\varepsilon_{io}/4\pi n_{io}e^2$ is the ion beam Debye radius and $\Delta\varepsilon_{io}$ is the energy spread (temperature) of the beam. The spatial change of the ion directed velocity can be taken from [5] and is

$$\tau u \frac{du}{dx} = -\alpha_{dr}(u)uP_o z_o; \quad \tau = \frac{T_i}{T_e}; \quad z_o = \frac{Z_{do}e^2}{aT_e}; \quad P_o = \frac{n_d Z_{do}}{n_{io}} \quad (1)$$

where Z_{do} is the electric charge of the dust in the absence of the beam and $u = u_b(M/KT_b)^{1/2}$. In the above equation, the effect of the change of the dust charge by beam ions is neglected. The coefficient α_{dr} is a known drag coefficient which has the following analytical form when

$$\alpha_{dr}(u) = \frac{1}{2u^3}\left[\ln\Lambda + \frac{\tau u^2}{z} + \frac{\tau^2 u^4}{z^2}\right] \quad (2)$$

where $\ln\Lambda$ is the Coulomb logarithm. In Eq. (2), two terms which do not contain the Coulomb logarithm are due to absorption of the beam ions on dust particles. The critical ion beam velocity U_{cr} where the absorption on dust particles starts to dominate is

$$U_{cr} \approx \sqrt{\frac{\ln(\Lambda)}{\tau}}$$

or in dimensional units

$$\frac{\varepsilon_{io,cr}}{T_e} = \ln\Lambda.$$

With the experimental T_e of 1eV and with an assumption that $\ln(\Lambda) = 10$, the critical beam energy becomes 10 eV. This value is close to the experimental value of 8 eV.

REFERENCES

1. Nakamura, Y., and Sarma, A., *Phys. Plasmas*, **8**, 3921 (2001).
2. Saitou, Y., and Nakamura, Y., in this proceeding.
3. Nakamura, Y., and Bailung, H., *Rev. Sci. Instrum.*, **70**, 2345 (1999).
4. Nakamura, Y., *Phys. Plasmas*, **8**, 5086 (2001).
5. Tsytovich, V. N., *Physics-Uspekhi*, **40**, 53 (1997).

Scattering of an Ion Beam by Charged Fine Particles with Coulomb Force

H. Amemiya* and Y. Nakamura[†]

*The Institute of Physical and Chemical Research (RIKEN) Hirosawa 2-1, Wako, Saitama 351-0198, Japan
[†]The Institute of Space and Astronautical Science Sagamihara, Kanagawa 229-8510, Japan

Abstract. Fine particles satisfying critical limits act as Coulomb forces and scatter charged particles like beams due to the long-range force. Otherwise, fine particles behave as tiny probes. The energy loss and broadening rates of an ion beam by particles having Coulomb fields are investigated where the Coulomb logarithm is taken as a variable. Dependence of the energy loss and broadening on the plasma density, dust charge and beam energy is obtained. A method for measuring the dust surface charge is also given.

INTRODUCTION

When fine particles are charged, energy loss and broadening due to Coulomb collisions are expected to occur. Usually, the particle has been dealt with as a tiny probe being surrounded by a sheath whose potential distribution is different from the Coulomb potential. No probe theory which accounts for the elastic recoil of particles exists. There must be some limits for which the dust has a Coulomb potential. An aim of this paper is to obtain them. Another aim is to obtain the energy change of an ion beam through interaction with a dusty plasma.

LIMIT OF THE VALIDITY OF THE PROBE MODEL

As the necessary condition for a particle to behave like an ion, no sheath should be formed around it. The conditions for no sheath formation are (a) the amount of charge on a grain $Z_d < 1$, (b) the floating potential difference V_f is less than the sheath edge potential difference V_s, (c) the ion current density at sheath edge is grater than the ion current density to the grain, and (d) the number of ions in the sheath is less than one. If a sheath exists, the orbital motion model [1] may hold for dust, though OML is not applicable [2]. But without sheath, OML may be applied under the condition

$$\frac{\partial}{\partial r}\{r^2[1+\eta(r)/\beta]\} \geq 0, \tag{1}$$

which is fulfilled for the Coulomb potential. For the negatively charged dust, it is necessary from the charge neutrality (called C.N.), $N_i = N_e + Z_d N_d$, that

$$(N_e + Z_d N_d) \sqrt{T_i/M_i} < N_e \sqrt{T_e/M_e}. \tag{2}$$

where N_d is the dust density, Z_d the dust charge, ($Z_d = 4\pi\varepsilon_\circ r_d V_f/e$), r_d is the dust radius, and T_e, T_i are the electron and ion temperatures. For the floating potential $\eta_f = -eV_f/\kappa T_e$

$$\left(1 + \frac{N_d}{N_e}\frac{4\pi\varepsilon_\circ r_d \kappa T_e \eta_f}{e^2}\right) \sqrt{\frac{\beta m_e}{M_i}} \left(1 + \beta\eta_f\right) = e^{-\eta_f}. \tag{3}$$

Equation (3), η_f and Z_d vs N_d/N_e with κT_e as a parameter for A$^+$ under the condition of (2), shows that η_f is almost constant up to $N_d/N_e \sim 10^{-3}$ then falls like $(N_d/N_e)^{-l}$ when $N_d/N_e > 10^3$. Depending on κT_e, Z_d is $10^2 - 10^3$ for $N_d/N_e/10^{-1}$ but decreases rapidly above $N_d/N_e = 10^{-1}$. Due to C.N., Z_d must decrease with N_d. The probe model is rejected when $Z_d < 1$. In what follows the case of $r_d = 10^{-5}$ cm is plotted, but the dependence on r_d vanishes, if the abscissa is changed to $Z_d N_d/N_e (=v)$.

Poisson's equation in the sheath is

$$\frac{\partial}{\partial \xi}\left(\xi^2 \frac{\partial \eta}{\partial \xi}\right) = \frac{1}{2}\left\{\sqrt{1 + \frac{\eta}{\beta}} + \sqrt{1 + \frac{\eta}{\beta} - \frac{4i}{\beta^{1/2}\xi^2}}\right\} - F(\eta), \tag{4}$$

where $i = I/I_\lambda$, $I_\lambda = 4\pi\lambda_{De}^2 \sqrt{2\kappa T_e/M_i} N_i e$,

$$F(\eta) = \frac{\exp(-\eta) + \exp(-\eta/\beta)v}{(1+v)}; \quad v = Z_d N_d/N_e \tag{5}$$

is extended from the negative ion containing case [3], $D_e = (\varepsilon_\circ \kappa T_e/N_e e^2)^{1/2}$ is the Debye length related to electrons. The ion current i for OML becomes

$$i = \frac{\beta^{1/2}\xi_d^2}{2\sqrt{\pi}}\left(1 + \frac{\eta_f}{\beta}\right), \quad \xi_d = r_d/\lambda_{De} \tag{6}$$

From the quasi-neutral condition from (4), the sheath edge current i_s becomes

$$i_s = \beta^{1/2}\xi_s^2 F(\eta_s)\left\{\sqrt{1 + \frac{\eta_s}{\beta}} - F(\eta_s)\right\}, \tag{7}$$

where ξ_s corresponds to the sheath edge where $\partial \xi/\partial \eta = 0$. Conditions (b),(c),(d) are

$$\eta_f < \eta_s, \quad i_s(\eta_d) > 1, \quad \frac{4\pi}{3}N_i r_d^3 \left(\frac{\xi_s^3}{\xi_d^3} - 1\right) < 1 \tag{8}$$

η_f, η_s, i and i_s have been calculated using (5), (6), (7), and $\partial xi/\partial \eta = 0$.

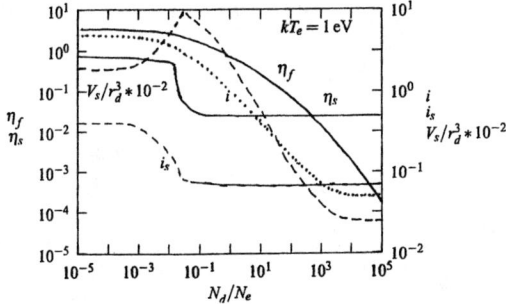

FIGURE 1. η_f, η_s, i, i_s and the sheath volume V_s vs N_d/N_e.

Figure 1 shows η_f, η_s, i, i_s and the volume of the sheath V_s/r_d^3 as a function of N_d/N_e ($\beta = 0.03$). At small N_d/N_e ($\eta_s \sim 0.6$) η_s is nearly constant drops and abruptly at $N_d/N_e \sim 2 \times 10^{-2}$. Near the same N_d/N_e i_s also drops to a small value. It happens that $\eta_f < \eta_s$ at $N_d/N_e = 5.6 \times 10^2$, and $i < i_s$ at $N_d/N_e = 1.8 \times 10^3$, i.e. no sheath formation is possible. The number of plasma ions in the sheath $N_i V_s$ does not exceed unity unless $N_i > 10^{13}$ cm^{-3}. Whene r_d is large $V_s N_i$ can exceed unity and the grain can behave as a tiny probe.

COULOMB SCATTERING BY DUST PARTICLES

The rate of energy loss and broadening of charged particles by Coulomb collisions have been investigated by several authors [4–8]. However, the previous work assumed that the Coulomb logarithm is constant. This doese not always hold. Recently, for more accuracy, the Coulomb logarithm has been treated as a variable [9]. Then the energy loss and broadening rates are given by

$$\frac{dE_\parallel}{dt} = -C\alpha \left\{ (L_\circ + 2) \nabla_\alpha \phi_1(\alpha) + \nabla_\alpha \phi_2(\alpha) \right\}, \tag{9}$$

$$\frac{dE_\perp}{dt} = -C (M/M_i) \left\{ L_\circ \phi_1(\alpha) + \phi_2(\alpha) \right\}, \tag{10}$$

where

$$\phi_1(\alpha) = -\int \frac{1}{(U/s)} f(\vec{v}) d\vec{v} \tag{11}$$

$$\phi_2(\alpha) = -\int \frac{\ln(U/s)^2 1}{(U/s)} f(\vec{v}) d\vec{v}; \tag{12}$$

$f(v)$ is the velocity distribution of scatterers, M: the reduced mass $M^{-l} = M_i^{-l} + M_d^{-l}$ (M_i: ion mass, M_d: dust mass), λ_D is the Debye length $\lambda_d^{-2} = \lambda_{De}^{-2} + \lambda_{Dd}^{-2}$, $\lambda_{Dd} =$

$(\varepsilon_\circ \kappa T_d/Z_d^2 N_d e^2)^{1/2}$, $r_\circ = Z_d Z_i e^2/(4\pi\varepsilon_\circ M U^2)$, $U = (V^2 + v^2 - 2Vv\cos\theta)^{1/2}$, ($\theta$: the angle of ion velocity V against dust velocity v). In refs.[4–8], $L_\circ + \ln(U/s)^2 = L_c$ ($s = 2\kappa T_d/M)^{1/2}$) was constant but here L_c is divided into $L_\circ$ and a variable part so $L_\circ$ can be rewritten as

$$L_\circ = \ln\left(\frac{\lambda_D}{\lambda_{Dd}}\right) + \ln(9N) - \ln\left(\frac{3}{2}Z_i\right), \tag{13}$$

where $N = (\pi/3)N_d\lambda_{Dd}^3$. In order for the present model to hold, $L_c > 0$ and $r_d < r_\circ < \lambda_D$ is equivalent to $L_c > 0$. $r_\circ > r_d$. When E_b is large $r_\circ < \lambda_D$ can be violated and besides Coulomb collisions two-body collisions can appear. The energy loss rate by the two-body collision

$$\frac{\Delta E_2}{dx} = N_d\left(\pi r_d^2\right)\frac{2M_i}{M_d}E_b = \frac{3LN_d M_i}{2r_d\rho}E_b, \tag{14}$$

is far below the values by (9) and (10). If $E_b = M_i V^2 \gg \kappa T_d$, ($E_b$: ion beam energy), asymptotic relations hold,

$$\frac{dE_\parallel}{dt} = \dot{E}_\parallel \supseteq -\frac{C}{\alpha}\{L_\circ + \ln(\alpha^2)\}\frac{L}{<v>}, \tag{15}$$

$$\frac{dE_\perp}{dt} = \dot{E}_\perp \supseteq -\frac{C}{\alpha}(M/M_i)\{L_\circ + \ln(\alpha^2)\}. \tag{16}$$

Then,

$$\Delta E_\parallel \supseteq \Delta E_\perp = \frac{C}{\alpha}\{L_\circ + \ln(\alpha^2)\}\frac{L}{<v>}, \tag{17}$$

where $<v>$ is the average ion velocity and L is the interaction length.

Consider a plasma with $T_e = 1$ eV, $T_d = 300$ K, $L = 1$ m, $\rho = 1$ g/cm^3, $r_d = 0.1\,\mu$m, $Z_d = 10^2 - 10^4$ and $N_d = 10^3 - 10^5$ cm^{-3}. Figure 2 shows $-\Delta E$ vs E_b with N_e as a parameter and ΔE decreases with E_b with weak dependence on N_e. When $E_b = 7.2$ eV $r_\circ = T_d$ is reached. At $N_e \sim 10^{15}$ cm^{-3}, $L_c < 0$ which is the limit of the model.

The plot of $-\Delta E$ ($L = l$ m) vs Z_d with N_e as the parameter for $E_b = 1$ eV, $\kappa T_e = 1$ eV, $N_d = 10^4$ cm^{-3} shows that $-\Delta E$ increases almost with Z_d^2. The plot of $-\Delta E$ ($L = l$ m) vs N_d with N_e as the parameter for the same E_b and κT_e for $Z_d = 10^2$ shows that ΔE increases almost proportionally to N_d.

METHOD FOR MEASURING THE CHARGE Z_D

An ion beam after passing the plasma can have a drift Maxwellian with the mean velocity $v_\circ$, $M_i v_\circ^2 = E_b - \Delta E$ and broadening $\kappa T_i = \Delta E$. The z-axis is the beam direction. Use a rotational analyzer with the surface parallel to the y-axis and the surface normal oriented to the direction θ from the z axis, the current $I(\theta)$ becomes

$$I(\theta) = \left(\frac{\kappa T_i}{2\pi M_i}\right)^{1/2}\{\exp(-\xi^2) + \sqrt{\pi}\xi[1 + \text{erf}(\xi)]\}, \tag{18}$$

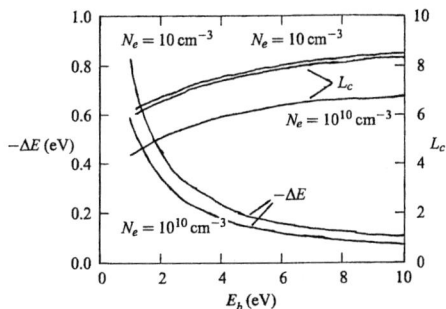

FIGURE 2. $-\Delta E$ ($L = 1$ m) and L_c vs E_b for some N_e.

or if the analyzer is directional,

$$I_d(\theta) = I_d(\theta) \exp\left[-\left(-\frac{v_\circ \sin\theta}{v_{\text{th}}}\right)^2\right], \quad (19)$$

where $\xi = v_\circ \cos\theta / v_{\text{th}}$, $v_{\text{th}} = (2\kappa T_i/M_i)^{1/2}$. If N_e and κT_e are known, e.g. $N_e = 10^8$ cm^{-3}, $\kappa T_e = 1$ eV and $E_b = 1$ eV as typical experimental conditions [10], we obtain $Z_d = 180$, 450 and 750 for $\Delta E = 0.05$, 0.2 and 0.5 eV respectively. The currents $I(\theta)$ and $I_d(\theta)$ are equal to each other at $\theta = 0°$ and $I(90°)$ corresponds to the beam's thermal current. The ratio $I(0°)/I(90°)$ is a function of Mach number vo/v_{th} and can be a diagnostic for Z_d where $I_d(\theta)$ is more sensitive to θ than (θ).

REFERENCES

1. Allen, J. E., *Physica Scripta*, **45**, 497 (1992).
2. Allen, J. E., et al. *J. Plasma Phys.*, **63**, 299 (2000).
3. Amemiya, H., et al., *Plasma Source Sci. Tech.*, **8**, 179 (1999).
4. Rosenbluth, M. N., et al., *Phys. Rev.*, **107**, 1–6
5. Trubnikov, B. A., *Reviews of Plasma Physics* vol. 1, pp.105–204 (1966).
6. Shkalovsky I. P., et al., *The Particle Kinetics of Plasmas*, 1966.
7. Poth, H., *Physics Reports*, **196**, 135 (1990).
8. Meshkov, I. N., *Phys. Part. Nucl.*, **25**, 631 (1994).
9. Amemiya, H., et al., *Riken Accel. Prog. Rep.*, **35**, 38 (2002).
10. Nakamura, Y., *Phys. Plasmas*, **8**, 5086 (2001).

Secondary Emission From Small Spherical Grains

Z. Nemecek*, J. Pavlu*, J. Safrankova*, I. Richterova* and I. Cermak[†]

Faculty of Mathematics and Physics, Charles University, Prague, Czech Republic
[†]*Technical University, Chemnitz, Germany*

Abstract. A number of phenomena connected with dust grains within the solar system can be explained by their electric charging. In the presented paper, we discuss the influence of the grain diameter on the changes of equilibrium potential of spherical grains from glass and pure SiO_2 in a laboratory experiment which is based on the levitation of a single dust grain in the AC electric field of a quadrupole. The yield of secondary emission derived from the measurements exhibits a strong dependence on the grain diameter. For larger grains, it reveals the properties of planar surfaces, whereas it can be higher for one micron grains and 10 keV of the primary beam energy. The rise of the secondary emission yield was attributed to the emission from the "back side" of the grain which starts when the penetration depth of primary electrons becomes comparable with the grain size. The results are discussed in view of present models of this process and several correction of the models are suggested.

INTRODUCTION

Electrons hitting a single dust grain may ionize the dust material thereby ejecting (secondary) electrons and producing a current. The yield δ_y, which is the ratio of emitted to incoming electrons, is a function of grain material and size and the kinetic energy, E of primary electrons in the inertial frame of the grain. According to Sternglass [1], it is often approximated by $\delta_y(E) = K\delta_M \frac{E}{E_M} \exp[-2(\frac{E}{E_M})^{1/2}]$ when the grain size is large compared to the stopping range of the incident electrons in the dust material. The yield has a maximum of $\delta_y = \delta_M$ (from 1 to 10) at an incident energy E_M ($\sim 100 \div 1000$ eV).

The energy dependence is in part due to the distance to the surface at which secondary electrons are produced. The proximity to the surface can also strongly enhance secondary electron emission from dust grains over that from macroscopic bodies while primary electrons may also pass through sufficiently small dust grains without producing many secondary electrons. Draine and Salpeter [2] give corrections to account for dust sizes. More recent works by Chow et al. [3] include detailed models for the corrections due to secondary electron production from dust grains of a variety of sizes and compositions that are in a good agreement with experimental results by Svestka et al. [4].

The initial decrease in equilibrium potential is due to secondary electron production deeper into the grain but eventually the production is closer to the exit side where the-mainly-forward produced secondary electrons can escape more easily. These size-dependent effects become important for electron energies of several keV in micron-sized dust grains. The motivation of the presented work is to investigate the effect of the grain size on its charging properties. Since we are not able to interpret our data using the Chow et al. model [3], we have developed own computer model of secondary emission.

EXPERIMENTAL SET-UP

In the present set-up, a single isolated dust grain levitates in the AC electric field of a quadrupole. The detail description can be found in [5] and [6]. The trapped grain is bombarded by the electron gun (200 eV ÷ 10 keV). In order to avoid the influence of the secondary electrons emitted from quadrupole electrodes, the electron beam as well as the quadrupole AC voltage are sampled [7]. The beam electrons are released only when the quadrupole voltage is zero, and thus the secondary electrons have small energies and cannot emit new secondaries. We would like to note that the measuring principles are similar to those used by the Svestka et al. [4] experiment but there are several important differences which will be discussed latter.

The shape and spread of diameters of used grains were checked using electron microscope but even in the case of "monospheres", a part of grains can be broken or they can create clusters and thus the determination of parameters of the grain is desirable. Since the experiment can provide only specific charge of the grain, an independent method is required for determination of other quantities. Charging of the grain by an ion beam can be used for determination of the surface potential and, under some assumptions, the mass and radius of the spherical grain [8, 9]. The change of the frequency of grain oscillation caused by one elementary charge can be determined for a sufficiently small grain. This change then allows us to determine the mass of the grain [9].

EXPERIMENTAL RESULTS

The measured specific charge, Q/m of three SiO_2 1.2 μm spherical grains are plotted in Figure 1a as a function of the primary electron beam energy. It attains a maximum at ~ 400 eV and then falls down as can be expected taking into account the formula [1] for the yield of secondary emission. However, when the beam energy exceeds 5 keV, the specific charge begins to rise with the increasing beam energy. The curves in Figure 1a are nearly identical. It suggests a very narrow distribution of grain diameters as well as a good stability of experimental conditions.

Figure 1b shows the dependence of the surface potential on the grain diameter. The data for one 1.2 μm grain are from Figure 1a; the data for a larger grain are from Svestka et al. [4]. Both curves exhibit a clear local minimum which shifts toward the higher beam energies with the grain diameter. It is interesting to note that this position seems to be roughly proportional to the grain diameter.

MODEL AND DISCUSSION

We have tried to find the parameters of the Chow et al. model [3] to fit our experimental data but without success. For this reason, we have developed a new model which follows

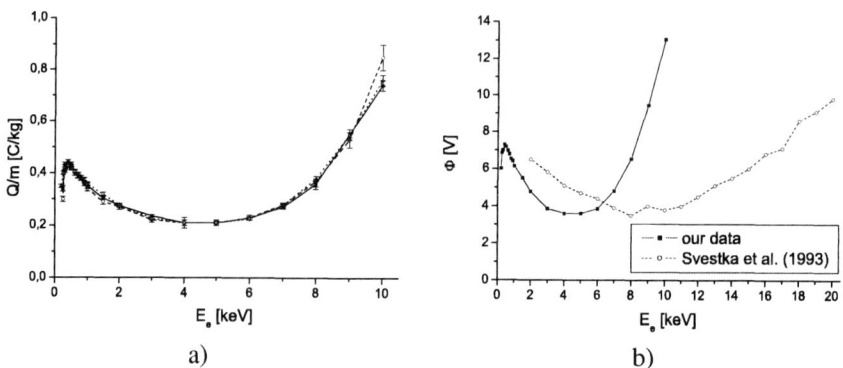

FIGURE 1. a) The specific charge as a function of the beam energy for three 1.2 μm grains. b) The comparison of the surface potentials for two grain diameters.

the trajectories of primary beam electrons inside the grain. The basic ideas are: The beam electrons penetrate into a grain and they undergo collisions. Each collision leads to the constant energy loss (ΔE) and to a random scattering of directions of further motion. The mean free path, λ between collisions is a linear function of the energy of primary electrons. Each collision generates an excited electron. The probability that the excited electron becomes secondary electron decreases exponentially with the distance of the particular collision from the surface. In order to test this simple approach, we have calculated the yield of secondary emission from a planar surface and compared it with the Sternglass formula in Figure 2a. The parameters of the model were taken from [1]. One can note a good agreement of the results of our simple model with the "universal" Sternglass curve. On the other hand, much larger disagreement between this curve and results of computation according to Chow et al. [3] is apparent from this figure.

The computed surface potential as a function of the primary beam energy is compared with the experimental data in Figure 2b. Our model can reproduce the grain potential in the whole range of energies, whereas Chow et al. [3] should split the experimental data into two parts and use two different sets of the model parameters. The principal difference of these models is that we use the random scattering of beam electrons (as used in [1]), whereas the primary electrons proceed into a grain along the straight lines in the Chow et al. model [3]. This assumption enormously increases the depth in which a majority of secondaries is generated.

Our model predicts that the minimum of the surface potential will shift nearly linearly toward the higher beam energies with the grain diameter and this fact is a good agreement with the experiment. However, the depth of the minimum would increase and even negative potentials can be expected for large grains. This behaviour seems to contradict to data in Figure 1b. We suppose that the measuring method used in Svestka et al. [4] had several features which can influence results. First, the quadrupole was supplied by an asymmetric source and thus the resulting electric field inside the quadrupole could affect the beam trajectory. Second, neither the beam nor quadrupole voltage were sampled and thus the primary electrons could hit electrodes and resulting secondary electrons would increase the minimum potential. However, a position of the minimum remains unchanged according to our simulation. Nevertheless, the disagreement between our model and data of Svestka et al. [4] will be a subject of further investigations.

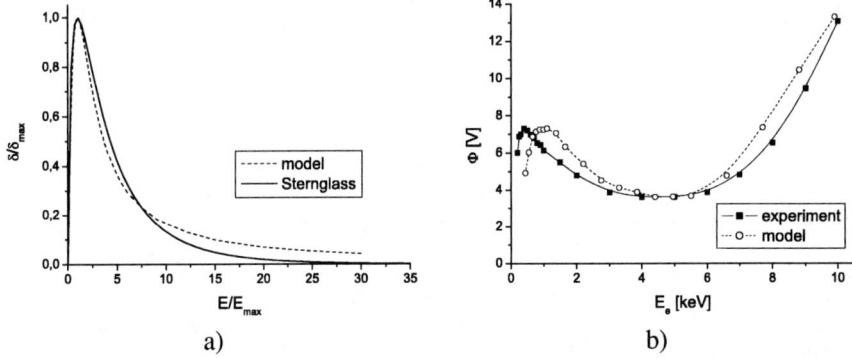

FIGURE 2. a) The normalized yield of secondary emission as a function of the normalized energy of primary electrons for a planar surface. b) Comparison of the model and experimental dependences of the grain potential on the beam energy.

ACKNOWLEDGMENTS

The present work was supported by the Charles University Grant Agency, Contract 176/01, and by the Research project, No. MSM 113200004.

REFERENCES

1. Sternglass, E. J., *Scientific Paper 1772*, Westinghouse Res. Lab., Pittsburgh, 1957.
2. Draine, B. T., and Salpeter, E. E., *Astrophysical Journal*, **231**, 1979, pp. 77–94.
3. Chow, V. W., Mendis, D. A., and Rosenberg, M., *IEEE Trans. Space Sci.*, **22** (2), 1994, pp.179–186.
4. Svestka, J., Cermak, I., Grun, E., *Adv. Space Res.*, **13** (10), 1993, pp.(10)199–(10)202.
5. Pavlu, J., Safrankova, J., Nemecek, Z., and Velyhan, A., *this issue*, 2002.
6. Zilavy, P., Sternovsky, Z., Cermak, I., Nemecek, Z., Safrankova, J., *Vacuum*, **50** No.1–2, 1998, pp. 139–142.
7. Sternovsky, Z., Zilavy, P., Nemecek, Z., and Safrankova, J., *Conf. on Contr. Fusion and Plasma Physics, Praha*, **ECA** Vol. 22C, 1998, pp. 2549–2552.
8. Cermak, I., Grun, E., Svestka, J., *Adv. Space Res.*, **15** (10), 1995, pp. (10)59–(10)64.
9. Zilavy, P., Nemecek, Z., Safrankova, J., *Proceedings of contributed papers WDS 1999 – part II*, Matfyzpress, Prague, 1999, pp. 252–257.

Charging Properties of Dust Grain Clusters

J. Pavlu*, J. Safrankova*, Z. Nemecek* and A. Velyhan*

Faculty of Mathematics and Physics, Charles University, Prague, Czech Republic

Abstract. The dust grains or their clusters can be frequently found in many space environments - interstellar clouds, tails of comets or planetary rings are only typical examples. Being surrounded by plasma, dust grains are charged by various processes as photoemission, collection of electrons and/or ions, secondary or field emissions. The determination of a charge, which can the grain or cluster of grains gain by these processes, is very important for an estimation of the dusty plasma properties because electrostatic forces highly dominate the gravitation. However, our knowledge of charging processes is inferred from measurements on large planar samples and thus it can be applied on micron-sized grains only with a care and their application on the clusters is under question.

Our experimental set-up allows us to trap a single dust grain or cluster of grains in an electro-dynamic quadrupole and to expose it by the electron and/or ion beams. Precise measurements of a stepwise change of charge-to-mass ratio induced by the elementary charge allow us to determine the charge and mass of the investigated grain. We have investigated SiO_2 spherical grains as well as their clusters. A systematic study reveals that the dependence of the charge acquired by an investigated object (grain or cluster of grains) by the particle beam of given energy depends not only on grain macroscopic characteristics but on the configuration of the grains in a particular cluster.

INTRODUCTION

The growth of a specific dust particle cluster proceeds through a sequence of randomly oriented and directed mutual collision processes that may result in sticking. The cluster consists of the same type of particle, usually spherical dust grains of a single size.

For most astrophysical applications, clustering effects are very important process. When glass particles in laboratory conditions were charged by ions, it was found that strong non-linear discharging starts at certain field strengths, independent of the vacuum pressure. The discharging current was attributed to field emission of ions from the particle's surface [1]. The threshold for this process lies at about 5.10^8 V/m; depends on the history of the particle's surface and is significantly lower than the expected (e.g., [2]). The large discrepancy is probably attributable to the fact that the theoretical calculations were based on electrically conductive particles. In the case of dielectric materials, the electric field penetrates into the particle and substantially alters the conditions on the particle surface.

Field emission prevents compact dielectric particles from electrostatic fragmentation. The highest attainable field strength of about 5.10^8 V/m produces and "electrostatic stress" of about 1 MPa which is insufficient to fragment compact particles (e.g., [3]).

Electrostatic fragmentation of loosely bound particles having diameters of $2 \div 20$ μm and made of Al_2O_3 and glass was studied. Their tensile strengths were found to range from 1 to 100 kPa, compatible with theoretical estimates, and to be proportional to the inverse square of the particle diameter. If this relation holds true, the onset of fragmentation depends on the electrostatic potential, Φ rather than the field strength, E_s. In addition, if Φ is practically independent of particle size, as is expected in many environments for particles within the size range studied, the onset of fragmentation is also independent of particle size [5].

In the present paper, we discuss results of charging of SiO_2 clusters and the influence of the cluster configuration on attainable value of the surface potential.

EXPERIMENTAL SET-UP

A single isolated dust grain is suspended in an electrodynamic quadrupole inside an ultra-high vacuum chamber and exposed to a primary beam of electrons or ions having energies up to 10 keV or 5 keV, respectively. A trapped grain is illuminated by a He-Ne laser; its image is projected by a simple optical system and the coordinates of the amplified light spot are determined. The coordinate signals are used to control the motion of the particle by the damping system. The current Q/m ratio is calculated from the grain oscillation frequency, f_z, the quadrupole supply voltage, V_{ac}, and its frequency f_{ac}. The expression, $\frac{Q}{m} = \pi^2 r_0 \frac{f_{ac} f_z}{V_{ac}}$, where r_0 stands for the inner radius of the middle quadrupole electrode, can be used in a first approximation.

The charging of a grain to a high surface potential by an ion source is usually used for the grain mass determination. This method cannot be applied for loosely bonded clusters and thus we are using the change of the frequency, f_c induced by one elementary charge for determination of the total grain charge.

EXPERIMENTAL RESULTS AND THEIR DISCUSSION

The current experimental set-up limits the lowest diameter of investigated grains to about 1 μm. In order to have a sufficient dynamic range, we have used the 1.2 μm SiO_2 spherical grains which tend to create clusters. For each cluster released from the container we have measured the equilibrium charging characteristics in the electron energy range from 0.25 to 10 keV and determined the total charge by the method of elementary charge. Altogether we have measured more than 200 clusters. Equilibrium surface potentials as a function of the beam energy for several of investigated clusters are plotted in Figure 1a. The curves differ by a number of grains in the particular cluster. The overall shape of the curve for a single grain was discussed in Nemecek et al. [6], and thus we will concentrate on differences caused by clustering. One can clearly see that, whereas the initial parts of all curves in Figure 1a are nearly the same, the slopes of high-energy tails decrease with a number of spheres in the cluster in a systematic manner. The exception from this rule will be discussed later.

The equilibrium potential of the grain or cluster is determined by a balance of incoming and outgoing electrons. If the beam energy is sufficiently low, the electrons are emitted only from the illuminated part of the cluster and the curves roughly follow the dependence found for planar surfaces. If the emission from the "back side" of particular grain becomes important, the surface potential is a decreasing function of a number of

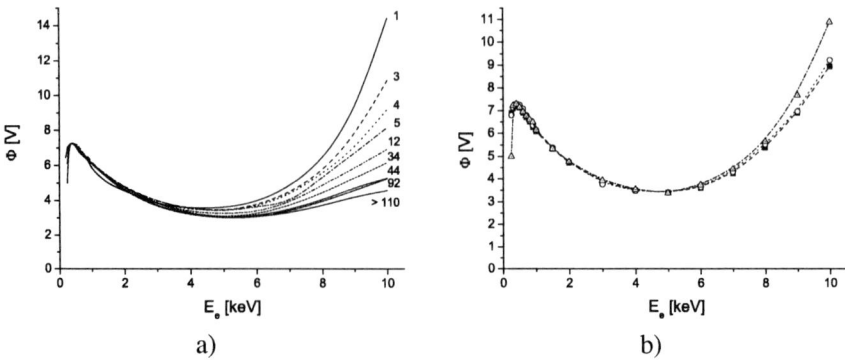

FIGURE 1. a) Equilibrium surface potentials of different clusters as a function of the primary electron beam energy. b) Equilibrium potentials of clusters consisted of 4 grains.

grains in the cluster. We suggest that it is caused by a shielding of secondary electrons by other grains in the cluster because secondary electrons escaping "inside" the cluster are captured by another surface. It is interesting to note that, even for a relatively high number of grains in the cluster, the charging characteristics differ from those measured for compact grains of a similar equivalent diameter. The lowest curve in Figure 1a was measured for one cluster consisted of ~ 110 grains. Its mass corresponds to a ~ 5 μm sphere on which any effect of enhanced emission would be observed [6]. We assume that the grains on the flank surfaces of a cluster, which are not shielded, are responsible for the enhanced cluster potential at high beam energies.

A better view on the effect of a cluster geometry on charging characteristics provides Figure 1b where equilibrium surface potentials of clusters containing 4 grains are plotted as a function of the primary electron beam energy. The different slopes of high-energy tails are caused by different spatial adjustment of grains in particular cluster and, consequently, by a different degree of shielding. This process is schematically illustrated in Figure 2 where two possible configurations of four grains are drown. In a compact configuration, the electrons escaping "inside" cannot leave the cluster even if they possess enough energy.

The clusters under study were probably bonded by hydrogen bonds. We have estimated the bonding force under assumption that two grains are covered by a monolayer of water. The corresponding bonding potential lies in order of ~ 10 V. This value is comparable with observed potentials. We assume that it can be a reason why we were not able to investigate clusters composed of two grains because the bonding force increases with a number of grains in a particular cluster.

This result can be probably applied on compact grains of an irregular shape. The surface potential of such grains would be determined by a part with smallest cross-section and thus it would be significantly higher than that derived in a spherical approximation. A higher potential can affect the clustering of micron-sized grains but it cannot lead to fragmentation of clusters because the potential attainable by secondary emission is too low.

As a cluster geometry in our experiment is unknown, we cannot reliably determine the tensile strength. However, the spherical approximation provides an upper limit of the order of 10 N/m^2, i.e., even below the value required for destruction of fluffy aggregates. However, the tensile strength increases with decreasing diameter as the surface potential rises and it could be sufficient to destroy clusters of nanoparticles.

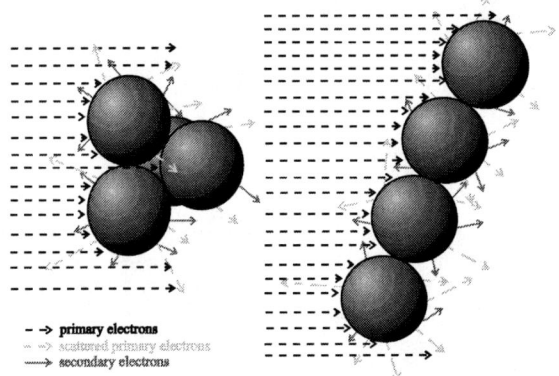

FIGURE 2. A sketch of shielding for two possible configurations of 4 grains in one cluster.

ACKNOWLEDGMENTS

The present work was supported by the Charles University Grant Agency, Contract 176/01, and by the Research project, No. MSM 113200004.

REFERENCES

1. Cermak, I., Grun, E., Svestka, J., *Adv. Space Res.*, **15** (10), 1995, pp. (10)59–(10)64.
2. Draine, B. T., and Salpeter, E. E., *Astrophysical Journal*, **231**, 1979, pp. 77–94.
3. Grun, E., in *The Giotto Spacecraft*, eds. E. Wolfe and B. Battrick, ESA SP-224, 1984, pp. 39–41.
4. Svestka, J., Cermak, I., Grun, E., *Adv. Space Res.*, **13** (10), 1993, pp. (10)199–(10)202.
5. Svestka, J., and Grun, E., in *Hypervelocity Impacts in Space*, ed. J. A. M. McDonnell, Canterbury (Univ. of Kent), 1992, pp. 139–143.
6. Nemecek, Z., Pavlu, J., Safrankova, J., Richterova, I., and Cermak, I., *this issue*, 2002.

Dust ion–acoustic solitons: Role of trapped electrons

S. I. Popel*, A. P. Golub'*, T. V. Losseva*, A. V. Ivlev[†] and G. Morfill[†]

*Institute for Dynamics of Geospheres RAS, Leninsky pr. 38, bld. 6, 117334 Moscow, Russia
[†]Max–Planck–Institut für Extraterrestrische Physik, D–85741 Garching, Germany

Abstract. The dust ion–acoustic solitons with trapped electrons are investigated. It is shown that the compressive soliton–like perturbations are damped and slowed down, but possess the main properties of solitons. There is a principal possibility to study experimentally the role of trapped electrons in the soliton formation.

The first study of dust ion–acoustic solitons in complex plasmas has been performed in [1]. The approximation has been used when it is possible to neglect variations in dust particle charge and absorption of electrons and ions by dust particles. In a typical complex plasmas these processes can result in the anomalous dissipation. This means that the existence of completely steady–state nonlinear structures is impossible in complex plasmas. In reality, this note is truth for any real system. However, in complex plasmas it leads to qualitatively new results. The usual consideration of a soliton assumes that the electrons are not trapped by the potential well formed by the soliton. However, this assumption is violated when the following inequality [2] is valid:

$$\tau \gg L/v_{Te}, \qquad (1)$$

where τ is the characteristic time of the variation of the soliton field, L is the characteristic spatial scale of the soliton, and v_{Te} is the electron thermal velocity. In complex plasmas the characteristic time τ of soliton damping is determined by the realtionship:

$$\tau^{-1} \sim \max\{v_{ch}, \tilde{v}\}, \qquad (2)$$

where the rate v_{ch}, at which the ions are absorbed by the dust grains, and the rate $\tilde{v}$, at which the ions lose their momentum as a result of their absorption on the grain surfaces and their Coulomb collisions with the grains, are given by Eqs. (4) and (5) in [3], respectively. The characteristic soliton width L is of the order of $10\lambda_{De}$, where λ_{De} is the electron Debye length. For such values of L the inequality (1) is fulfilled always for $Z_{d0}d \geq 1$, where $q_d = -Z_d e$ is the grain charge, $-e$ is the electron charge, $d = n_{d0}/n_{e0}$, n_d is the dust density, n_e is the electron density, and the subscript 0 stands for the unperturbed plasma parameters. The latter inequality is the typical one for the most of dusty plasmas. Thus in the dusty plasmas it is necessary to take into account the effect of adiabatically trapped electrons.

This effect influences the spatial electron distribution. If the electrons are under the action of the soliton electric field corresponding to the *positive* electrostatic potential

φ (the potential well for the electrons), then their distribution is determined by the Gurevich's formula

$$n_e = n_{e0}\left\{\exp\left(\frac{e\varphi}{T_e}\right)\left[1-\mathrm{erf}\left(\sqrt{\frac{e\varphi}{T_e}}\right)\right]+2\sqrt{\frac{e\varphi}{\pi T_e}}\right\}, \qquad (3)$$

where T_e is the electron temperature, $\mathrm{erf}(x)$ is the error function.

The equations for the description of the dust ion–acoustic solitons with the positive electrostatic potential (compressive solitons) are analogous to those of the ionization source model (see Eqs. (1)–(7) in [3]) with the following exceptions: instead of the Boltzmann distribution for the electrons (Eq. (3) in [3]) it is necessary to use the Gurevich distribution (3); we do not take into account the kinetic pressure term (containing the ion temperature T_i) in Eq. (2) in [3]. The form of the ionization source intensity S_i depends on the plasma parameters. In our calculations we use the parameters similar to those of the experiments on dust ion–acoustic soliton excitation [4]: the argon ion density $n_{i0} = 3 \cdot 10^8$ cm^{-3}; the dust particle size $a = 4.4$ μm; the electron and ion temperatures $T_e = 1.5$ eV and $T_i = 0.1$ eV, respectively. The experiments [4] were performed on the same installation (double plasma device) as in the experiments [5]. As it has been shown (see [3]), in this case it is possible to consider the source S_i as a constant.

We use the following normalization:

$$\frac{e\varphi}{T_e} \to \varphi, \quad \frac{v}{c_s} \to v, \quad \frac{n_{i,d}}{n_{e0}} \to n_{i,d}, \quad \frac{tc_s}{\lambda_{De}} \to t, \text{ and } \frac{x}{\lambda_{De}} \to x, \qquad (4)$$

where v is the velocity, $c_s = \sqrt{T_e/m_i}$ is the ion–acoustic speed, m_i is the ion mass, and n_i is the ion density.

We have performed the following investigations.

1) We have studied the steady–state compressed solitons, which propagate with a constant speed (Mach number) M and are the solutions of the set of Eqs. (1) – (2) (with the right–hand sides equal to zero) and Eq. (6) from [3] as well as Eq. (3). Here we neglect the dissipation processes related to the dust particle charging and absorption of plasma particles by dust grains (below we use the notion "steady–state soliton" for the solitons satisfying equations which do not take into account these dissipation processes). The solitons obey the equation

$$\partial_x^2 \varphi = -d_\varphi V(\varphi), \qquad (5)$$

where the subscript x (φ) denotes derivative with respect to x (φ), and the Sagdeev potential $V(\varphi)$ is given by

$$V(\varphi) = 1 - \exp(\varphi) - \frac{2\sqrt{\varphi}}{\sqrt{\pi}} - \frac{4\varphi^{3/2}}{3\sqrt{\pi}} - \varphi Z_d d$$
$$+ M(1+Z_d d)\left(M - \sqrt{M^2 - 2\varphi}\right) + \exp(\varphi)\mathrm{erf}(\sqrt{\varphi}). \qquad (6)$$

2) We have considered the evolution of the initial steady–state soliton (from the previous item) in complex plasmas with taking into account the dissipation processes related to the dust particle charging and absorption of plasma particles on dust.

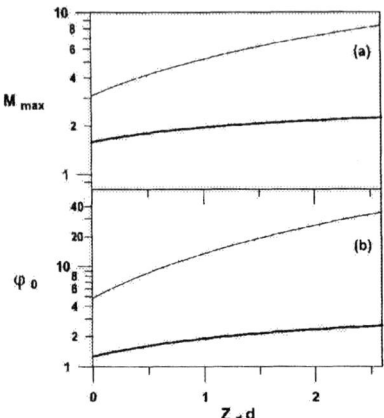

FIGURE 1. Dependencies of the maximum value of Mach number M_{max} (a) and the maximum soliton amplitude φ_0 (b) on $Z_d d$ for the compressive soliton with the trapped electrons (thin lines) and that with Boltzmann electrons (bold lines).

3) We have studied the interaction of two different compressive dust ion–acoustic solitons with the trapped electrons taking into account their damping due to the anomalous dissipation.

The main results of the investigation are the following.

1) The properties of the compressive solitons with the trapped electrons are very different from those with the Boltzmann electrons. In particular, the maximum possible amplitude of the soliton with the trapped electrons is much larger than that of the "Boltzmann" soliton, while the region of allowable Mach numbers for the former is much wider than for the latter (see Figure 1). This shows the principal possibility to study experimentally the role of trapped electrons in the soliton formation.

2) The evolution of the initial perturbation in the form of the steady–state compressive soliton with the trapped electrons occurs in the following manner. The soliton is damped due to the dissipation originating from the dust particle charging processes. The speed of the perturbation (the Mach number M) decreases. However, at any time the form of the evolving perturbation is similar to that of the steady–state compressive soliton with the trapped electrons corresponding to the Mach number at this moment of time. This fact is related to small variations in the dust particles charges (less than several per cent from the equilibrium value).

3) After the interaction of two damped solitons (see Figure 2), each perturbation has the form, which is close to that of the same soliton perturbation propagating individually from the beginning (not subjected to the interaction). This property is the property inherent in solitons.

Thus there is a possibility of the existence of the dust ion–acoustic compressive solitons which are damped and slowed down, but their form corresponds to the soliton one for the running value of their speed. After their interaction they conserve the soliton form. The role of trapped electrons in such solitons is significant.

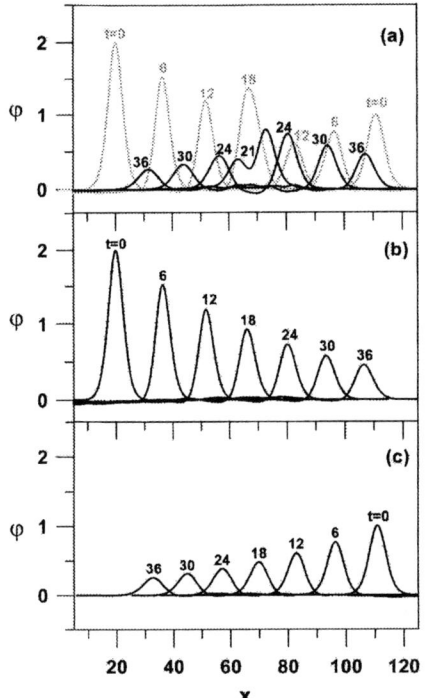

FIGURE 2. Time evolution of compressive solitons with the trapped electrons. The interaction (a) of two individual solitons (b) and (c). Grey and black lines correspond to the soliton profiles respectively before and after the interaction at $t = 0, 6, 12, 18, 21, 24, 30, 36$. At $t = 0$ all the perturbations have the form of the steady–state solitons. The initial ($t = 0$) amplitude of the left soliton (a) and the individual soliton (b) is $\varphi_0 = 2$, the Mach number is $M = 2.8$. The initial parameters of the steady–state right soliton (a) and the individual soliton (c) are $\varphi_0 = 1$ and $M = 2.435$. The remaining plasma parameters are as follows: the argon ion density $n_{i0} = 3 \cdot 10^8$ cm^{-3}, $a = 4.4$ μm, $T_e = 1.5$ eV, $T_i = 0.1$ eV, and $Z_d d = 2$.

ACKNOWLEDGMENTS

This work was supported by RFBR (grant no. 02–02–26678).

REFERENCES

1. Popel, S. I. and Yu, M. Y., *Contrib. Plasma Phys.*, **35**, 103 (1995).
2. Lifshitz, E. M., and Pitaevskii, L. P., *Physics Kinetics*, Nauka, Moscow, 1979, pp. 182–185.
3. Popel, S. I., Golub', A. P., and Losseva, T. V., "Shock wave–like structures in complex plasmas: Theory and experiments," in *ICPDP–2002 Conference Proceedings* (see this Volume).
4. Nakamura, Y. and Sarma, A., *Phys. Plasmas*, **8**, 3921 (2001).
5. Nakamura, Y., Bailung, H., and Shukla, P. K., *Phys. Rev. Lett.*, **83**, 1602 (1999).

Dusty plasma structures under the external influences

V.E. Fortov, V.I. Molotkov, V.P. Efremov, A.P. Nefedov, M.Y. Poustylnik, V.M. Torchinsky

*Institute for High Energy Densities, Russian Academy of Sciences
Izhorskaya 13/19, 125412, Moscow, Russia*

Abstract. Expeiments on different external influences on the dusty plama structures in the dc glow discharge are presented. The possibility of their application for diagnostic purposes is shown. Gas dynamic influence on the dusty plasma structures was considered for the first time. It was used for the excitation of nonlinear dust acoustic waves.

Investigations of different external influences on the dusty plasma structures are of great interest. First of all, the influences that make negligible perturbation in the background plasma may serve as diagnostic means. External influences may be also used to control the spatial position and the order of the dusty plasma structures. Moreover, external influence may be employed to drive energy into the system of charged dust grains in a plasma and observe the dynamics of the response.

This work represents the investigations of three different external influences in the dc glow discharge striations. The following influences are: the effect of focused laser light, neutral gas temperature gradient and the gas-dynamic influence.

The experiments were carried out in a vertically positioned dischrage tube [1], in which the glow discharge with cold electrodes was created. The tube is filled with neon up to the pressure 0.1 - 2 torr. The discharge current varies from 0.1 to 4 mA. In these range of regimes the standing striations existed. The dust grains were held above the discharge area in a container with the grid bottom. When shaking the container the dust grains fall down through the grid and form the ordered structures in the striations. The dust particles were highlighted with a laser diode 50 mW power. The scattered light was observed by a CCD camera at a frame rate of 25 fps and in some cases by a CMOS camera at frame rate up to 300 fps. The observations were conducted in the first striation above the cathode.

The influence of the focused laser light on the dust grains was used for the measurements of the charge of dust grains levitating in striations [2]. The light beam from an Ar^+ laser was focused onto a single particle in the structure. The beam power was up to 200 mW, the waist thickness was about 60 μm. Under the effect of the laser light the particle moves 1.5 - 3 mm out of the structure, then leaves the beam and returns back into the structure. The velocity of a dust particle has a maximum on the returning trajectory. In the point of the maximum the neutral drag force equals to the radial electric force. From this condition the charge is found.

As it was shown in a number of experiments [3, 4] thermophoresis may have a great

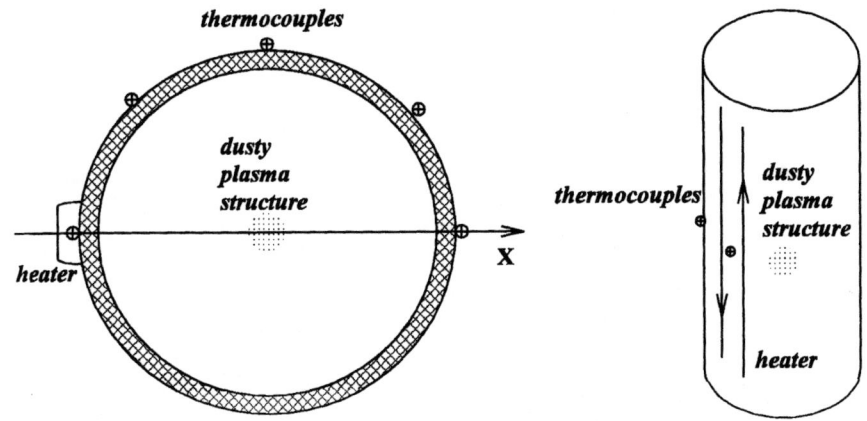

FIGURE 1. Scheme of the experiment on the thermophoretic influence.

influence on the dust grains levitating in the discharge. In this work this influence was used to measure the value of the radial force acting on a dust grain in the discharge. To create the gradient of the gas temperature a 16 cm long heating wire was maintained on a tube parallel to its axis (see Fig.1). Five thermocouples were set to measure the temperatre profile on the tube circumference $T_w(\phi)$, ϕ is the polar angle. Since the length of the heater is much longer than the size of a dusty plasma structure the temperature distribution in the volume of the discharge could be considered as two-dimensional. The temperature variation in this experiment is several K so the dependence of the gas heat conductivity on the temperature may be neglected. Therefore the temperature distribution is determined by the stationary linear two-dimensional heat conductivity equation

$$\Delta T(\rho, \phi) = 0 \tag{1}$$

and boundary condition

$$\lambda_{glass} \frac{T_w(\phi) - T_r(R, \phi)}{d} = \lambda_{gas} \left. \frac{dT}{d\rho} \right|_{R, \phi}, \tag{2}$$

where $d = 2$ mm is the tube wall thickness, $R = 18$ mm is the tube radius and ρ is the radial variable. Solving this equation we aquire the temperature distribution, which allows to calculate the thermophoretic force in each point inside the tube. The formula for the thermophoretic force acting onto a spherical particle with the radius a is given in [5]:

$$F_{th} = -\frac{8}{45}\sqrt{\pi} p a^2 l \frac{\nabla T}{T}, \tag{3}$$

where p is the neutral gas pressure and l is the mean free path of gas atoms. The experiments were done with vertical chains of dust particles. Fig. 2a shows the effect of thermophoresis on such a chain. It is seen that the chain is distorted in such a way that the unknown interparticle force has the constituent in the radial direction for all

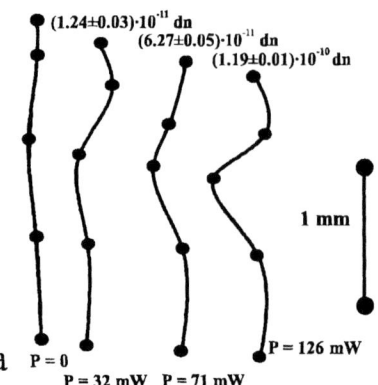
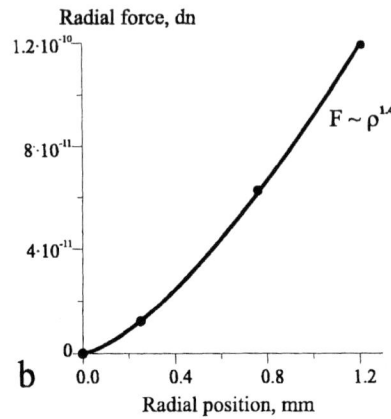

FIGURE 2. a. Influence of the thermophoresis on a vertical chain of dust grains. **P** denotes the power released in the heater, numbers above each configuration of a chain show the corresponding value of the thermophoretic force; b. Profile of the radial force, obtained from the positions of the lowest particle. Discharge conditions: neon pressure 0.8 torr, current 1.66 mA.

the particles with the exception of the lowest one, which experiences only the radial trapping force and the thermophoresis. The value of thermophoresis is determined by the reconstructed temperature distribution and the position of a dust grain. In this way the radial trapping force profile is obtained (Fig. 2b).

Gas-dynamic influence was employed for the excitation of nonlinear waves in the dusty plasma structures. The scheme of the experiment is shown on Fig 3a. The cathode was made in the form of a hollow cylinder. Below the cathode a plunger was set into the tube. It was moved manually with the help of a magnet at a velocity of 30-40 cm/s. A grid kept under the floating potential was inserted 7 cm above the cathode. The discharge regime was chosen in such a way that the first striation was formed exactly below the grid (pressure 0.3 torr, current 0.1 mA). Upper edge of the dusty plasma structure was 4 mm separated from the grid. Melamineformaldehyde dust grains 1.03 μm diameter were used in these experiments.

After moving the plunger downward the structure was again for some time streaming downward, then it stopped and began moving towards its initial equilibrium and when it returned to the stable position a disturbance propagating through it appeared (Fig. 3b). The disturbance consists of three parts: first compresion, rarefaction and the second compression. The amplitudes of the dust density for these three parts respectively are 1.6 and 1.3 respectively and 0.65 for the rarefaction. This means that the wave is strongly nonlinear. The velocities of the compressions are about 2.5 cm/s, but the second compression always runs a bit faster reaching the speed of 3 cm/s. The dust acoustic speed is estimated using the following formula [6]:

$$C_{da} = \sqrt{\frac{Z_d^2 T_i n_d}{m_d n_i}}, \qquad (4)$$

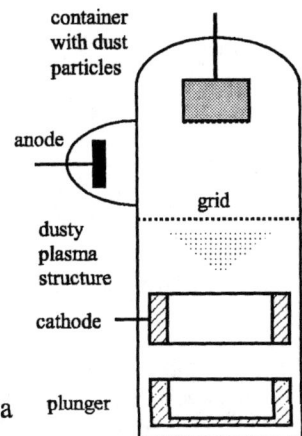
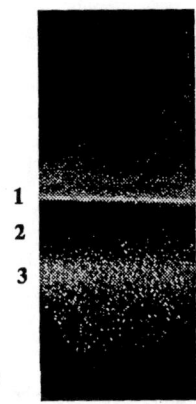

FIGURE 3. Scheme of the experiment on the gas-dynamic influence on the dusty plasma structures (a) and the sequence of videoimages, representing the disturbance (b). Zone 1 is the first compression, zone 2 - the rarefaction and zone 3 - the second compression. Vertical size of the image is 14 mm.

where n_i, n_d are the ion and dust density, T_i is the ion temperature and Z_d, m_d are the dust grain charge and mass respectively. The value of the dust particle charge was calculated by the extrapolation of experimental dependences obtained in [2] and was assumed to lie in the range 400 - 750 electrons. The ion density is estimated to be equal to $4 \cdot 10^7 - 10^8$ cm^{-3}. The dust density also slightly changes inside the initial dusty plasma structure around the average value of $3 \cdot 10^4$ cm^{-3}. Thus C_{da} = 1.8 - 5.2 cm/s.

The conditions of our experiment are charactarized by comparatively high neutral gas pressure and consequently great damping of the waves due to the neutral drag. If we use the *Epstein* formula for the neutral drag force which is applicable in our case (mean free path of Ne atoms at the pressure p = 0.3 torr is 170 μm and particle radius 0.5 μm), we obtain the frictional damping rate 75 s^{-1}. We work at a frame rate of 300 fps and consequently the disturbance must be damped within 4 frames if it has no energy source other than the initial impulse. However dust acoustic instability which is a typical phenomenon for the dc discharge striations could serve as such a sorce [7].

REFERENCES

1. V.E. Fortov, A.P. Nefedov, V.M. Torchinsky et. al., *Phys. Lett. A* **229**, 317 (1997)
2. V.E. Fortov, A.P. Nefedov, V.I. Molotkov, M.Y. Poustylnik, V.M. Torchinsky, *Phys. Rev. Lett.* **87**, 205002 (2001)
3. G.M. Jellum, J.E. Daugherty, D.B. Graves, *J. Appl. Phys.* **69**, 6923 (1991)
4. V.V. Balabanov, L.M. Vasilyak, S.P. Vetchinin et.al., *JETP* **92**, 86 (2001)
5. T. Nitter, *Plasma Sources Sci. Technol.* **5**, 93 (1996)
6. P.K. Shukla, A.A. Mamun, *Introduction to Dusty Plasma Physics*, IOP Publishing, London, 2001
7. V.E. Fortov, A.G. Khrapak, S.A. Khrapak, V.I. Molotkov, A.P. Nefedov, O.F. Petrov, V.M. Torchinsky, Phys. Plasmas, **7**, 1374 (2000)

Dusty plasmas in a dc glow discharge

V.E. Fortov, A.P. Nefedov, V.I. Molotkov, O.F. Petrov, M.Y. Poustylnik,
V.M. Torchinsky, A.G. Khrapak

*Institute for High Energy Densities, Russian Academy of Sciences
Izhorskaya 13/19, 125412, Moscow, Russia*

Abstract. The striations proved to be a very useful tool to study various phenomena taking place in complex plasmas. In this work we present results of observations within a wide range of formations from an ordinary chain of several charged dust grains up to complex structures consisting of regions with different dusty plasma states. Plasma crystals, liquid-like structures with a short-range order and with a convective motion of dust particles, a development of dust waves have been studied at different plasma conditions and with spherical dust particles varying in size from 1.9 μm up to 63 μm. Experiments with 300 μm long cylindrical dust grains discovered the formation of the ordered structure similar to the liquid crystal.

Recently, quasicrystalline dusty structures have been found in the dc glow discharge plasma. The main difficulty here is that in the near-cathode region of a dc discharge the electric field increases simultaneously with a decrease of the electron density. This leads to a sharp decrease of the particle charge and a reduction of the electrostatic force balancing gravity. In the stratified discharge the situation is qualitatively different. Here, a sharp increase of the field in the striation head is accompanied by only an insignificant decrease of the electron density.

The experiments were carried out in a vertically positioned glass tube filled with neon gas, in which the glow discharge with cold electrodes was created [1]. The upper electrode was the anode, the lower electrode was the cathode. In the experiments we observed levitation of separate particles as well as chains consisting of several particles arranged on the axis of the discharge. In this case the particles were located not equidistantly and the maximum distance was in the lower part of the chain. The addition of separate particles into the chain was possible only when the number of particles $n <$

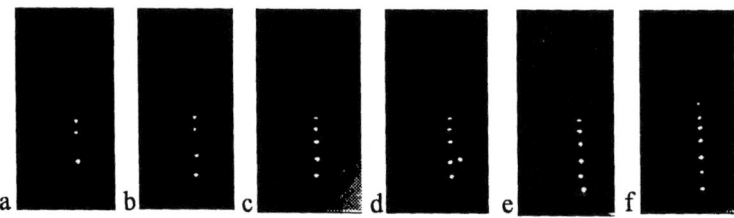

FIGURE 1. Formation of a vertical chain of melamineformaldehyde dust particles 10.24 μm diameter. Vertical size of each frame is 10 mm

10. When a number of particles is greater than the number mentioned above, the subsequent particles found a place at the side of the central part of the chain. As an example in Fig.1 we present chains of monodisperse melamine formaldehyde particles of diameter 10.24 μm ($\rho = 1.5$ g/cm^3), obtained at $I = 0.5$ mA, $p = 0.3$ Torr. In all the frames in Fig.1 except Fig.1d there are stationary states and Fig.1d gives the view of dynamics of incorporating the dust particles into the chain during their injection: the particle present in the frame at the side of the chain finally did not incorporate into the structure and fell down to the bottom of the discharge tube.

In this work we performed the molecular-dynamic simulation of the system of dust particles present in the glow discharge plasma. The parameters of the electric field have been chosen to be close to the experimentally measured values [2]. The radial component of the potential was given by the expression: $\phi_w = \phi_t(r/R_t)^\alpha$, where $R_t = 1.5$ cm is the discharge tube radius, $\phi_t = \phi_0 + \phi_1/(1+((z-z_1)/d_1)^2)$ is the wall potential at the height z. α, z_1 and d_1 are parameters. The particle charge Z_d was in a rigid connection with the floating potential ϕ_t and its diameter d, $Z_d = d\phi_t/2$. The interaction of the dust particles was taken into account with the help of the screened Coulomb potential. A dust particle was subjected along the vertical axis z to gravity and a random Langevin force $\xi(t)$, which satisfied to the condition $<\xi^2> = 2\beta T/\Delta t$, where β is the friction force of the buffer gas, T is the gas temperature and Δt - is the step of integration on time. Besides we took into account the ion drag force [4]. Its role increases with decreasing the particle size. For the integration of the equation of motion the Shtermer method of 8-th order was used [3]. The number of dust particles was varied in the range from 1 to 800. For more details of the simulation see [5].

It is possible to achieve the quantitative agreement of particle arrangements in the chain obtained in the experiment with the molecular dynamic simulation only at the particular choice of parameters when along the striation in the direction of the anode first the maximum of the electron density takes place, then the maximum of the floating potential (the particle charge) and last the maximum of the electric field. This is in an agreement with measurements of the parameters distribution in striations in the neon discharge [2] and testifies to the adequacy of the chosen model to the real experimental conditions. The analysis of the arrangement of charged particles in the chain can be used as an indirect means for diagnosing properties of the glow discharge positive column.

In the experiments we used four types of particles: particles of borosilicate glass ($\rho = 2.3$ g/cm^3), in the form of thin-walled, hollow spheres of diameter 50-63 μm with wall thickness 1-5 μm; Al_2O_3 particles ($\rho = 4$ g/cm^3) with diameter 3-5 μm, monodisperse melamine-formaldehide particles of two diameters 10.24 μm and 1.87 μm. Masses and charges of particles were in the range from $5 \cdot 10^{-12}$ - 10^{-7} g and 10^4 - $5 \cdot 10^5$ e, respectively.

In the case of small particles an increase of their number leads at the specific discharge parameters to formation of structures where different regions coexist: the regions of the strong ordering (plasma crystal), the regions with convective and oscillatory motion of particles (dusty plasma liquid). Fig.2 shows a vertical section of the structure of monodisperse particles with diameter $d = 1.87$ μm which is obtained at $I = 5.0$ mA, $p = 0.3$ Torr. In the lower part of the structure presented the oscillatory motions of particles in the vertical direction (waves of particles' density) are observed. The frequency of these

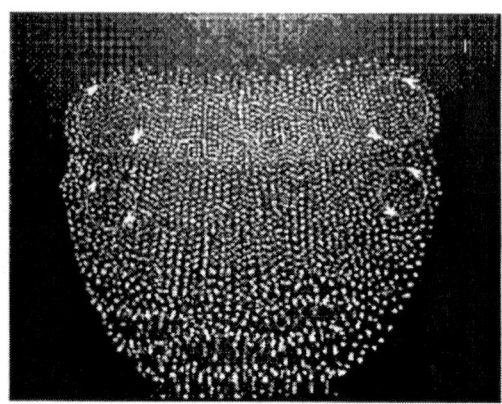

FIGURE 2. Videoimage of a complicated dusty plasma structure, formed of 1.87 μm diameter melamineformaldehyde dust grains. The central part of the structure is close to crystalline, side parts are liquid like and contain convective cells, in the lower part the self-excited dust acoustic waves are observed. The size of the image is 14x10.6 mm

oscillations is about 25-30 Hz and the wavelength is about 1000 μm with an average interparticle distance of about 200 μm. These waves are caused by the dust acoustic instability which is often observed in the dc discharge dusty plasma [6]. Practically in the whole central part there is a plasma crystal with a distinct chain-like ordering. At the periphery of the upper part of the structure there are regions with the convective motion of particles which weakens towards the structure center. The very interesting peculiarity of the structure presented in Fig.2 is the coexistence of several regions with a convective motion of dust particles. Each of the regions is a toroid. The toroid occupies the periphery part of the structure. In the vertical section these regions of the structure shown in Fig.2 present two zones which are symmetrical with respect to the axis of the discharge tube. In the zones there is a circular motion of dust particles. As it is shown by the arrows in Fig.2, in the left part of the structure dust particles move in the clockwise direction and in the right part there is a motion in the counterclockwise direction. Besides, let us note the boundary line between the upper and lower regions. The line is clearly seen in Fig.2. At the boundary of these regions the counter motion of dust particles is observed. Such a complicated picture is associated with the peculiar distribution of forces acting on the dust particles: the ion drag force, the electrical force and distribution of plasma parameters along the striation.

The experiments with cylindrical nylon particles ($\rho = 1.1$ g/cm^3) were also performed [7]. The length of a cylinder was 300 μm and a diameter 15 μm. It was not possible to realize levitation of such heavy particles in striations of the positive column of the neon glow-discharge. However, in the case of glow-discharge in a mixture of hydrogen with neon when a ratio of luminous part of the striation to its total length decreases and, therefore, the electric field in the striation head increases, cylindrical dust particles of the relatively great mass ($m_p = 6 \cdot 10^{-8}$g) levitate forming one or two layers in the horizontal plane. The video image of the layer mentioned above is presented in Fig.3. The particles align parallel to each other and form the ordered structure which is like the liquid crystal

 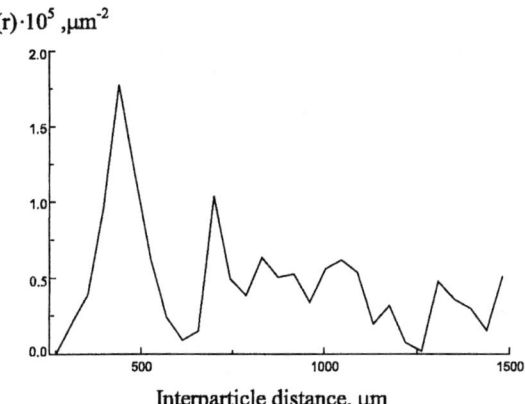

FIGURE 3. Videoimage of the horizontal section of a structure formed by cylindrical dust grains and a distribution function of their centers of masses.

of the nematic thread-like type with a mean distance between particles equals to 750 mm. In some regimes the interparticle distance was greater than 1 mm.

It could be supposed *a priori* that in the case of levitation of these cylindrical dust particles they will follow, in one or other form, the cylindrical symmetry of the discharge. Nevertheless, as the experiment revealed the preferential direction in the arrangement of particles has no connection with the discharge symmetry and depends very slightly upon a discharge current. The direction mentioned above is not an accidental one and is reproduced when the experiment is repeated.

REFERENCES

1. V.E. Fortov, A.P. Nefedov, V.M. Torchinsky, V.I. Molotkov, O.F. Petrov, A.A. Samarian, A.M. Lipaev, A.G. Khrapak, *Phys. Lett. A* 229, 317 (1997)
2. Y.B. Golubovsky, S.U. Nisimov, *J. Tech. Phys.* 65, 46 (1995)
3. V.V. Zhakhovsky, S.I. Anisimov, *JETP*, 84, 743 (1997)
4. M.S. Barnes, J.H. Keller, J.C. Foster, I.A. O'Neil and D.K. Coultas, *Phys. Rev. Lett.* 68, 313 (1992)
5. V.E. Fortov, V.I. Molotkov, V.M. Torchinsky in *Frontiers in Dusty Plasmas* ed. by Y. Nakamura, T. Yokota, P.K. Shukla, Elsevier, 2000, p. 445
6. V.E. Fortov, A.G. Khrapak, S.A. Khrapak, V.I. Molotkov, A.P. Nefedov, O.F. Petrov, V.M. Torchinsky, *Phys. Plasmas* 7(5), 1374 (2000)
7. V.I. Molotkov, A.P. Nefedov, M.Y. Poustylnik, V.M. Torchinsky, V.E. Fortov, A.G. Khrapak and K. Yoshino, *JETP*, 71, 102 (2000)

Oscillations of Few Particle Vertical Structures

N. J. Prior*, L. W. Mitchell* and A. A. Samarian[†]

*School of Chemistry, Physics, and Earth Sciences,
Flinders University of South Australia.
GPO Box 2100, Adelaide, SA 5001, Australia.
[†]School of Physics, University of Sydney NSW 2006 Australia.

Abstract. Experiments on the excitation of vertically aligned particle resonances are presented. A two particle string was driven vertically and horizontally by electrostatically and laser driven oscillations through a range of frequencies. The laser excitation technique allows us to drive one particle and observe the response of another, whereas in earlier experiments single particle systems were excited to vertical resonance. From the close agreement between our model and the experimental resonance curves, the charges of the top and bottom particles were established and a value for the Debye screening length and interparticle forces were estimated.

A complex plasma is a low-temperature plasma, consisting of the neutral gas, ions, electrons, and micron-sized particles of solid matter (dust particles). One of the basic problems in any complex plasma study is determining the value of the particle charge. In typical experimental situations the particles are trapped in the sheath, above the lower electrode of an RF discharge, where the electric field force on the particles balances their weight. There the particles interact by means of their Coulomb repulsion and form an ordered lattice, the so-called plasma crystal. Hence, the charge on the particles is a crucial parameter for particle trapping as well as for the formation of ordered plasma crystals.

Several theories concerning dust charging in a plasma environment exist, however even the applicability of the most common orbital motion limited (OML) charging theory [1] (in the plasma sheath) is still an open question [2, 3]. A self-consistent theory of particulate charging in the plasma sheath does not appear to currently exist. Under this circumstance the experimental measurement of the value of the charge on the dust particle becomes very important. This poster presents an overview of the experiments undertaken at the Flinders University of South Australia and possible theories which allow estimation of the particle charge and interparticle interaction force.

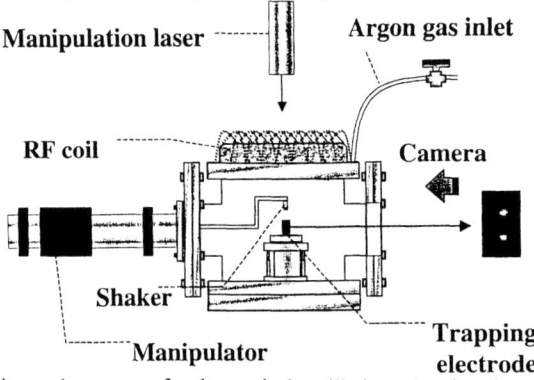

FIGURE 1. Experimental apparatus for the vertical oscillations showing the manipulation laser above the chamber. The camera on the side of the chamber was used to record the vertical oscillations. The particles were illuminated by a further laser through one of the side windows (not shown).

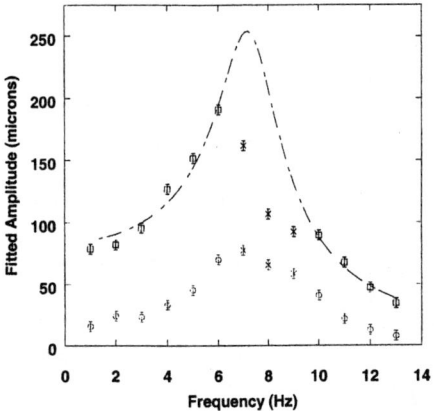

FIGURE 2. Fitted amplitude as a function of frequency for the laser driven oscillations. The circles, (o), represent the amplitude of the top particle and the squares, (□), represent the amplitude of the bottom particle. The crosses, (×), are amplitudes which exhibit vibration absorber behaviour [5]. The bottom particle data set is fitted with a natural resonance frequency curve, that excludes the particles affected by vibration absorber behaviour.

The experimental arrangement was a cylindrical chamber, as shown in figure 1 and described in detail elsewhere [4]. The experiments were undertaken in an inductively coupled argon plasma at a pressure of 7 Pa, with an electron density of 2.2×10^9 cm^{-3} and an electron temperature of 1.9 eV. The particles were 6.21 μm diameter melamine formaldehyde spheres. The dust particles were stored in a small shaker with a ~ 50 μm exit hole, to allow only a few particles into the plasma upon agitation of the manipulator. Once in the plasma, two particles were trapped in the viewing region using a 5.6 mm diameter pad held at a potential of ~ 16 V relative to the grounded lower electrode. The structure was illuminated with a helium-neon laser and the oscillations were recorded using a video camera attached to a microscope. For the vertical oscillations, there were two techniques used to oscillate the particles, after they were trapped above the electrode (Fig. 1). The first method was to vary the trapping voltage sinusoidally (pad driven oscillation) which involved the addition of a 100 mV p-p AC signal to the DC trapping voltage, while the second employed a 35 mW diode laser which was focussed onto individual particles and pulsed. This led to square wave driven oscillations (laser driven oscillations).For the horizontal oscillations, the laser was repositioned on the side of the chamber and used to drive each particle in the two particle string inturn. This was the only way that horizontal oscillations were driven. Both the vertical and the horizontal oscillations were recorded on a video recorder. The recording of the behaviour of the oscillating particles was digitised and the vertical and horizontal positions of every particle in every frame were obtained. The software used a centroid extraction algorithm which returned an x and y corresponding to the centroid of the particle image. This in turn allowed extraction of the amplitude of the oscillation to an RMS accuracy of ± 0.5 pixel. Once the position of the particle was known the amplitude could be approximated and a series of plots of amplitude as a function of frequency were generated.

The goal of the vertical oscillation experiments was to determine the charge of two particles in a vertical string using a variation of the resonance technique. Consequently it is important to note here that when the structure was driven by the pad, we can neglect the interparticle coupling. This was due to the fact that the pad excites the system as a whole, oscillating both particles in the structure simultaneously, preventing the excitation due to the interparticle coupling. This is in contrast to the laser driven case, where the bottom particle only was driven and the top particle allowed to respond.

With this in mind, for the pad driven results, the analysis is identical to that for the one particle case [6]. Using the parabolic sheath approximation, the sheath electric field gradient was determined to be 3.3×10^5 Vm^{-2} and the dust particle mass is given as 1.9×10^{-13} kg. This one particle analysis can be used because the particles were not allowed to interact, essentially behaving as one particle with twice the mass and twice the charge. Evaluating Ze in this manner gave the average charge of the two particles. From the graph of pad driven oscillation amplitude as a function of frequency the resonance value was determined to be $f_0 \sim 6.5$ Hz, indicating that the average particle charge was $\sim 6.1 \times 10^3$ times the charge on the electron.

For the laser driven results, the plot of oscillation amplitude as a function of frequency (figure 2) was clearly not a simple resonance curve, rather a resonance curve with a dip in the tail. This is similar to the resonance plot of a vibration absorber [5], and so a new model was required for this case. A coupled oscillator model with both particles having linear coupling to the plasma was chosen. Such an approach is adequate for the parabolic sheath potential, where the electrostatic restoration force is linear, ($F_{el} \sim E'Zer$). The simplest approximation for particle interaction is the screened Coulomb-Debye potential, although we note that for particles levitating in the plasma sheath the interaction potential in the vertical direction is asymmetric, because of the ions flowing towards the electrode. However, it is still interesting to consider the case for the Debye interaction.

Using the model we were able to extract two parameters describing the experimental system, the debye length, λ_D, and the charge ratio, Q_r, From the ideal fit of the modelled data to the experimental data the charge ratio indicated that the bottom particle carried 0.7 times the charge of the top particle. The charge on the top particle was $(7.83 \pm 0.18) \times 10^3 e$ which resulted in a calculated charge on the bottom particle of $(5.48 \pm 0.26) \times 10^3 e$. These charges when averaged compare well with the pad driven case which indicated that the average charge was $6.1 \times 10^3 e$.

For the horizontal experiments there were two cases, hitting the top particle with the laser and hitting the bottom particle. The two cases showed very different responses. For the experiments driving the top particle, the particles oscillated in phase at low frequency and then as the frequency was increased, started to swap phase, see figurephasetop. When the bottom particle was driven the results were quite different. The particles started out of phase and then as the frequency was increased the phase difference decreased between them, see figure 4. This is an example of the asymmettry of the vertical system. In order to investigate the interparticle force, we used each particle's instantaneous positions in each frame to obtain equations for the velocity and acceleration of each particle. Following this the equation of motion of each particle was used to determine the interparticle force. This analysis is preliminary but the force was found to be attractive in both cases indicating that the dust particle had molecular properties. In all cases the forces were of the order of 10^{-14} N.

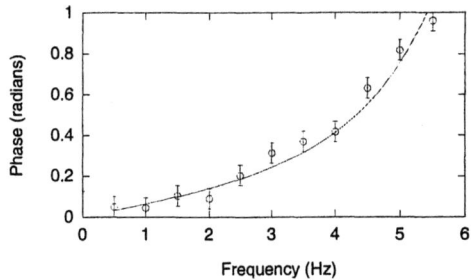

FIGURE 3. Phase as a function of frequency for the horizontal laser driven oscillations In this case the top particle was driven. The data set is fitted with a natural resonance phase curve.

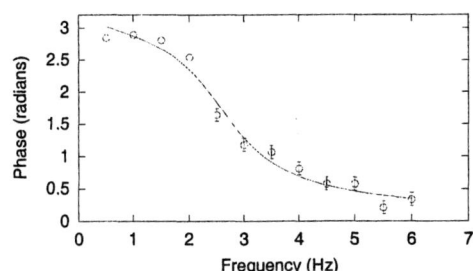

FIGURE 4. Phase as a function of frequency for the horizontal laser driven oscillations In this case the bottom particle was driven. The data set is fitted with a natural resonance phase curve.

In conclusion, we have shown that oscillations are a promising tool for obtaining information about both the plasma system and multiple particle interaction.

Acknowledgments

The authors would like to thank the Science Foundation for Physics within the University of Sydney and Dr. S. Vladimirov for useful discussions. NJP acknowledges the support of a Flinders University Research Scholarship. AAS was supported by U2000 Fellowship of the University of Sydney.

REFERENCES

1. Whipple, E. C., *Rep. Prog. Phys.*, **44**, 1197 (1981).
2. Allen, J. E., *Physica Scripta*, **45**, 497 (1992).
3. Tomme, E. B., Annaratone, B. M., and Allen, J. E., *Plasma Sources Sci. Technol.*, **9**, 87 (2000).
4. Prior, N. J., Mitchell, L. W., Samarian, A. A., and James, B. W., *Phys. Rev. E (in preparation)* (2002).
5. Palm, W. J., *Modelling, Analysis and Control of Dynamic Systems*, Wiley, 2000.
6. Homann, A., Melzer, A., and Piel, A., *Phys. Rev. E*, **59**, R3835 (1999).

Lunar and Martian dust charging on surfaces

Zoltan Sternovsky, Mihaly Horanyi and Scott Robertson

Department of Physics, University of Colorado, Boulder, CO 80309-0390 USA

Abstract. The contact charging of two planetary analog dust samples are investigated in a simple experiment. Dust particles of lunar (JSC-1) and Martian (JSC-MARS-1) regolith simulants are brought into contact with metals of known work function (Co, Ni, Au, Pt) and the resulting contact charge is measured. The dust charge is typically 10^5 elementary charges for a ~100 micron dust particle and increases with repeated agitation of the surface. The dust charge scales linearly the work function of the contacting surface. The effective work functions of the planetary analogs are determined by extrapolation to be 5.8 eV and 5.6 eV for the lunar and Martian dust simulants, respectively. Additional measurements are made with oxidized metal surfaces and a glass surface.

INTRODUCTION

The dynamics of charged dust particles is strongly influenced by electromagnetic forces in various space environments [1]. The charging of dust particles suspended in plasmas is primarily due to collection of plasma electrons and ions and due to UV radiation. Often, however, dust dynamics on or near surfaces of large solid bodies is of interest where the charging is from dust-dust or dust-surface contact. Such circumstances exist in Martian or terrestrial dust storms, in Saturn's ring system, or on the surfaces of airless planetary objects. The charging of dust particles then can be affected or even dominated by triboelectric or contact charging processes. Contrary to dust charging in plasma, contact charging is determined by the material properties of the dust and the contacting surface. Earlier, we have developed a simple experimental method to investigate dust charging on surfaces [2] by either contact or induction in an applied electric field. The dust particles studied were both metallic and insulating in the 40 – 200 μm size range. Here, we use the same experimental technique to investigate the contact electrification of two planetary analog dust samples: JSC-1 lunar and JSC-Mars-1 Martian dust simulants.

CONTACT CHARGING

Charge transfer between materials of different electronic structure occurs when brought into contact. Electrons are transferred from the material with a lower work function, W_1, to the material with higher work function, W_2, until thermodynamic equilibrium is reached. Upon separation, the residual contact charge is often found proportional to the contact potential, $V_c = (W_1 - W_2) / e$, where e is the elementary charge [3,4]. For breaking metal - metal contacts, the contact charge is strongly reduced by electron tunneling current, which is drawn by the increasing potential difference between the separating species. Due to the absence of this backflow current, the contact charging is considerably greater when at least one of the contacting materials is an insulator [3,4].

The electronic structure of insulator materials is complicated. Localized electron energy levels inside the forbidden gap exist due to impurities, crystal defects, or the presence of the surface. These energy levels can be filled up to a certain level, that is customarily assigned as the "effective" work function of the insulator [3,4]. Charge accumulation is often observed with repeated contacts of insulators that may be caused by increased contact area or increased charge penetration depth.

In the contact charging process the properties of the surfaces are important. Several metal surfaces used in the experiments rapidly oxidize in air. A model by Harper [5] assumes the presence of an oxygen atom layer on the oxidized surface. The acceptor levels of atomic oxygen are close to 5.5 eV and will be occupied by electrons from the metal. Thus the oxidized metals behave like materials with 5.5 eV work function in contact charging.

EXPERIMENTS AND RESULTS

The experimental set-up is similar to that used in earlier investigations [2]. In a small vacuum chamber, a 5 cm diameter disc of carefully cleaned thin metal or of glass is mounted horizontally. A small amount of the dust sample is loaded onto the surface. The surface is mechanically agitated by a small electromagnet acting on a pair of permanent magnets attached to the disc. Due to the agitation, dust particles drop through a 1 mm central hole into a Faraday cup beneath. The Faraday cup is connected to a sensitive and calibrated electrometer that measures the net charge on the dust particles with $\pm 2 \times 10^4$ e accuracy.

The experiments are performed in high vacuum conditions (pressure < 10^{-6} Torr) with two regolith dust samples: JSC-1 and JSC-Mars-1 simulants. These samples imitate the chemical composition, mineralogy, density, and other properties of the lunar mare soil and the oxidized soil of Mars, respectively. Dust samples are size selected by using standard sieves. The surfaces were made of Zr, V, W, Co, Ni, Au, Pt, and stainless steel, in the form of high purity thin foils with thickness in the range 0.02 to 0.2 mm. Additionally, a glass disc is used.

An example of experimental data is shown in Fig. 3 for the JSC-1 dust charged from clean Ni surface. The contact charge on the dust increases during the course of the experiment, i.e. a dust particle dropped later in the queue had, on average, larger charge than one dropped earlier. This behavior is typical for both dust samples used and it is due to the repeated contact effect. Dust charges up to 10^6 e are found that correspond to about −20V of surface potential assuming spherical shape and homogeneous surface charge distribution [1]. Most probably, however, the charge on the dust is localized to the spot of contact and thus the local potential differences can be considerably larger. The estimated surface electric field intensity E is on the order of $10^5 - 10^6$ V/m depending on the size of the charged spot.

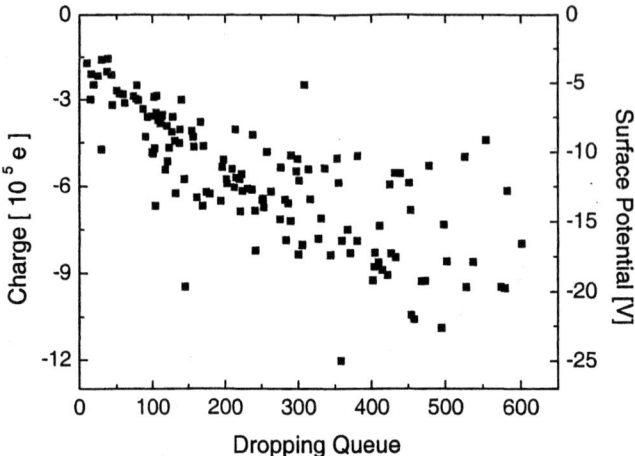

Figure 1. The evolution of the charge on JSC-1 lunar dust simulant particles after falling from a Ni surface in a queue. The particle diameters were 125 – 150 μm. The right axis shows the equivalent dust surface potential assuming spherical shape. Dust particles drop less frequently than on every agitation.

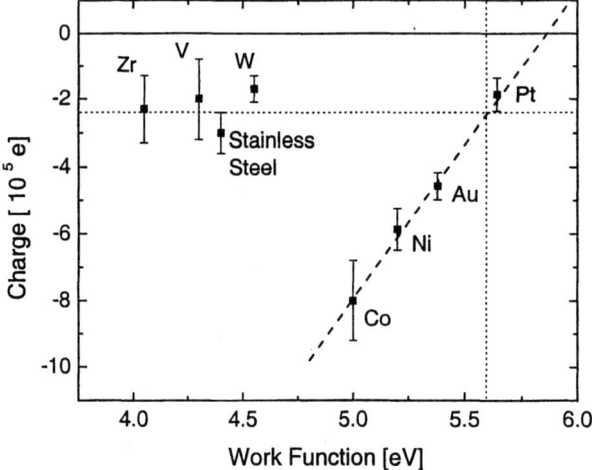

Figure 2. The charging of JSC-1 from different metals. The dust sample size range is 125 – 150 μm. The points represent the average dust charge from six individual measurements with the standard deviations indicated. The dashed line is the linear fit for the non-oxidized surfaces: Co, Ni, Au, and Pt. The horizontal dotted lines represent the average dust charge from oxidized surfaces. The vertical dotted line indicates the effective work function of oxidized metals.

The data in Fig. 2 show the average charge on the JSC-1 dust particles from different metal surfaces as a function of their work function. For surfaces that oxidize instantly on air (Zr, V, W and stainless steel) the dust charge shows little variation with the work function. On the other hand, the dust charge varies linearly with the work function of metals resistive to oxidation (Co, Ni, Au and Pt). The effective work function of the dust sample is determined extrapolating the linear fit to zero dust charge giving 5.8 eV for JSC-1. The average contact charge from oxidized surfaces (dotted horizontal line) matches the linear fit close to 5.5 eV that is in agreement with the model by Harper [5]. The work function of JSC-Mars-1 was determined similarly to be 5.6 eV. In additional experiments it was also shown that the contact charge on the dust particles varies linearly with size [2,6].

Glass, a nonconductor, represents a more realistic situation for dust charging on the surfaces of planetary objects. Both the JSC-1 and JSC-Mars-1 charged negatively on glass. Figure 3 shows the average dust charge on the simulants together with two dust samples that have been investigated in an earlier study [2]: alumina and silica. The effective work functions are 5.25 and 5.5 eV for Al_2O_3 and SiO_2, respectively. The intensity of charging of the planetary simulants from the glass surface is similar to charging from metal surfaces.

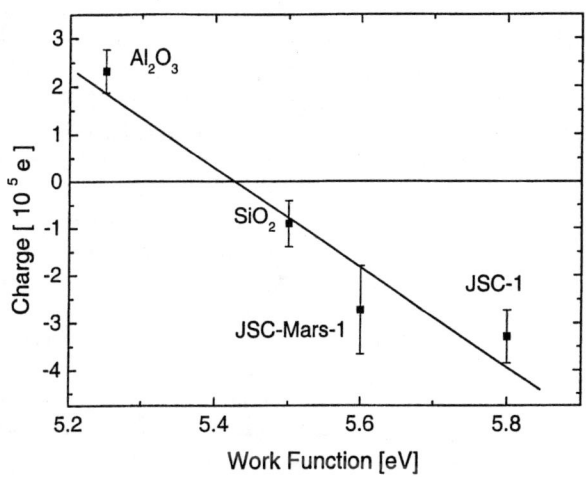

Figure 3. Contact charge from a glass surface on different dust compositions (alumina, silica, JSC-1 and JSC-Mars-1). The dust samples were from a 125 – 150 μm size range.

ACKNOWLEDGMENTS

The authors acknowledge the support from the National Aeronautics and Space Administration (Microgravity Sciences).

REFERENCES

1. M. Horányi, *Annu. Rev. Astron. Astrophys.* **34**, 383 (1996).
2. Z. Sternovsky, M. Horányi, and S. Robertson, *J. Vac. Sci. Technol. A*, **19**, 2533 (2001).
3. J. Lowell and A. C. Rose-Innes, *Advances in Physics* **29**, 947 (1980).
4. J. Cross, *Electrostatics: Principles, Problems and Applications* (Bristol, Hilger, 1987), Chap. 2.
5. W. R. Harper, *Contact and frictional electrification* (Clarendon Press, Oxford, 1967), Chaps. 1, 2.
6. Z. Sternovsky, A. Sickafoose, J. Colwell, S. Robertson, and M. Horányi, submitted to *J. Geophys. Res.* (2002).

Dust Vortex in Complex Plasma

A.A. Samarian and O. Vaulina

School of Physics, University of Sydney, NSW 2006, Australia

Abstract. The results of experimental observations of self-excited dust vortex motions in dc and rf-discharge plasma are presented. We found vertical vortices of dust particles in inductively coupled plasma rotated with an angular frequency ~ 0.2-1.5 1/sec and similar motion of the dust structures formed in striation region of a dc-discharge. Two types of the dust vortex were also excited in the presence of an additional electrode in a planar radio frequency discharge. The first is vertical rotation of macroparticles in bulk dust clouds; the second occurs in the horizontal plane for a monolayer structure of particles.

Problems associated with the formation and development of vortices in dissipative systems of interacting macroparticles are under discussion in various fields of science (plasma physics, molecular biophysics, hydrodynamics etc.). Dusty plasma provides a good experimental model for the study of vortices in these systems.

The theoretical analysis and numerical simulation [1-3] have shown that in plasma with gradients of dust charge, various dust particle motions could be excited. The vertical vortices of dust particles were found to be rotating in inductively coupled plasma with an angular frequency of 0.2–$1.5\,\text{s}^{-1}$. Similar motion of dust structures was observed in striation region of a dc-discharge. There are two basic reasons which prevent the development of the dissipative instability in planar capacitive rf-discharges. The first is the homogeneity of plasma, and the second reason is concerned with the small number of observed layers of macroparticles and, accordingly with the small shift parameter γ_o [3]. This explains the absence of experimental observations of vortex motion in the usual circumstances. Nevertheless, the formation of bulk dust clouds, for example, in microgravity, or the introduction of an extra electrode, leads to the convection of dust particles [4,5]. Below, we analyse the conditions for generation of dust vortex motion which was observed in an rf-discharge plasma with the different auxiliary electrodes.

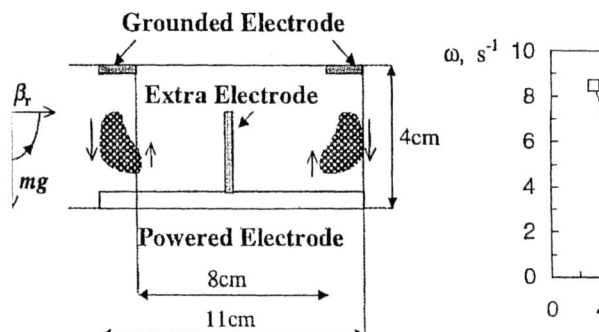
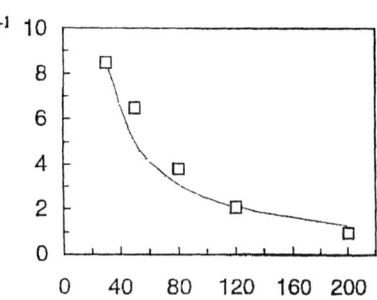

FIGURE 1. Experimental set up for vertical vortex production formation of horizontal dust vortex

FIGURE 2. The dependency of rotation frequency ω on the pressure. The points are the experimental values, solid line corresponds to $\omega = g/4P\,[\text{mTorr}]$

Experiments were carried out in argon (Ar) at the pressures in the range $P = 0.06 - 0.2$ Torr, with the melamine formaldehyde particles ($a \cong 2.8\,\mu m$, $\rho = 1.5\,g/cm^3$). The presence of extra electrodes in the centre of the system allows displacement of the dust particles to the extreme region of the system, where the gradient β_r of dust charges is greater. This increase in β_r is connected with the fast growth of the radial potential $\phi(r)$ of the electric field $E(r)$ near the edge of the powered electrode. It should be noted that the decrease of electron temperature near the edge of the electrode could also play an important role in decreasing the dust charges in this case. A simplified schematic diagram of the experiment on the formation of bulk dust cloud in the presence of an extra electrode is shown in Fig. 1. The direction of dust rotation is in accordance with the theoretical forecast ($\Omega < 0$). Dependency of the rotation frequency ω on pressure is presented in Fig. 2. For our experimental conditions the friction frequency can be written, using a free-molecular approach, as $v_{fr}[s^{-1}] \cong 200P$ [Torr]. Assuming $F_{non} = m_p g$ and $\omega = \Omega/2 \sim g\beta_r/2 < Z > V_{fr}$, we obtain the value $\beta_r/<Z> \approx 0.1\,cm^{-1}$.

The presence of additional electrodes in the rf-plasma can not only destroy the homogeneous conditions in the plasma, but can also lead to additional forces F_{non}, which in conjunction with the electric forces leads to rotation of the macroparticles. A simplified schematic diagram of the experiment on the formation of horizontal vortex in the presence of an extra electrode is shown in Fig. 3.

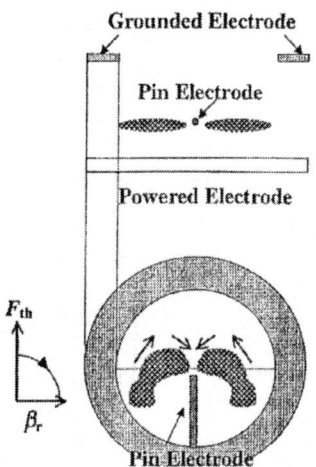
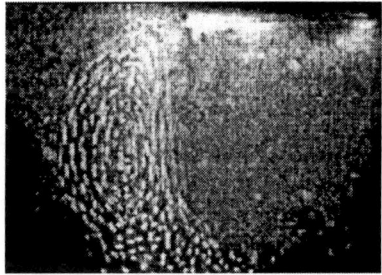

FIGURE 3. Simplified schematic diagram of the apparatus used for the formation of horizontal dust vortex

FIGURE 4. Typical video image of the dust vortex

The typical images of induced dust rotation in the horizontal plane for the few layers of macroparticles in a rf-plasma source with a heated pin electrode are presented in Fig. 4. To achieve heating by electron current a positive potential U_1 (with respect to the surrounding plasma) was applied to the pin electrode.

We have studied the dependence of rotation frequency ω on the pressure P for $U_1 = 40$ V, and on the potential U_1 of the extra electrode for $P = 0.2$ Torr. The direction

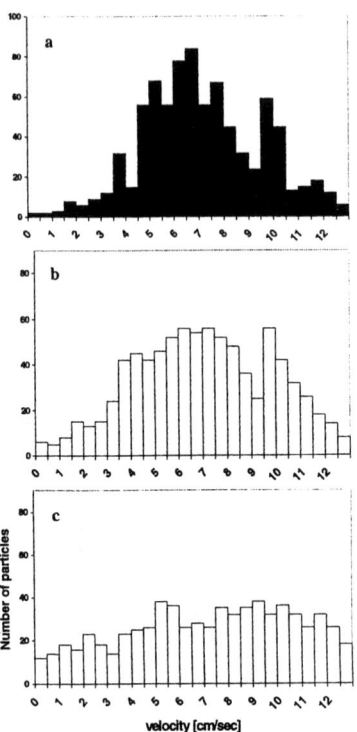

FIGURE 5. Velocity distribution function for different input power. a- 100 W b- 70 W

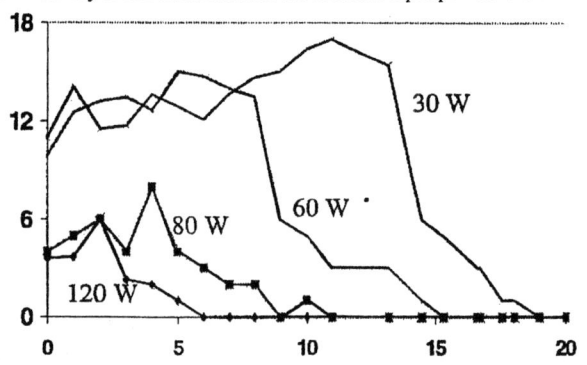

FIGURE 6. The effect of power on the z-component of the velocities of the particles

of dust rotation is in accordance with the direction of the thermophoretic force and the decrease of dust charges from the pin electrode ($\Omega < 0$). There are two basic reasons that can lead to an increase in dust charges towards a positive pin electrode. The first is the reduction of ratio n_i/n_e, and the second is connected with the growth of electron temperature in the region of high electric field. A lower limit estimate of angular velocity ω_{min}

of dust rotation can be obtained from $\omega_{\min}[s^{-1}] = \Omega_{\min}/2 \approx \beta_r F_{th}/2m_p <Z> v_{fr} \approx$ $50\beta_r[\text{cm}^{-1}]\partial T/\partial y[\text{Kcm}^{-1}]/<Z>v_{fr}[s^{-1}]$. In turn a lower limit estimate of gas temperature gradient can be obtained from a calculation of $\partial T/\partial y$ for a heated sphere with radius R equal to the radius of the extra electrode ($\sim 50\,\mu\text{m}$): $\partial T/\partial y = R\Delta T/\Lambda^2 \sim 0.1\Delta T$, where $\Lambda \sim 0.5\,\text{cm}$ is the characteristic distance from the extra electrode to the edge of dust cloud, and ΔT is the difference between the electrode and gas temperatures, respectively T_1 and T_0. The value of T_1 can be estimated by equating the input power density $P_1/\pi h R^2$ to the heat losses $\chi_1 \Delta T/2\pi Rh$ through the electrode, where h is the length of electrode, and $\chi_1 = 0.8$ W/cmK is the thermal conductivity of steel. Thus for $P_1 = 1.1$ W ($U_1 = 40$ V), we have $\Delta T \sim 100$ K and $\beta_r/<Z>\sim 0.6$ cm^{-1}.

In addition to the angular velocities of the vortices the distribution of linear velocities in the vortices was obtained. The velocity distribution was measured for different input rf powers. The distribution is computed by binning particles in steps of 0.5 cm/s from 0 to 14 cm/s.

A comparison of the velocity distributions for three different input rf powers is shown in Fig. 5. It is observed that at the lowest power the velocity distribution is fairly broad. As input rf power is increased the velocity distribution becomes peaked around 6 cm/s. The broadening of the velocity distribution function with decrease in input power was attributed to the expansion of the vortex in the vertical direction and occurrence of a vertical (z) component of particle velocity (see Fig.6).

The results of experimental observation of two types of self-excited dust vortex motions (vertical and horizontal) in a planar rf-discharge are presented. The first is vertical rotation of macroparticles in bulk dust clouds. The second is formed in the horizontal plane for a monolayer structure of particles. We attribute the occurrence of these vortices to the development of dissipative instability in a dust cloud with a dust charge gradient, which has been produced by an extra electrode. The presence of an additional electrode also produces an additional force which in conjunction with the electric forces leads to rotation of dust structure in the horizontal plane.

ACKNOWLEDGMENTS

This work was supported by Australian Research Council, the Science Foundation for Physics within the University of Sydney and Russian Foundation for Fundamental Research, project No.01-02-16658. AAS was supported by University of Sydney U2000 Fellowship.

REFERENCES

1. Vaulina, O.S., Nefedov, A.P., Petrov, O.F. and Fortov, V.E., *JETP* **91**, 1063 (2000).
2. Vaulina, O.S., Samarian, A.A., Nefedov, A.P., and Fortov, V.E., *Phys. Lett. A*. **289**, 240 (2001).
3. Vaulina, O.S., Samarian, A.A., Petrov, O.F, James, B.W., and Fortov, V.E.,*Phys. Rev. E.* (to be published).
4. G. Morfill, H. Thomas, U. Konopka, *et al.*, *Phys. Rev. Lett.* **83**, 1598 (1999).
5. D.A. Low, W.H. Steel, B.M. Annaratone and J.E. Allen, *Phys. Rev. Lett.* **80**, 4189 (1998).

Charging of Different Size Dust Particles in the Plasma Sheath

A.A. Samarian, S.V. Vladimirov, and S.A. Maiorov

School of Physics, University of Sydney, NSW 2006, Australia

Abstract. In this work, we investigate particle charges as a function of the particle size by studying the particle levitation position experimentally. These data were compared with those obtained on the basis of the model calculation of the particle levitation, involving the OML approximation for charging plasma currents, as well as with the results obtained by a particle-in-cell (PIC) numerical simulation performed for a simplified model of a dust particle in the plasma flow. The observed discrepancy between experimental and theoretical, simulation and modelling results was explained by the non-Maxwellian character of the electron distribution in the sheath.

One of the most important problems associated with the presence of macroscopic objects (``dust grains") in a plasma is the value of the particle charge. For the typical conditions of a radio-frequency (rf) discharge plasma, the charging of a micron-size dust grain is mainly due to plasma currents onto the grain surface, with the resulting net charge being negative and large, as related to the charges of plasma electrons and ions (such that the dimensionless charge Z_d is of order 10^3-10^4). In the simplest approximation, the charge is determined by the floating potential and therefore related to the capacity of the grain that, in the case of spherical particles, is directly proportional to the particles' radius.

$$Q = R_d(1 + \frac{R_d}{\lambda_D})\varphi_s , \text{ for } R_d << \lambda_D \quad Q = R_d \varphi_s \quad (1)$$

In the laboratory, the problem of the charge acquired by dust grains is directly related to dust levitation. For the majority of experiments, dust grains levitate in the sheath region where strong inhomogeneities of the plasma field and density distributions exist. One of the major forces in the balance of forces acting on the levitating particles in the plasma sheath is the electrostatic force, and therefore the value of the dust charge for the levitating particle is strongly related to the levitation height (and vice versa). Because of the inhomogeneities of the plasma field and density distributions, the surface potential of the particle levitating in the plasma sheath is a function of the levitating position $\varphi_s = f^\varphi(r_{eq})$. Actually the surface potential determined by the ion electron density ratio, electron temperature and the ion velocity $\varphi_s = f^\varphi(n_e/n_i, T_e, v_i)$, which in the sheath depends on the distance from the electrode $n_e/n_i = f^n(r)$, $T_e = f^T(r,)$, $v_i = f^v(r)$. Thus the charge of the particle levitating in the plasma sheath can be expressed as follows

$$Q = R_d f^\varphi(f^n(r_{eq}), f^T(r_{eq}), f^v(r_{eq})) \quad (2)$$

In the OML approximation, f^φ can be expressed as a product of $f^\varphi{}_n f^\varphi{}_T f^\varphi{}_v$, where

$f^\varphi{}_n \sim 1 + \beta \ln \frac{n_e}{n_i}$, $f^\varphi{}_T \sim T_e$, $f^v \sim$ f(v_i). In this case charge will be proportional to

$$Q \sim R_d T_e (1 + \beta \ln \frac{n_e}{n_i}) \quad (3)$$

Considering Eq. (2), we should mention that the levitating position r_{eq} is strongly dependent on particle size such as the bigger and therefore more massive particles tend to levitate deeper in the sheath, with reduced electron plasma density and larger sheath fields, while lighter particles levitate closer to the sheath edge with more plasma electrons and weaker sheath electric field. Therefore the charge is a complex function of R_d:

$$Q \sim R_d T_e (f^{eq}(R_d)) \left[1 + \beta \ln \frac{n_e(f^{eq}(R_d))}{n_i(f^{eq}(R_d))} \right], \qquad (4)$$

and we can see that charge of the particle levitating in the plasma sheath is a nonlinear function of the particle size even if the simplest OLM charging model is assumed.

Here, we report the experiments to find the dependence of the dust charge as a function of its size in an rf-discharge plasma. We obtain the experimental value of the charge from the balance of forces method [1] and vertical resonance technique [2]. Furthermore, we compare the experimental results with those obtained on the basis of PIC simulations using the code developed in [3] as well as with those found on the basis of the levitation model developed [4].

Figure 1 shows the charges on different size dust particles are presented for $P = 65$ mTorr, $W = 64$W. For comparison, the values of dust charge calculated using the plasma parameters obtained from probe measurements 3 mm above the sheath edge (T_e=2.1, V_p=35V, V_{sb}=78V, n_e=5.7 10^8 cm^{-3}) are also given (solid line). For this calculation we take into account that when the emission processes are unimportant, the equilibrium charge Z_d of dust particles can be obtained from the OML theory under the condition $a \ll D \ll \lambda_{mfp}$, where λ_{mfp} is a mean free path for electron-neutral or ion-neutral collisions. If we assume that the drift electron (ion) currents are less than thermal ones, $T_i \approx 0.03$eV, and $n_e \approx n_i$, we have [5]

$$<Z> = C_z\, a\, [\mu m]\, T_e\, [eV], \qquad (5)$$

where the value of C_z and is dependent on the type of background gas and equal 1.8 10^3 for Ar.

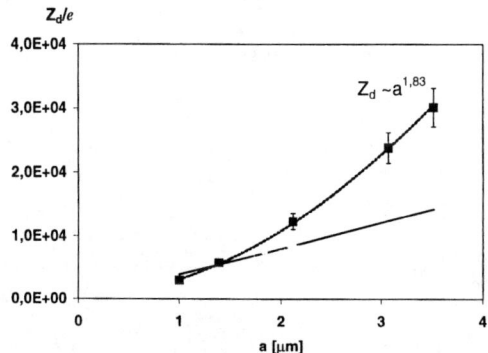

FIGURE 1. Experimental dependence of the particle charge on its size

We consider the results of PIC modelling of the particle charging in a plasma without the ion flow. This allows us to elucidate the character of the dust charging for various dust sizes when the plasma environment does not change. As was mentioned above that in the laboratory, the dust particles levitate above the horizontal electrode, and the levitation height strongly depends on the particle charge and mass and therefore on the particle size: the bigger particles levitate lower, i.e. closer to the electrode, where the plasma parameters are different than those for lighter particles. However, it is instructive to check the dependence of the dust particle charge on its size in the case of the same plasma environment. That is why we have performed the special simulation runs.

For the simulation we have chosen an argon plasma with the electron temperature is $T_e=2$ eV and the ion temperature is $T_i=0.025$ eV. The ion Debye length for such plasma normalized to the average distance between ions is $r_{Di}/n_i^{1/3}=1.06$, while for electrons we have $r_{De}/n_e^{1/3}=6.6$. Figure 2 demonstrates the dependence of the charge accumulated on the dust grain on its size. For better comparison with other results, we have chosen the same particle sizes. The fitting of the power function of size gives the characteristic index 1.38. For bigger grain sizes, we see that the grain charge is almost directly proportional to the grain size with good accuracy. On the other hand, for relatively small grains the dependence demonstrates the nonlinear character: the charge grows faster than linear with the increasing particle size.

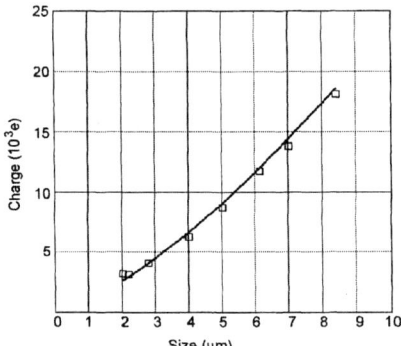

FIGURE 2. Experimental chamber and image of test gains levitated above the electrode.

Here, we also present the results of a self-consistent hydrodynamic model of the dust levitation and equilibrium in a collisional plasma sheath taking into account plasma ionization. We treat the sheath problem self-consistently and investigate dust levitation at various positions corresponding to different particle sizes, with not only supersonic, but also subsonic ion flow velocities at the levitation position of the dust grain in a collisional plasma with an ionization source. For more details of the model, see [4].

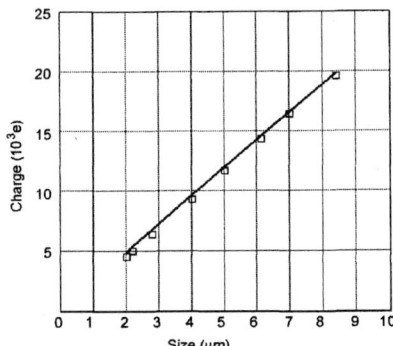

FIGURE 3. Dependence of the levitating charge on its size (hydrodynamic model + OLM).

Figure 3 shows the charges of the levitating particles of various sizes. By fitting this data, we see that the actual power index (0.96) is (although very close to unity) less than 1. This effect demonstrates the discussed above nonlinear dependence of the levitating particles on the grain size when bigger and therefore heavier particles levitate deeper into the sheath (and closer to the electrode). On the other hand, the deviation from the linear dependence is very small.

The observed discrepancy between experimental and theoretical, simulation and modelling results can be explained by the non-Maxwellian character of the electron distribution in the sheath, e.g., by the presence of "hot" electrons (for example due to secondary emission), that were not taken into account in the above theoretical consideration. In this case the electron temperature can be expressed as

$$T_e^* = T_e + \ln(1 + \gamma \exp\left(\frac{e\varphi_s(T_e^h - T_e)}{T_e^h T_e}\right)), \quad \gamma = \frac{n_e^h}{n_e}, \quad (6)$$

and the variation of γ along the sheath can lead to the variation of charge. We can expected the increasing value of γ towards the electrode which, according to (4), explains the observed non-increasing of the dust particle charge.

ACKNOWLEDGMENTS

This work was supported by Australian Research Council and partially by the Science Foundation for Physics of the University of Sydney. AAS was also supported by the University of Sydney U2000 Fellowship.

REFERENCES

1. Samarian, A.A. and James, B.W., *Phys. Lett.* **A287**, 125 (2001)
2. Melzer A., T. Trottenberg T., A. Piel A., *Phys. Lett.* A**191**, 301 (1994)
3. Vladimirov SV, Cramer NF, Maiorov S.A. *25th ICPIG, Nagoya, Japan, 2001, vol.3, p.21-2*.
4. Vladimirov S.V., and Cramer N.F., *Phys .Rev. E*, **62**, 2754 (2000).
5. Vaulina O.S. *et al, Phys. Rev. E* (to be published)

Nanoscale SiO_2 Particles at High Temperatures: Size Dependent Properties

I.V. Schweigert*, K.E.J. Lehtinen[†], M.J. Carrier** and M.R. Zachariah[‡]

Institute of Theoretical and Applied Mechanics, Novosibirsk 630090, Russia
[†] *University of Helsinki, Finland*
**National Institute of Standards and Technology, Gaithersburg, USA*
[‡] *Department of Mechanical Engineering and Chemistry, University of Minnesota and the Minnesota Supercomputer Institute, Minneapolis, USA*

Abstract. The properties of silica clusters at temperatures of 1500 to 2800 K have been investigated using classical molecular dynamics simulations for particles containing up to 1152 atoms. We found that the atoms in the cluster were arranged in a shell like structure at low temperatures and that the radial density profile peaked near the outer-edge of the particle. Our computed surface tension did not show any significant size dependent behavior. Finally our computed diffusion coefficients in the liquid state are larger than bulk computed diffusivities. Smaller clusters have much higher pressures and lower the temperature of melting.

One of the most important methods of synthesizing nanoparticles is through vapor phase nucleation and growth. Several methods are used, including combustion, plasmas, thermal reactors, and evaporation–condensation [1, 2, 3].

For nanoparticle synthesis, one of the most fundamental and important kinetic properties of interest in nanoparticle growth is the sintering or coalescence rate between particles during vapor–phase growth. Knowledge of these rates and their dependence on process parameters ultimately impact the ability to control primary particle and agglomerate growth, which are of critical importance to nanoscale particles whose properties depend strongly on size, morphology and crystal structure [4].

In this work the structural and dynamical properties of silica nanoclusters were studied in MD simulations with Tsuneyuki's pair potential of inter–atom interaction [5].

At lower temperature $T = 1500$ K in a solid 'glass' state all clusters exhibit a oscillating density distribution. With increasing temperature there seems to be a tendency to dampen out the radial density gradient, though some peaks still remain. The radial dependence of mass density of different size clusters is shown in Fig. 1 at a temperature $T = 2080$ K.

In general we observe an enhanced density for smaller clusters, small thermal expansion, and oscillating density profile.

In order to study the diffusion in a finite system such as a cluster it is necessary to separate the transport of atoms into three regime. The first stage of diffusion is characterized by motion of atoms near their equilibrium sites or 'cages' created by neighbor atoms. In the second regime the atoms jump from one site to another and the mean square displacement increases linearly with time. In the third stage we observe a saturation in the mean square displacement of atoms because the characteristic diffusion

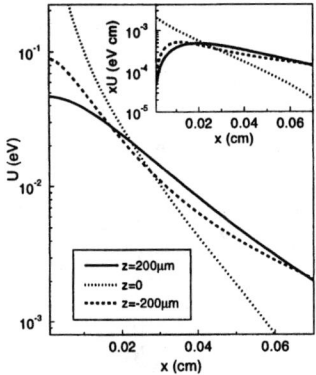

FIGURE 1. The radial density distribution for the clusters with 288 atoms (dashed line), 576 atoms (solid line) and 1156 atoms (dotted line) at $T = 2080$ K. The sketch of surface structure is shown in the insert.

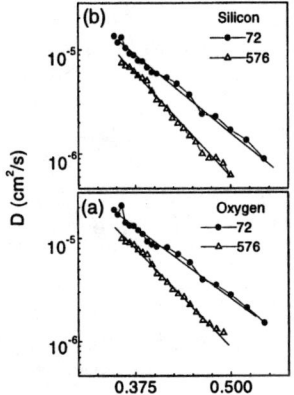

FIGURE 2. Diffusion coefficients of oxygen (a) and silicon atoms (b) for the clusters with 72 atoms (circles) and 567 atoms (triangles) as a function of inverse temperature.

length is of the order of the cluster size. The diffusion coefficient D for the cluster can be calculated from the second regime. The computed D is presented in Fig. 2 for silicon and oxygen atoms in two clusters. The diffusion coefficients are presented in an Arrhenius plot and give an activation energy $E_A = 15000$ K for the cluster with 72 atoms and $E_A = 16100$ K for the cluster with 576 atoms. This is considerably lower that that reported from MD results for bulk silica ($E_A = 35000$ K [6]). The diffusion coefficient in a small cluster is higher than that for a larger cluster as one would expect based on

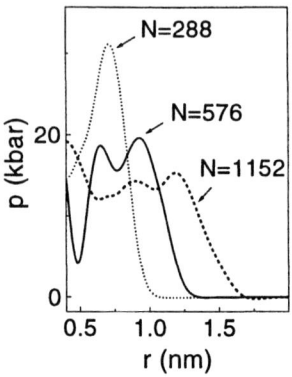

FIGURE 3. The smoothed radial pressure distribution for clusters with $N = 288$ atoms (dotted line), $N = 576$ atoms (solid line) and $N = 1152$ atoms (dashed lined line) at $T = 2000$ K.

surface to volume ratio considerations. Moreover, it has been shown theoretically [7] and experimentally [8] that an increase of the internal pressure (as it takes place for smaller clusters and will be discussed below) enhances the diffusion process in silica.

The internal pressure of a cluster includes a kinetic part, which is determined by the temperature, and an electrostatic contribution associated with the interactions of the atoms. To obtain the electrostatic pressure within a particle we compute the Irving–Kirkwood pressure tensor [9] by extension to a spherically symmetric system using the method described by Thompson et al [10]. In Fig. 3 the smoothed distribution of pressure over radius is shown for three clusters with $N = 288, 576, 1152$ atoms. It is seen that the smaller clusters have the higher internal pressure. The internal pressure based on Laplace's equation $P = \sigma/2R$ at constant surface tension predicts a significant increase as a particle shrinks in size, that qualitatively agrees with our MD results.

We calculate the surface tension σ within the mechanical approach, following the algorithm described by Thompson et al in Ref. [10].

One can see in Fig. 4 that the small cluster with $N = 72$ atoms and larger cluster with $N = 576$ atoms have the same value of surface tension at $T = (1920 - 2500)$ K.

At higher temperature the smaller cluster losses the spherical shape and the probability of dissociation of SiO_2 fragment quickly increases. Therefore the surface tension of the 72 atoms cluster quickly decreases at $T > 2500$ K. In our calculations we have obtained the surface tension which is equal to 0.67 J/m^2, that is higher than the plane surface tension $\sigma = 0.3$ J/m^2. We attribute this difference to the inter–atomic potential chosen for our calculations.

We found that smaller clusters have a larger density and a higher internal pressure, which matches qualitatively and in magnitude the Laplace–Young equation. Most interesting was that the surface tension did not show any significant size dependent effects over the range of cluster sizes studies $72 - 1152$ atoms.

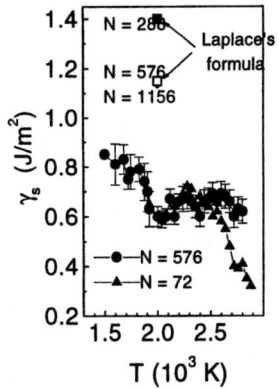

FIGURE 4. Surface tension as a function of the temperature for the clusters with $N = 72$ atoms (triangulars) and $N = 576$ atoms (open circles). The three points above denote the surface tension from Laplace's formula for the clusters with 288 atoms (squares), 576 atoms (diamonds), and 1156 atoms (open squares).

ACKNOWLEDGMENTS

This work is supported by NSF grant CTS–9802998 (Dr. Mike Roco – Program Manager) and one of the authors (I.S.) is partly supported by NATO grant SfP 974354. The authors would like to thank the Supercomputer Institute of University of Minnesota for a grant of computer time.

REFERENCES

1. Zachariah, M.R., Aquino-Class, R.D.S.M, and Steel, E., *Nanostructured Materials*, **5**, 383 (1995).
2. Steffens, K.L., Zachariah, M.R., and Axelbaum, R. L., *Chem. Mater.*, **8**, 1871 (1996).
3. Kortshagen, U.R., Bhandarkar, U.V., and Girshick, S. L., *Pure Appl. Chem.*, **71**, 1871 (1999).
4. Siegel, R. W., *Springer Series in Materials Science 27 Springer–Verlag, Berlin* (1994).
5. Tsuneyki, S., Tsukuda, M., and Matsui, Y., *Phys. Rev. Lett.*, **61**, 869 (1988).
6. Valle, R. G. D., and Anderson, H. C., *J. Chem. Phys.*, **97**, 2682 (1992).
7. Angell, C.A., and Tamaddon, S., *Science*, **218**, 885 (1982).
8. Shimizu, N., and Kushiro, I., *Geochim. Cosmochim. Acta*, **48**, 1295 (1984).
9. Irving, J. H., and Kirkwood, J. G., *J. Chem. Phys.*, **18**, 817 (1950).
10. Thompson, S.M., *et al.*, *J. Chem. Phys.*, **81**, 530 (1984).

Dynamical Phase Transition in Dust Crystals

V.A. Schweigert*, I.V. Schweigert†, V. Nosenko** and J. Goree**

*Institute of Theoretical and Applied Mechanics, 630090 Novosibirsk, Russia
†Institute of Theoretical and Applied Mechanics 630090 Novosibirsk, Russia
**Department of Physics and Astronomy, The University of Iowa, Iowa City Iowa 52242

Abstract. Experiments and simulations are reported for a monolayer plasma crystal that is disturbed by an extra particle moving in a plane below the monolayer. Numerical simulations and experiments are performed to find an explanation for the motion of the extra particle. The simulations take into account the ion wakefield downstream of the monolayer of particles, in the presence of ion flow. In the experiment, the orbit is straight at low gas pressures, but with higher damping it is crooked and less energetic. The same trend is observed in the simulations, supporting a conclusion that the wakefield is responsible for the particle acceleration. The simulation reveals that the energy of the extra particle exhibits distinctive transitions, between three regimes.

In the experiments of Samsonov et al. [1], a kind of spontaneous particle motion was observed. The experiments were performed using a monolayer of particles in the electrode sheath of rf discharge, with a few extra particles in an incomplete second layer 200 μm below the monolayer. These extra particles moved about spontaneously, and because they were charged, they disturbed the particles in the main layer. Samsonov et al. discovered that this motion resulted in the generation of a Mach cone.

The mechanism that accelerates the particles must be a persistent force, because it overcomes the constant friction experienced by the extra particle as it moves in gas. Presumably the accelerating mechanism is an electric force, but the mechanism behind it until now has not been identified.

Here we report results of simulations of plasma crystal, in which a single lattice layer and an incomplete lower layer, are considered. We observe lower particle orbits like those in experiment of Samsonov et al. In the present paper we also report experiments, which allows us to draw a conclusion about the acceleration mechanism by comparing to the simulations.

We used molecular dynamics (MD) simulations to observe the motions of all particles in a monolayer, plus an extra particle below it. We modeled a section of a suspension that included $N = 1021$ particles, and we applied periodic boundary conditions. The primary difference between our MD simulation, and the simulation in Ref. [1], is the inter-particle potential. Both simulations were intended to model the experimental results of Ref. [1]. However, in the simulation of Ref. [1], the inter-particle potential was a simple Yukawa repulsion, the extra particle in that simulation was not spontaneously accelerated; it was necessary to artificially move it. Here we use a different particle interaction, which includes the effects of the ion wakefield.

The inter-particle interaction was determined accurately by performing a particle-in-cell Monte-Carlo (PIC MC) simulation including the particles in the monolayer, electrons, and ions.

For calculation of forces acting on the particles, we approximate the potential from the "ab-initio" PIC MC calculations by the following analytic expressions. In the expressions below, the potentials have units of electron volts and the distances are measured in centimeters. The particle potential U_i in the crystal lattice is assumed to be a Debye-Huckel type

$$U_i(\vec{p}_i - \vec{p}_j) = \frac{Ze^2}{|\vec{p}_i - \vec{p}_j|} \exp(-\kappa_i |\vec{p}_i - \vec{p}_j|/a),$$

where the effective screening length $\kappa_i = 1.64$ and $|\vec{p}_i - \vec{p}_j|$ is the inter-particle distance between the particles belonging to the monolayer. The extra particle acts on a particle above it, in the monolayer, with a force determined by the following inter-particle potential

$$U_u(\vec{p}_i - \vec{p}_e) = \frac{Z_1 e^2}{|\vec{p}_i - \vec{p}_e|} \exp(-\kappa_e |\vec{p}_i - \vec{p}_e|/a),$$

where $Z_1 = 12930$, $\kappa_e = 0.916$ and $|\vec{p}_i - \vec{p}_e|$ is the distance between a particle in the monolayer and the extra particle. For computing the force acting on the extra particle from a particle in the monolayer we use the potential

$$U_l(\vec{p}_e - \vec{p}_i) = Z_2 e^2 \exp(-\kappa_l |\vec{p}_e - \vec{p}_i|/a) \left(-\frac{1}{|\vec{p}_e - \vec{p}_i|} + \frac{0.156 a^2}{|\vec{p}_e - \vec{p}_i|^3} - \frac{1.1 a^4}{|\vec{p}_e - \vec{p}_i|^5} \right),$$

where $Z_2 = 4905$, $\kappa_l = 0.559$ and $|\vec{p}_e - \vec{p}_i|$ is the distance between the extra particle and a particle in the monolayer. For illustration we show in Fig. 1 the absolute values of the particle potential in different horizontal planes. U_i (dotted line) and U_u (solid line) are the repulsive, positive potentials and U_l (dashed line) is the attractive potential which has the negative value. The insert in Fig. 1 gives xU(x) values to show a discrepancy as compared to the Yukawa potential.

Note that the inter-particle potential is not symmetric, between the particle in the monolayer and a particle in the lower layer. The attractive potential U_l is larger than the repulsive U_u by about a factor of two when the extra particle sits in the vertically aligned position. In other words, $U_u \neq U_l$. This situation arises because the system is not closed. This phenomenon was reported previously [2, 3] for the analysis of instabilities in bilayer crystals.

The experimental part of this work was carried out using the same apparatus as used in [1]. In the experiment we varied the gas pressure from $P = 2.7 - 11.7$ Pa, thereby adjusting the damping rate. We adjusted the gas pressure until we observed orbits that resembled those in the simulation, for all three regimes.

As one of our main results, we found in both the experiment and the MD simulations that the extra particle moves about, and that its trajectory depends on the gas pressure. More precisely, it depends on the dimensionless friction coefficient ν/ω_{pd} which is the ratio of the friction coefficient and the dust crystal frequency $\omega_{pd} = \sqrt{Z^2 e^2 / \varepsilon_0 M_p a^3}$. The range of the dimensionless friction coefficient ν/ω_{pd} in the experiment and in MD simulation coincides.

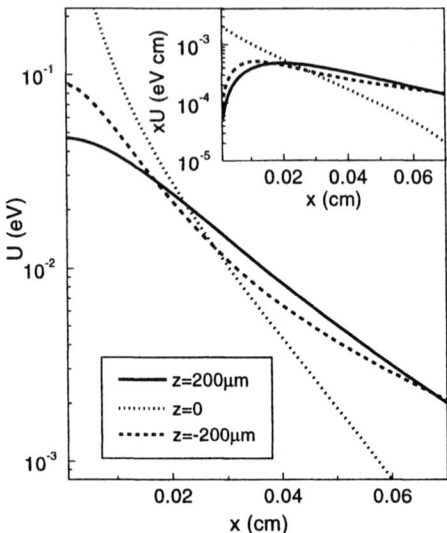

FIGURE 1. Potential distributions around a particle from the PIC-MC simulation as a function of the *x*-coordinate, i.e. the coordinate perpendicular to the ion flow. Three curves refer to the horizontal lines in Fig. 1. The insert shows xU(x) as a function of the x.

In the simulation, we found that as we varied the gas damping, the motion of the extra particle had three distinct regimes, with transitions between them. This is shown in Fig. 2. Within each regime, the particle motion was generally about the same. However, near a critical value of gas damping, a small change in gas pressure resulted in a significantly different kind of motion of the extra particle. At a higher pressure (regime I), an extra particle and a particle from the monolayer remain trapped. However, near a critical value of gas damping, a small change in gas pressure resulted in a significantly different kind of motion of the extra particle. The first transition occurs with a small decrease of the friction coefficient from $\nu = 0.3\ \omega_{pd}$ to $\nu = 0.27\ \omega_{pd}$. This results in a large increase in the kinetic energy of the extra particle from 5 eV to 80 eV, causing the couple to dissociate. The motion in regime II of Fig. 2 is diffusive.

The second transition occurs also with a small change of the friction coefficient, this time at $\nu = 0.08\ \omega_{pd}$. This results in another large increase in the energy of the extra particle, up to 1000 eV. This kinetic energy is higher than the attraction energy, so that the extra particle is not tightly coupled to a single particle above it. Now the particle trajectory is undisturbed straight line (regime III of Fig. 2).

In this work one of our main results is that the orbit of the extra particle has the same shape, in the simulation and the experiment, and the shape has the same trend of becoming straighter as the gas damping is reduced. However in the experiment we did

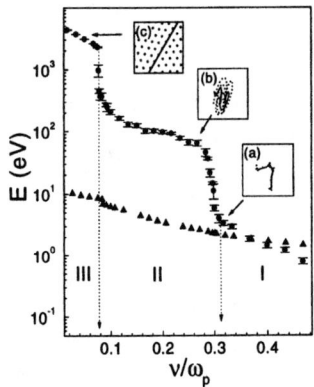

FIGURE 2. Regimes of motion of the extra particle: (I) "molecular" state, (II) diffusion, (III) straight-line motion. The mean kinetic energy of particles in the monolayer (triangles) and the extra particle (circles) are plotted as a function of the dimensionless friction coefficient. The energy of the extra particle undergoes two transitions, where a small change in friction results in a large change in particle energy. The insets show the enlarged trajectories of the particles (a) at $v = 0.3\ \omega_{pd}$, (b) at $v = 0.27\ \omega_{pd}$, and (c) at $v = 0.03\ \omega_{pd}$. The size of the inset boxes 0.1×0.1 mm^2 for (a) and (b), and 2×2 mm^2 for (c). The trajectory of the extra particle in the incomplete lower layer is shown by a solid line, and that of the upper layer by a dashed line in (a) and (b). In (c) the particles in the upper layer are shown in a snapshot.

not observe the rapid transition between different regimes of particle motion.

Another of our main results is that the mechanism of accelerating the extra particle must arise from the ion wakefield. In this simulation the extra particle initially has a kinetic energy equivalent to room temperature, and then it gradually gains energy with time. The acceleration of the particle ceases when its velocity reaches some value, for example, 4.2 cm/s at a pressure $P = 5$ Pa which is surprisingly close to the experimental one for the same pressure Ref. [1].

The fact that our experiment and simulation agree in the motion of the extra particle suggests that the simulation incorporates the physics responsible for the acceleration. As compared to the MD simulation of Ref. [1], where the particle did not accelerate spontaneously, the primary difference is that our inter-particle potential takes into account the ion wakefield. Thus, we attribute the acceleration of the particles to the wakefield.

REFERENCES

1. Samsonov, D., *et al*, *Phys. Rev. Lett.*, **83**, 3649 (1999).
2. Schweigert, V.A., *et al*, *Phys. Rev. E*, **54**, 4155 (1996).
3. Schweigert, V.A., *et al*, *Phys. Rev. Lett.*, **80**, 5345 (1998)

Electrostatic Response of a Dusty Plasma with a Grain Size Distribution to a Moving Test Charge

M. Shafiq* and Michael A. Raadu*

Alfvén Laboratory, Royal Institute of Technology, SE-100 44, Stockholm, Sweden

Abstract. The presence of a dust component in a plasma leads to many new effects that depend on the physics of the interaction between the dust grains and the ambient plasma. In particular the electrostatic response to a moving test charge is modified. In general, natural or industrial dusty plasmas are expected to have a distribution of particle sizes. The linear kinetic behaviour is then quite different to the case with single sized grains. Thermodynamic equilibrium for a hot dust component would lead to a Maxwellian velocity distribution with a unique temperature for all masses. Low mass grains will then dominate the tail of the velocity distribution. For reasonable choices of the size distribution the contribution of the dust component to the dispersion relation has been found to be equivalent to that for a kappa (generalised Lorentzian) distribution of mono-sized particles (Raadu, 2001). Using this dispersion relation the response to a moving test charge can be investigated both analytically and numerically. Explicit analytical expressions, correct to second order in the velocity, are found for the potential in the case of a slowly moving test charge.

INTRODUCTION

While many papers have tried to account in someway for charge fluctuations, and energy loss of test particle when it propagates through dusty plasma [1], little has been done for dust grains, with a distribution in mass and size. Even if thermal equilibrium is assumed so that the velocity distribution is Maxwellian, the electrostatic response when there is a distribution of sizes is effectively the same as for non-Maxwellian distribution of mono-sized particles. In particular for a mass distribution which is exponential for large masses and a power law for small sizes, the dispersion relation is the same as that of a kappa (generalized Lorentzian) distribution of single sized particles [2]. This case is particularly interesting since the dispersion properties have been extensively investigated [3]. Here these results will be used in the analysis of the dusty plasma response to a moving test charge.

ELECTROSTATIC RESPONSE OF DUST WITH A SIZE DISTRIBUTION

A basic modification to conventional plasma theory is that the dust grains can be expected to have a distribution of sizes [4, 5]. If the dust approaches thermodynamic equilibrium, the tail of the velocity distribution should be dominated by small low-mass grains. As a consequence, the properties of the plasma will be significantly different from the mono-sized Maxwellian distribution. In order to have a specific example of the size distribution function $h(a)$, the following form will now be assumed:

$$h(a)\,da = An_d a^\beta \exp\left(-\alpha^3 a^3\right) da \tag{1}$$

Assuming a thermal equilibrium for the dust grains with a unique temperature T, the distribution over size and velocity is

$$f_0(v,a)\,dvda = h(a)[m(a)/2\pi kT]^{1/2}\exp\left(-m(a)v^2/2kT\right) dvda \tag{2}$$

Following [2] a kinetic analysis of electrostatic perturbations leads to a Vlasov equation with the particle radius a as a free parameter. In a simplified charging model the charge $q(a)$ is proportional to the vacuum capacity and therefore to a. A standard Landau treatment may be used, modified by an integration over a when Poisson's equation is used. The dispersion function $D(K, \omega)$ is found to be given by

$$K^2 D(K,\omega) \equiv K^2 + K_{De}^2 + K_{Di}^2 - \int_{-\infty}^{+\infty} \frac{1}{[v-\omega/K]} \frac{\partial}{\partial v}\left[\int_0^\infty \frac{q(a)^2}{\varepsilon_0 m(a)} f_0(v,a)\,da\right] dv \tag{3}$$

where the ions and electrons only give Debye shielding for low frequencies. Comparison with standard results shows that the dust responds with an effective zero-order velocity distribution

$$f_{d,0}^{eff}(v) \equiv \frac{n_d}{\omega_{pd}^2} \int_0^\infty \frac{q(a)^2}{\varepsilon_0 m(a)} f_0(v,a)\,da \tag{4}$$

For the choice of $h(a)$ used here this leads to an effective distribution [2]:

$$f_{d0}^{eff}(v) = Cn_d \left[\frac{3kT\alpha^3}{2\pi\rho}\right]^{(\beta+1)/3} (v^2 + 3\alpha^3 kT/2\pi\rho)^{-(5/6+\beta/3)} \tag{5}$$

which is a kappa distribution (generalized Lorentzian distribution [6, 7]). This somewhat surprising but convenient relation allows the known dispersion relations for kappa distributions to be directly translated to the present case of Maxwellian velocity distributions combined with the mass distribution $h(a)$. The corresponding effective dust plasma frequency is given by

$$\omega_{pd}^2 \equiv \int_0^\infty \frac{q(a)^2}{\varepsilon_0 m(a)} h(a)\,da \tag{6}$$

ELECTROSTATIC RESPONSE TO A MOVING TEST CHARGE

We now consider the response to a moving test charge of an unmagnetised dusty plasma characterized by the number density n_j, temperature T_j (in energy units), mass m_j and charge q_j of the j-th species (j equals e for electrons, i for ions and d for the dust). For the dust there is a size distribution $h(a)$, and m_d and q_d are weighted averages defined by the equivalence to a kappa distribution [2].

The expression for the electrostatic potential of a moving test particle is

$$\varepsilon_0 \phi_1(\mathbf{r}) = \frac{q_t}{8\pi^3} \int \frac{\exp[i\mathbf{K}\cdot\mathbf{r}]}{K^2 D(K, \mathbf{K}\cdot\mathbf{V}_t)} d\mathbf{K} \tag{7}$$

where $D(K, \mathbf{K}\cdot\mathbf{V}_t)$ is the dielectric response function given by the equivalent kappa distribution with $\kappa = (2\beta+5)/6$. The inverse of the dielectric response function up to second order in $\hat{K}\cdot\mathbf{V}_t/\alpha$ is

$$\frac{1}{D(K,\mathbf{K}\cdot\mathbf{V}_t)} = \frac{K^2}{\left(K^2+K_{eff}^2\right)} \left[1 + \frac{iA(\kappa)K_D^2}{K^2+K_{eff}^2}\right.$$
$$\left. \times \left(\frac{\hat{K}\cdot\mathbf{V}_t}{\alpha}\right) - \frac{B(\beta)K_D^2}{K^2+K_{eff}^2}\left(\frac{\hat{K}\cdot\mathbf{V}_t}{\alpha}\right)^2\right]^{-1} \tag{8}$$

where $K_{eff} = \sqrt{K_{De}^2 + K_{Di}^2 + K_D^2}$ is the effective Debye wave-number, $A(\kappa) = \sqrt{\pi/\kappa}\,\gamma(\kappa+1)/\gamma(\kappa+1/2)$, $B(\kappa) = (2+1/\kappa)$ and κ is related to β, the power law index of the size distribution for small radii by the relation $\kappa = (2\beta+5)/6$. The resulting expression for ϕ_1 can be written as follows to highlight the dependence on the angle λ between the velocity and the radial vector $\mathbf{r}$,

$$\phi_1(r,\lambda) = \frac{q_t}{4\pi\varepsilon_0 r}\left[g_{00} - \left(\frac{2}{\pi}\right)\right.$$
$$\left. \times \frac{V_t}{\alpha}(rK_D)^2 A(\kappa) g_{11}\cos(\lambda) + \frac{r^4}{2\pi^2\alpha^2}(g_{20}+g_{22}\cos^2(\lambda))V_t^2\right]$$

where the strength functions are given by the following functions of the radial distance $r = |\mathbf{r}|$,

$$g_{00} = \exp(-rK_{eff})$$

$$g_{11} = \left[\frac{1}{2r^2 K_{eff}^2} - \frac{Ei(rK_{eff})}{4rK_{eff}\exp(rK_{eff})}\left(1+\frac{1}{rK_{eff}}\right)\right.$$
$$\left. + \frac{1}{r^2 K_{eff}^2}\right) - \frac{E_1(rK_{eff})}{4rK_{eff}\exp(-rK_{eff})}\left(1-\frac{1}{rK_{eff}}+\frac{1}{r^2 K_{eff}^2}\right)\right]$$

$$g_{20} = \left[\left\{\frac{\pi^2 K_D^4 A^2(\kappa)}{4r^3 K_{eff}^3 \exp(rK_{eff})} + \frac{2\pi^2 K_D^2\left(-B(\kappa)K_{eff}^2 + K_D^2 A^2(\kappa)\right)}{r^6 K_{eff}^6}\right.\right.$$
$$\left.\left. \times \left(1-\exp(-rK_{eff})\right)\left\{1+rK_{eff}+\frac{r^2 K_{eff}^2}{2}\right\}\right)\right\}\right]$$

$$g_{22} = \left[\left\{-\frac{\pi^2 K_D^4 A^2(\kappa)}{4r^2 K_{eff}^2 \exp(rK_{eff})} - \frac{6\pi^2 K_D^2\left(-B(\kappa) K_{eff}^2 + K_D^2 A^2(\kappa)\right)}{r^6 K_{eff}^6}\right.\right.$$
$$\left.\left. \times \left(1 - \exp(-rK_{eff})\right)\left\{1 + rK_{eff} + \frac{r^2 K_{eff}^2}{2} + \frac{r^3 K_{eff}^3}{6}\right\}\right)\right\}\right]$$

DISCUSSION

Here we have presented a model for the electrostatic response to a moving charge of a dusty plasma with a grain size distribution. The effect of different choices for the size distribution is contained in the parameters $A(\kappa)$ and $B(\kappa)$, where $\kappa = (2\beta + 5)/6$. The first order term has a strength function g_{11} that is essentially the same as for the Maxwellian case, although the effective Debye wave number K_{eff} depends through K_D on the form of the size distribution. Explicitly we have

$$K_D^2 \equiv \int_0^\infty \frac{q(a)^2}{\varepsilon_0 kT} h(a) da$$

This applies for a general choice of the grain size distribution function $h(a)$. The second order term depends on both $A(\kappa)$ and $B(\kappa)$, and the structure of the strength functions depends on β, the power law index of the size distribution $h(a)$ used here (1). Formally the Maxwellian case can be recovered by setting $A(\kappa) = \sqrt{\pi/2}$ and $B(\kappa) = 1$.

Although the work presented here concerns the response of a dusty plasma for a class of grain size distributions together with a Maxwellian velocity distribution, it is clear that the analysis would be identical for a monosized dust component with a generalised Lorentzian distribution.

ACKNOWLEDGMENTS

The authors would like to thank their colleagues at the Alfvén laboratory for useful discussions and suggestions. This work was partly supported by the Swedish Research Council.

REFERENCES

1. Nasim, M.H., Shukla, P.K., and Murtaza, G., *Phys. Plasmas*, **6**, 1409 (1999).
2. Raadu, M.A., *IEEE Trans. Plasma Sci*, **29**, pp. 182–185 (2001).
3. Mace, R.L., and Hellberg, M.A., *Phys. Plasmas*, **2**, 2098 (1995).
4. Havnes, O., Aanesen, T.K., Melandsø, F., *J. Geophys. Res.*, **95**, pp. 6581–6585 (1990).
5. Aslaksen, T.K., and Havnes, O., *J. Geophys Res.*, **97**, pp. 19175–19185 (1992).
6. Summers, D., and Thorne, R.M., *Phys. Fluids B*, **3**, pp. 1835–1847 (1991).
7. Summers, D., and Thorne, R.M., *Phys. Fluids B*, **3**, pp. 2117–2123 (1991).

Chaotic Behavior of Electron-Positron Dusty Magnetoplasma with Equilibrium Flows

Arshad M. Mirza*, M. Shafiq[†], Michael A. Raadu[†] and Khalid Khan*

Department of Physics, Quaid-i-Azam University, Islamabad 45320, Pakistan
[†]*Alfvén Laboratory, Royal Institute of Technology, SE-100 44, Stockholm, Sweden*

Abstract. The nonlinear dynamics of low-frequency (in comparison with electron gyro-frequency) electrostatic disturbances in nonuniform dusty electron-positron magnetoplasma with equilibrium sheared flow is investigated. In the linear limit, a local dispersion relation has been derived and analyzed. On the other hand, in the nonlinear case, the temporal behavior of the nonlinear dissipative system can be expressed as a set of three coupled nonlinear equations, which are a generalization of Lorenz-Stenflo equations that admit chaotic trajectories. The results of our present investigation should be very helpful to understand plasma turbulence and wave phenomena in several laboratory and astrophysical plasma systems.

INTRODUCTION

Electron-positron pair plasmas are generally produced at high-energies and typically exists in some astrophysical system such as in solar flares, pulsars, jets from active galactic nuclei, in the early universe, near the polar cap region of fast rotating neutron star and black holes. Experiments show that nonrelativistic electron-positron pair plasma can be produced in laboratory conditions by the interaction of an electron beam with high-Z target [1], lepton-ion interactions [2] and by ultrashort laser pulses in plasmas [3]. Propagation of intense short laser pulses in a plasma can also lead to pair production resulting in a three-component electron-positron-ion plasma. In fact, such three-component plasmas have been seen in several laboratory experiments intending to use positrons as a probe to study transport in tokamaks [4].

In this paper, we study the linear as well as nonlinear properties of low-frequency electrostatic waves in a nonuniform magnetized dust-contaminated electron-positron pair plasma. We show that nonlinearly coupled electrostatic modes in nonuniform, resistive magnetoplasma with shear flows can have a chaotic behavior. It is shown that new set three mode coupling equations can be represented as a generalization of Lorenz-Stenflo type equations [5, 6], which admit chaotic behavior under certain conditions.

NONLINEAR DYNAMIC EQUATIONS

Consider an electron-positron pair plasma with positively charged dust grains embedded in an external magnetic field $B_0\hat{z}$, where B_0 is the strength of the external magnetic field.

The dusty plasma also has equilibrium density and velocity gradients that are maintained by some external sources.

For low-frequency electrostatic waves, the perpendicular components of the electron and positron velocity perturbations may be given using the drift-approximation as

$$\mathbf{v}_{e\perp} \approx \frac{c}{B_0}\hat{\mathbf{z}} \times \nabla\phi + \frac{c}{B_0\omega_c}\left[\partial_t + \nu_e + \mu_e \nabla_\perp^2 + \frac{c}{B_0}\hat{\mathbf{z}} \times \nabla_\perp \phi \cdot \nabla\right]\nabla_\perp \phi, \quad (1)$$

and

$$\mathbf{v}_{p\perp} \approx \frac{c}{B_0}\hat{\mathbf{z}} \times \nabla\phi - \frac{c}{B_0\omega_c}\left[\partial_t + \nu_e + \mu_p \nabla_\perp^2 + \frac{c}{B_0}\hat{\mathbf{z}} \times \nabla_\perp \phi \cdot \nabla\right]\nabla_\perp \phi, \quad (2)$$

where $\mu_e(\mu_p) = 0.51\rho_e^2 \nu_e$ is the electron (positron) gyroviscosity, ν_e is the electron collision frequency and ρ_e is the electron gyroradius, ϕ is the electrostatic potential.

The parallel components of the electron and positron velocities can be expressed as

$$D_t v_{ez} = \frac{e}{m}\partial_z\phi + \frac{c}{B_0}\left(\partial_x v_{ez0}(x)\right)\partial_y\phi \quad (3)$$

and

$$D_t v_{pz} = -\frac{e}{m}\partial_z\phi + \frac{c}{B_0}\left(\partial_x v_{pz0}(x)\right)\partial_y\phi \quad (4)$$

where $D_t = \partial_t + \nu_e + (c/B_0)\hat{\mathbf{z}} \times \nabla\phi \cdot \nabla = d_t + \nu_e$.

If we take a difference of positron and electron continuity equations together with Poisson's equation, viz. $\nabla^2 \phi = 4\pi e(n_e - n_p - Z_d n_d)$, then we have

$$D_t\left(\nabla^2 + \frac{\omega_p^2}{\omega_c^2}\nabla_\perp^2\right)\phi + \frac{\omega_p^2}{\omega_c}\kappa_d \partial_y\phi + \mu_e \frac{\omega_p^2}{\omega_c^2}\nabla_\perp^4 \phi - \nu_e \nabla^2\phi$$
$$-4\pi e\left(n_{p0}\partial_z v_{pz} - n_{e0}\partial_z v_{ez}\right) = 0 \quad (5)$$

where $\omega_p^2 \equiv \omega_{pe}^2 + \omega_{pp}^2 \equiv 4\pi(n_{e0} + n_{p0})e^2/m$, $\kappa_d = (n_{e0} + n_{p0})^{-1}\partial(Z_d n_{d0})/\partial x$, $\mu_e(\mu_p) = 0.51\rho_e^2 \nu_e$ is the electron (positron) gyroviscosity, and ρ_e is the electron gyroradius. The dust dynamics has been ignored since we are looking for wave phenomena on a time scale much shorter than the dust plasma and dust gyroperiods.

LINEAR DISPERSION RELATION

The local dispersion relation for electrostatic waves can be derived by neglecting the nonlinear terms and Fourier transforming the resultant equations. We readily obtain the following dispersion relation in terms wave frequency ω and wave number $\mathbf{K}$,

$$(k^2 + ak_y^2)\omega^{*2} + \omega^*\left[\omega_d - i(\nu_e k^2 + a\mu_e k_y^4)\right] - \omega_p^2 k_z^2\left[1 + \frac{k_y}{(n_{e0}+n_{p0})\omega_c k_z}\right.$$
$$\left. \times \left(n_{e0}\partial_x(v_{ez0}) - n_{p0}\partial_x(v_{pz0})\right)\right] = 0, \quad (6)$$

where $\omega^* = (\omega + i v_e)$, $a = \omega_p^2/\omega_c^2$, $k^2 = k_y^2 + k_z^2$ and $\omega_d = (\omega_p^2/\omega_c)\kappa_d k_y/(k^2 + k_y^2 a)$ is the dust convective cell frequency.

In the collisionless case, Eq. (3) predicts an instability and the threshold condition is

$$\left[n_{p0}\partial_x\left(v_{pz0}\right) - n_{e0}\partial_x(v_{ez0})\right] > \frac{k_z\omega_c(n_{e0}+n_{p0})}{k_y}\left[1 + \frac{\omega_d^2\left(k^2 + ak_y^2\right)}{4\omega_p^2 k_z^2}\right] \quad (7)$$

An inspection of Eq. (4) reveals that the dust convective cells are driven at nonthermal levels on account of the free energy stored in the magnetic field aligned velocity gradients of electron and positron plasma system.

NONLINEAR CHAOTIC BEHAVIOR

We follow Lorenz and Stenflo [5] and derive a set of equations in a dusty electron-positron dissipative magnetoplasma. We introduce the ansatz:

$$\phi = \phi_1(t)\sin(K_x x)\sin(K_y y) \quad (8)$$
$$v_{ez} = v_1(t)\sin(K_x x)\cos(K_y y) - v_2(t)\sin(2K_x x), \quad (9)$$
$$v_{pz} = v_3(t)\sin(K_x x)\cos(K_y y) - v_4(t)\sin(2K_x x), \quad (10)$$

where K_x and K_y are constant parameters, and $\phi_1, v_{ez}, v_{pz}, v_1, v_2, v_3$ and v_4 are amplitudes. Substituting above into (1) and (2), we obtain a system of five nonlinear equations, which may be put in the normalized form of a 5×5 matrix which is similar to Lorenz and Stenflo type equations [6]

$$\begin{pmatrix} d_\tau X \\ d_\tau Y \\ d_\tau Z \\ d_\tau U \\ d_\tau T \end{pmatrix} = \begin{pmatrix} -\sigma & \sigma & 0 & s & 0 \\ r-Z & -1 & 0 & 0 & 0 \\ Y & 0 & -1 & 0 & 0 \\ -1 & 0 & 0 & -1 & X \\ -U & 0 & 0 & 0 & -1 \end{pmatrix} \begin{pmatrix} X \\ Y \\ Z \\ U \\ T \end{pmatrix} \quad (11)$$

where $\sigma = a[v_e - \mu_e K_\perp^2]/v_e(1+a)$, $s = 4\pi e n_{p0} K_y f(x) a_4/(1+a) a_1 v_e K_\perp^2$, $r = 4\pi e f(x) n_{e0} K_y^2 F_1 / a v_e (v_e - \mu_e K_\perp^2) K_\perp^2$, with $K_\perp^2 = K_x^2 + K_y^2$, $\tau = t/t_0$, and $t_0 = v_e^{-1}$. The variables X, Y, Z, U, T are the normalized forms of ϕ_1, v_1, v_2, v_3 and v_4.

Here, we have taken into account small magnetic shear effects such that $\partial_z V_{jz} \approx (\partial_x f(x))\partial_y V_{jz}$ and assumed that $|\partial_t + (c/B_0)\hat{z} \times \nabla\phi \cdot \nabla| \gg v_{jz0}\partial_z$; and $\nabla_\perp^2 \gg \partial_z^2$ with $F_1 = e(\partial_x f(x) + \omega_c^{-1}\partial_x(v_{ez0}))/m$, $F_2 = e(\partial_x f(x) - \omega_c^{-1}\partial_x(v_{pz0}))/m$, and $K_\perp^2 = K_x^2 + K_y^2$. The characteristic equation which governs the linear stability of the stationary state is

$$(\lambda+1)^3\left[\lambda^2 + (1+\sigma)\lambda + \sigma(1-r) + s\right] = 0, \quad (12)$$

The eigenvalues are $\lambda = -1$, $\left[-(1+\sigma)\pm\sqrt{(1-\sigma)^2 - 4(s-r)\sigma}\right]/2$. For $s = 0$ and $0 < r < 1$, the eigenvalues become $\lambda = -1, -\sigma$ thus the equilibrium point is stable because all the eigenvalues are negative. At $r = 1$ we have $\lambda = 0, -(1+\sigma)$ i.e., the fixed point is marginally stable. It is stable for $1 < r < r_1$ where r_1 is any arbitrary value. Finally, at $r_1 < r_H$ two of the eigenvalues become complex i.e., two limit cycles result which are stable as long as the real part of the complex eigenvalues are negative, where $r_H \equiv 2(7\sigma^3 + 56\sigma^2 + 104\sigma - 30)/\sigma(\sigma^2 + 11\sigma + 27)$. On the other hand, above r_H the limit cycles becomes unstable and chaos sets in.

CONCLUSIONS

We have investigated the nonlinear properties of low-frequency electrostatic disturbances dust-contaminated electron- positron magnetoplasma with equilibrium sheared flows. When linearly excited finite amplitude electrostatic modes interact among themselves may lead to chaotic state of the plasma. A generalization of Lorenz-Stenflo type equations admitting chaotic trajectories is found. In conclusion, we stress that the present investigation should be useful to have a better understanding of plasma turbulence and wave phenomena in several laboratory and astrophysical plasma systems.

ACKNOWLEDGMENTS

This research was partially supported by the Pakistan Science Foundation Project No. PSF/Res /C-QU/Phys.(111), the Pakistan Atomic Energy Commission Project No. PAEC /D & D– 1(103)/97–98, and the Quaid-i-Azam University Research Fund URF (2001-2002).

REFERENCES

1. V. Tsytovich and C. B. Wharton, Comm. Plasma Phys. Cont.Fusion, **4**, 91 (1978).
2. E. P. Liang, S. C. Wilks and M. Tabak, Phys. Rev. Lett. **81**, 4887 (1998).
3. V. I. Berezhiani and I. G. Murusidze, Phys. Lett. A, **148**, 338 (1990).
4. C. M. Surko and T. Murphy, Phys. Fluids B, **2**, 1372 (1990).
5. L. Stenflo, Phys. Scr. **53**, 83 (1996).
6. A. M. Mirza and P.K. Shukla, Phys. Lett.A. **229**, 313 (1997).

Effect of Grain Charging Dynamics on the Response of a Dusty Plasma to a Moving Test Charge

Michael A. Raadu* and M. Shafiq*

Alfvén Laboratory, Royal Institute of Technology, SE-100 44, Stockholm, Sweden

Abstract. A large number of new physical processes related to the presence of charged dust grains in a plasma have been discovered both theoretically and in experiments. The dynamics of the charging of the grains adds a new degree of complexity to the physics of these dusty plasmas. New phenomena such as damping due to the phase shift between a plasma wave and the dust charge have been found. The response of a plasma to a moving test charge reveals many characteristic aspects of collective effects, such as charge shielding, the excitation of dispersive waves and the formation of wave fronts. The effect of even a slow test charge velocity is significant. The exponential shielding, which limits the influence of a stationary charge to a few Debye lengths, is modified so that there is a potential disturbance with an asymptotic power law dependence. Here we include the influence of grain charging in the plasma dielectric function. A standard description of the charging dynamics is used. Improved models would lead to more realistic forms of the dielectric function, that could be directly incorporated in our analysis. We find analytical expressions for the potential structure using a power series expansion in the test particle velocity. The modifications to the potential are clearly evident from extra terms which depend on the parameters for the dust charging. Experimentally, the form of the potential could therefore be used to diagnose the charging dynamics and as an indirect test of the model.

INTRODUCTION

In this paper we investigate the response of plasma including grain charging dynamics using linearized dielectric theory. Dynamical charging is known to lead to a new type of wave damping [1]. We compute the contribution of this effect to electrostatic potential of a test charge passing through a dusty plasma. The linearized potential of a test particle moving through a collisionless plasma was given by Rostoker [2] in terms of the dielectric function. Cooper [3] found an explicit expression for the potential, up to second order in velocity, of a slowly moving charge for the case of thermal electrons. More recently, Neufeld and Ritchie [4] computed the potential of a moving test charge also with ions forming a fixed background. This work was further extended by taking into account the ion dynamics [5]. One of the main objectives of the present investigation is to incorporate dust charge fluctuations in a multicomponent dusty plasma so as to determine their effect on the response to a test charge. We consider a weakly coupled dusty plasma in which dust sizes and the inter-grain spacings are assumed to be much smaller than the dusty plasma Debye length. The electrons and ions are assumed to obey a Boltzmann distribution.

MATHEMATICAL FORMULATION OF THE PROBLEM

The dusty plasma is characterized by number density n_j, mass m_j and temperature T_j (j equals e for electrons, i for ions and d for dust). Considering the low phase velocity (in comparison to the electron and ion thermal velocities), the number density perturbations of electrons and positive ions follow a Boltzmann distribution.

The dust charge fluctuation (q_{d1}), which arises due to the wave motion induced oscillations in the electron and ion currents that flow onto the grain surface, is a new dynamical variable.

The linearized equation for the dynamical charging is then found to be [1],

$$\frac{\partial q_{d1}}{\partial t} = -\Omega_{u0} q_{d1} - 4\pi\varepsilon_0 a_d \Omega_{v0} \phi_1 \tag{1}$$

where $C = 4\pi\varepsilon_0 a_d$ is the capacitance of the dust grain, Ω_{u0} is the charging relaxation rate determined from the equilibrium electron and ion currents flowing towards the dust grain surface. The Fourier component of the dust charge fluctuation may be written as [6]

$$n_{d0} q_{d1}(\mathbf{K}, \omega) = -\frac{i\beta}{(\omega + i v_0)} \left(\varepsilon_0 \kappa^2 \phi_1\right) \tag{2}$$

where $v_0 = \Omega_{u0}$, $\beta = (C n_{d0}/\varepsilon_0 \kappa^2) \Omega_{v0}$ and κ is the combined Debye wave number for the electron and ion shielding. The expression for the electrostatic potential, in the frame of reference of the moving test charge, is as follows [2],

$$\varepsilon_0 \phi_1(\mathbf{r}) = \frac{q_t}{8\pi^3} \int \frac{\exp[i\mathbf{K}\cdot\mathbf{r}]}{K^2 D(K, \mathbf{K}\cdot\mathbf{V}_t)} d\mathbf{K} \tag{3}$$

where $D(K, \mathbf{K}\cdot\mathbf{V}_t)$ is the dielectric response function having the following form

$$D(K, \mathbf{K}\cdot\mathbf{V}_t) = 1 + \frac{K_1^2}{K^2} + \frac{W(\hat{\mathbf{K}}\cdot\mathbf{V}_t) - 1}{K^2} + \frac{1}{K^2 v_0} \frac{i\kappa^2 \beta (\mathbf{K}\cdot\mathbf{V}_t)}{(v_0 - i\mathbf{K}\cdot\mathbf{V}_t)} \tag{4}$$

with $K_1 = \sqrt{1 + \kappa^2(1 + \beta/v_0)}$, the effective Debye wave number, and v_0 the charge relaxation rate [6] determined from the equilibrium electron and ion currents flowing towards the dust grain surface. The dust thermal speed V_{td} and dust Debye length, are both normalized to unity for convenience. W is the plasma dispersion function [7]. By introducing the expression for the dielectric response function (4) in (3), expanding as a power series in the test particle velocity $\mathbf{V}_t$ to second order, and simplifying, we get

$$\phi_1(r, \lambda) = \frac{q_t}{8\pi^3 \varepsilon_0} \left[g_{00} + V_t g_{11} \cos\lambda + V_t^2 \left(g_{20} + g_{22} \cos^2\lambda\right)\right]$$

where λ is the angle between the test particle velocity $\mathbf{V}_t$ and the radial vector $\mathbf{r}$ and the strength functions are given by the following relations

$$g_{00} = \frac{2\pi^2}{r}\exp(-K_1 r)$$

$$g_{11} = \left[\frac{\pi^2\kappa^2\beta}{v_0^2\exp(K_1 r)} - \frac{\sqrt{2}\pi^{\frac{3}{2}}}{rK_1^2}\left\{1 + \frac{\Phi}{2} - \frac{(1+r^2K_1^2)\Psi}{2rK_1}\right\}\right]$$

$$g_{20} = \left[\left\{-\frac{\pi^2\kappa^4\beta^2}{4K_1 v_0^4} + \frac{\pi^2\kappa^2\beta}{v_0^3 r} - \frac{\pi^3}{8K_1^3} - \frac{\pi^2(2K_1^2 - \pi)}{K_1^6 r^3}\right.\right.$$
$$\left.\times\left(1 + K_1 r + \frac{K_1^2 r^2}{2}\right)\right\}\exp(-K_1 r) + \frac{\pi^{\frac{3}{2}}\kappa^2(2K_1^2 r^2 + 3)\beta}{2\sqrt{2}K_1^4 r^2 v_0^2}$$
$$\left.\times\left(\Phi - \frac{\Psi}{K_1 r}\right) + \frac{3\pi^{\frac{3}{2}}\sqrt{2}\kappa^2\beta}{2 r^2 v_0^2 K_1^4} + \frac{\pi^2(2K_1^2 - \pi)}{K_1^6 r^3}\right]$$

$$g_{22} = \left[\left\{\frac{\pi^2\kappa^4\beta^2 r}{4v_0^4} - \frac{\pi^2\kappa^2(1+K_1 r)\beta}{v_0^3 r} - \frac{\pi^3 r}{8K_1^2} + \frac{3\pi^2(2K_1^2 - \pi)}{K_1^6 r^3}(1 + K_1 r\right.\right.$$
$$\left.+ \frac{K_1^2 r^2}{2} + \frac{K_1^2 r^2}{6}\right)\bigg\}\exp(-K_1 r) - \left\{\Phi - \frac{\Psi\left((K_1^2 r^2 + 3)^2 - K_1^2 r^2\right)}{K_1 r(2K_1^2 r^2 + 9)}\right\}$$
$$\left.\times\left(\frac{(2K_1^2 r^2 + 9)\kappa^2\beta}{2\sqrt{2}\pi^{-\frac{3}{2}}K_1^4 r^2 v_0^2}\right) - \frac{1}{2}\frac{\pi^{\frac{3}{2}}\sqrt{2}\kappa^2(9+K_1^2 r^2)\beta}{r^2 v_0^2 K_1^4} - \frac{3\pi^2(2K_1^2 - \pi)}{K_1^6 r^3}\right]$$

where we have used the following relations for the exponential integrals in the above expressions to define the functions Φ and Ψ,

$$E_1(K_1 r) = Ei(1, K_1 r) = \frac{1}{2}[\Psi(K_1, r) + \Phi(K_1, r)]\exp(-K_1 r)$$
$$Ei(K_1 r) = -[Ei(1, -K_1 r) + i\pi] = \frac{1}{2}[\Psi(K_1, r) - \Phi(K_1, r)]\exp(K_1 r)$$

and where the exponential integrals are related to alternate standard expressions [8] as

$$E_i(1, K_1 r) = E_1(K_1 r)$$
$$E_i(1, -K_1 r) = -[E_i(K_1 r) + i\pi]$$

DISCUSSION

The results shown here clearly show the effect of charging dynamics on the dusty plasma response to a moving test charge. Without the charging terms a comparison may be made with the early results of Cooper [3], but some corrections must be made for errors in his work. The modification of the first order term due to charging is seen to fall exponentially. A further effect of the charging dynamics is an increase in effective Debye wave vector K_1, so that the shielding of the test charge is enhanced. In the present work the convective term in the equation for the dynamical charging has been ignored. Further work will include this and will take into account the velocity distribution of the dust grains.

ACKNOWLEDGMENTS

The authors would like to thank their colleagues, at the Alfvén laboratory for useful discussions and suggestions. This work was partly supported by the Swedish Research Council.

REFERENCES

1. Melandsø, F., Aslaksen, T. K. and Havnes, O., *Planet. Space Sci.*, **41**, 321 (1993).
2. Rostoker, N., *Nuclear Fusion*, **1**, 101 (1960).
3. Cooper, G., *Phys. Fluids*, **12**, 2707 (1969).
4. Neufeld, J. and Ritchie, R. H., *Phys. Rev.*, **98**, 1632 (1995).
5. Sanmartin, J. R. and Lam, S. H., *Phys. Fluids*, **14**, 62 (1971).
6. M. H. Nasim, P. K. Shukla, and G. Murtaza, *Phys. Plasmas*, **6**, 1409 (1999).
7. S. Ichimoru, *Basic Principles of Plasma Physics, A Statistical Approach*, W. A. Benjamin, P 56, 1973.
8. M. Abramowitz and I. A. Stegun, "*Handbook of Mathematical Functions*", (National Bureau of Standard, 1964).

Two-Dimensional Strongly Coupled Plasma on a Liquid Surface

T. Shoji*, K. Shinohe[†], H. Tomita[¶] and Y. Sakawa*

Department of Energy Engineering and Science, Nagoya University, Nagoya 464-8603, Japan
[†]*Peripherals Engineering Department, Fujitsu Co.,1405 Ohmaru, Inagi-shi, Tokyo 206-8503, Japan*
[¶]*Department of Nuclear Engineering, Nagoya University, Nagoya 464-8603, Japan*

Abstract. A simple system of a two-dimensional (2D) strongly coupled plasma has been developed by using charged fine particles in an oil suspension. The fine silica (SiO_2) particles placed on the bottom of a concave electrode where silicon oil is filled are charged up by applying a voltage with respect to the upper glass electrode, which is coated by electro-conductive film (ITO). The particles are confined on the oil surface by the surface tension and the electric field between the electrodes. The long time evolution of Coulomb crystal growth was observed.

INTRODUCTION

When the charge of the particles confined in a finite space is large enough that their potential energy exceeds the kinetic energy, a long range Coulomb interaction dominates the collective behavior of the particles and such a plasma is called a strongly coupled plasma (SCP). Several experiments have been conducted to study SCP in laboratories by charging up small particles in a liquid [1], air [2] and in low density plasmas [3-4]. In dust plasma experiments, the charge of the dust particles is determined by ion and electron flux to the particles in a sheath region, so the interaction between the particles and the plasma makes control and measurement of the charge difficult. In colloid experiments, the particles are charged chemically and easily measured, but the control of the charge of the particles is difficult in the liquid suspension. We have developed a simple 2D SCP system by charging particles in an oil suspension. This new system has several advantages in studying the Coulomb solidification process and the dynamic behavior of SCP in a liquid phase compared with the conventional colloidal suspension in a water: the charge of the particles can be easily controlled by the voltage, the number of impurity ions which shield out the Coulomb potential can be reduced in the insulating oil.

2D SCP SYSTEM CONFINED ON AN OIL SURFACE

Silica (SiO_2) particles of the diameter d=3, 5,10,15µm are charged and confined within parallel plate electrodes (Fig. 1). The accuracy of the diameters is within 1%.

The stainless steel lower electrode (50mm in diameter) has a hollow (3.5mm in diameter and 0.5mm in depth at the center). The upper electrode is made of glass and coated by transparent electro-conductive material ITO on the lower side.
The Silicon oil (TSF451-10, GE Toshiba Silicones Co., Ltd.) of $\varepsilon=2.6\varepsilon_0$ and $\rho_o=0.935\times10^3$ kg/m^3 are filled in the hollow, where ε and ρ_o are the specific dielectric constant and density of the oil, and ε_o is a vacuum dielectric constant.

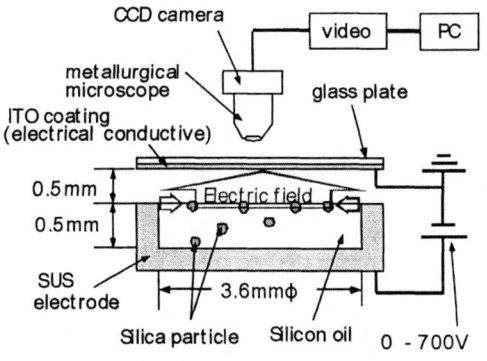

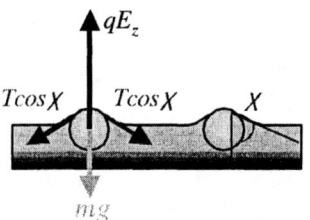

FIGURE 1. Experimental set-up.

FIGURE 2. Force balance on particles at oil surface, where T and χ are surface tension and the angle between oil and particle boundary.

The particles are put into the oil and charged by the application of the voltage on the electrodes ($\leq$700V). When the voltage increases above the critical voltage V_C, the particles move upward and finally stop at the oil surface as a result of the surface tension (Fig.2). Due to the concave shape of the hollow electrode, the radial component of the electric field compresses particles toward the center of the hollow on the oil surface. The charge q is precisely measured by adjusting the voltage of the electrodes to balance the forces on the particle in the oil and it is found that $q \propto d^{5/2}$, where d is the diameter of the particles. The charge can be controlled easily by the diameter and the specific mass of the particles in this system.

SOLIDIFICATION OF THE CHARGED PARTICLES ON THE OIL SURFACE.

We show an example of the 2D hexagonal crystals of the charged particles confined on the oil surface (Fig. 3). The particles are compressed toward the center of the oil surface by the radial electric field in the hollow electrode. The time variation of the surface density of the particles at the center on the oil surface is shown in Fig. 4. When the charged particles first appeared on the oil surface, they spread over the whole area on the oil surface. Then the compression of the particles in the radial direction proceeds, so the density of the particles at the center increases in early time and finally saturate to the equilibrium value. The force balance that governs the radial motion of the particles is expressed as

$$m\dot{v}_r = F_{Coulomb} + F_{conf} + F_{fric}, \tag{1}$$

where $F_{Coulomb}$, F_{conf}, F_{fric}, are Coulomb, electrostatic confinement and friction forces, respectively.

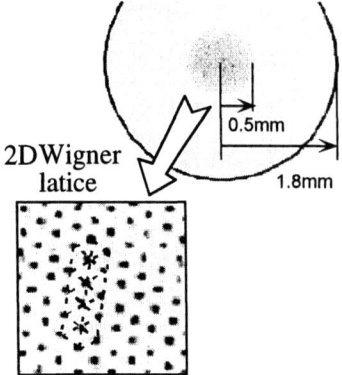

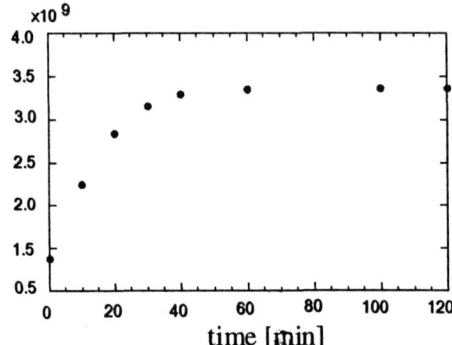

FIGURE 3. View of charged particles (one layer) confined on oil surface. $d=10\mu m$.

FIGURE 4. Time variation of radial surface density (m^{-2}) at the center after the charged particles appear on the oil surface. $d=10\mu m$.

In the early stage of the compression, particles are uniformly spread over the oil surface and move inward balanced by the friction force, $F_{conf} + F_{fric} \approx 0$. When the average density $\bar{n}$ is defined as, $N = \bar{n} \pi R^2$, where N and R are the total number and the radius of the boundary of the particles (R=5x10^{-4}m in Fig. 4), $\bar{n}$ can be deduced as,

$$\frac{\dot{\bar{n}}}{\bar{n}} \approx \frac{2qK}{3\pi\eta d}, \tag{2}$$

where K and η are the coefficient of the parabolic confinement potential and viscosity of the oil, respectively. K is calculated as $K \approx 5.6 \times 10^4 V_{conf}$ [V/m^2] in this experiment, where V_{conf} is the voltage applied on the electrodes. In the equilibrium stage where $F_{Coulmb} + F_{conf} \approx 0$, the density profile is no more uniform and observed to be parabolic. Assuming that the particles form a continuum surface charge and using the variation method to obtain the radial density profile [5], one finds

$$n(r) = \frac{\varepsilon K R}{2q}(\frac{\lambda_D}{R})\left\{\left(\frac{R}{\lambda_D}\right)^2 - \left(\frac{r}{\lambda_D}\right)^2\right\} \tag{3}$$

where λ_D is the effective shielding (Debye) length due to the existence of small counter ions diffused in the oil. The equilibrium density profiles obtained in this experiment well fit the parabolic shape one as expected from Eq. (3). From the measured equilibrium density at the center in Fig. 4, λ_D is calculated from Eq. (3) as $\lambda_d \cong 4$ for the particles of d=10 μm.

The time evolution of the crystal growth during the compression of the particles on the oil surface is examined by taking the statistics of the voronoi polygons of the particles (Fig. 5). In about 60 minutes after the appearance of the particles on the oil surface, more than 80% of the polygon cells become hexagonal. Because of the cylindrical symmetry of the confinement potential, there are some irregularities in the cells like pentagon and heptagon cells around the periphery of the crystal. The radial distribution function g(r) is measured to evaluate the morphology of the crystal in the compression process of the particles. Even in the early stage of the compression, the ordered structures in the particles are observed and the peaks of g(r) well agree with the hexagonal one.

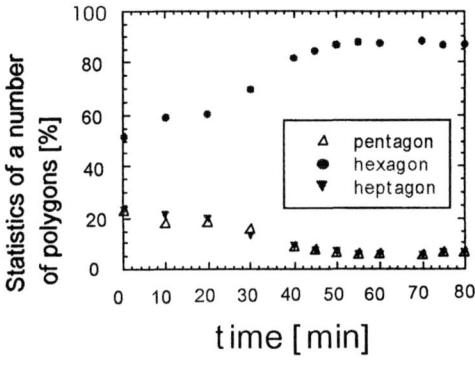

FIGURE 5. Statistics of voronoi polygons in time.

CONCLUSIONS

A new colloid type strongly coupled plasma system has been developed by charging silica particles in oil. In this system, the charge on the particles is well defined and controlled by the diameter of the particles and the DC voltage on the parallel electrodes. The charged particles are confined on the oil surface and the 2D hexagonal crystal formed are found to be stable for more than a day. The time evolution of Coulomb crystal growth is observed on a long time scale of minutes.

REFERENCES

1. Sakaki, T., *J. Chem. Phys.* **82**, 996 (1985).
2. Wuerker, R., Shelton, H., and Langmuir, R., *J. Appl.Phys.* **30**, 342 (1959).
3. Chu, J., et al., *J Phys.* **D27**, 296 (1994).
4. Y. Hayashi, and Tachibana, K., *Jpn. J. Appl. Phys.* **33**, 804 (1994).
5. Totsuji, H., *Phys.Plasmas* **8**, 1856 (2001).

Levitation and Transport of Charged Dust Over Surfaces in Space

Joshua E. Colwell, Mihály Horányi, Scott Robertson and Amanda A. Sickafoose

LASP, University of Colorado, Boulder, CO 80309-0392 USA

Abstract. Dust in planetary regoliths may become charged and levitated in plasma sheaths and photoelectron sheaths near the surface [1,2]. This provides an explanation for the observations of the lunar horizon glow [3]. Horizontal electric fields or inhomogeneities in the sheath may lead to net transport of dust on the surface. Electrostatic levitation of dust may also explain observations of regolith deposits in craters on the asteroid 433 Eros by the NEAR spacecraft [4]. We present the results of a simple model of dust transport in a photoelectron sheath across a surface with simple topographical forms. We find a net deposition of particles launched in random directions at photoelectron sheath boundaries such as might occur in the terminator region. Topographic boundaries such as blocks and craters provide an additional sink for particles moving horizontally across the surface in a sheath.

INTRODUCTION

Dust is abundant on the surfaces of the Moon, Mars, asteroids, comets, planetary ring particles, and in protoplanetary disks. The dusty regoliths on the surfaces of these airless planetary objects are generated by the continual bombardment by the interplanetary micrometeoroid flux. The surfaces of airless bodies in the solar system are directly exposed to the solar wind and to the full solar spectrum. Direct impact of solar wind protons onto a surface ejects electrons from the surface. Similarly, solar ultraviolet photons produce photoelectrons from the surface. Both processes can lead to an electron sheath near the surface resulting in strong local electric fields. In the absence of topography the electric field near the surface is directed upwards and can lift positively charged dust grains off the surface. At a certain height in the sheath, the electric force balances gravity depending on the charge to mass ratio of the particle and gravity of the object. Particles lifted off the surface may become stably levitated in the sheath, accelerated to escape velocity, or follow modified ballistic trajectories to other points on the surface. An example of dust levitation was observed above the lunar surface by the Surveyor spacecraft [3]. Lunar horizon glow imaged by the Clementine spacecraft may be explained by dust accelerated to escape velocity in the lunar photoelectron sheath [5].

Another example of transport of charged dust in planetary environments is the spokes in Saturn's rings, believed to be composed of small dust particles lifted off the surface of larger ring particles by impact-generated plasma [6]. Recent high resolution observations of the surface of 433 Eros by the NEAR spacecraft show smooth deposits in the floors of some craters. The gravitational acceleration on Eros is too small (less than 10^{-3} g) to result in downslope flow of dust particles without some other force to break interparticle surface forces and mobilize the dust particles. The sizes of particles in these crater bottoms, called "ponds", are only constrained to be smaller than cm-sized. Mobilization of micron-sized dust grains in a photoelectron sheath is a possible explanation for these ponds [4,7]. Experimental [8] and theoretical [1,2] work has demonstrated that dust can be levitated in plasma and photoelectron sheaths in planetary environments. Here we report on simulations of horizontal transport in a photoelectron sheath to study net transport of dust in planetary regoliths.

MODEL DESCRIPTION

We follow the trajectories of individual dust grains using a direct integrator that simultaneously solves for the charge on the particle and its trajectory. The forces on the particle are gravity downward and the electric force resulting from the grain charge and the upwad electric field normal to the model surface that is produced by the sheath. We follow trajectories of the particles in 2 dimensions, and impose simple topography on the surface. We do not attempt a physically realistic duplication of a planetary surface or an inhomogeneous sheath; rather we start with a simple model that illustrates the dynamics of dust in a sheath and the net transport of dust if the sheath has breaks or the surface has topography. Particles are launched with a variety of initial velocities from 100 evenly spaced points along the 1-D "surface"at an initial angle of 45 degrees. Particles are tracked until they come into contact with the surface or a maximum simulation time has passed. The positions where the particles reimpact the surface are recorded and used to calculate a transport matrix whose coefficients determine the transport from one location on the model surface to another. The vertical resolution of the model sheath is fixed at 0.05 Debye lengths, where the Debye length is calculated at the surface. The spacing of the grid and the scale of model topography is much larger than a Debye length.

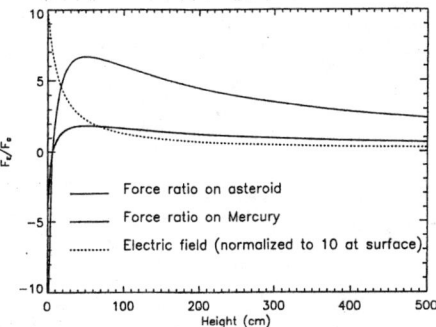

FIGURE 1. The electric field as a function of height above the surface, and the ratio of the electric force (positive upward) to the gravitational force (negative upward) are plotted for a $0.1 \mu m$ radius grain above a surface with $g = 360 \, cm\text{-}s^{-2}$ and $g = 1 \, cm\text{-}s^{-2}$. Forces balance where the ratio equals one, and the electric force was calculated using the equilibrium charge on the grain. For our dynamical simulations reported below, grain charge is integrated simultaneously with grain position, so the particle is not necessarily in charge equilibrium with the sheath.

The electron density and electric field are calculated based only on the vertical position above the surface, so the sheath model is one-dimensional (Figure 1). A shadowed region can be imposed by using topography and specifying a solar illumination angle. The electric field and electron density are set to zero with the shadow. We also test the effect of introducing a break in the sheath without the use of topography, such as would occur at the terminator.

RESULTS

We conducted simulations above surfaces with gravitational accelerations matching that of Mercury ($g = 360$ cm-s^{-2}) and a medium-sized asteroid ($g = 1$ cm-s^{-2}). Particles were launched at speeds between 1 and 10^3 cm/s. Particles tend to accumulate at topographic boundaries because their trajectories intersect crater or block walls (Figures 2–3).

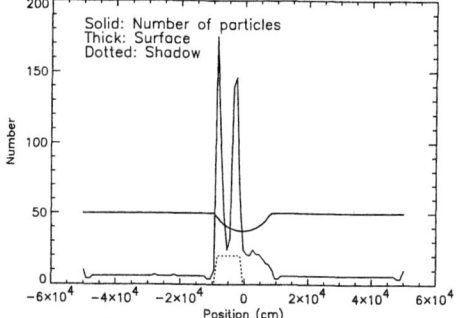

FIGURE 2. The distribution of final dust positions for a simulation with a shallow crater (thick line) and sunlight incident from the left resulting in a shadow on the crater floor (dotted line). Particles suspended in the photoelectron sheath fall to the surface when they enter the shadow and the electric field vanishes, and also accumulate near the crater walls (solid line).

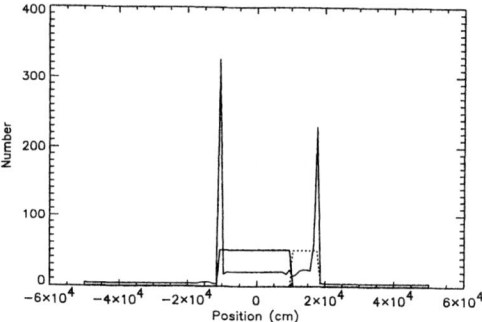

FIGURE 3. Distribution of dust positions like Figure 2, but for a surface with a rectangular block (thick line) instead of a crater. Sunlight is incident from the left, and the dotted line indicates the region of the block's shadow. Particles accumulate at the block boundary and at the shadow boundary (solid line).

Particles also accumulate at sheath boundaries where the electric force vanishes. Particles that are levitated in the sheath move across the surface with their initial horizontal velocity. This initial velocity could be the result of an impact on the surface, for example, or by levitation due to a large surface bias. When the particle crosses the local terminator, the photoelectron sheath vanishes and gravity is the only remaining force on the particle and it falls to the surface. The distribution of particle positions on the surface show this as large accumulations at the boundaries of shadows produced by topography (Figures 2–3). Future work will include a full three-dimensional model of the sheath above realistic surfaces with a time-variable terminator.

ACKNOWLEDGMENTS

This research was supported by NASA Office of Biological and Physical Research.

REFERENCES

1. Nitter, T., and Havnes, O., *Earth, Moon, and Planets*, **56**, 7–34 (1992).
2. Criswell, D.R., and De, B.R., *J. Geophys. Res.*, **82**, 1005–1007 (1977).
3. Rennilson, J.J., and Criswell, D.R., *The Moon*, **10**, 121–142 (1974).
4. Robinson, M.S., Thomas, P.C., Veverka, J., Murchi, S., and Carcich, B., *Nature*, **413**, 396–400 (2001).
5. Zook, H. A., Potter, A.E., and Cooper, B.L., *Lunar Planet. Sci.*, **26**, 1577–1578 (1995).
6. Goertz, C. K., *Reviews of Geophysics*, **27(2)**, 271–292 (1989).
7. Tepliczky, I., and Kereszturi, A., *Lunar Plan. Sci. Conf.*, **33**,Abs. #1656), (2002).
8. Sickafoose, A.A., Colwell, J.E., Horányi, M., and Robertson, S., *J. Geophys. Res.*, submitted, (2002).

Dust-acoustic Waves with a Non-thermal Ion Velocity Distribution

S.V. Singh*, G.S. Lakhina*, R. Bharuthram[†] and S.R. Pillay**

*Indian Institute of Geomagnetism, Colaba, Mumbai-400005, India
[†]University of Natal, Durban- 4041, South Africa
**Physics Department, University of Durban-Westville, Durban-4000, South Africa

Abstract. The effect of a non-thermal ion velocity distribution is studied on linear and nonlinear dust-acoustic waves in an unmagnetized three component dusty plasma. For the linear theory, electrons are Boltzmann distributed, ions are treated kinetically with a non-thermal ion velocity distribution and the dust particles are considered to be streaming with respect to ions. It is found that growth of the waves is possible when streaming velocity of the dust particles is larger than the phase velocity of the dust-acoustic wave. The model is used to study nonlinear dust acoustic waves.

INTRODUCTION

Charged dust grains are important constituents of matter in space and astrophysical systems such as the planetary rings, asteroid zones, cometary tails, interstellar clouds, as well as the lower parts of the Earth's ionosphere. The charged dust grains react to electromagnetic and gravitational fields, and give rise to low–frequency collective phenomena in dusty plasmas.

Dusty plasmas open up an ultra low–frequency regime for the existence of novel types of wave modes, electrostatic acoustic–like mode called the Dust–Acoustic Wave (DAW). The DAW mode was first predicted theoretically by Rao et al.[1] by including the dust collective dynamics. Their model consists of Boltzmann distributed thermal electrons and ions which provide the restoring force, while the inertia arises due to the heavier dust component. In recent years, a large number of authors have theoretically investigated linear and nonlinear wave propagation in dusty plasmas. Nonlinear propagation of the low–frequency DAWs leads to the formation of coherent structures, such as solitons. Depending on the parameter regime, it is possible to have rarefactive as well as compressive electrostatic potential structures associated with the nonlinear DAW solitons in three component dusty plasmas. Mamun et al[2]. (1996) have studied the dust-acoustic solitons with non-thermal distribution of ions in a two component dusty plasma consisting of negatively charged dust grains and positive ions.

Here, we study the effect of a non-thermal ion velocity distribution on the linear and nonlinear dust-acoustic waves in an unmagnetized three component dusty plasma. For the linear theory, electrons are Boltzmann distributed, ions are treated kinetically with a non-thermal velocity distribution[3] and the dust particles are streaming relative to the ions. Thereafter, the model is used to study nonlinear dust-acoustic waves.

BASIC EQUATIONS

We consider a three component, unmagnetized dusty plasma with Boltzmann distribution for electrons, streaming dust particles and non-thermal velocity distributed ions. The governing equations for such a system are given by

$$n_e = n_{e0} \exp(e\phi/T_e), \tag{1}$$

$$\frac{\partial n_d}{\partial t} + \frac{\partial}{\partial x}(n_d v_d) = 0, \tag{2}$$

$$\frac{\partial v_d}{\partial t} + v_d \frac{\partial v_d}{\partial x} = \frac{Z_d e}{m_d} \frac{\partial \phi}{\partial x} \tag{3}$$

$$\frac{\partial^2 \phi}{\partial x^2} + 4\pi e [n_i - n_e + Z_d n_d] = 0. \tag{4}$$

For the ions we chose a non-thermal velocity distribution[3] given by

$$f_{i0}(v) = \frac{n_{io}}{\sqrt{2\pi v_{ti}^2}} \frac{(1 + \alpha v^4/v_{ti}^4)}{(1+3\alpha)} \exp(-v^2/2v_{ti}^2). \tag{5}$$

In the equations (1)-(5), e is the electronic charge, and $n_{e0}(n_{io})$ is the equilibrium electron (ion) number density and Z_d is number of charges on the dust grain, $v_{ti} = \sqrt{T_i/m_i}$ is the ion thermal velocity; $T_i(m_i)$ is the temperature (mass) of the ions, and α is a parameter which determines the population of the non-thermal ions. We note that $\alpha = 0$ corresponds to a Maxwellian distribution. In the next section, we solve the above equations for linear nonresonant and resonant modes in dusty plasmas.

ANALYSIS OF LINEAR MODES

Linearizing equations (1)-(4) and using (5), we obtain the following linear dispersion relation

$$1 = \frac{k^2 \lambda_{de}^2}{1 + k^2 \lambda_{de}^2} \left[\frac{\omega_{pd}^2}{(\omega - kv_{do})^2} + \frac{\omega_{pi}^2}{2(1+3\alpha)k^2 v_{ti}^2} \left\{ Z'(\xi) + \sqrt{\frac{2}{\pi}} \frac{\alpha}{v_{ti}^3} \left(\frac{d^2 I}{d\mu^2} \right) \right\} \right], \tag{6}$$

where $\lambda_{de} = \sqrt{T_e/4\pi n_{eo} e^2}$, is the electron Debye length, $\omega_{pi} = \sqrt{4\pi n_{i0} e^2/m_i}$ and $\omega_{pd} = \sqrt{4\pi Z_d^2 n_{d0} e^2/m_d}$ are the plasma frequencies of the ion and dust particles, respectively and $Z'(\xi)$ is the plasma dispersion function with argument $\xi = \omega/\sqrt{2}kv_{ti}$, $I = \sqrt{\pi \mu} Z'(\xi)$ and $\mu = 1/2v_{ti}^2$. We obtain a nonresonant mode by expanding the plasma dispersion function in the limit $\xi \gg 1$, to obtain

$$1 = \frac{k^2 \lambda_{de}^2}{1 + k^2 \lambda_{de}^2} \left[\frac{\omega_{pd}^2}{(\omega - kv_{do})^2} + \frac{\omega_{pi}^2}{\omega^2} \left(1 + \frac{3k^2 v_{ti}^2}{\omega^2} \frac{1+15\alpha}{1+3\alpha} \right) \right], \tag{7}$$

where v_{do} is the dust particles drift speed. It can be seen from the dispersion relation (7) that the effect of energetic ions will be small on the nonresonant mode which is of the Buneman type mode. For the resonant mode we expand the plasma dispersion function in the limit $\xi \leq 1$ and obtain the following expressions for the real frequency and growth rate for dust acoustic wave,

$$\omega_r = kv_{d0} \pm kc_d / \sqrt{1+k^2\lambda_d^2}, \tag{8}$$

$$\gamma = \mp \sqrt{\frac{\pi}{8}} \omega_r \frac{\left(1 - 4\alpha \frac{\omega_r^2}{k^2 v_{ti}^2}\right)}{(1+3\alpha)} \left\{ \frac{Z_d(m_i/m_d)^{1/2} T_e^{3/2} (n_{do} n_{io}^2)^{1/2} \exp(-\omega_r^2/2k^2 v_{ti}^2)}{[n_{eo}T_i(1+k^2\lambda_{de}^2) + n_{io}T_e(1-\beta)]^{3/2}} \right\}, \tag{9}$$

where $c_d = \omega_{pd}\lambda_d$; $\lambda_d = \sqrt{T_e T_i / 4\pi e^2 \{n_{eo}T_i + n_{io}T_e(1-\beta)\}}$. It can be seen from equation (9) that for $v_{do} = 0$, the growth rate is negative and waves are damped. It is expected as there is no source of free energy to excite waves. Considering the lower signs in equations (8) and (9), it is seen that the dust acoustic mode become unstable provided $v_{do} > c_d/(1+k^2\lambda_d^2)^{1/2}$. The energetic ions reduce the growth rate. In the next section we discuss the effect of non-thermal ion distributions on nonlinear (solitons) dust-acoustic waves. It is assumed that the waves have already reached the saturation stage.

STUDY OF NONLINEAR DUST-ACOUSTIC WAVES

Following Cairns et al.[3] (1995), integrating (5) over velocity space, the ion density can be obtained as

$$n_i = n_{io}(1 + \beta\phi + \beta\phi^2)e^{-\phi}, \tag{10}$$

where $\beta = 4\alpha/(1+3\alpha)$. Solving the set of equations (1)-(4) and (10) and applying appropriate boundary conditions[2], we obtain an energy integral

$$\frac{1}{2}(d\phi/d\eta)^2 + V(\phi) = 0, \tag{11}$$

where $\eta = x - Mt$; M is Mach number and the Sagdeev potential is given by

$$V(\phi) = \frac{n_{eo}}{\sigma}(1 - e^{\sigma\phi}) + n_{io}\left\{1 + 3\beta - (1 + 3\beta + 3\beta\phi + \beta\phi^2)e^{-\phi}\right\}$$
$$+ (M - v_{do})^2 - (M - v_{do})\sqrt{(M - v_{do})^2 + 2\phi}, \tag{12}$$

where $\sigma = T_i/T_e$ and equations (10)-(12) are normalized equations, i.e., potential is normalized with T_i/e, electron and ion densities with $Z_d n_{do}$ and dust density by n_{do}, distance with $\sqrt{T_i/4\pi Z_d n_{do} e^2}$, time with ω_{pd}^{-1} and velocities with $\sqrt{Z_d T_i/m_d}$.

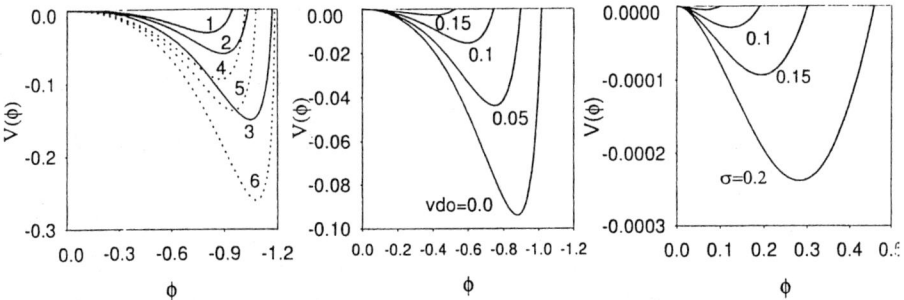

FIGURE 1. Rarefactive solitons: Plot of $V(\phi)$ with ϕ for the parameters $\alpha = 0.2$, $v_{do} = 0$, $\sigma = 1.0$, $n_{eo} = 0.0$ (solid curves) and $n_{eo} = 0.1$ (dashed curves) for $M = 1.45$ (curves 1,4), 1.48 (curves 2,5) & 1.54 (curves 3,6)

FIGURE 2. Rarefactive solitons: variation of $V(\phi)$ with ϕ for various values of v_{do} as indicated on the curves for $M = 1.45$, $\sigma = 1.0$, $n_{eo} = 0.1$

FIGURE 3. Compressive solitons: variation of $V(\phi)$ with ϕ for various values of σ as indicated on the curves for $M = 1.35$, $v_{do} = 0.0$, $n_{eo} = 0.1$

We have numerically analyzed (12) to obtain both rarefactive as well as compressive solitons. The typical chosen parameters are $\alpha = 0.2, \sigma = 1.0, n_{eo} = 0.1, v_{do} = 0$. Figure 1 shows the variation of Sagdeev potential $V(\phi)$ with potential, ϕ for rarefactive solitons. It is observed that with the inclusion of electrons in the two-component (dust and ions) plasma the range of Mach number for which solitons are observed is much wider but shifts towards lower Mach numbers. For example, for above parameters, we could get solitons for $1.24 < M < 1.54$ but for $n_{eo} = 0$ it is possible to obtain soliton solutions for $1.41 < M < 1.6$. Also, the peak amplitude is large and potential well is deeper for three component dusty plasma than the two component case considered by Mamun et al.[2] (1996). Figure 2 shows the variation of $V(\phi)$ with ϕ for $M = 1.45$ for various values of v_{do} as indicated on the curves. It is noted that for three component case soliton solution is possible for wider range of dust beam drift speed than the two component case. The peak amplitude decreases with increase dust beam speed.

In Figure 3, we show the variation of $V(\phi)$ with ϕ for compressive solitons for the parameters, $M = 1.35, n_{eo} = 0.1, v_{do} = 0$ for various values of σ as shown on the curves. It is observed that with the inclusion of electrons in two component dusty plasma, the range of Mach numbers for which compressive solitons are observed gets narrower.

REFERENCES

1. Rao, N.N., Shukla, P.K. and Yu, M.Y., *Planet. Space Sci.*, **38**, 543 (1990).
2. Mamun, A.A., Cairns, R.A. and Shukla, P.K., *Phys. Plasmas,* **3**, 2610 (1996).
3. Cairns, R.A., Mamun, A.A, Bingham, R., Dendy, R., Bostrom, R., Nairns, C.M.C. and Shukla, P.K., *Geophys. Res. Lett.*, **22**, 2709 (1995).

Production Mechanism and Chemical Structure of Dust Particles in Fluorocarbon Plasmas

Kazuo Takahashi, Kouichi Ono, and Yuichi Setsuhara

*Department of Aeronautics and Astronautics, Kyoto University,
Yoshida-Honmachi, Sakyo-ku, Kyoto 606-8501, Japan*

Abstract. Dust particles were found to be produced in octafluorocyclobutane (c-C_4F_8) plasmas used for reactive ion etching of silicon dioxide films and chemical vapor deposition of low dielectric constant films. The dust particles contained the ultrahigh mass polymers of 100 000 in molecular weight, which were compounds carbon-rich compared with deposited films and were formed from gas phase products related to C_2F_4. The ultrahigh mass polymers could work as nuclei, and were grown to be spherical micrometer-sized dust particles. The dust particle included crystalline grain of a few nanometer scale which corresponded to the nuclei size estimated form the molecular weight.

INTRODUCTION

Surface processing using low pressure plasmas is one of the most important techniques in semiconductor manufacturing. Especially, reactive ion etching and chemical vapor deposition (CVD) using plasmas with reactive gases are indispensable for fabrication of semiconductor devices. Since, recently, highly selective and anisotropic etching is required to obtain contact holes with high aspect ratio in ultralarge-scale integrated circuits. On the other hand, low permittivity intermetal dielectrics are investigated for the purpose of reducing the resistance capacitance time delay, which is conspicuous due to shrinkage of the spacing between metal lines in high density multilayer integrated circuits. From these industrial requirements, fluorocarbons, e.g., C_4F_8, are being applied to plasma etching and CVD processes. In the fluorocarbon plasmas, it was shown that the highly polymerized molecules were formed, which were detected in measurements with electron attachment mass spectrometry [1, 2, 3]. Mechanisms analogous to cluster formation in silane discharges [4], polymerized molecules may have high sticking probability, and may cause formation of dust particles in the gas phase as well as thin film on the surface [5]. Since such dust particles can damage semiconductor devices on silicon wafers, investigation of dust particle production is significant for purification of the processing plasmas for semiconductor fabrication. In the present paper, chemical characterization of the dust particles with gel permeation chromatography (GPC) and x-ray photoelectron spectroscopy (XPS) is reported, and the production mechanism is discussed based on the results of gas phase measurements with Fourier transform infrared (FTIR) spectroscopy and observation with scanning electron microscope (SEM) and high resolution transmission electron microscope (HRTEM).

CHEMICAL STRUCTURE OF THE DUST PARTICLES

A conventional parallel plate type reactor was used [6]. Octafluorocyclobutane (c-C_4F_8) was chosen as a kind of fluorocarbon compounds. For the plasma generation, a rf (13.56 MHz) power supply was connected to the upper electrode where the power density was maintained at 0.15 W/cm^2. The gas pressure was varied between 23 and 250 mTorr.

In the whole tested gas pressure range, amorphous fluorocarbon film was deposited irrespective of the surface condition. The dust particles, however, were observed only at the pressure higher than 50 mTorr, whose diameter measured in SEM observation ranged between 0.5 and 2.3 µm [7]. Chemical characterization of the dust particles was performed with GPC and XPS. From the GPC results, it became clear that the dust particles produced in gas phase contained polymers of ultrahigh mass around 100 000. In addition, species with molecular weights less than 2000 were also found in an extract from the dust particles as in the case of the deposited film. In the first XPS measurement, the XPS spectrum of the dust particles was almost the same as that of the film, because the dust particles were possibly covered with the fluorocarbon films after dropping on the surface. In order to remove the film on the dust particles, Ar ion sputtering was performed in the XPS system using a low-energy ion beam so as to avoid the transformation of the chemical composition. The XPS spectrum of the dust particles so obtained is obviously different from that of the film. The content of C-CF$_x$ bond in the dust particles was more than that in the film. This result indicates that the branching of carbon network was promoted in the precursor molecules composing the dust particles. This is also consistent with the GPC result that the dust particles are composed of the ultrahigh mass polymers.

PRODUCTION MECHANISM OF THE DUST PARTICLES IN FLUOROCARBON PLASMAS

Many fluorocarbon molecules, CF_4, C_2F_4, C_2F_6 and other products, were found to be produced from the C_4F_8 plasma in the measurements with FTIR spectrometer [8]. The production amount of C_2F_4 molecule and the products including the bond components related to the molecule, e.g., -CF=CF- and -CF=CF$_2$, increased with the source gas pressure. These products appeared just after plasma initiation and increased rapidly with discharge time. More noteworthy is that the products were consumed with the dust particle production rather than decreased with the depletion of the source gas molecule. The fact was shown by the correlation between time evolution of the product pressure and time scale of the dust particle production. Therefore we can safely agree that the species related to C_2F_4 molecule and the molecule itself are candidates for the dust particles

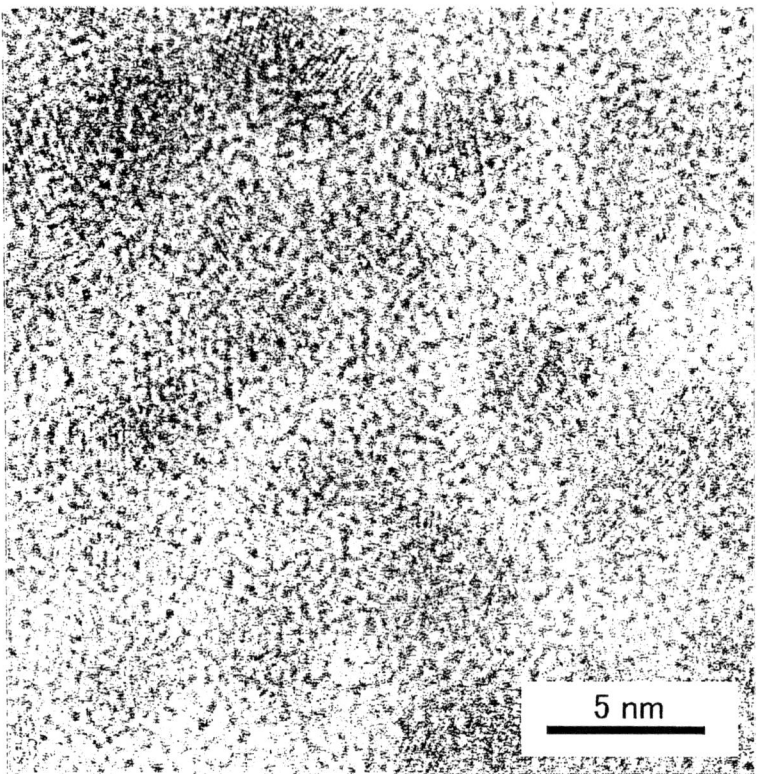

FIGURE 1. HRTEM image of the dust particle nuclei extracted in THF solution.

produced in the fluorocarbon plasmas.

The dust particle precursor of the species related to C_2F_4 molecule can form the ultra-high mass polymers as the nuclei. The geometrical dimension of the nuclei is estimated to be 9.7 nm from molecular weight of 100 000. The nuclei was tried to be observed in tetrahydrofuran (THF) solution for GPC measurements with HRTEM. It is noted that many crystalline grains were found in the nanoscopic observation (Fig. 1). The electron beam diffraction pattern showed that the ordered atomic arrangements corresponded to diamond structure. The crystal structure did not continue to grow on the nuclei, since a few hundreds nanometer dust particles were covered with amorphous fluorocarbon layer. The nuclei, some of them are crystalline and other are amorphous, must be charged up with a few electrons in plasmas. Since the charge of the nuclei increase with their size, a larger nucleus tends to be trapped more easily in discharge. The trapped nuclei continue to grow up to micrometer-sized dust particles by the mutual aggregation and by deposition of film precursor around the nuclei. From the point of view of material science, it is significant that such crystalline structure is synthesized in the non-heated plasma. In the gas phase, the reactions excited by energetic electrons of a few electron volts and bombarding ions may cause the crystallization.

CONCLUSION

The dust particle production was confirmed in the c-C_4F_8 plasmas. In the plasmas, the C_2F_4 molecule played an important role in gas phase polymerization and nucleation of the dust particles. The molecule was activated and transformed to highly reactive species, -CF=CF- and -CF=CF$_2$. The species formed ultrahigh mass polymers working as nuclei of the dust particles. The nuclei of 100 000 in molecular weight was estimated to be 9.7 nm in their size, and many of the nuclei were crystalline grains with diamond structure. The nuclei can be charged by a few electrons and trapped in discharge. Then, they continue to grow covered with amorphous fluorocarbon layer. The growth is proceeded until the nuclei become micrometer-sized dust particles.

It is noted that crystalline nuclei of diamond structure could be obtained in the nonheated plasma. The fluoro-diamond nuclei will attract many interests of scientist and engineer as one of the nanometer-sized materials for industrial applications. Now the properties of the crystalline nuclei are under investigation to explore possibilities as a functional nanometer-sized material.

ACKNOWLEDGMENTS

The present work was supported in part by a grant from the Murata Science Foundation. The authors would like to thank Mr. Takashi Kouzaki and Mr. Tetsuyuki Okano of Matsushita Technoresearch, Inc. for the HRTEM observation.

REFERENCES

1. Stoffels, E., Stoffels, W. W., and Tachibana, K., *Rev. Sci. Instrum.*. **69**, 116-122 (1997).
2. Stoffels, W. W., Stoffels, E., and Tachibana, K., *J. Vac. Sci. Technol. A,* **16**, 87-95 (1998).
3. Imai, S., and Tachibana, *Jpn. J. Appl. Phys.,* **38**, L888-L891 (1999).
4. Veprek, S., Schopper, K., Ambacher, O., Rieger, W., and Veprek-Heijman, M. G. J., *J. Electrochem.,* **140**, 1935-1942 (1993).
5. Anderson, H. M., and Radovanov, S. B., *J. Vac. Sci. Technol. A,* **14**, 608-614 (1996).
6. Takahashi, K., and Tachibana, K., *J. Appl. Phys.,* **89**, 893-899 (2001).
7. Takahashi, K., and Tachibana, K., *J. Vac. Sci. Technol. A,* **19**, 2055-2060 (2001).
8. Takahashi, K., and Tachibana, K., *J. Vac. Sci. Technol. A,* **20**, 305-312 (2002).

Relativistic Dust Particles Approaching the Earth as Highest Energy Cosmic Rays

V.N.Tsytovich* and R.Bingham[†]

*General Physics Institute, Moscow, 119991, Russia
[†]Rutherford Appleton Laboratory, Chilton Didcot, OX11 0QX, UK

Abstract. The biggest observational mystery is the presence in cosmic rays particles with energies about $10^{21} eV$ which should be destroyed while travelling through $3^0 K$ black body radiation. One possibility is that they are not protons but dust particles containing 10^{10} or more protons with moderate $\gamma = (1 - v_d^2/c^2)^{-1/2}$ about 10-100 (v_d being the dust velocity). By analyzing the process of charging for fast grains we obtained that the dust grains travelling through interstellar media will loose their charge during less than 10^4 s when the dust velocities exceed the thermal electron velocity of interstellar plasma and can not be charged if $v_d > 0.1 \times c$. This result excludes dust acceleration by a random walk in magnetic shocks to create relativistic neutral atoms in cosmic rays by a decay of accelerated grains [1]. The grains surrounding pre-supernova can be accelerated up to $\gamma = 100$ by radiation pressure of the light emitted in supernova explosions. Relativistic dust grains will approach the Earth as showers but not as flux constant in time. When approaching the Earth the relativistic dust particles can be destroyed by solar radiation [2] since the solar radiation in the rest frame of relativistic grain corresponds to X-ray frequency range . For energy larger than critical the δ- electrons created inside the dust particles by X-ray photons will escape the grain and will continuously charge the dust positively until it is disrupted by the large positive dust charge. The critical value γ_{cr} such that for $\gamma > \gamma_{cr}$ the dust particle is destroyed by solar radiation is calculated in the present paper numerically and indicates that for rigid crystalline grains of the size about $1 \mu m$ γ_{cr} is rather large 4.7×10^4. The obtained results can be used to distinguish the source of cosmic ray showers in the planned measurements of cosmic ray showers from space by a system of detectors on satellites. The disruption of relativistic grains by solar radiation produces small explosions which can be detected.

PLANNING EXPERIMENTS

It is expected that in the space around the Earth the relativistic dust particles with $\gamma = 1/\sqrt{1 - u^2/c^2} \gg 1$ can be present and can cause effects which can be detected experimentally: 1) a flux of relativistic neutral atom cosmic rays approaching the Earth [1], 2) cosmic ray showers imitating the showers of single proton cosmic rays of the highest energy [2] (such particles should contain about $10^{10} - 10^{11}$ nucleons (several μm size) and should have a Lorentz factor γ about 10) when approaching the Earth atmosphere. In the planning experiments the difference between the cosmic ray showers produced by dust grains and the showers produced by a single high energy proton can be observed by set of counters on a system of satellites. Two main physical problems exists: 1)whether the relativistic grains approaching the solar system could be largely charged and 2) whether they can survive not being disrupted by solar optical radiation which in they frame of reference corresponds to ultraviolet or X-ray radiation and is charging the grains positively until they are disrupted before reaching the Earth atmosphere [2].

CHARGING OF FAST GRAINS IN THE INTERSTELLAR MEDIA

The thermal electron velocity v_{T_e} in the interstellar media is less than $\approx (1-3) \times 10^8$ cm/s, ($T_e < (3-10)$ eV. In the grain frame with $v_d \gg v_{T_e}$ the electrons and ion move with equal drift velocity $-v_d$ and therefore the cross-section of electron capture is larger than the geometric cross-section on a factor $1 + q/am_e v_d^2$ (a being the grain radius), while the ion capture cross-section is practically coinciding with the geometrical cross-section. The charging equation takes the form

$$\frac{d}{dt}\left(\frac{q}{e}\right) = I_{ph} + I_{sec} - v_d n_e \pi a^2 \left(1 + \frac{qe}{am_e v_d^2}\right) + v_d n_i \pi a^2$$

$$I_{ph} + I_{sec} - \left(\frac{q}{e}\right) \pi a^2 v_d \left(n_d + n_e \frac{e^2}{am_e v_d^2}\right); n_i = n_e - \frac{q}{e} n_d \quad (1)$$

where I_{ph} and I_{sec} are the photo-electron current and secondary electron emission per unit charge respectively. In absence of the latter (1) describes a fast grain charge decay while travelling in the interstellar media on the typical timescale of the order of 10^4s. In presence of photo-effect and secondary emission the small equilibrium charge is always positive. A constrain for describing the charging by the equation (1) is the presence of non-zero capture coefficient of electrons and ions by grains (in (1) they were assumed to be 1). For grain velocities larger than $0.1c$ both electrons and ions will pass freely the grains of the size of $1\,\mu$m and the capturing coefficient vanishes.

GRAIN ACCELERATION IN SPACE

The acceleration by subsequent passing of the interstellar shocks is not possible since the grains loose their charges, the only mechanism is the light pressure acceleration in supernova explosions [3]. The time of acceleration up to γ of the order of 10 for $1\,\mu$m dust grain can be estimated about $(3-10) \times 10^5$s with the heating not sufficient to melt the pre-supernova grains. The energy reached will be proportional to a^2 and for grain power-law distribution with $n_d(a) \propto 1/a^s$ and $s \approx 3.5$ (the value obtained from observations, here the normalization used for dust distribution in sizes is $n_d = \int n_d(a)da$. We get the distribution of accelerated grains in their energies ε_d as $n_d(\varepsilon_d) \propto 1/\varepsilon_d^\nu, \nu = (s+1)/2 \approx 2.25$ (the normalization used is $n_d = \int n_d(\varepsilon_d)d\varepsilon_d$). This energy distribution is close to that observed for the high energy part of the cosmic rays at the Earth. This consideration shows that one should expect that the flux of relativistic grains approaching the Earth corresponds to grain "showers" but not to a constant in time flux.

DISRUPTION OF THE GRAINS IN VICINITY OF THE EARTH BY SOLAR RADIATION

The thermal solar radiation in vicinity of the Earth operates in the frame of relativistic grain as ultraviolet and X-ray radiation, produces δ- electrons which can leave the grain and increase continuously its positive charge. The intensity of solar radiation is inverse proportional to square of the distance from the Sun and the possibility of disruption is determined by the closest distance to the Sun that the grain will reach. We assume this distance to be the astronomical unit. The maximum value of the electric field strength at the surface of the grain where the disruption starts is 10^{10} V/m for crystal hard grain and is 10^8 V/m for not rigid dust grain. The electric field strength and the grain charge increase continuously when the grain approaches the vicinity of the Earth. Since the grain becomes positively charged it automatically creates a potential barrier for δ-electrons to escape and for further increase of the positive charge. The corresponding critical grain energy ε_{cr} where this potential barrier cannot prevent the further grain charging is $\varepsilon_{cr}/m_e c^2 \approx (q/q_{max})10^2 a$ (cm) where the q_{max} is the maximum grain charge where the disruption starts. Another restriction is related with energy losses of δ electrons in the grain and gives the second inequality $a < (q/q_{cr}^2)\,\mu$m. For the case the both restriction are satisfied we get the equation for γ_{cr} such that the grains reaching the vicinity of the Earth from the side opposite to the Sun with $\gamma < \gamma_{cr}$ will be not destroyed by solar radiation

$$40 E_{cr} \gamma_{cr} a(\mu m) \exp\left(-\frac{10^5 (a(\mu m))^{1/4}}{\gamma_{cr}^{3/4}}\right) = 1 \qquad (2)$$

where the constant E_{cr} is determined by the value of the critical electric field and is

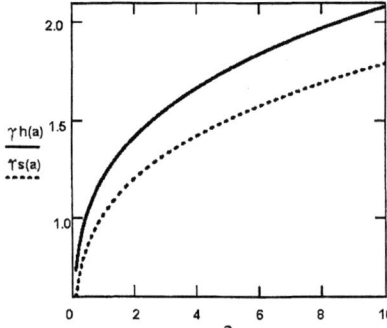

FIGURE 1. Dependence of $\gamma = \gamma_{cr} 10^{-5}$ on the dust size a in μm, the solid line γh correspond to the hard crystalline dust, the dotted line γs correspond to not rigid grains

normalized in the way to be equal to 1 for hard crystalline grain and to be 100 for not rigid grains. Figure 1 presents the numerical results of solution of (2) for γ_{cr} as function of a:

These results show that one can expect that in vicinity of the Earth there can be present a much more energetic relativistic dust grains than that expected from the previous investigation [2] and can produce the cosmic ray showers. The non-relativistic fast grain are not destroyed by solar radiation and can occupy the near Earth space. Of special interest is to distinguish the disruptions of grains with $\gamma > \gamma_{cr}$ in the near-solar space and the observed γ ray bursts and X-ray bursts of radiation which are believed to be produced by remote cosmic explosions.

REFERENCES

1. Istomin, Ya., and Barabach,S., Proceedings workshop *"Neutral cosmic Atoms"*, Kiruna, Sweden, 1997.
2. Berezinsky, V.S., and Prilutsky, O.F., *Astroph.and Space Sci.*, **21**, 475 (1973).
3. Spitzer L., *Phys. Rev.*, **76**, 583 (1949).

Instability Caused by Dust Drift and the Observed Polar Mesospheric Summer Echoes (PMSE's)

V.N.Tsytovich* and O.Havnes[†]

General Physics Institute, Moscow, 119991, Russia
[†] *University of Tromso, Tromso, Norway*

Abstract. Observations show that the Polar Mesopause Summer Echoes (PMSE's) radar scattering layers near the Earth mesopause are strongly linked to the dust layer in the same hight region [1]. No single mechanism has been found which is capable of explaining the PMSE over large frequency range and different atmospheric conditions at which it has been observed. One possible explanation in the high frequency range around 1 GH is the non-coherent scattering on the non-linear plasma screening clouds around dust particles [2]. This requires that $a < \lambda_D$ (a is the size of the grains, λ_D is the Debye radius) and that $P = n_d Z_d / n_i$ is close to 1. New simultaneous and common volume observations by radars at 50 MHz and by rockets show that PMSE can also occur for $P \ll 1$ and that the reflected signal rise rapidly with dust charge density as P increases in the range below about 0.1. This may be an indication that some instability caused by dust is developing. In the regions of PMSE's the ion-dust mean free path $\lambda_{id} = \lambda_D^2 / aP$ is 6 orders of magnitude larger than the ion-neutral mean free path λ_{in} and most instabilities are damped by high rate of ion-neutral collisions. We investigate the role of gravity forced dust drift in typical conditions ($n_n \approx 10^{14} \text{cm}^{-3}$, $T \approx 100°$ K, $a \approx 0.01 \mu$m with the falling velocity ≈ 20 cm/s thermal dust velocity of the order of 50 cm/s). The gravity balance, the condition of charge neutrality and the balance of ionization and absorption on dust is taken into account. In the dispersion equation we include account the changes in dust charges, the ion-neutral collisions, the ion and electron pressures, the dust-ion drag and the dust inertia. We find an instability of dissipative type caused by the dust drift, dust charge variations and dust drag. The instability develops in broad range of wave lengths with threshold for typical parameters being estimated as $\lambda > 10 \text{cm}(1/\sqrt{P})$. For $P = 0.01$ (approximately the lowest value observed) this corresponds to 1 m which is much shorter than the wavelength (6 m) at which the observations of PMSE for the low P-value were done. The growth rate is estimated to be developing in ≈ 200 s during which the dust falls $50m$ while the thickness of PMSE layer is from several hundred meters to several kilometers. The instability operates till $\lambda_{max} \approx 30$ m.

EQUILIBRIUM CONDITIONS IN THE DUST LAYER

We will use the dimensionless units $n = n_i/n_{i,0}$, $n_e = n_e/n_{i,0}$; $P = n_d Z_d/n_{i,0}$, $z = Z_d e^2/aT$; $\mathbf{u} = \mathbf{v}_i/\sqrt{2} v_T$; $\mathbf{v} = \mathbf{v}_d/\sqrt{2} v_*$; $v_* = (3\sqrt{pi}/4\sqrt{2})(n_{0,i}/n_n) v_T$; $v_T = \sqrt{T/m_i}$, $T_e = T_i = T_n = T$; $m_i = m_n$; $v_{d,0} = (v_T/v_*)m(m_i g \lambda_{id}/T)$; $m = (m_d/m_i)(v_*/v_T)(e^2/aT)$; $\mathbf{r} \rightarrow \mathbf{r}/\lambda_{id}$; $t \rightarrow t\sqrt{2} v_T/\lambda_{id}$; $\lambda_{id} = \lambda_D^2/aP$ where a is the dust size, λ_{di} is the ion-dust mean free path, g is the gravitational constant, $v_{d,0}$ is the dimensionless effective dust falling velocity and m is the effective dimensionless dust mass. The rate of ionization will characterized by coefficient α_i which gives the number of dimensionless n created in the unit of dimensionless time. The rate of absorption on dust particles in this units is $\alpha_c P n$ where the coefficient of ion attachment to dust grain in Orbit Motion Limited Approach is $\alpha_c = (1/2\sqrt{\pi})(1+z)/z$.

The equilibrium state in the dust layer is formed for dust grains by balance of the gravity force and friction in neutral gas $\mathbf{v}_d = v_{d,0}\mathbf{z}$ where $\mathbf{z}$ is the unit vertical vector, for other parameters it is formed by the charge balance $n_{e,0} = 1 - P_0$ ($n_i = n_{i,0}, n = 1$) and by the balance of ionization and absorption on dust $\alpha_i = \alpha_c P_0$, we use the parameter P_0 as the main parameter characterizing the state where the balance is established. The charging equation is sufficient to find the equilibrium dust charge as function of P_0. The equilibrium value of dust drift was estimated in the abstract. For the conditions in upper ionosphere $v_{d,*} \ll v_T, m \ll 1$.

DISPERSION RELATION FOR PERTURBATIONS AND ITS SOLUTIONS

The dimensionless frequency and the dimensionless wave vectors of the perturbations of the equilibrium state will be introduced according to the above given normalization of the space coordinates and time. In the perturbation of the balance equations for dust we take into account the change in gravity force due to change of dust velocity, the dust inertia, the electric fields appearing in perturbations and the drag force from ions (the drag coefficient is chosen as $2/3\sqrt{\pi}\ln\Lambda$ where $\ln\Lambda$ is the Coulomb logarithm and takes into account only the multiple Coulomb scattering on ions on dust charges). In the perturbations of the electron motion we take into account only the electron pressure variations and electric field perturbations. In perturbation of ion motion we take into account the friction related with dust drag, ion inertia, ion pressure and collisions with neutrals. In the continuity equation we take into account the changes of absorption on dust due to density of ion variations, variations of dust charges and diffusion in neutral gas. The variation of dust charges is found from charging equation and is expressed through the ion and electron density variations. The ion-neutral collisions are characterized by additional dimensionless parameter $s = (\sigma_{in}/\pi a^2)(16\sqrt{2}/3)(aT/e^2)$ where σ_{in} is the cross-section of ion-neutral collisions. The simplifications of the dispersion equation is obtained by using small parameters $1/k^2 \ll 1; v_{d,*}/v_T \ll 1, 1 \ll \omega \ll 1/m$ when the dispersion relation takes the form

$$-z_0(s\omega + i\kappa^2) = (\omega - \omega_0)\left[\left(\frac{1}{P_0} + \frac{z_0+1}{z_0(z_0+2)}\right) + \frac{s\omega + i\kappa^2}{i\kappa^2}\left(\frac{1-P_0}{P_0} + \frac{z_0+1}{z_0(z_0+2)}\right)\right] \quad (1)$$

where

$$\kappa = k^2\frac{v_*}{v_T}; \quad \omega_0 = (\mathbf{k}\cdot\mathbf{v}_0)\sqrt{\frac{v_*}{v_T}} \quad (2)$$

The real part of this relation gives the value of the real part of the frequency ω_R:

$$-z_0 s\omega_R = (\omega_R - \omega_0)g \quad g = \frac{2-P_0}{P_0} + \frac{2(z_0+1)}{z_0(z_0+2)} \quad (3)$$

where
$$\omega_R = \omega_0 \frac{g}{g+sz_0} \quad \omega_R - \omega_0 = \frac{sz_0}{g+sz_0}\omega_0 \tag{4}$$

For $s \ll 1$ the difference between the real part of the frequency ω and ω_0 is small. We then find the imaginary part of the frequency putting

$$\omega = \omega_R + i\Gamma \tag{5}$$

for $\Gamma > 0$ we have an instability and for $\Gamma < 0$ we have the damping of the perturbations. The imaginary part of (2) gives

$$-\frac{s\omega_R(\omega_R-\omega_0)}{\kappa^2}\left(\frac{1-P_0}{P_0}+\frac{z_0+1}{z_0(z_0+2)}\right)+\Gamma\left(g+\frac{s\Gamma}{\kappa^2}\left(\frac{1-P_0}{P_0}+\frac{z_0+1}{z_0(z_0+2)}\right)\right)$$
$$= -z_0(s\Gamma+\kappa^2) \tag{6}$$

To go further we can estimate the parameter s defined above, namely $\sigma_{in} \simeq 10^{-14}$ cm^2 and $\pi a^2 \simeq 3 \cdot 10^{-12}$ and $aT/e^2 \simeq 1/25$, thus $s \simeq 3 \cdot 10^{-3}$. We consider therefore the case

$$s\Gamma \ll \kappa^2 \tag{7}$$

$$\Gamma = -z_0\kappa^2 + \frac{s^2 z_0 \omega_0^2}{g^2\kappa^2}\left(\frac{1-P_0}{P_0}+\frac{z_0+1}{z_0(z_0+2)}\right) \tag{8}$$

We can make some estimates of the growth rate dependencies. Since $\omega_0 \sim \kappa$ we find that the second term is independent on the wavelength. For instability to exist we need

$$\kappa^2 = k^2 \frac{v_*}{v_T} \ll s^2 v_0^2 \frac{v_*}{v_T}\frac{1}{g^2}\left(\frac{1-P_0}{P_0}+\frac{z_0+1}{z_0(z_0+2)}\right) \tag{9}$$

We used here the estimate

$$\omega_0^2 \simeq \kappa^2 v_0^2 \frac{v_*}{v_T} \tag{10}$$

For $P_0 \ll 1$, $g \sim 1/P_0$ and the right-hand side of (8) is proportional to P_0. Thus, with a decrease of the parameter P_0 the disturbances excited should have larger wavelength. But for P_0 of the order of 1 the factor $(1/g^2)((1-P_0)/P_0 + (z_0+1)/(z_0(z_0+2)))$ is of the order of 1. An estimate of v_0 depends much on the size of dust particles. For $m_d \simeq 4 \cdot 10^{-18}$g which corresponds to dust material density of the order of 1 g/cm^3 and corresponds to the size $a \sim 10^{-6}$ cm^{-3} for $n_{i,0} \sim 10^3$ cm^{-3} we find for the value of the dimensionless dust drift velocity $v_0 \simeq 3 \cdot 10^7$ (the dimensional drift velocity is then ~ 20 cm/sec). Thus $v_0 s \simeq 10^5$ and the threshold condition (8) gives approximately $k \ll 10^5$. Having in mind that k is measured in the ion dust mean free path $\lambda_{id} \sim 10^6$ cm we find the threshold for the critical wavelength λ_{cr}

$$\lambda_{cr} > 10 \text{ cm} \cdot \frac{1}{\sqrt{P_0}} \tag{11}$$

which corresponds to the sizes which will affect the scattering. The starting condition $k \gg 1$ means $\lambda \ll 10^6$ cm. Thus, in the conditions (7) the growth rate exists in the broad range of wavelength with an estimate $\Gamma \sim s^2 v_0^2 \frac{v_*}{v_T}$. Since $v_*/v_T \simeq 10^{-11}$ and $v_0^2 \sim 10^{15}$, $s^2 = 10^{-5}$ we have $\Gamma \sim 1/10$. The units of Γ are $\omega_{pi} = 2 \cdot 10^4$ sec^{-1}; $\omega_d \simeq 4 \cdot 10^{-2}$ sec^{-1}. Thus the development of instability needs 200 s. The dust is falling during the instability development approximately on 50 m. The threshold wavelength existed for $P_0 \ll 1$ is estimated as following: for $P_0 \simeq 0.01$ still $\lambda_{cr} > 1$ m. For dimensionless mass we have an estimate $m = (m_d/m_i)(v_*/v_T)(e^2/aT) \simeq 10^{-3}$, $1/m \sim 10^3$ and the condition (10) $\Gamma \ll 1/m$ is well satisfied. It means that dust inertia is unimportant. The left-hand side of the inequality (10) means $\omega_0 \gg 1$ or $\lambda_{di} \frac{v_*}{v_T} v_0 \gg \lambda$. Since $\lambda_{Di} = 10^6$ cm $\cdot (v_*/v_T) \sim 10^{-11}$, $v_0 \sim 3 \cdot 10^7$ it means $\lambda \ll 10^3$ cm ~ 10 m $\sim \lambda_{max}$. The heavier the particles the larger is λ_{max} since $v_0 \sim a^2$; $\lambda_{di} \sim 1/a$, thus $\lambda_{max} \sim a$.

In the limit $\omega \ll 1 \ll 1/m$ the dispersion relation has the form

$$-i\kappa^2 z_0 = (\omega - \omega_0) \left[\frac{\left(\frac{1}{P_0} + \frac{1}{z_0}\frac{(z_0+1)}{(z_0+2)}\right)\left(\kappa^2 - \frac{(z_0+1)s}{2\sqrt{\pi}z_0}\left(1 - P_0 + \frac{1}{z_0(z_0+2)}\right)\right)}{\kappa^2 + \frac{(z_0+1)s}{2\sqrt{\pi}z_0}\left(1 + P_0 + \frac{1}{z_0(z_0+2)}\right)} + \left(\frac{1-P_0}{P_0} + \frac{(z_0+1)}{z_0(z_0+2)}\right) \right] \quad (12)$$

The analysis of (12) gives an estimate of the imaginary part of the frequency $\Gamma \sim -\kappa^2 P_0$ which means that all perturbations are damped.

REFERENCES

1. Aslaksen, T., and Bratti, A., *Physica Scripta*, **T89**, 133–137 (2001).
2. Havnes, O., de Angelis, U., Bingham, R., Goertz, C.K., Morfill, G.E., and Tsytovich, V.N., *Journ. Atmosph. Terrestr. Physics*, **52**, 637–643 (1990).

Strong Damping and Universal Instability of Dust-Ion Sound Waves(DISW) and Dust-Acoustic Waves(DAW)

V.N.Tsytovich* and K.Watanabe[†]

*General Physics Institute, Moscow, 119991, Russia
[†]Institute for Fusion Science, Toki, Gifu, 509-5292, Japan

Abstract. In laboratory experiments the balance between the plasma absorption on dust and plasma ionization is established on time scales much shorted than the period of DISW and DAW. The dispersion, damping and instability of these waves is therefore determined not only by deviations from quasi-neutrality as treated before [1] but mainly by deviations from the ionization and absorption balance in the basic balanced state. The proper formulation of the ground state is necessary for formulation of any dispersion relation for waves. In the present consideration the ionization source which compensate the absorption of plasma particles on dust in the ground state is assumed to be proportional to the electron density (as it is in most existing experiments). We found at the first time the correct dispersion of DAW which is in agreement with experiments [2]. The largest change of DISW and DAW is the appearance of rigid restriction in wave numbers for waves to exist ($k > k_{cr,d}$), a cardinal change of dispersion, appearance of strong collective damping and appearance of new collective modes and universal instabilities. The restrictions show that in present plasma crystal experiments the DISW and DAW cannot exist and that they should exist in other experiments [3]. For $k_{cr,i} < k < k_{cr,d}, k_{cr,i} \ll k_{cr,d}$ the DISW are strongly damped and the dispersion of DAW strongly deviates from the linear dispersion and the frequency reaches constant value. Both critical values $k_{cr,d}$ and $k_{cr,i}$ can be reached in the experiments. For $k < k_{cr,i}$ universal instability in absence of any gradients or flows is found both for DISW and DAW. The growth rate of instability is sufficient to be detected experimentally. The instability can be related with formation of dust structures or with a start of the phase transition to the plasma crystal state. The threshold of instability in the range of frequencies and wave numbers corresponding to DAW corresponds to dust temperatures about $T_{d,cr} = T_i(Z_d T_e/T_i)^{4/3}(n_i/Z_d n_d)^{5/3}(m_i/m_d)^{2/3}$, close to that observed for plasma crystal condensation.

INTRODUCTION. THE BASIC STATE

To define the basic state we use the dimensionless notations

$$n = \frac{n_i}{n_0}; n_e = \frac{n_e}{n_0}; P = \frac{n_d Z_d}{n_0}; \tau = \frac{T_i}{T_e} \ll 1; z = \frac{Z_d e^2}{a T_e}; Q = \frac{dn_i}{dt}\frac{\sqrt{T_i m_i}}{4\sqrt{2}\pi n_0^2 e^2 a}; \quad (1)$$

where tn_0 is the ion density in the basic state, T_i, T_e, T_d are the ion, electron and dust temperatures respectively, assumed to be constant, n_i, n_i, n_d are the ion, electron and dust densities respectively, Z_d is the dust charge in units of electron charge, m_i is the ion mass and a is the dust size. By α_{ch} we denote the charging coefficient and by α_{dr} the drag coefficient. They have a particular simple expressions $\alpha_{ch} = 1/2\sqrt{\pi}$, $\alpha_{dr} = (2/3\sqrt{\pi})\ln\Lambda$

(where $\ln\Lambda$ is the Coulomb logarithm) for Orbit Motion Limited Approach model and multiple Coulomb scattering models respectively. These coefficients are left arbitrary not to be restricted by any models. In these dimensionless units the ion absorption on dust is $Q_{abs} = \alpha_{ch}Pn$ and the ionization production of ions is assumed to be proportional to electron density $Q_{ion} = \alpha_{ion}n_e$. Contrary to standard definition of the basic state using only the quasi-neutrality condition we use also the condition of the balance of absorption and ionization (the value of the basic state are denoted here by a subscript $_0$): $n_{e0} = 1 - P_0$; $Q_{abs,0} = Q_{i,0}$ or $\alpha_i = P_0\alpha_{ch}/(1-P_0)$; the charging equation allows to find the dust charge z_0 of the ground state as function of P_0 (z_0 depends also on two other fixed parameters τ and m_e/m_i). The ground state depends on single parameter P_0. The deviations of the ground state are related both with violation of quasi-neutrality and with violation of the balance of ionization and absorption. The waves and other perturbations can be considered as the one in homogeneous infinite media if the characteristic size of experimental device is larger than the Debye length (for deviations from quasi-neutrality) and if the characteristic size is larger than the ion absorption length (for deviations of absorption and ionization). Both conditions are well satisfied in existing experiments.

DISPERSION RELATION FOR DISW

The perturbations of the basic state we consider in hydrodynamic approach assuming that in the electron force balance equation only the electric field and electron pressure perturbations are important, that in ion force balance equation the electric field, ion inertia, ion pressure and friction on dust due to the dust drag by ions are important. For DISW we neglect the dust motion and assume that that the perturbations have the phase velocities larger than both ion thermal and dust thermal velocities. In the continuity equation we take into account both the change of ionization due to electron density perturbation and the change of absorption due to the ion and the dust charge perturbations. Since the dust is considered unmovable $\delta P/P_0 = \delta z/z_0$ and the value of the dust charge variation δz is found from the charging equation:

$$\frac{\delta z}{z_0} = \frac{\alpha_{ch}}{(1+z_0)\alpha_{ch} - i\omega}\left(\frac{\delta n_e}{n_{e0}} - \delta n\right) \qquad (2)$$

In the Poisson equation which gives the dielectric permittivity $\varepsilon_{\mathbf{k},\omega}$ we take also into account the dust charge variations in the perturbations:

$$\varepsilon_{\mathbf{k},\omega} = 1 + \frac{\tau}{k^2\lambda_{Di}^2}\left\{\left(1 - P_0 + \frac{\alpha_{ch}P_0}{(1+z_0)\alpha_{ch} - i\omega}\right) + \left(1 + \frac{\alpha_{ch}P_0}{(1+z_0)\alpha_{ch} - i\omega}\right)\frac{\left[1 - \frac{\alpha_{ch}^2 P_0 z_0(\alpha_{dr}P_0 z_0 - 2i\tau\omega)a^2}{((1+z_0)\alpha_{ch} - i\omega)k^2\lambda_{Di}^2}\right]}{\left[\tau + \frac{a^2}{k^2\lambda_{DDi}^2}\left(\frac{P_0 z_0 \alpha_{ch}^2}{(1+z_0)\alpha_{ch} - i\omega} - i\omega\right)(\alpha_{dr}P_0 z_0 - 2i\tau\omega)\right]}\right\} \qquad (3)$$

where ω is the dimensionless frequency in units $\omega_{pi}a/\sqrt{2}\lambda_{Di}$ and ω_{pi} is the ion plasma frequency. According to (3) the DISW exist only in the limit $\omega \gg \alpha_{dr}P_0 z_0/2\tau$ which means physically that the frequency should be much larger than the ion-dust absorption frequency or in other notations that

$$k\lambda_{Di} \ll k_{cr,d}\lambda_{Di} = \sqrt{\frac{2}{\pi}} \ln \Lambda P_0 z_0 \sqrt{(1-P_0)} \frac{a}{\lambda_{Di}\sqrt{\tau}} \quad (4)$$

which means also that for $\tau \ll 1$ the DISW exist only for frequencies larger than the charging frequency. In this limit their phase velocity is $\sqrt{T_e/m_i(1-P_0)}$. When k approaches $k_{cr,d}$ the damping of the waves become strong $\gamma_{dam} \approx -P_0 \omega_{pi} \alpha_{dr} z_0/2\sqrt{2\tau}\lambda_{Di}$. For $k < k_{cr,i} \ll k_{cr,d}$ the perturbations became universally unstable

$$k_{cr,i}^2 \lambda_{Di}^2 = \frac{a^2}{\lambda_{Di}^2} \frac{\alpha_{dr}\alpha_{ch} P_0^3}{1+z_0+P_0} \quad (5)$$

The maximum growth rate corresponds to the k values less but of the order of $k_{cr,i}$ and the growth rate is of the order of that found for structurization instability [4].

DISPERSION RELATION FOR DAW

We consider the range of phase velocities larger than the dust thermal velocity but less than both the ion and electron thermal velocities, take into acount the dust charge variations and dust density variations and consider the frequency to be much less than the charging frequency and the frequency of ion absorption on dust grains. Then the perturbations of the ground state defined above lead to the following expression for the dielectric permittivity

$$\varepsilon_{\mathbf{k},\omega} = 1 + \frac{\tau}{\lambda_{Di}^2 k^2} \frac{1}{\left(\tau + \frac{\alpha_{dr}\alpha_{ch}P_0^2 z_0^2 a^2}{k^2\lambda_{Di}^2(1+z_0)}\right)} \left\{ 1 + \frac{P_0}{1+z_0} + \tau\left(1 - \frac{P_0 z_0}{1+z_0}\right) - \right.$$

$$\left. \frac{\alpha_{ch}\alpha_{dr}P_0^3 z_0^2 a^2}{k^2\lambda_{Di}^2(1+z_0)} + \frac{P_0\left(\tau + \frac{\alpha_{ch}\alpha_{dr}z_0^2 P_0(P_0-\tau)a^2}{k^2\lambda_{Di}^2(1+z_0)}\right)\left(\tau + \frac{\alpha_{dr}\alpha_{ch}P_0 z_0(1+P_0)a^2}{k^2\lambda_{Di}^2}\right)}{\left[\left(\tau_d - \omega^2 \frac{\tau a^2}{\mu k^2 \lambda_{Di}^4}\right)\left(\tau + \frac{\alpha_{dr}\alpha_{ch}P_0^2 z_0^2 a^2}{k^2\lambda_{Di}^2(1+z_0)}\right) - \frac{P_0 z_0 \alpha_{ch}\alpha_{dr}\tau a^2}{k^2\lambda_{Di}^4}\right]} \right\} \quad (6)$$

In conditions where the ion-neutral collisions dominate in the equation for the ion motion the friction of ions on dust should be substituted for the drag coefficient. The neutral dust collisions change the dielectric permittivity in a way that one should substitute $\omega(\omega + iv_{di})$ for ω^2 in the expression (6). The DAW according to (6) exist only if the inequality (4) is satisfied which requires $a^2 \lambda_{Di}^2 \ll \tau$. The latter relation is not satisfied in most experiments on plasma crystals and is satisfied for experiments [3], where the DAW were observed. The dispersion relation for DAW found from (6) in the range of their existence is

$$\omega^2 \approx \omega_{daw}^2 \left(1 + \frac{\alpha_{dr}\alpha_{ch}P_0 z_0(1+P_0)a^2}{k^2\lambda_{Di}^4 \tau} - k^2\lambda_{Di}^2\right); \omega_{daw}^2 = \frac{P_0 Z_d T_i}{m_d} \frac{1}{1+\frac{P_0}{1+z_0}} \quad (7)$$

The factor $1/(1+P_0/(1+z_0))$ in the frequency ω_{daw}^2 is related with dust charge variation which changes of effective Debye screening, and the first dispersion term in (7) is related with collective effects in perturbations of the ground state. The curvature of the dispersion curve introduced by the first correction of (7) is opposite to the curvature of the curve introduced by the second correction describing the deviations from quasi-neutrality. In the present experiments one can easily distinguish the two deviations from the linear law and that already detected in [3] corresponds to the first type of corrections. This is an indications that the effect described here is already observed and can be investigated experimentally in more details. The condition that the dispersion corrections due to the perturbations of the basic state dominate is $k^2 \lambda_{Di} \gg \frac{a}{\lambda_{Di}\sqrt{\tau}}$. In the range $k_{cr,d} \gg k \geq k_{cr,i}$ the wave are converted into another mode with constant frequency $\omega_{mathrmcoll}$

$$\omega_{mathrmcoll}^2 = \omega_{pd}^2 \frac{a^2}{\tau \lambda_{Di}^2} \frac{\alpha_{ch}\alpha_{dr}z_0 P_0(1+P_0)(1+z_0)}{1+P_0+z_0} \tag{8}$$

with dumping $\gamma/\omega \approx P_0 a \omega_{pd}/\lambda_{Di}\sqrt{\tau}\omega_{pi}$. The whole range $0 < k < k_{cr,i} - \Delta k_{cr,i}; \Delta k_{cr,i} \ll k_{cr,i}$ corresponds to unstable perturbations but the most unstable are those with k close to $k_{cr,i}$ (almost single periodic mode excitation)

$$\frac{\gamma_{max}}{\omega \frac{a}{\sqrt{2}\lambda_{Di}}} = \frac{\sqrt{3}}{2}\left(\frac{Z_d m_i}{2m_d \tau}\alpha_{ch}\alpha_{dr}P_0^2(1+P_0)(1+z_0)\right)^{1/3} \tag{9}$$

$$\frac{\Delta k_{cr,i}}{k_{cr,i}} \approx \frac{\sqrt{3}}{4} \frac{\alpha_{ch}\alpha_{dr}P_0^3 z_0^2}{(1+z_0+P_0)^2} (\alpha_{ch}\alpha_{dr}P_0^2(1+P_0)(1+z_0))^{1/3}\left(\frac{Z_d m_i}{2m_d \tau}\right)^{1/3} \tag{10}$$

This mode can turn the system to a crystal state. The stabilization of this instability can be shown to be only related with the dust pressure effect. The criteria of stabilization is

$$\frac{T_d}{T_e Z_d} > \frac{2(1+z_0+P_0)}{\sqrt{3}\alpha_{ch}\alpha_{dr}P_0^3 z_0^2}(\alpha_{ch}\alpha_{dr}P_0^2(1+P_0)(1+z_0))^{2/3}\left(\frac{Z_d m_i}{m_d \tau}\right)^{2/3} \tag{11}$$

For the charging, the drag coefficients and z_0 of the order of 1 the dependence of the critical temperature on the parameters of the system is

$$T_{d,cr} \approx T_i \frac{Z_d^{4/3}}{\tau^{4/3} P_0^{5/3}}\left(\frac{m_i}{m_d}\right)^{2/3} \tag{12}$$

By using the data which are usually found in the present experiments [5] $\tau = 0.02$, $Z_d = 10^3, P_0 \approx 1; m_d/m_i \approx 10^{10}$ we get from (12) $T_d > (1.5) \times T_i$. The obtained criterion is close to that observed in experiments for phase transition to the plasma dust crystal state. The instability considered here can serve as the most probable candidate for a start of formation of the plasma crystal state.

REFERENCES

1. Shukla, P.K., and Silin V.P., *Physics Scripta*, **45**, 508 (1992).
2. Tsytovich, V.N., and Watanabe K., *Contr. Plasma Phys.*, (accepted), preprint NIFS-720 (2002).
3. D'Angelo, N., *Physics of Plasmas*, **4**, 3422 (1997).
4. Morfill, G.E., and Tsytovich, V.N., *Plasma Physics Reports*, **26**, 727–736 (2000).
5. Thomas, H.M., and Morfill, G.E., *Nature*, **379**, 806 (1996).

Theory of Small Atomic-Like 2D Dust Clusters

Sh. G. Amiranashvili*, N.G. Gousein-zade* and V.N.Tsytovich*

*General Physics Institute, Moscow, 119991, Russia.

Abstract. In several experiments atom-like dust clusters with parabolic confining potential were observed [1-3]. Here we present a general theory of 2D clusters confined by $(1/2)m_d\omega_0^2 r^2$ potential with *arbitrary pair interaction potential* depending on the inter-dust distance. It describes the equilibrium conditions, normal modes, their frequencies and possible instabilities of clusters with arbitrary N number of grains. The mono-layer clusters can have $2N$ frequencies of oscillations in the cluster plane among which 3 modes are trivial ($\omega = 0$ and double degenerate frequency of oscillation in the potential well). The $2N - 3$ non-trivial modes are considered. For example, for square dust cluster with potential $V(r)$ the equilibrium is described by $\omega_0^2 = -(4/m)\left[V'(\sqrt{2}R) + V'(2R)\right]$, the frequency of radial oscillations is $\omega^2 = (16R^2/m)\left[V''(\sqrt{2}R) + 2V''(2R)\right]$, the two single modes frequencies are $\omega^2 = (32R^2/m)V''(2R); \omega^2 = (16R^2/m)V'''(\sqrt{2}r)$ and one double degenerated mode frequency is $\omega^2 = (1/m)\left[V'\sqrt{2}R) - V'(2R) + 4R^2 V''(\sqrt{2}R)\right]$ where ' corresponds to the differentiation of the potential $V(r)$ with respect to $\sqrt{r}$. The general stability criterion was found and investigated for $N \geq 4$. The pair interaction potential $V(r)$ is considered as a sum of different attraction and repulsion terms, including that which describe the non-screened collective and non collective attraction, the screened non-Coulomb interaction and the non-screened repulsion. The collective non-screened potential causes the absence of equilibria at certain dust cluster sizes. For screened Coulomb potential $V_c(r) = (Z_d^2 e^2 \alpha_{scr}/r)\exp(-r/\lambda_{scr})$ the clusters with the size R are considered. The pentagon cluster is found to be stable for $R < 3.3\lambda_{scr}$ and the clusters with $N \geq 6$ are found to be always unstable. The measurements of the frequencies of the cluster modes, the thresholds of cluster equilibria and the stability of the clusters can be used for detection of the dust-dust interactions and for comparison of the theory of 2D dust clusters and the observations.

THEORY OF CLUSTERS WITH ARBITRARY POTENTIAL INTERACTION

We consider 2D clusters of N grains confined by external parabolic potential $V_{conf} = m_d \omega_0^2 r^2/2$ and arbitrary pair dust potential of interaction, $V(r) = f(r^2)$, which defines the function $f(x)$, through which all results can be expressed. It is found that for equilibrium stationary polygon to exist the following restriction should be valid

$$\omega_0^2 = -\frac{4}{m_d}\sum_{k=1}^{N-1}\sin^2\left(\frac{\pi k}{N}\right)f'\left[4R^2\sin^2\left(\frac{\pi k}{N}\right)\right] > 0 \qquad (1)$$

where $f'(x)$ is the derivative of $f(x)$ with respect to its argument x. Equation (1) can be used to determine the equilibrium size R of the polygon in the given trapping field. For equilibrium to exist it is necessary that the right hand side of (1) be positive. The criterion of stability for small perturbations of the existing equilibrium is described by

$$(A_1 - A_{l+1} + B_{l+1})(A_1 - A_{l-1} + B_{l-1}) \geq (B_1 - B_l)^2 \quad (2)$$

where

$$A_l(N) = -\frac{4}{m_d} \sum_{k=1}^{N-1} \sin^2\left(\frac{\pi k l}{N}\right) f'\left[4R^2 \sin^2\left(\frac{\pi k}{N}\right)\right];$$

$$B_l(N) = \frac{4}{m_d} \sum_{k=1}^{N-1} \sin^2\left(\frac{\pi k l}{N}\right) 4R^2 \sin^2\left(\frac{\pi k}{N}\right) f''\left[4R^2 \sin^2\left(\frac{\pi k}{N}\right)\right] \quad (3)$$

For given N the criterion (2) should be met for all possible l ranging from 0 to $N/2$.

DUST-DUST BINARY INTERACTIONS

Different type of potentials can operate between dust grains in clusters: 1) the screened Coulomb potential which only for rather large distances can be approximated by Yukawa potential $\propto \alpha_{scr} exp(-r/\lambda_{scr})/r$ with a screening length λ_{scr} being different from Debye length (due to dust charge variations in screening and due to collective effects), 2) the repulsive long range non-screened potential $\propto 1/r^2$, 3) the shadow non-collective non-screened attraction potential $\propto -1/r$ 4) collective attraction potential $\propto (1/r)\cos(r/\lambda_{coll})$ where λ_{coll} is the collective length $\propto 1/\sqrt{n_d Z_d/n_i}$. The general expression of the potential which takes into account these interactions is

$$V(r) = \frac{Z_d^2 e^2}{r}\left[\alpha_{scr} \exp\left(-\frac{r}{\lambda_{scr}}\right) + \frac{\alpha_{fl}}{r} - \alpha_{ncoll} + \alpha_{coll} \cos\left(\frac{r}{\lambda_{coll}}\right)\right] \quad ((4)$$

The introduced by (4) coefficients vary in the ranges $1/\eta_{coll} < \alpha_{scr} < 1$; $\eta_{coll} \approx (a^2/\lambda_{Di}^2 \tau)$; $\tau = T_i/T_e$; $\sqrt{1/\eta_{coll}} < \lambda_{scr} < \sqrt{1/(1+P_0/(1+z_0))}$; $P_0 = n_d Z_d/n_i$; $z_0 = Z_d e^2/aT_e$; $\alpha_{fl} \approx a/2$; $\alpha_{ncoll} > a^2 \lambda_{Di}^2$; $\eta_{coll} < \alpha_{coll} < 1$; $\lambda_{Di}/\lambda_{Di}\sqrt{1/P_0 \tau \eta_{coll}} > \lambda_{coll} > \lambda_{Di}/\sqrt{P_0 \tau}$ where a is the dust size and λ_{Di} is the ion Debye screening length. Mention that in existing dust plasma crystal experiments the parameter η_{coll} is not small ranging from 2 to 4. These coefficients determine the possibility of forming equilibrium clusters, frequencies of normal modes and the stability thresholds. From these thresholds one will be able to determine experimentally the strengths of different listed binary interaction which were proposed so far [4-12].

NUMERICAL RESULTS FOR CLUSTER EQUILIBRIA

For the simplest screened Yukawa potential $V \propto (1/r)\exp(-r/\lambda_{scr}$ and for the simplest case of pure collective interaction pair potential $V \propto (1/r)\cos(r/\lambda_{coll})$. For the first case the potential depends on the effective strength coefficient α_{scr} and on the effective screening length λ_{scr} (which usually for large dust charges in the present experiments differs from standard expressions $\alpha_{scr} = 1$ and $\lambda_{scr} = \lambda_{Di}$ due to non-linear effects, charging and collective effects). In this case we normalize the **square of frequency** ω_0^2 on $Z_d^2 e^2 \alpha_{scr}/m_d R^3$ and denote this value as $YN(R)$ where R is the distance normalized on λ_{scr}. For the case of collective interaction potential we normalize the **square of the frequency** ω_0^2 on $Z_d^2 e^2 \alpha_{coll}/m_d R^3$ and denote this value as $CN(R)$ where R is then the distance normalized on λ_{coll}; in these notations N stays for the number of grains in the clusters. Numerical results for dependence of YN on its dimensionless R for Yukawa type of potential are shown on Fig.1a, while for the collective potential the numerical results the dependence of CN on its R is shown on Fig.1b. On can see that in the second case the square of the frequency becomes negative at certain sizes of the clusters which means the absence od equilibria. Also it should be noticed (see Fig.1a) that at large distances the equilibria can be again established. The Yukawa clusters can have equilibria at all distances (Fig.1b). If one adds to the Yukawa interaction the non-collective attraction

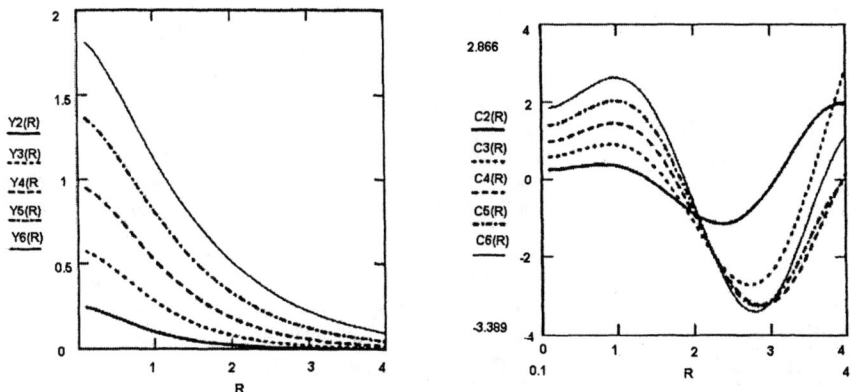

FIGURE 1. Fig.1a — dependence of the square of normalized equilibrium frequency YN on its normalized cluster size R for Yukawa interaction potential $V \propto (1/r)exp(-r/\lambda_{scr})$; Fig.1b — dependence of the square of the normalized equilibrium frequency CN on its normalized cluster size R for collective dust interaction potential $V \propto (1/r)cos(r/\lambda_{coll})$.

then its influence can be to destabilize the equilibria. For example for triangle and square clusters

$$Y3(R) = \frac{1}{\sqrt{3}}\left((1+\sqrt{3}R)\exp(-\sqrt{3}R) - \frac{\alpha_{ncoll}}{\alpha_{scr}}\right) \quad (5)$$

$$Y4(R) = \frac{1}{\sqrt{2}}(1+\sqrt{2}R)\exp(-\sqrt{2}R) + \frac{1}{4}(1+2R)\exp(-2R) - \frac{\alpha_{ncoll}}{\alpha_{scr}}\left(\frac{1}{\sqrt{2}}+\frac{1}{4}\right) \quad (6)$$

Here R is normalized on λ_{scr}.

By measuring the threshold of the equilibria one can make conclusions for the strength of non-collective dust-dust attraction. Spatial interest are the cases where $\omega_0^2 = 0$ or on Fig.1b the distances where $CN(R) = 0$ and for examples (5)(6) the distances at which $YN(R) = 0$. Those cluster do not need an external parabolic potential to exist and can be called as boundary free clusters. Their stability is of special interest.

NORMAL MODES AND STABILITY OF CLUSTERS

The frequency of the modes is determined by dispersion relation. Among $3N$ modes of the cluster tree are trivial (zero frequency and shifts of the whole cluster); left are $3N - 3$ non-trivial modes. The Fig. 2a shows the displacements in modes for simplest clusters. Some of the modes are degenerate, such as radially symmetric polygon oscillations. The

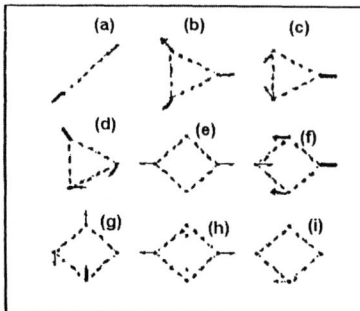
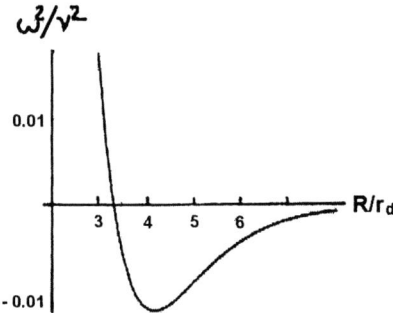

FIGURE 2. Fig.2a — Sets of normal displacements for simplest clusters, the arrows indicate the direction of the grain motion; Fig.2b — The squared frequency of pentagon oscillations against normalized distance $R = r/\lambda_{scr}$; the range of R where the square of the frequency is negative corresponds to instability, on Fig.2b $r_d = \lambda_{scr}$

stability of clusters was investigated numerically. For $N \leq 4$ the Yukawa clusters are stable. The pentagon Yukawa clusters are unstable for $R > 3.32$ (see Fig.2a) where the squared normalized frequency of pentagon oscillations is shown as function of R being negative for $R > 3.32$ (as before R is normalized on λ_{scr}).

REFERENCES

1. Goree, J., Samsonov, D, Ma, Z., Bhattacharjee, A., Thomas, H., Konopka, U. and Morfill, G.E., *Frontiers in Dusty plasmas*, Elsevier B.V. Proc. of Second Conf. on Dusty Plasma Physycs -ICPDP-99, 91 (2000).
2. Lin,I., Yin-Ju Lai, Wen-Tau Juan,and Ming-Heng Chen *Frontiers in Dusty plasmas*, Elsevier B.V. Proc. of Second Conf. on Dusty Plasma Physycs-ICPDP-99, 75 (2000).
3. Meltzer, A., Schweigert, V. A., and Piel, A., *Frontiers in Dusty plasmas*, Elsevier B.V. Proc. of Second Conf. on Dusty Plasma Physycs-ICPDP-99, 115 (2000).
4. Tsytovich, V. N. , Khodataev, Y. and Bingham R., *Comments on Plasma Physics and Controlled Fusion*, **17**, 249 (1996).
5. Ignatov, *Comments P.N.lebedev Inst.A*. **58**, 1 (1995).
6. Khodataev, Ya. K., Morfill, G., and Tsytovich, V. N., *J. Plasma Phys.*, **65**, 257 (2001).
7. Lampe, M., Gavrishchaka, V., Ganguli, B.N., and Joyce, G., *Phys. Rev. Lett.*, **86**, 5278 (2001).
8. Tsytovich, V.N., de Angelis, *Phys. Plasmas*, **7**, 554 (2000).
9. Tsytovich, V.N., and Morfill G., *Fhyz. Plasmy*, (in press) (2002).
10. Kompaneetz, R., and Tsytovich, V., *Fhyz. Plasmy*, (in press)(2002).
11. Tsytovich, V., *Contr. to Plasma Phys.*, (submitted) (2002).

Influence of Charge Variations on Dust Dynamics in a Planar rf Discharge

O.S. Vaulina*, A.A. Samarian, and B.W. James

School of Physics, University of Sydney, NSW 2006, Australia
*Institute of High Energy Densities, Russian Academy of Sciences, Moscow, Russia

Abstract. The influence of charge variations on the dynamics of mono layered structure levitating above the powered electrode in planar rf discharge have been considered. We analysed the random fluctuations of the particle charges, caused by the stochastic variations of charging current, and also the fluctuations of charges due to their random motion in the presence of charge gradient in the vertical direction. We found that the oscillation kinetic energy acquired is mostly due to the fluctuations of the charging currents.

In a complex plasma the dust particles achieve electrostatic equilibrium with respect to the plasma by acquiring negative charge. This charge is extremely large compared to the ionic charge. In addition, the particle charge is not fixed but is coupled self-consistently to the surrounding plasma parameters. There are two mechanisms that can lead to random fluctuations of particle charge relative to its equilibrium (time averaged) value [1]. The first is related to the random nature of ionic and electronic currents which charge the particles. The second is a result of random motion of the particles (for example due to their Brownian motion) in the spatially inhomogeneous plasma.

Here we consider the influence of charge variations on the dynamics of a mono-layered structure levitated above the powered electrode in a planar rf discharge. Vertical oscillations in such a structure were first reported in [2]. In the present experiments, the 6.28 μm diameter melamine formaldehyde particles have also been used. The dust particles suspended in the plasma are illuminated using a Helium-Neon laser. The laser beam enters the discharge chamber through a 40-mm diameter window. A window mounted on a side port in a perpendicular direction allows a view of the light scattered at 90° by the suspended dust particles and provides a vertical cross-section of dust structure. In addition, we use the top window to view the horizontal dust-structure. The laser beam can be expanded in the vertical and horizontal directions into sheets of light by a system of cylindrical lens. Images of the illuminated dust cloud are obtained using a charged-coupled device (CCD) camera with a 60 mm micro lens, and a digital camcorder (focal length: 5-50 mm). The camcorder was operated at 25 to 100 frames/s. The video signals are stored on a videotape recorder or are transferred to a computer via a frame-grabber card with an 8-bit gray scale and 640X480-pixel resolution. The coordinates of particles were measured in each frame, and individual particles were traced from one frame to the next.

The experiment used a one-layer structure. It was found that when the pressure was decreased bellow a critical value the dust particles began to oscillate spontaneously in the vertical direction. Fig. 1 shows typical time dependence of the vertical oscillation amplitude and the frequency spectrum of the oscillations. The amplitude and frequency of the oscillations are several millimetres and about ten Hz, respectively. When the rf input power was decreased the oscillation amplitude was found to increase. Fig. 2 shows the dependence of oscillation amplitude on gas pressure and rf power. For pressures below 35 mTorr the oscillation amplitude increased dramatically. This increase is greater for lower rf powers. For the 6.28 mm particles saturation and decrease of the oscillation amplitude is observed.

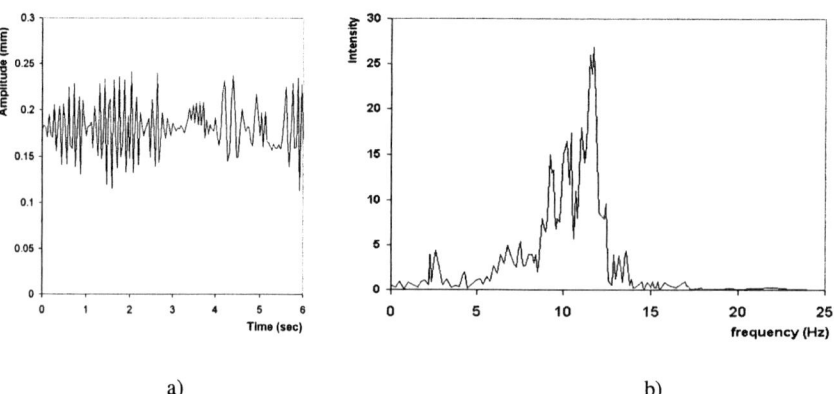

a) b)

FIGURE 1. Oscillation amplitude time dependence (a) and frequency spectrum (b).

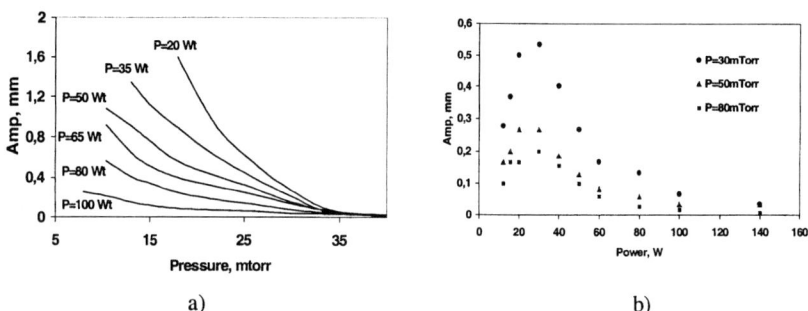

a) b)

FIGURE 2. Experimental chamber and image of test gains levitated above the electrode.

For explanation of the observed self-excited oscillations we have taken into account the random fluctuations of the particle charges caused by the stochastic variations of charging current, and also fluctuations of charges due to their random motion in the presence of gradient of particle charge in the vertical direction. In this 2-D case the dynamics of dust particles levitated in the plasma sheath are described by the following equations

$$m_p y'' = -m_p v_{fr} y' - (\alpha_y e \langle Z \rangle + \beta_y E_y) y + e \tilde{Z}_f E_y + F_y^{br}, \qquad (1a)$$

$$m_p r'' = -m_p v_{fr} r' - \alpha_r e \langle Z \rangle r + e(\tilde{Z}_f + \tilde{Z}_s) E_r + F_r^{br}, \qquad (1b)$$

where r'', y'', r', y' are derivatives with respect to time, and $\alpha_y = dE_y/dy$, $\alpha_r = dE_r/dr$ are spatial gradients of the electric field $\mathbf{E} = (E_y, E_r)$. The charge variations $\tilde{Z}_f$, $\tilde{Z}_s$ and their mean square displacements $\langle \tilde{Z}_f \rangle = \sqrt{\langle \tilde{Z}_f^2 \rangle_t}$, $\langle \tilde{Z}_s \rangle = \sqrt{\langle \tilde{Z}_s^2 \rangle_t}$, $(\langle \tilde{Z}^2 \rangle_t$ are given by

$$\frac{d\tilde{Z}_f}{dt} = -\eta \tilde{Z}_f + \tilde{F}_f, \quad \langle \tilde{Z}_f \rangle = \xi \sqrt{\langle Z \rangle}, \qquad (2a)$$

$$\tilde{Z}_s = \beta_y y, \quad \langle \tilde{Z}_s \rangle = \beta_y \sqrt{\langle y^2 \rangle}, \qquad (2b)$$

where η and ξ are parameters determined by the plasma condition. Assuming no correlation between $\tilde{F}_f, F_y^{br}, F_r^{br}$ the total kinetic energy of a dust particle can be defined as $2\langle K \rangle = 3T_o + \Delta^f T + \Delta^s T$, where T_o is gas temperature, and $\Delta^{f(s)} T$ depends on the nature of the charge fluctuation. By solving the equations (1)-(2a) we found an expression for the value of oscillation kinetic energy acquired due to fluctuations of the charging currents:

$$\Delta^f T_{y(r)} = \frac{e^2 Z \xi^2 E_{y(r)}^2}{m_p v_{fr}(v_{fr} + \eta)}, \qquad (3)$$

where the value of E_r is $\sim (eZ/l)^2$. Estimating $\Delta^f T$ for typical experimental conditions yields $\Delta^f T = 2\Delta^f T_r + \Delta^f T_y < 10 T_o$. This is less than experimentally observed values which are about 100eV.

The expression for the kinetic energy acquired due to random motion in the presence of a charge gradient can be obtain from equations (1)-(2b):

$$\Delta^s T_r = (T_o + \Delta^f T_y + \Delta^s T_y)\theta_1, \qquad (4)$$

where

$$\theta_1 \approx \left(\frac{\beta_y}{e\langle Z \rangle}\right)^2 \frac{2\alpha \omega_y^2 e^4 \langle Z \rangle^4}{m_p^2 g(\alpha_y/E_y + \beta_y/\langle Z \rangle)((\omega_r^2 - \omega_y^2)^2 + 2(\omega_r^2 - \omega_y^2)v_{fr}^2 + 4\omega_y^2 v_{fr}^2) l_p^4} \qquad (4a)$$

And $\omega_r^2 = e\langle Z \rangle \alpha_r/m_p \sim (e\langle Z \rangle)^2 n_p$, $\omega_y^2 = (\beta_y E_y + e\langle Z \rangle \alpha_y)/m_p$. For $(\omega_r^2 - \omega_y^2) \ll v_{fr}^2$:

$$\theta_1 \approx \left(\frac{\beta_y}{e\langle Z \rangle}\right)^2 \frac{\alpha e^4 \langle Z \rangle^4}{2m_p^2 g(\alpha_y/E_y + \beta_y/\langle Z \rangle)v_{fr}^2 l_p^4}. \qquad (4b)$$

Taking into account that $\Delta^f T_y \approx \Delta^f T/3$ $l_p^2 \cong l^2 + \langle y^2 \rangle$, where

$$\langle y^2 \rangle = \frac{(T_o + \Delta^f T_y + \Delta^s T_y)}{m_p \omega_y^2},$$

$\Delta^f T_y \approx \Delta^f T/3$, $\Delta^s T_y = \gamma \Delta^s T_r$, and γ is the parameter of kinetic energy redistribution, for $l^2 \gg \langle y^2 \rangle$ and $\gamma = 1$ ($\Delta^s T = \Delta^s T_y \equiv \Delta^s T_r$) we finally obtain:

$$\Delta^s T = \frac{(T_o + \Delta^f T/3)}{1 - \theta_1} \quad (5)$$

Estimates of the maximum value of kinetic energy and oscillation amplitude using a linear approximation in the case $l^2 \ll \langle y^2 \rangle$ give us

$$\Delta^s T^{max} = \theta_1 l^2 \omega_y^2 m_p, \qquad A_y^{max} \approx l\sqrt{2\sqrt{\theta_1}} \quad (6)$$

where $A_y^{max} = \sqrt{2\langle y^2 \rangle}$.

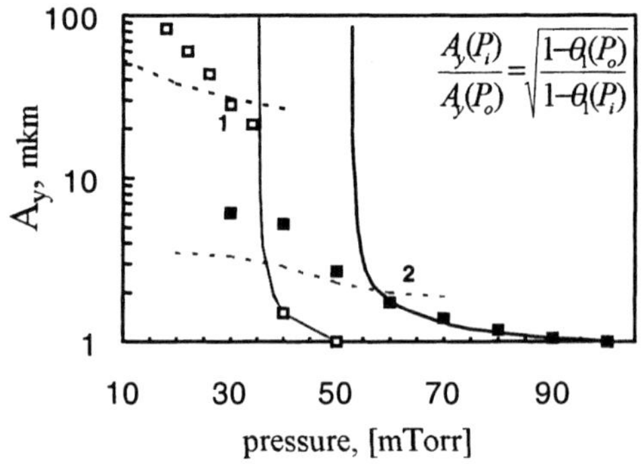

FIGURE 3. Normalised amplitude versus pressure. 1-for a=1.05 μm, 2- for a=3.07 μm. Solid lines are simulation results. Dotted lines $A_y \approx n_p^{-1/3}$. $A_y(P_i)$ is oscillation measured for different pressures P_i.

The dependencies of the ratio $A_y^{max}(P_i)/A_y(P_o)$ on pressure is shown in Fig.3. Taking into account $\theta_1(P_i) = \theta_1(P_o)(P_o/P_i)^2$, the value of $\theta_1(P_o)$ can be obtained using equation (Fig.3) by fitting the simulation data to the experimental results. Using this routine $\theta_1 \approx 0.5$ for $P_o = 50$ mTorr ($a=1$μm) and $\theta_1 \approx 0.28$ for $P_o = 100$ mTorr, ($a=3.07$ μm) were obtained.

ACKNOWLEDGMENTS

This work was supported by Australian Research Council, the Science Foundation for Physics within the University of Sydney. ASS was supported by University of Sydney U2000 Fellowship.

REFERENCES

1. Khrapak S.A., Nefedov A.P., Petrov O.F., Vaulina O.S., Phys.Rev. E **59**, 6017 (1999).
2. Samarian, A., James, B., Vladimirov, S., and Cramer, N., *Phys. Rev.* **64**, 025402(R) (2001).

Criteria for Phase-Transitions in Yukawa Systems (Dusty Plasma)

O. S. Vaulina, S. V. Vladimirov*, O. F. Petrov, V. E. Fortov

Institute of High Temperatures, RAS, Moscow, Russia
*School of Physics, University of Sydney, Australia

Abstract. Phenomenological criteria for various phase transitions in the Yukawa system are considered including the melting of cubic lattices and the transitions between the *bcc*- and *fcc*-structures.

Weakly ionized dusty plasma can be considered as a dissipative system of macroparticles interacting with the screening Yukawa potential $\varphi = eZ\exp(-l/\lambda)/l$ (where eZ is the particle charge, and λ is the screening lengths). Under certain conditions, the dust immersed in plasma may form the stable structures similar to the liquid or the solid [1-5]. For analyses of phase state of dust systems the values of Coulomb coupling parameter are often used: $\Gamma = (eZ)^2/Td$, were T is the kinetic dust temperature in energy units, $d=n^{-1/3}$, and n is the dust concentration. In the case of Coulomb interactions the critical Γ_c value is about 106 at the line of fluid-solid transition. Nevertheless this rule is not suitable for Yukawa systems. Assumption of screening interactions ($\kappa = d/\lambda \neq 0$) leads to increasing of critical value Γ_c [6-9].

There are two known empirical rules for the first-order fluid-solid phase transition in three dimensions. The first one is the Hansen criterion for the freezing point, which determines the first maximum of the liquid structure factor as a value less than 2.85. The second rule is the Lindemann criterion, which determines the ratio of the root-mean-square displacement Δ_o of particle from its equilibrium position to the interparticle distance d on the melting line of the solid as a value close to 0.15. Thus the ratio $\delta_c = \sqrt{2}\Delta_o /d$ must be equal to ~ 0.21 on the melting line of system. It should be noted that the particle displacement $\Delta = \sqrt{2}\Delta_o$ from the center of mass of the system, measured in numerical simulations, is usually less than 0.2 d. Theoretical estimations show that the value of δ_c tends to 0.2 with increasing number of particles in the simulated systems [7]. The condition analogous to the Lindemann criterion can be obtained on the assumption that the average volume of thermal fluctuations $V_{tf} \cong 4\pi \Delta^3/3$ for the *bcc*-lattice should not exceed $(1-\pi\sqrt{2}/8)V \approx 0.32V$, where $V = n^{-1} \equiv (\alpha\, a_{W-S})^3$, $\alpha = (4\pi/3)^{1/3}$, and $a_{W-S} = (4\pi n/3)^{-1/3}$ is the Wigner-Seitz radius. For the stable *fcc*-structure to exist, it is necessary to satisfy the condition $V_{tf} < (1-\pi\sqrt{2}/6)V \approx 0.26V$ [10]. Taking into account the possibility of the counter displacement of grains, we find that the value of the ratio Δ/d must exceed either 0.211 ($\Delta_o/d > 0.15$) to melt the *bcc*-structure or 0.198 ($\Delta_o/d > 0.14$) to melt the *fcc*-lattice. The criterion for the transition between *bcc*- and *fcc*-structure in Yukawa system can be obtained on the assumption that for the change of the *bcc*-symmetry of lattice, the interparticle distances should exceed λ (the intergrain interaction is in this case similar to that of "hard spheres" when formation of *fcc*-structures is possible):

$$2(1-\pi/(3\sqrt{2}))^{-1/3}\Delta_o \approx (a_{W-S} - \lambda). \quad (1)$$

Values of Δ_o/d for various phase transitions are presented in Fig. 1. We can see that the range of κ between 5.8 and 6.8 defines the region of the triple phase transition (*bcc-fcc*–liquid) and the values of Δ_o/d on the melting lines of *bcc*- and *fcc*- lattices are in accordance with the Lindemann Criterion.

Here we present a new empirical rules which determine the normalized coupling parameter $\Gamma_n = K_n \exp(-\kappa) \Gamma$ as a values close to the constants C_p at the line of different phase transitions (including the melting of cubic lattices and the transitions between the *bcc*- and *fcc*- structures): $\Gamma_n = K_n \exp(-\kappa) \Gamma \equiv C_p \approx$ const. The values of normalized coefficients K_n and constants C_p can be obtained from the relationship for the harmonic oscillator:

$$\Delta_o^2 = 3T/m_p\omega_c^2, \quad (2)$$

where m_p is the mass of the dust particle, and ω_c is the characteristic grain oscillation frequencies in lattices of different types. The value of $\omega_c = \omega_{bcc}$ in the *bcc*-lattice can be obtained from expression $F = (eZ)^2 \exp(-l/\lambda)(1+ l/\lambda)/l^2$ for the intergrain force assuming that the electric fields of all particles except the nearest ones are fully compensated [9]: $\omega_{bcc} = eZ(4n/\pi m_p)^{1/2} (1+\kappa +\kappa^2/2)^{1/2} \exp(-\kappa/2)$. Substitution of ω_{bcc} in Eq. (2) gives

$$\Gamma_n = \Gamma^* \equiv (1+\kappa+\kappa^2/2) \exp(-\kappa) \Gamma \quad (3)$$

and $C_p \cong 3\pi/(2\delta_c^2) \approx 106$ n accordance with [9,11], here $\delta_c \equiv \sqrt{2}\Delta_o/d = (1-\pi\sqrt{2}/8)^{1/3}/2\alpha \approx 0.211$ at the melting line of *bcc*- structure.

Under the assumption that the characteristic frequency for the *fcc*–structure $\omega_c = \omega_{fcc} \propto dF/dl$, the value of the parameter Γ^* on the crystallization line for lattices of both types should be constant which contradicts to the results of numerical simulations, see Fig. 2. Suitable approximation $\omega_{fcc}^2 \approx 2\alpha^3 n (eZ)^2 \exp(-\kappa)(\kappa-\alpha)/m_p$ for the characteristic grain frequency in the *fcc*–structure can be obtain for the homogeneous system, where the gradient dF_Σ /dl of the sum F_Σ of the electrical forces can be estimated as $dF_\Sigma /dl \propto n (eZ)^2 \exp(-\kappa)(\kappa-\alpha)$. Thus, assuming that $\delta_c = (1-\pi\sqrt{2}/6)^{1/3}/2\alpha$ on the melting line of the *fcc*–lattice, we determine from Eq. 2:

$$\Gamma_n = (\kappa-\alpha) \exp(-\kappa) \Gamma, \quad C_p \cong 6/(2\alpha^3\delta_c^2) \approx 18.5,$$

and the parameter Γ^* for this phase transition as

$$\Gamma^* \approx 18.5\,(\kappa-\alpha)^{-1}(1+\kappa+\kappa^2/2). \qquad (4)$$

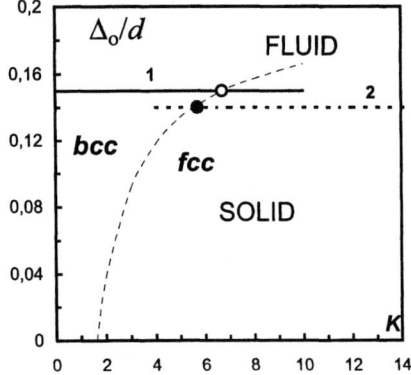

FIGURE 1. Dependencies of Δ_o/d on κ for different phase transitions:
1 – $bcc \to$ fluid, 2 - $fcc \to$ fluid; 3 - $bcc \leftrightarrow fcc$;
● - $\kappa = 5.8$; ○ - $\kappa = 6.8$.

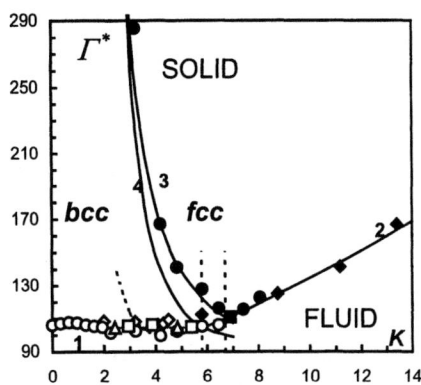

FIGURE 2. Dependencies of Γ^* on κ for different phase transitions:
○ - [6]; □ - [7]; ◇ - [8]; ∇ - [9]
1 – $\Gamma^* =106$ ($bcc \to$ fluid) (Eq. 3), 2 - $fcc \to$ fluid (Eq. 4); 3 – $bcc \to fcc$ (Eq. 5); 4 – $fcc \to bcc$ (Eq. 6).

The use of the modified parameter Γ^* Eq.(2) allows us to illustrate the behavior of the melting curves for the transitions fcc–lattice – liquid and bcc–lattice – liquid for the linear (not logarithmic) scale, Fig. 2. The difference between condition (4) and previous results [7,11] of modeling the fcc-lattice -liquid phase transition does not exceed 2% for $\kappa > 6.8$.

Normalized coupling parameter Γ_n and modified parameter Γ^* at the line of transition of system from bcc- to fcc-structure can be obtain from Eqs. 1-2 with $\omega_c = \omega_{bcc}$

$$\Gamma^* \approx 64\,\kappa^2(\kappa-\alpha)^{-2}. \qquad (5)$$

Suitability of condition (5) as a criterion for the bcc-fcc-transition was checked using data of Ref. [6]. We obtained that deviation of the calculated values of Γ^* for the bcc-fcc-transition from Eq. (5) is within the limits ±2%, as illustrated on Fig. 3 (curve 3). Taking into account that the possibility of the reverse transition from the fcc- to bcc-structure is defined by the frequency ω_{fcc}, as a criterion for this transition we can consider the expression

$$\Gamma^* \approx 9.8\,\kappa^2\,(\kappa-\alpha)^{-3}(1+\kappa+\kappa^2/2) \qquad (6)$$

Note that condition (6) depends on the approximation of the frequency ω_{fcc} and therefore can be incorrect for small values of $\kappa \to \alpha$. However, calculations on the basis of expressions (5)-(6) (curves 3 and 4 on Fig. 2) fully determine the region of the triple phase transition (κ=5.8-6.8) and agree with the results of calculation of Δ_o/d (see Fig. 1). We add also that the difference in the positions of curves described by Eqs. (5) - (6) is capable to explain disagreements of the results of numerical simulations on finding the position of the triple point of system in Refs. [6,7].

To conclude, we considered the existent and proposed new phenomenological criteria for various phase transitions in the Yukawa system. Algorithm presented for the calculations of phase diagrams may be used for the systems with the another types of lattices or intergrain interactions in the case of $T \gg \theta_d$, where θ_d is the Debye temperature. The values of characteristic frequencies ω_c can be obtained by the direct calculations, the estimations of gradients of intergrain interaction forces and also by the analytical approximations of the numerical simulation results or the experimental data. Presented results are independent on viscosity of surrounding gas (a friction produced by impacts of the gas atoms or molecules) and can be applied for analysis of particle dynamics in binary colloidal systems of different types, for example, in solutions of viruses, or for studies of diffusion controlled processes in physics of polymers, where Yukawa-type potentials are used extensively.

ACKNOWLEDGMENTS

This work was partially supported by the Russian Foundation for Fundamental Research, project nos. 01-02-16658, by INTAS Grant No. 2000-0522, and the Australian Research Council.

REFERENCES

1. Thomas, H., Morfill, G.E., Demmel, V., et al., *Phys. Rev. Letts.*, **73**, 652 (1994).
2. Chu, J.H., and L. I, *Phys. Rev. Letts.*, **72**, 4009 (1994).
3. Fortov, V.E., Nefedov, A.P., Petrov, O.F., et al., *Phys. Rev. E: Rapid Comm.*, **54**, 2236 (1996).
4. Rosenberg, M., and Mendis, D.A., *IEEE Trans. Plasma Sci.*, **23**, 177 (1995).
5. Fortov, V.E., Nefedov, A.P., Torchinsky, V.M., et al., *Phys. Letts. A*, **229**, 317 (1997)
6. Hamaguchi, S., Farouki, R., and Dubin, D., *Phys. Rev. E*, **56**, 4671 (1997).
7. Stevens, M., and Robbins, M., *J. Chem. Phys.*, **98**, 2319 (1993).
8. Meijer, E., and Frenkel, D., *J. Chem. Phys.*, **94**, 2269 (1991).
9. Vaulina, O., Khrapak, S.., *JETP* **92**, 228 (2001).
10. Hofman, J.M.A., Clercx, Y.J.H., Schram, P.P.J., *Phys. Rev. E*, **62**, 8212 (2000).
11. Vaulina, O.S., Vladimirov, S.V., *Plasma Phys.*, **9**, 835 (2002)

Rotational modes of oscillation in a chain of rod-shaped particles in a plasma

S.V. Vladimirov, M.P. Hertzberg, and N.F. Cramer

School of Physics, University of Sydney, New South Wales 2006, Australia

Abstract. Three dimensional rotatory modes of oscillations in a one-dimensional chain of rod-like charged particles or dust grains in a plasma are investigated. The dispersion characteristics of the modes are analyzed. The stability of different equilibrium orientations of the rods, phase transitions between the different equilibria, and a critical dependence on the relative strength of the confining potential are analyzed. The relations of these processes with liquid crystals are discussed.

The dynamic properties of the particle motion, formation of colloidal crystals and phase transitions in plasma-dust systems are important fundamental questions related to the general theory of self-organization in open dissipative systems [1]. The cases already studied, experimentally and theoretically, mostly correspond to spherical dust grains, but there is growing interest in the properties of colloidal structures composed of elongated (cylindrical) particles [2, 3] levitating in the sheath region of a gas discharge plasma.

Unlike point-like or spherical particles, elongated rotators exhibit a number of additional oscillations related to the new (rotational) degrees of freedom [4, 5]. The excitation and interactions of all these modes will strongly affect the lattice dynamics, leading in particular to new types of phase transitions, as well as affecting those phase transitions already existing in lattices composed of spherical grains. Here, we present a full three-dimensional analysis of the rotatory modes in the chain of rod-like particles, and analyse the critical dependence of the equlibrium and stability of such a chain.

Each rod-like particle is modeled as a rotator having two charges (and masses) concentrated on the ends of the rod, the upper charge being Q_a and the lower charge Q_b; the charges are constant and the masses are equal. The rod of length L, connecting these two charges, has zero radius and mass. We consider a one-dimensional infinite linear chain of rods, with their centers of mass evenly separated by the distance d in equilibrium, along the x-axis. The relevant forces are due to the external potentials and the interparticle interactions. The external potential Φ_{ext} is a combination of the potentials due to both gravitation and the external electrodes. The interparticle force is Coulombic in nature, and there is an exponential decay of the interparticle potential with the distance λ_D (the plasma Debye length). The Lagrangian is given by

$$L = \frac{m}{2}\sum_n \left(\frac{d\mathbf{R}^n}{dt}\right)^2 + \frac{I}{2}\sum_n \left[\left(\frac{d\phi^n}{dt}\right)^2 (\sin\theta^n)^2 + \left(\frac{d\theta^n}{dt}\right)^2\right]$$
$$- \sum_n (Q_a\Phi_a + Q_a\Phi_b + Q_b\Phi_a + Q_b\Phi_b + Q_a\Phi_{\text{ext}} + Q_b\Phi_{\text{ext}}), \qquad (1)$$

where the sum is over all the particles, and I is the common rotational inertia of each particle. Here θ is the angle the rod makes with the z (vertical) axis, and ϕ is the angle the projection of the rod onto the $x-y$ (horizontal) plane makes with the x-axis (the direction of the chain of particles).

The external potential can be approximated by a parabolic potential for small oscillations, whose minimum lies at the center of mass of a rod ($y=0, z=0$). The assumption of an infinite chain in the x-direction removes the need for a confining potential in that direction. The external potential is therefore assumed to act in both the z (vertical) and y directions, $Q_a \Phi_{\text{ext}}(\mathbf{a}^n) + Q_b \Phi_{\text{ext}}(\mathbf{b}^n) = k_y y^2 + k_z z^2$, where (y,z) are the coordinates of the upper charge Q_a.

The equations of motion are extremely nonlinear; moreover, the θ and ϕ behaviour is coupled. In order to proceed we must *linearise* these equations about an equilibrium position. We shall call the equilibrium colatitudinal and azimuthal angles, about which we consider small perturbations, θ_0 and ϕ_0 respectively, which are common to all particles. The analysis then proceeds as follows: let ε be a small perturbation of θ, so that $\theta^n = \theta_0 + \varepsilon^n$; and similarly for ϕ by letting η be a small perturbation from ϕ_0.

We next determine the common *equilibrium* orientation of the particles. The result is that equilibria exist in only the following orientations (unless $k_y = k_z$): $(\theta_0, \phi_0) = (0, \phi_0)$ (with ϕ_0 arbitrary) or $(\pi/2, 0)$ or $(\pi/2, \pi/2)$, which we draw successively as shown in Fig. 1 (a), (b), and (c), respectively. If $k_y = k_z$, equilibrium exists for the orientation

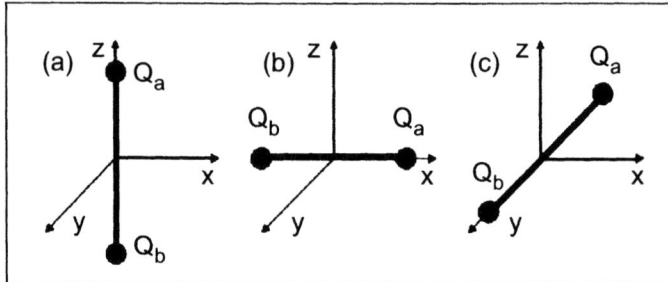

FIGURE 1. The three equilibrium orientations of the rod-like dust grains.

$(\theta_0, \pi/2)$, where θ_0 is arbitrary. Oscillations in θ about the second equilibrium have already been considered in Ref. [5], and oscillations in ϕ about that equilibrium can be obtained simply by exchanging k_y and k_z. Since ϕ is undefined for a vertically oriented rod, we concentrate here on the horizontal equilibrium case $(\pi/2, \pi/2)$.

The perturbation equation describes small oscillations in azimuthal angle ϕ^n about the equilibrium Fig. 1(c), $\theta = \pi/2$, $\phi = \pi/2$. To investigate the existence of an oscillatory solution we compute its Fourier transform and obtain:

$$I\omega^2 = +L^2 \left[Q_a \Phi_a''(d) - \frac{Q_a}{L_d} \Phi_b'(L_d) \right] \sin^2(kd/2)$$
$$+ \frac{d^2 L^2}{L_d^2} \left[Q_a \Phi_b''(L_d) - \frac{Q_a}{L_d} \Phi_b'(L_d) \right] \cos^2(kd/2) + [Q_a \leftrightarrow Q_b] - \frac{L^2}{2} k_y, \qquad (2)$$

where $L_d^2 = L^2 + d^2$. For the Debye-Coulomb potential in equation (2), the product $Q_a \Phi'_a$ is negative, while $Q_a \Phi''_a$ is positive, as is true for any potential that falls off with distance. The result is that the coefficients of the oscillatory sine and cosine terms in Eq. (2) are always positive. Thus the dust particles would always exhibit stable oscillations, except for the presence of the term $-k_y L^2/2$ from the external potential, which acts to pull the grains away from this equilibrium to the x-axis (where $y = 0$). These competing terms may then give rise to regions of stable behaviour and regions of unstable behaviour. Recall that the moment of inertia is $I = mL^2/2$. Hence the factor L^2 cancels throughout and is only present, implicitly, through the quantity L_d. We may plot the dispersion relation, by selecting some typical values of the parameters involved: $m = 10^{-15}$kg, $Q = 10^3 e$ to $10^4 e$, and $\lambda_D = 300 \mu$m; the resulting dispersion relation is shown in Fig. 2. This plot clearly shows stable regions, corresponding to $\omega^2 > 0$, and unstable regions

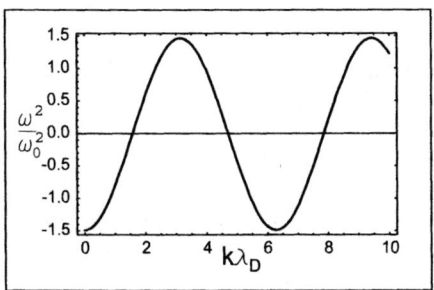

FIGURE 2. Normalised frequency squared versus normalised wavenumber for perturbations in the angle ϕ, travelling in the x direction. The equilibrium orientation is $(\theta_0 = \pi/2, \phi_0 = \pi/2)$. Here ω_0 is the dust plasma frequency, $d = \lambda_D$ and $k_y = 10^{-10}$kgs^{-2}.

corresponding to $\omega^2 < 0$ (i.e ω imaginary.)

We now investigate the behaviour of the perturbation ε of the colatitudinal angle θ, recalling that ε and η completely decouple in the linear approximation. The Fourier transformed equation of motion leads to the following dispersion relation:

$$I\omega^2 = L^2 \left[\frac{Q_a}{d} \Phi'_a(d) - \frac{Q_a}{L_d} \Phi'_b(L_d) \right] \sin^2(kd/2) + [Q_a \leftrightarrow Q_b] + \frac{L^2}{2}(k_z - k_y). \quad (3)$$

Once again note the oscillatory dependence on wavenumber. The first term on the right hand side is this time *negative*, since $Q\Phi'(\mathbf{r}) < 0$ and $d < L_d$. Thus the dust particle's mutual repulsion causes instability. Note the competing terms from the external potential. In the horizontal case one needs $k_z > k_y$ for the possibility of stability (of course wavenumber gaps are still possible). Note that similar dependence exhibit pairs of unbound spherical particles levitating in the confining potential in x- and z-directions [6]. We also recall that the interchange $k_z \leftrightarrow k_y$ gives the vertical equilibrium case. By selecting $k_z > k_y$ the dispersion relation will be qualitatively the same as in Fig. 2.

Instability implies that the rods may switch from the horizontal to the vertical equilibrium (or vice versa) depending on the sign of $k_z - k_y$. However, our analysis is only valid for small perturbation angles, and breaks down at large amplitudes. In order to examine what ensues if the rods lie near an unstable orientation we must consider what will happen physically: the rods will in fact move in opposite directions to move away from one another. Thus on the average, the even rods, say, will move clockwise and the odd rods counter-clockwise (or vice versa). Hence it may be that there exists some intermediate value of stability between the vertical and horizontal. Our analysis is different now from earlier, since we are allowing alternate rods to have opposite equilibrium orientations. By considering the vertical case, this corresponds to equilibria at θ_0 and $-\theta_0$ alternately. Let the even rods be described by perturbations ε^n in θ as before, and the odd neighbours by ζ^{n-1} and ζ^n. The first order perturbation equation is

$$I\ddot{\zeta}^n = \frac{L^2}{4}\left[-\frac{Q_a\Phi'_a(R_1) + Q_b\Phi'_b(R_1)}{R_1} + \frac{2Q_a\Phi'_b(R_2)}{R_2}\right]$$
$$\times \left(2\cos(2\theta_0)\zeta^n - \eta^n - \eta^{n-1}\right) - \frac{L^2}{4}(k_y - k_z)\cos(2\theta_0)\zeta^n. \tag{4}$$

Thus equilibrium occurs if $\theta_0 = 0$ (the vertical case), or $\theta_0 = \pi/2$ (the horizontal case), or if the term in brackets vanishes. In the case where $k_z = k_y$, and when $Q_a = Q_b$, an equilibrium occurs at $\theta_0 = \pi/4$, when the charges are furthest away from one another, and neighbours are out of phase by $\pi/2$. In this case the right hand side of the perturbation equation (4) vanishes; this is because for an external potential symmetric about the x axis, an arbitrary rotation of θ_0 may be made, as long as neighbours are at right angles to each other. If the potential is not symmetric ($k_y \neq k_z$), equilibria may occur at intermediate angles different to $\theta_0 = \pi/4$.

Hence the described system of rods can undergo phase transitions from one state to another i.e vertical to horizontal and vice versa, or to an intermediate equilibrium, dependent on the easily adjustable external potentials. In fact it is a natural extension to see that this is also true of the third equilibrium case, corresponding to the equilibrium at $(\pi/2, 0)$. This process has a relation to processes in liquid crystals [7], where a state of matter exists between the solid and liquid phases, wherein rod shaped molecules exhibit a partial alignment, rather than a rigid array seen in crystals. The direction of this partial alignment (and phase) can be altered by an external influence.

ACKNOWLEDGMENTS

This work was supported by the Australian Research Council.

REFERENCES

1. Thomas, H.M., and Morfill, G.E., Nature **379**, 806 (1996).
2. Mohideen, U, Rahman, H.U., Smith, M.A., *et al*, Phys. Rev. Lett. **81**, 349 (1998).
3. Annaratone, B.M., Khrapak, A.G., Ivlev, A.V., *et al*, Phys. Rev. E **63**, 036406 (2001).
4. Vladimirov, S.V., and Nambu, M., Phys. Rev. E **64**, 026403 (2001).
5. Vladimirov, S.V., and Tsoy, E.N., Phys. Rev. E **64**, 035402(R) (2001).
6. Vladimirov, S.V., and Samarian, A.A., Phys. Rev. **65**, 046416 (2002).
7. Kumar, S., *Liquid Crystals: Experimental study of Physical Properties and Phase Transitions* (Cambride University, Cambridge, 2001).

Particle Flows and Ambipolar Electric Field in Dusty Partially Ionized Plasmas with Temperature Gradients

V.S. Tsypin*, S.V. Vladimirov[†], R. Galvão*, I. Nascimento*, M. Tendler**
and A. de Assis[‡]

*Institute of Physics, University of São Paulo, 05508-900, Brasil
[†]School of Physics, University of Sydney, New South Wales 2006, Australia
**Royal Institute of Technology, 10044 Stockholm, Sweden
[‡]Fluminense Federal University, Niterói, Rio de Janeiro, Brazil

Abstract. In this paper the plasma flows and ambipolar electric field induced due to the temperature stresses on ions are estimated. These flows are similar to the plasma "residual" rotations in tokamaks, but are related to dust contamination of tokamak plasmas. They can also entrain neutrals in a multi-component partially ionized plasma.

It is well-known that the sheared radial electric field E_r plays an important role in tokamak plasmas, e.g., being responsible, first, for creation of transport barriers [1] and, second, for squeezing particle banana orbits in the core of large tokamaks [2]. In the first case, simple estimates to find the required radial electric field shear are

$$\frac{cB_\theta R}{B} \frac{\partial}{\partial r} \frac{E_r}{RB_\theta} \geq \gamma_{max}, \qquad (1)$$

where c is the light speed, r and R is the torus minor and major radius, respectively, B is the magnetic field, γ_{max} are the increments of most dangerous instabilities in edge tokamak plasmas. In the second case, there is the so-called squeezing parameter S [2]

$$S = 1 - \frac{e_i B_\zeta^2}{M_i \omega_{ci}^2 B_\theta^2} \frac{dE_r}{dr}. \qquad (2)$$

Here, B_θ and B_ζ are the poloidal and toroidal components of the magnetic field, respectively, e_i and M_i are the ion charge and mass, respectively, $\omega_{ci} = e_i B/(cM_i)$ is the ion cyclotron frequency. The approximate value of the radial electric field E_r can be found from the radial projection of the ion momentum equation, and is given by

$$E_r \approx \frac{1}{c}\left(-B_\zeta U_{i\theta} + B_\theta U_{i\zeta} + BU_{pi}\right), \quad U_{pi} = \frac{c}{e_i n_i B}\frac{\partial p_i}{\partial r}, \qquad (3)$$

where $U_{i\theta}$ and $U_{i\zeta}$ are the poloidal and toroidal "physical" projections of the ion velocity $\mathbf{V}_i$ respectively, p_i is the ion pressure, n_i is the ion density. The important role of ion viscosity in inducing residual plasma flows and radial electric field in fully ionized plasmas was first demonstrated in [3]. In the presence of impurities or dust, plasma flows can be induced by thermal forces affecting ions mainly as a result of their collisions with other heavy particles [4]. These thermal forces can exceed ion viscous forces.

In this paper the plasma flows and ambipolar electric field induced due to the temperature stresses on ions are estimated. These flows are similar to the plasma "residual" rotations in tokamaks, but are related to dust contamination of tokamak plasmas. They can also entrain neutrals in a multi-component partially ionized plasma.

For brevity, we consider the simplest case when the quasineutrality condition has the form $-en_e + eZ_i n_i - eZ_d n_d \simeq 0$, where $-e \equiv e_e$ is the electron charge, n_e and n_d are the electron and dust densities, respectively, and $eZ_i \equiv e_i$ and $-eZ_d \equiv e_d$ are the ion and dust effective charges. In addition, we assume that $Z_d n_d \ll \{n_e, Z_i n_i\}$, $M_i \ll M_d$, M_d is the dust particle mass, and dust particles are motionless (dust temperature is equal to zero). To find ion flows we start from the summed ion and electron momentum equations [5]

$$-\nabla p - \nabla \cdot \pi_\parallel^i + \frac{1}{c}[\mathbf{j} \times \mathbf{B}] + \mathbf{R} \simeq 0. \tag{4}$$

Here $p = p_e + p_i$ is the plasma pressure, $\pi_\parallel^i$ is the ion viscosity tensor, $\mathbf{j}$ is the plasma current, $\mathbf{B}$ is the magnetic field vector, $\mathbf{R}$ is the friction force between ions and dust including thermal forces $\mathbf{R}_T$ which is defined by the relation $\mathbf{R} = M_i \int \mathbf{w} C d\mathbf{w}$, $\mathbf{w} = (\mathbf{v} - \mathbf{V}_i)$, $\mathbf{V}_i$ is the ion mean velocity, and C is the collision integral in the ion kinetic equation [5].

As far as usually the ion mass M_i is negligibly small in comparison with the dust particle mass M_d, to find the friction force $\mathbf{R}$ we can use results of papers [5] and [6] with replacements in corresponding equations $e_e \to e_i$, $e_i \to e_d$, $M_e \to M_i$, $M_i \to M_d$, $T_e \to T_i$, $T_d = 0$. As a result, we arrive at

$$\mathbf{R} \simeq -M_i n_i \nu_{id} \mathbf{V}_i + \mathbf{R}_T, \tag{5}$$

$$\mathbf{R}_T = -2.69 \frac{n_d Z_d^2}{Z_i^2} \mathbf{b} \nabla_\parallel T_i + \frac{3}{2} \frac{\nu_{id} n_i}{\omega_{ci}} [\mathbf{b} \times \nabla T_i], \tag{6}$$

$$\nu_{id} = \frac{4\sqrt{2\pi} \lambda e^4 Z_i^2 Z_d^2 n_d}{3\sqrt{M_i} T_i^{3/2}}, \tag{7}$$

where $\mathbf{b} = \mathbf{B}/B$, $\nabla_\parallel = \mathbf{b} \cdot \nabla$, $\mathbf{V}_{i\parallel} = \mathbf{b}(\mathbf{b} \cdot \mathbf{V}_i) \equiv \mathbf{b} V_{i\parallel}$, and $\mathbf{V}_{i\perp} = [\mathbf{b} \times [\mathbf{V}_i \times \mathbf{b}]]$. Obtaining equations (5) and (6), we assumed that $\nu_{id} \ll \omega_{ci}$, $Z_d \to \infty$ and $\nu_{id} \gg v_{Ti}/qR$, where $v_{Ti} = \sqrt{2T_i/M_i}$ is the ion thermal velocity and $q = rB_\zeta/RB_\theta$ is the safety factor. Thus we restricted ourselves by the case of collisional plasma relevant for the tokamak edge. To get qualitative results we also restricted ourselves by the case of an axially-symmetric, large aspect-ratio, circular cross-section tokamak.

Now we employ the parallel and ζ-covariant projections of equation (4). In the first case, we have

$$\frac{1}{qR}\frac{\partial p}{\partial \theta} + \mathbf{b} \cdot \left(\nabla \cdot \pi_\parallel^i\right) + M_i n_i \nu_{id} V_{i\parallel} + 2.69 \frac{n_d Z_d^2}{Z_i^2 qR}\frac{\partial T_i}{\partial \theta} \simeq 0, \tag{8}$$

using here for the contravariant projection of the vector **b** an approximate expression $b^\theta \simeq 1/qR$. We average equation (8) over θ:

$$\left\langle \mathbf{b} \cdot \left(\nabla \cdot \pi^i_\parallel\right)\right\rangle_\theta + M_i n_i v_{id} \left\langle V_{i\parallel}\right\rangle_\theta \simeq 0, \qquad (9)$$

where $\langle...\rangle_\theta = \int_0^{2\pi} (...) d\theta$. The vector $\nabla \cdot \pi^i_\parallel$ has the form [6]:

$$\nabla \cdot \pi^i_\parallel = \frac{3}{2}\mathbf{b}(\mathbf{b}\cdot\nabla)\pi^i_\parallel - \frac{1}{2}\nabla\pi^i_\parallel + \frac{3}{2}\pi^i_\parallel\left[\mathbf{b}(\nabla\cdot\mathbf{b}) + (\mathbf{b}\cdot\nabla)\mathbf{b}\right], \qquad (10)$$

where the ion parallel viscosity scalar [6, 7] is $\pi^i_\parallel = -1.92\frac{p_i}{rv_i}\left(U_{i\theta} + 1.83U_{Ti}\right)\frac{\partial \ln B}{\partial \theta}$, and $v_i = \frac{4\sqrt{\pi}\lambda e^4 Z_i^4 n_i}{3\sqrt{M_i}T_i^{3/2}}$ is the ion-ion collision frequency, $U_{Ti} = \left(1/M_i\omega_{ci}\right)\partial T_i/\partial r$. From (10) one finds

$$\left\langle \mathbf{b} \cdot \left(\nabla \cdot \pi^i_\parallel\right)\right\rangle_\theta = -\frac{3}{2qR}\left\langle \pi^i_\parallel \frac{\partial \ln B}{\partial \theta}\right\rangle_\theta. \qquad (11)$$

As a result, one has from (9) and (11)

$$U_{i\theta} + 1.83U_{Ti} + 1.96\frac{qR^2}{\varepsilon\lambda_i^2}\frac{n_d Z_d^2}{n_i Z_i^2}\left\langle V_{i\parallel}\right\rangle_\theta \simeq 0, \qquad (12)$$

where $\lambda_i = v_{Ti}/v_i$ is the ion mean free path and $\varepsilon = r/R$ is the inverse aspect ratio. In the absence of dust, $n_d = 0$, we arrive from (12) at the well-known result [7] (see, e.g., [8])

$$U_{i\theta}^{(1)} = -1.83U_{Ti}. \qquad (13)$$

Now we turn to the ζ-covariant projection of equation (4). For an axially-symmetric tokamak, $\partial/\partial\zeta = 0$, we obtain from this equation

$$-\left(\nabla\cdot\pi^i_\parallel\right)_\zeta + \frac{\sqrt{g}}{c}j^r B^\theta + R_\zeta \simeq 0, \qquad (14)$$

where

$$R_\zeta \simeq -M_i n_i v_{id} RU_{i\zeta} - 2.69\frac{n_d Z_d^2}{Z_i^2 q}\frac{\partial T_i}{\partial \theta} - \frac{3}{2}v_{id}n_i M_i \sqrt{g}b^\theta U_{Ti}, \qquad (15)$$

and g is the determinant of the metric tensor, j^r is the r-contravariant projection of **j**. From the plasma current continuity equation $\nabla\cdot\mathbf{j}\simeq 0$ we find the ambipolarity condition $\int_0^{2\pi} j^r \sqrt{g}d\theta = 0$. Using (10), it can be shown that

$$\left\langle \sqrt{g}\left(\nabla\cdot\pi^i_\parallel\right)_\zeta\right\rangle_\theta = 0. \qquad (16)$$

Multiplying (14) by $\sqrt{g}$ and using (16) and the Maxwell equation $\partial\left(\sqrt{g}B^\theta\right)/\partial\theta = 0$, we arrive at

$$\left\langle \sqrt{g}\left\{RU_{i\zeta} + 1.91\frac{1}{v_i qM_i}\frac{\partial T_i}{\partial \theta} + \frac{3}{2}\sqrt{g}b^\theta U_{Ti}\right\}\right\rangle_\theta \simeq 0. \qquad (17)$$

The term $\partial T_i/\partial \theta$ can be found from the ion heat transfer equation [5]

$$\frac{1}{qR}\frac{\partial q_{i\|}}{\partial \theta} + \nabla \cdot \mathbf{q}_{i\perp} + \frac{3M_e n_0 v_e}{M_i}\left(T_i - \langle T_i \rangle_\theta\right) = 0, \tag{18}$$

where $q_{i\|} = -3.91\frac{n_0 T_i}{M_i v_i qR}\frac{\partial T_i}{\partial \theta}$, and $\mathbf{q}_{i\perp} = \frac{5}{2}\frac{cn_i T_i}{e_i B}\left[\mathbf{h} \times \nabla T_i\right]$. Solving equation (18) leads to

$$\frac{\partial T_i}{\partial \theta} = 2.55 \frac{bT_i}{rv_i d(b)} U_{Ti} \frac{\partial^2 \ln B}{\partial \theta^2}, \tag{19}$$

where $d(b) = 1 + 2.2b\sqrt{M_e/M_i}$ and $b = q^2 R^2/\lambda_i^2$. Substituting (19) into (17), we arrive at

$$\langle V_{i\|}\rangle_\theta \simeq \left(B_\zeta U_{i\zeta} + B_\theta U_{i\theta}\right)/B = -1.5\frac{\varepsilon}{q}U_{Ti}\left[1 + 1.62\frac{q^2}{d(b)}\right]. \tag{20}$$

It follows from (12) and (20) that

$$U_{i\theta} = -1.83 U_{Ti}\left\{1 - 1.60\frac{R^2}{\lambda_i^2}\frac{n_d Z_d^2}{n_i Z_i^2}\left[1 + 1.62\frac{q^2}{d(b)}\right]\right\}\left(1 + 1.96\frac{R^2}{\lambda_i^2}\frac{n_d Z_d^2}{n_i Z_i^2}\right)^{-1}. \tag{21}$$

We see from equations (3) and (21) that in the presence of dust in tokamak plasma edges the ion poloidal rotation velocity $U_{i\theta}$ can change sign because of thermal forces resulting from interaction of ions with dust. It occurs at a very small density of dust, $n_d > 0.63\left(\lambda_i^2/R^2\right)n_i Z_i^2/Z_d^2$. This velocity can also substantially exceed the conventional drift velocity U_{Ti}. When the parameter $\alpha = \left(R^2/\lambda_i^2\right)n_d Z_d^2/\left(n_i Z_i^2\right)$ satisfies the condition $\alpha \gg 1$, we arrive at

$$U_{i\theta}^{(2)} = 1.50 U_{Ti}\left[1 + 1.62\frac{q^2}{d(b)}\right]. \tag{22}$$

Thus poloidal velocity $U_{i\theta}$ with the growth of parameter α monotonically changes from $U_{i\theta}^{(1)}$ to $U_{i\theta}^{(2)}$. Correspondingly, the radial electric field can also change sign and achieve large values substantially affecting tokamak edge plasma dynamics.

REFERENCES

1. Burrell, K.H., Phys. Plasmas, **4**, 1499-1518 (1997).
2. Shaing, K.C. and Hazeltine, R.D., Phys. Fluids B, **4**, 2547-2551 (1992).
3. Tamm, I.E., Theory of a magnetic thermonuclear reactor, in Plasma Physics and the Problem of Controlled Thermonuclear Reactions, ed. M.A.Leontovich (Oxford: Pergamon) **1**, 1-20 (1961).
4. Tsypin, V.S., Vladimirov, S.V., Tendler, M., Elfimov, A.G., de Assis, A.S. and de Azevedo, C.A., Phys. Plasmas, **4**, 3436-3438 (1997).
5. Braginskii, S. I., in *Reviews of Plasma Physics*, ed. by M. A. Leontovich (Consultants Bureau, New York, 1965), **1**, 205-265.
6. Mikhailovskii, A.B. and Tsypin, V.S., Contrib. Plasma Phys., **24**, 335-354 (1984).
7. Mikhailovskii, A.B. and Tsypin, V.S., Sov. Phys. JETP, **56**, 75-79 (1982).
8. Claassen, H.A., Gerhauser, H., Rogister, A. and Yarim, C., Phys. Plasmas, **7**, 3699-3706 (2000).

Nonlinear Periodic Waves in Dusty Plasma with Variable Dust Charge

Lakhan Lal Yadav* and R. Bharuthram[†]

*Department of Physics, Kigali Institute of Education, P.O. Box 5039, Kigali, Rwanda
[†]University of Natal, Durban- 4041, South Africa

Abstract. Using the reductive perturbation method, we present a theory of nonlinear periodic waves, viz. the cnoidal waves, in a dusty plasma consisting of electrons, ions, and cold dust grains with charge fluctuations, which in the limiting case reduce to dust acoustic solitons. It is found that the frequency of the dust acoustic cnoidal wave increases with its amplitude. The dust charge fluctuations are found to affect the characteristics of the cnoidal waves.

INTRODUCTION AND FORMULATION OF THE PROBLEM

There has been a significant interest in the study of dusty plasmas [1-4 and references therein]. Some authors have shown that the dust charge fluctuations affect the properties of the nonlinear localized structures [2-4]. The aim of this paper is to present a theory of nonlinear periodic waves in a dusty plasma with variable dust charge, which in the limiting case reduce to dust acoustic solitons.

We consider a collisionless unmagnetized dusty plasma consisting of electrons, ions and negatively charged cold dust grains with charge fluctuations. For many laboratory and astrophysical dusty plasmas, the dust hydrodynamical time scale is much greater than the dust charging time scale, therefore, on the hydrodynamical time scale, the dust charge can reach local equilibrium and the currents from electrons and ions to the dust grains are balanced [4]. In this situation the current balance equation of ref. [2] becomes

$$I_e + I_i \approx 0, \qquad (1)$$

where $I_e = -\pi e R^2 (8T_e/\pi m_e)^{1/2} n_e exp(e\psi/T_e)$ and $I_i = \pi e R^2 (8T_i/\pi m_i)^{1/2} n_i (1 - e\psi/T_i)$, are the electron and ion currents [2], respectively, $m_e(m_i)$, $T_e(T_i)$ and $n_e(n_i)$ are the mass, temperature and density of the electrons (ions), respectively, $\psi = -q_d/R$ is the dust grain surface potential relative to plasma potential ϕ, q_d and R are the magnitude of charge and radius of the dust grain, respectively. The dynamics of the one-dimensional low-frequency dust acoustic waves in this plasma is governed by the following dimensionless equations:

$$\frac{\partial n_d}{\partial t} + \frac{\partial}{\partial x}(n_d u_d) = 0, \qquad (2)$$

$$\frac{\partial u_d}{\partial t} + u_d \frac{\partial u_d}{\partial x} = q_d \frac{\partial \phi}{\partial x}, \qquad (3)$$

$$\alpha_i \mu_1 (1 - \sigma\psi)exp(-\sigma\phi) - exp(\sigma_1\sigma(\psi+\phi)) = 0, \tag{4}$$

$$\frac{\partial^2 \phi}{\partial x^2} = \frac{1}{Z_{d0}}(n_e - n_i) + q_d n_d, \tag{5}$$

where $n_{e,i} = n_{e0,i0} exp(\beta_{e,i}\phi)$, $n_{e0}(n_{i0})$ is the equilibrium electron (ion) number density, $\beta_e = \sigma\sigma_1$, $\beta_i = -\sigma$. n_d, u_d and m_d are the density, fluid velocity and mass of the dust grains respectively. $\alpha_i = (\sigma_1 m_e/m_i)^{1/2}$, $\mu_1 = n_{i0}/n_{e0}$, $\sigma = T_{eff}/T_i$, $\sigma_1 = T_i/T_e$ and the effective temperature, $T_{eff} = Z_{d0} n_{d0} T_e T_i / (n_{e0} T_i + n_{i0} T_e)$. In equations (2)-(5), densities, time t, space coordinate x, dust fluid velocity u_d, charge $q_d = eZ_d$, potentials ϕ and ψ are normalized by n_{d0}, $\omega_{pd}^{-1} = (m_d/4\pi n_{d0} Z_{d0}^2 e^2)^{1/2}$, $\lambda_{Dd} = (T_{eff}/4\pi Z_d n_{d0} e^2)^{1/2}$, $C_d = (Z_d T_{eff}/m_d)^{1/2}$, unperturbed grain charge $q_{d0} = eZ_{d0}$, and T_{eff}/e, respectively. In obtaining equation (4), we used current balance condition (1).

DERIVATION OF THE KDV EQUATION

To solve equations (2)-(5), we introduce the stretched coordinates $\xi = \varepsilon^{1/2}(x - v_0 t)$, $\tau = \varepsilon^{3/2} t$, where ε is a small parameter and v_0 is the phase velocity of the wave, and expand the variable quantities about their equilibrium values in powers of ε. From the first order equations, we obtain

$$n_d^{(1)} = -(1+\gamma_1)\phi^{(1)}, \tag{6}$$

$$u_d^{(1)} = -(1+\gamma_1)^{1/2}\phi^{(1)} + C^{(1)}(\tau), \quad q_d^{(1)} = \gamma_1 \phi^{(1)}, \tag{7}$$

$$v_0 = -(1+\gamma_1)^{-1/2}, \tag{8}$$

where $C^{(1)}(\tau)$ is an integration constant that may depend on τ and $\gamma_1 = -1/\psi_0$. Using first order solutions, we find that the second order equations give

$$q_d^{(2)} = \gamma_1 \phi^{(2)} + \gamma_2 \phi^{(1)2}, \tag{9}$$

$$n_d^{(2)} = \frac{\partial^2 \phi^{(1)}}{\partial \xi^2} - \phi^{(2)} + \left\{ \frac{(\mu_1 - \sigma_1^2)\sigma^2}{2(\mu_1 - 1)} + \gamma_1(1+\gamma_1) \right\} \phi^{(1)2} - q_d^{(2)}, \tag{10}$$

$$u_d^{(2)} = \frac{1}{2(1+\gamma_1)^{-1/2}} \left[\frac{n_d^{(2)}}{(1+\gamma_1)} - (\frac{1}{2}+\gamma_1)\phi^{(1)2} - \phi^{(2)} + \frac{\partial C^{(1)}}{\partial \tau}\xi \right] + C^{(2)}(\tau), \tag{11}$$

where $\gamma_2 = \sigma^2/\{(1-\sigma\psi_0)(1+\sigma_1)\}$ and $C^{(2)}(\tau)$ is the second integration constant that may depend on τ. The periodic boundary condition implies that $\partial C^{(1)}/\partial \tau$ in eqtaion (11) should be equal to zero, therefore, $C^{(1)}$ is independent of ξ and τ. Using equations (7)-(11), the second-order momentum equation gives the following KdV equation

$$\frac{\partial \phi}{\partial \tau} + a\phi\frac{\partial \phi}{\partial \xi} + C^{(1)}\frac{\partial \phi}{\partial \xi} + b\frac{\partial^3 \phi}{\partial \xi^3} = 0, \tag{12}$$

where $b = \frac{1}{2}(1+\gamma_1)^{-3/2}, a = b\left[\frac{(\mu_1 - \sigma_1^2)\sigma^2}{(\mu_1 - 1)} - 3(\gamma_1 + 1) - 2\gamma_2\right]$, and $\phi = \phi^{(1)}$.

NONLINEAR PERIODIC WAVE SOLUTION OF KDV EQUATION

For the steady state solution of the KdV equation (12), we consider $\eta = \xi - u_1\tau$, where u_1 is a constant velocity. Integrating equation (12) twice with respect to η, we obtain

$$\frac{1}{2}\left(\frac{d\phi}{d\eta}\right)^2 + V(\phi) = 0, \tag{13}$$

where Sagdeev potential $V(\phi)$ is given by

$$V(\phi) = \frac{a}{6b}\phi^3 - \frac{u}{2b}\phi^2 + \rho_0\phi - \frac{E_0^2}{2}, \tag{14}$$

ρ_0 and E_0 are, respectively, the charge density and electric field where ϕ vanishes, and $u = u_1 - C^{(1)}$. A cnoidal wave solution of equation (13) is given by [5]

$$\phi = \alpha_2 + (\alpha_1 - \alpha_2)cn^2\{D\eta, m\}, \tag{15}$$

where cn is the Jacobi elliptic function. The parameter m, called modulus, and D can be expressed in terms of the three real zeros of the Sagdeev potential, α_1, α_2 and α_3 as:

$$m^2 = \frac{\alpha_2 - \alpha_1}{\alpha_3 - \alpha_1}, \quad D = \sqrt{\frac{(\alpha_1 - \alpha_3)a}{12b}}. \tag{16}$$

The $\alpha_{1,2,3}$ are such that $\alpha_1 < \alpha_2 \leq \alpha_3$ for the plasma considered here for which $a < 0$ and

$$u_1 = C^{(1)} + \frac{a}{3}(\alpha_1 + \alpha_2 + \alpha_3). \tag{17}$$

The wavelength λ of the cnoidal wave is defined as

$$D\lambda = 2K(m), \tag{18}$$

where $K(m)$ is the first kind of complete elliptic integral. Using the conservation condition of the number density of the dust grains

$$\int_0^\lambda (n_d - 1)d\eta = 0, \tag{19}$$

we can find the three real zeros of the Sagdeev potential in terms of modulus m. To evaluate (19), we assume that $n_d^{(2)}$ and higher order terms are very much less than $n_d^{(1)}$. The velocity, $V = v_0 + u_1$ of the cnoidal wave in the laboratory frame is obtained by using equations (6), (8), (15), (17) and (19)

$$V = (1+\gamma_1)^{-1/2} + C^{(1)} - \frac{Aa}{3}\left\{\frac{1}{m^2}(2 - 3H(m)) - 1\right\}, \tag{20}$$

where $H(m) = E(m)/K(m)$ is the ratio of the complete elliptic integral of the second kind $E(m)$ to that of the first kind $K(m)$. The frequency $f = V/\lambda$, of the cnoidal wave can be obtained from equations (6),(15),(16),(18) and (19), as

$$f = \frac{V}{2K(m)m}\left[-\frac{aA}{12b}\right]^{1/2}. \qquad (21)$$

DISCUSSION

The coefficients a and b in the KdV equation (12) are functions of γ_1 and γ_2, therefore, equations (20) and (21) show that the amplitude and the frequency of the cnoidal wave are modified by the dust grain charge variations. From equation (21), it is clear that the frequency of the cnoidal wave increases with the amplitude of the cnoidal wave. From equation (20), we find that for values of modulus m above (below) a critical value $m_c \approx 0.9486$, the velocity V of the cnoidal wave increases (decreases) with amplitude.

In the limiting case, $m \to 1, \alpha_2 = \alpha_3 = 0, \alpha_1 = -A = 3u_1/a$ and $cn \to sech$. This situation happens when $\rho_0 = E_0 = 0$. Therefore, in this case, the cnoidal wave solution (15) is reduced to the dust acoustic soliton solution of ref. [4] with single ion species

$$\phi = \phi_m sech^2(\eta/\delta), \qquad (22)$$

where the amplitude of the soliton $\phi_m = 3u_1/a$ and width of the soliton $\delta = (4b/u_1)^{1/2}$.

It is plausible that the periodic signals may appear in astrophysical dusty plasmas. We infer that the present theory may be useful to explain the periodic signals as well as localized solitary structures in the space environments, where the dusty plasmas exist.

ACKNOWLEDGMENTS

One of the authors (LLY) thanks the Local Organizing Committee of ICPDP-2002 for providing the financial support to present this work at ICPDP. He also thanks the Management of KIE for the interest shown in this work. Useful discussion with Dr. Satyavir Singh is also thankfully acknowledged.

REFERENCES

1. Shukla, P.K., *Phys. Plasmas*, **8**,1791–1803 (2001).
2. Rao, N.N., and Shukla, P.K., *Planet. Space Sci.*, **42**, 221–225 (1994).
3. Popel, S.I., Yu, M.Y., and Tsytovich, V.N., *Phys. Plasmas*, **3**, 4313–4315 (1995).
4. Xie, B., He, K., and Huang Z., Phys. Plasmas **6**, 3808–3816 (1999).
5. Yadav, L.L., Tiwari, R.S., Maheshwari, K.P., and Sharma, S.R., *Phys. Rev. E*, **52**, 3045–3052 (1995).

Instability of Dust Lattice Waves due to Periodically Varying Charges of Dust Particles

Victoria V. Yaroshenko* and Gregor E. Morfill[†]

Institute of Radio Astronomy of National Academy of Science of Ukraine, Chervonoprapoma 4, Kharkov, Ukraine 61002
[†]*Max-Planck-Institut für Extraterrestrishe Physik, D-85740, Garching, Germany*

Abstract. The instability of the longitudinal and transverse dust lattice waves as a result periodical modulations of equilibrium dust charges is analyzed. The criterion and the maximum growth rate of the parametric instability is obtained for the case of a small depth modulation. The parametric resonance of the DL modes can provide a useful tool for determination of dusty plasma parameters.

INTRODUCTION

A consistent modelling of wave processes in dusty plasmas runs into interesting peculiarities that do not occur in standard plasma theory. Besides dealing with a medium in which massive charged particles can be weakly or strongly coupled, and thus sustain quite different waves, the dust charges are not fixed and independent from variations of the plasma parameters and potentials.

Indeed, a dust particle immersed in a plasma acquires a charge, which is completely specified by the electron-ion plasma and/or radiative environment. If the latter conditions are outwardly varying in time or in space, this gives a prerequisite for modulation of the equilibrium dust charges. One can expect that such charge variations might be easily produced in laboratory dusty plasmas. Although technical details of possible experiments and direct applications are beyond the scope of this theoretical paper, such charge modulation might arise due to the variations in time of the rf/dc discharge power or as a result of repeatedly illumination of the dust particles by an UV laser.

An idea of an external charge modulation may be especially important for laboratory dusty plasmas from a standpoint of excitation of the dust lattice (DL) modes and determination of unperturbed complex plasma parameters. We investigate in this paper the possibility of specific parametric instability of a condensed plasma, when equilibrium dust charges are subjected to periodic variations in time.

PARAMETRIC INSTABILITY OF THE DL WAVES

To investigate the DL wave propagation in a plasma crystal, formed by the particles with charges, periodically varying in time, we consider the simplified model of the 1D horizontal particle string [1], when each particle has the charge $Q(t)$, mass M and equilibrium separation Δ. The electrostatic potential of the particle is assumed to be the screened Coulomb potential with a screening length $\lambda_D \leq \Delta$, so that we take into account only the nearest neighbour particle interactions. Following standard procedure [2,3], we introduce the dimensionless particle displacement from the equilibrium for the n-th particle as $y_n = (y_{0n} - n\Delta)/\Delta$ along the string and the transverse one as $z_n = z_0 n/\Delta$. Assuming that $|y_n - y_{n\pm 1}|, |z_n - z_{n\pm 1}| \gg 1$ and $y_n, z_n \sim \exp(ikn\Delta)$ (a wavenumber k obeying $-\pi \leq k\Delta \leq \pi$), allows us to write basic equations, describing longitudinal and transverse oscillations as

$$\ddot{y}_n + 2\gamma \dot{y}_n + 4\sin^2\left(\frac{k\Delta}{2}\right)\Omega_\parallel^2(t)y_n = 0 \qquad (1)$$

$$\ddot{z}_n + 2\gamma \dot{z}_n + \Omega_v^2 z_n - 4\sin^2\left(\frac{k\Delta}{2}\right)\Omega_\perp^2(t)z_n = 0 \qquad (2)$$

Here γ is the damping rate due to the neutral gas friction [4]. The characteristic DL frequencies are defined through

$$\Omega_\parallel^2 = \frac{Q^2}{M\Delta^3}\left(2 + 2\frac{\Delta}{\lambda_D} + \frac{\Delta^2}{\lambda_D^2}\right)\exp\left(-\frac{\Delta}{\lambda_D}\right), \qquad (3)$$

$$\Omega_\perp^2 = \frac{Q^2}{M\Delta^3}\left(1 + \frac{\Delta}{\lambda_D}\right)\exp\left(-\frac{\Delta}{\lambda_D}\right), \qquad (4)$$

$$\Omega_v^2 = \frac{Q}{M}\frac{\partial E(z)}{\partial z}, \qquad (5)$$

where $E(z)$ is a sheath electric field.

Keeping in mind possible applications of the results to real experiments, a model of small periodic variations of the equilibrium particle charges seems rather reasonable. Let, for example, $Q(t)$ be varying in time with a frequency $\omega_\circ$ as

$$Q(t) = Q_\circ(1 - \beta\cos\omega_\circ t), \qquad (6)$$

where a depth modulation $\beta < 1$ insures that charge variations are small. In what follows we expand (3)–(5) in powers of small β, restricting our consideration to the lowest order. The squared frequencies (3)–(5) can be then represented in the form

$$\Omega_j^2(t) = \Omega_{\circ j}^2(1 - N\beta\cos\omega_\circ t) \qquad (7)$$

with $\Omega_{\circ j}^2$ being defined by corresponding relation (3)–(5) and N takes the value 2 for $\Omega_\parallel^2$ and $\Omega_\perp^2$ but $N = 1$ in the case of Ω_v^2.

Introducing a new function x either through $y_n = x\exp(-\gamma t)$ or $z_n = x\exp(-\gamma t)$ and inserting (7) into (1) or (2), reduces this to the canonical form of Mathieu's equation for the new function x

$$\ddot{x} + (a - 2q\cos 2\tau)x = 0 \qquad (8)$$

where $\tau = \omega_\circ t/2$. As is well known, Mathieu's equation has both stable and unstable solutions according to the coefficients a and q [5]. For small values of q the instability appears when $a = n^2$, $n = 1, 2, 3\ldots$. We are mostly interested in the first zone of parametric instability ($n = 1$, $a \to 1$, $q < 1$), which leads to the highest growth rate

$$\mu = \frac{1}{2}[q^2 - (a-1)^2]^{1/2} \qquad (9)$$

LONGITUDINAL DL WAVES

First we consider the longitudinal DL waves, which propagate along the string and are described by (1), from which it follows that the coefficients of Mathieu's equation (8) are given by

$$a = 4(4\Omega_{0\|}^2 \sin^2(k\Delta/2) - \gamma^2)\omega_0^{-2}, \quad q = 16\beta\Omega_{0\|}^2 \sin^2(k\Delta/2) - \gamma^2)\omega_0^{-2}. \qquad (10)$$

The maximum increment (9) is equal to $\mu = (beta/2)(1 + 2\gamma^2/\omega_o^2)$. The condition for the parametric resonance of the DL modes $a \to 1$, in the case of small damping reduces to $4\Omega_{o\|} \sin(k\Delta/2)\omega_o^{-1} \to 1$. Therefore in the short-wavelength limit, when $k\Delta \to \pi$, this yields $\omega_o = 2\Omega_{o\|}$. By contrast, the long-wavelength range, $k\Delta \ll \pi$, requires $\omega_o \ll 2\Omega_{o\|}$.

TRANSVERSE DL WAVES

For generality, we now consider the vertical DL modes (2), when the dust charges are modulated in time as (6). In the same way one easily gets Mathieu's equation (8) with the coefficients

$$a = 4\left(\Omega_{ov}^2 - 4\Omega_{o\perp}^2 \sin^2(k\Delta/2) - \gamma^2\right)\omega_0^{-2}, \quad q = 2\beta\left(\Omega_{ov}^2 - 8\Omega_{o\perp}^2 \sin^2(k\Delta/2)\right)\omega_o^{-2}$$

Hereby the resonance conditions $a \to 1$, $q < 1$ can be satisfied only if $\Omega_{ov} > 2\omega_o \sin(k\Delta/2)$. In experiments the frequency Ω_{ov} for the particles to oscillate in the plasma sheath is of the order of $60 - 200$ Hz while longitudinal $\Omega_{o\perp} \sim 100 - 500$ Hz [2]. This means that conditions of the parametric instability can be realized only for the long-wavelength DL, when $k\Delta \ll 1$. At small damping rate $\gamma < \Omega_{ov}$ the instability conditions imply $2\Omega_{ov}/\omega_o \to 1$, providing the maximum growth rate $\mu = \beta/2$.

Since the full solutions of the equations (1) and (2) are proportional to $\exp(\mu\omega_o t/2 - \gamma t)$, the excitation of the DL modes occurs when $1 > \beta > 4\gamma/\omega_o$. The latter determines a critical damping rate $\gamma_{cr} = \beta\omega_o/4$. The unstable DL modes to occur γ needs to be smaller than the threshold value γ_{cr}. This imposes the restriction on the possible values of the neutral gas pressure.

CONCLUSIONS

To summarize, we have investigated the possibility of parametric instability in a dusty plasma crystal resulting from a periodic modulation of the equilibrium particle charges.

We have discussed both the longitudinal and transverse dust lattice modes and formulated the conditions for the parametric resonance of these waves. Starting with the respective basic equations, we have shown that the possible parametric instability for the DL modes can be described by Mathieu's equation in the case of small depth modulation.

The dust charge has been evaluated until now only through model calculations. For this reason, every new experiment suggesting evaluation of the particle charges would be of importance. One such experiment might be observation of the parametric excitation of the DL modes due to the periodic variations in equilibrium dust charges. The identification of the maximum increment (9), corresponding to the condition $a \to 1$, immediately gives values of the characteristic frequencies Ω_{oj}^2 and thus determine the equilibrium dust charge Q_o (for given M, Δ, k and λ_D).

ACKNOWLEDGMENTS

This work was supported by the Alexander von Humboldt Foundation through the research fellowship (V. Y.). Fruitful discussions with A. Ivlev and H. Thomas are kindly acknowledged.

REFERENCES

1. Melanc, F., *Phys. Plasmas*, **3**, 3890–3897 (1996).
2. Ivlev A.V., and Morfill, G.E., *Phys. Rev. E*, **63**, 016409 (2001).
3. Morfill, G.E., Ivlev A. V., and Jokipii J.R., *Phys. Rev. Lett.*, **83**, 971-975 (1999).
4. Epstein, P., *Phys. Rev.*, **61**, 710-715 (1924).
5. McLachan, N.W., *Theory and Applications of Mathieu functions*, Clarendon Press, Oxford, 1951, pp. 39–104.

Planetary Ring and Spoke Simulation Experiment in Fine Particle Plasmas

Toshiaki Yokota

Department of Physics, Faculty of Science, Ehime University, Bunkyo-cho 3, Matsuyama, Ehime 790-8577, Japan

Abstract. The origins of planetary rings and the spokes of Saturn have been the subjects of much recent study. This paper describes a simulation laboratory experiment using a two component fine size particle plasma. The confinement ring is created around the rotating small insulator sphere in a fine particle plasma using unipolar induction. The spoke formation is performed by a shot of Nd-YAG pulse laser light.
Particle size and plasma density were in the order of 0.5 μm and 10^7 cm^{-3}, respectively, and particle charge was in the order of $\pm 25e$. All processes of ring formation and spoke creation were recorded on the VTR camera and analyzed in detail.

THEORETICAL FOUNDATION

The major component of the force for the existence of planetary rings is a gravitational force; however the structure and dynamics of rings have not been determined by only the gravitational force. In reality, the rings are constructed by a set of charged particles which move through a magnetic field [1]. It is well known that Jupiter, Uranus and Neptune also have rings similar to those of Saturn. These planets have dipole moments and rotate around their axes. This indicates that charged particles suffer electro-magnetic forces from planets, which is a key factor in our simulation experiments [2].

The force of gravity affects uncharged and charged particles in equal manner. For a particle of charge Q, the electric force is $F_e = QE$ where E is the strength of the electric field. If a planet has magnetic dipole moment M located at the gravity center and oriented along the rotation axis, a unipolar induction field will be created around a planet [3]. The simulating condition for the generation of ring is written as,

$$\frac{E_S}{E_M} = \frac{\frac{P_{m_S}}{2r_S^2 T_S}}{\frac{P_{m_M}}{2r_M^2 T_M}} = \frac{P_{m_S}}{P_{m_M}}\left(\frac{r_M}{r_S}\right)^2 \frac{T_M}{T_S} = 1$$

where the suffixes s and M mean Saturn and miniature sphere. If the values for Saturn and the miniature sphere are put into this equation, the relation between rotation frequency $n = 1/T_m$ and magnetic moment P_m become $P_m = 1.265 \times 10^5/n$.

We are interested in the role of meteorite collisions with the rings concerning in the creation of the spokes of Saturn. One hypothesis is that the expansion and ionization of vapor created by meteorites impact on Saturn's ring will trigger the formation of spokes

[4] [5]. This mechanism reveals some radial alignment of spokes, subsequent Keplerian shear, a fine structure, and properties of light scattering. Planetary meteorites impact on the Saturn's rings with velocity between 20 and 30 km/s. When a projectile impacts on solid target, a forward and a reverse shock is produced which compresses the projectile producing heating and subsequent vaporization. Thereafter the heated gas is converted to a plasma cloud with two components; an impa
ct component and a secondary component. These plasmas expand with sub-Alfvenic velocity V_d and create 'Spokes', see Fig.2.

EXPERIMENTAL SETUP

Two components of aluminum fine particle plasmas were generated by irradiation of UV-light on fine particles which were created by evaporation and cooling of aluminum vapors by argon gas flows (boat method). The particle size and density were controlled b
y boat temperature and gas flow rate. Typical particle size and plasma density were $\sim 0.5\mu m$ and $\sim 10^7 cm^{-3}$, respectively, (obtained by the laser light scattering intensities), and the particle charge was about ± 25 electrons obtained by splitting of fine particle beams in electrostatic fields [6].

A small insulated sphere adhered to a small permanent magnet in its inside was hung in the fine particle plasma and connected to a small motor by an insulating rod, and was rotated around the axis of the magnetic dipole moment. The rotating frequency was monitored by chopping of laser diode light (Fig.1). The process of ring formation and motion around the sphere were observed by using VTR and still camera and the pictures were digitally analyzed by a computer.

The spoke simulation experiment, at first, was investigated by shooting a Nd-YAG pulse laser light on the created fine particle ring (Fig.1). The pulse laser light melts the small part of fine particles and creates an expanding plasma. In this experiment high speed charged particles, accelerated by the electric field which is constructed with coaxial condensers, interact with the generated ring.

EXPERIMENTAL RESULT AND DISCUSSION

As stated, the ring formation experiment was performed under the condition of particle density $\sim 10^7 cm^{-3}$, particle size $\sim 0.5\mu m$, and particle charge $\sim \pm 25e$. The generated fine particle plasma was made of two component
(mixtures of positively and negatively charged particles). The Coulomb coupling parameter, Γ is estimated to be in $1 < \Gamma < 10$ for these conditions [7] [8].

A ring of fine particles around the magnetized sphere was clearly created when the rotation frequency agreed with the estimated value for the simulation conditions ($n=10$ Hz). The ring was located on the outskirts, 2 – 3 times the sphere radius and its thickness was very thin. The fine particles appeared to rotate slowly around the sphere in the same direction as the sphere rotation, as shown in Fig.3. According to the calculation, the

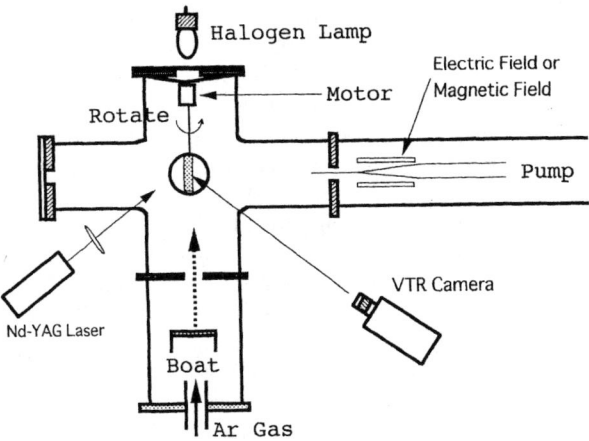

FIGURE 1. Experimental setup for simulation of planetary ring and spoke. Aluminum fine particles were generated by boat method and fine particle plasmas were created by irradiation of UV-light. Magnetized small sphere was rotated in plasmas. Plasma density was measured from the scattering intensity of laser light, and particle charge was obtained from the separation of plasma beam after pass through the static electric fields or static magnetic fields. Nd-YAG laser light shot the created ring to simulate the spokes. All of images were recorded on VTR camera and steal camera.

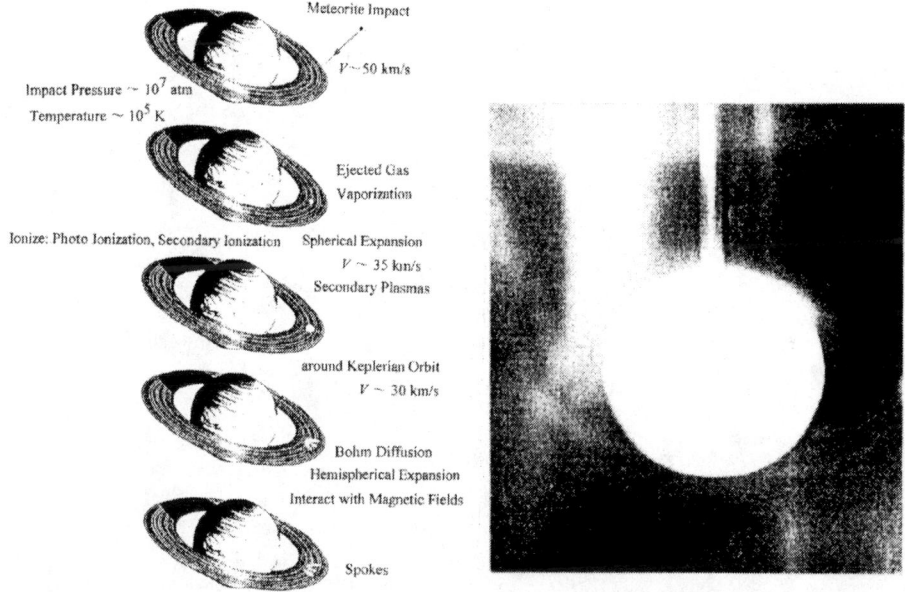

FIGURE 2. Spoke creation by the meteorite impact.

FIGURE 3. Typical image around the rotating small sphere. Ring was created when rotating frequency agreed with the simulated value (n=10 Hz).

electric field E_r generated by unipolar induction takes positive values over a distance twice the satellite radius in the equatorial plane, as shown in Fig.3. In this experiment, the charged fine particles were trapped in this region and co-rotated in the equatorial plane. This indicates the fact that the unipolar induction played an important role for ring creation of outer planets such as the Saturn in the early stage of solar system formation.

The next focus of the simulation experiment is the origin of the spokes of Saturn rings, according to the meteorite impact hypothesis [4]. A Nd-YAG laser light is focused on the fine particle ring by using focusing lens, and melts the partial regi
on of ring particles in order to generate a local expanding gaseous plasma, just like an impact plasma as shown in the experimental setup (Fig.1). A small spot of gaseous plasma is created on the simulated ring by shooting a laser light, but the "spoke" expansion has not yet been observed. More experiments are being performed, and also, an electrically accelerated fine particle impact experiment is planned.

CONCLUSION

We can simulate planetary ring formation in a laboratory by applying unipolar induction. Our inital results indicate that unipolar induction played a key role in the early formation stage of the solar system. Further experiments are designed to study the charge and polarity of the rings, as well as the "spoke" formation.

ACKNOWLEDGMENTS

This work was partially supported by a Grant-in Aid for Scientific Research from the Ministry of Education, Science and Culture, Japan (No.10680461).

REFERENCES

1. Grun, E., and *et.al*, H. K., *Geophys. Res. Lett*, **24**, 2171–2174 (1997).
2. Juhasz, A., and Horanyi, M., *J. Geophys. Res.*, **102**, 7237–7246 (1997).
3. Bliokh, P., Sinitsin, V., and Yaroshenko, V., *Dusty and Self-Gravitational Plasmas in Space*, Kluwer Academic Publishers, Dordrecht, 1995.
4. Goertz, C. K., and Morfill, G. E., *Icarus*, **53**, 219–229 (1983).
5. Morfill, G. E., and Goertz, C. K., *Icarus*, **55**, 111–123 (1983).
6. Yokota, T., and Manabe, S., *J. Quant. Spectrosc. Radiat. Transfer*, **61**, 219–225 (1999).
7. Yokota, T., *Physica Scripta*, **T84**, 175–177 (2000).
8. Yokota, T., *Frontiers in Dusty Plasmas*, pp. 321–328 (2000).

Thermophoretic and Ion Drag Forces Acting on Free-Falling Charged Particles in an RF-Driven Plasma

C. Zafiu, A. Melzer, A. Piel

Institut für Experimentelle und Angewandte Physik, Christian-Albrechts-Universität Kiel, 24098 Kiel, Germany

Abstract. The thermophoretic force and ion-drag acting on micro-particles falling through a long RF plasma has been measured here under two different experimental situations: either an unperturbed plasma characterized by its symmetric potential and temperature distribution or a plasma with an asymmetric temperature and potential profile. The dominant effect of the gravitation force is reduced here due to the dynamic equilibrium with the friction force of the particles with the neutral gas.

The reduction or even suppression of the dominant influence of the gravitation force on the complex plasma systems has allowed the study of "hidden" effects like voids in three-dimensional dust structures [1, 2, 3, 4], which occur due to smaller forces like thermophoretic force and ion drag. For typical particle diameters of $1\ldots 10\mu$m, these forces are about one order of magnitude smaller than the gravitation force.

Simulation of experiments under micro-gravity [2] show that an ad-hoc enhancement of the ion drag force of at least a factor of 10 is necessary to simulate voids like those observed in experiments [1]. Therefore, quantitative measurements of these two forces are necessary in order to check or improve the theoretical calculations and simulations.

Here, experiments to measure the ion drag and thermophoresis under laboratory conditions are presented.

In our experiment, the micro-particles are dropped through a long plasma discharge. Within several centimeters, the particles reach a regime of constant vertical velocity due to the balance of the gravitation force by the friction force of the falling particles with the neutral gas.

$$m\vec{g} - m\beta\vec{\dot{y}} = 0, \qquad (1)$$

Thus, this experiment provides a dynamic equilibrium, thereby effectively reducing the influence of gravity.

In the following, the forces acting on the particles on the horizontal direction (see Fig. 1d) are summarized. In a discharge tube, the electric field structure is determined by the positive polarization of plasma with respect to the walls. This typically provides a linearly increasing radial electric field $E(x)$. The electric force on a particle of charge Q is then defined by

$$\vec{F}_{electric} = Q\vec{E}(x). \qquad (2)$$

For the particles with a horizontal velocity, the friction force is

$$\vec{F}_{frict\_x} = m\beta\vec{\dot{x}}. \tag{3}$$

The ions are accelerated towards the walls due to the same electric field $E(x)$ in the plasma and act on the negatively charged particles by the so-called ion drag [2, 5, 6]. It has two components: a direct one, which is due to the momentum transport of the ions which hit the particle surface

$$\vec{F}_{collect} = n_i m_i \vec{v}_i v_s \pi b_c^2, \tag{4}$$

and a second one, the Coulomb component, which is produced by exchange of momentum of the ions whose trajectories are bent in the electric field of the particles

$$\vec{F}_{Coulomb} = 4\pi n_i m_i \vec{v}_i v_s b_{\frac{\pi}{2}}^2 \Gamma. \tag{5}$$

Here, v_s is the mean speed of ions approaching the particle, b_c is the maximal collection impact parameter, $b_{\frac{\pi}{2}}$ is the impact parameter corresponding to a deflection angle of $\pi/2$, Γ is the Coulomb logarithm. n_i, m_i, T_i, q_i and v_i are the number density, mass, temperature, charge and drift velocity of ions, and λ_D is the Debye length.

The thermophoretic force due to a temperature gradient of the neutral gas pushes the particle to colder regions of the plasma and is given by [6]

$$\vec{F}_{thermo} = -\frac{32}{15}\frac{r_p^2}{v_{th}}\lambda\vec{\nabla}T_n, \tag{6}$$

where λ is the translation part of heat conductivity of the gas, and v_{th} is the thermal velocity of the neutral gas.

In summary, the horizontal equation of motion of the falling particles is described by

$$m\vec{\ddot{x}} = \vec{F}_{electric} + \vec{F}_{frict} + \vec{F}_{collect} + \vec{F}_{Coulomb} + \vec{F}_{thermo} \tag{7}$$

The influence of the ion drag and thermophoresis have been studied here both for a symmetric and an asymmetric temperature and potential distribution in the plasma. This has been done to establish plasma regions with opposing ion drag and thermophoretic forces.

The experiments have been performed in a long RF discharge tube (see Fig. 1a) in He at a gas pressure between 100 and 200 Pa, for plasma power between 0.8 and 2.8 W. The particles are observed at a certain level in the middle of the plasma column in the scattered light of a thin vertical laser sheet, using a high frame rate video camera. The laser stroboscopically illuminates the particles at 400 Hz. This method provides a high time resolution, even when using a low speed camera.

The particles used here are of 9.55 µm and 20.02 µm diameter and are made of Melamine Formaldehyde (MF) and Polymethylmethacrylate (PMMA) respectively.

The forces on the particles are measured here in the following way: From the particle trajectories, the acceleration and friction force (in horizontal direction) are directly

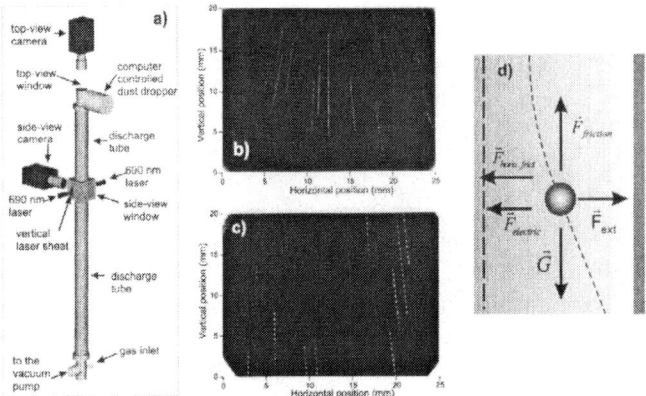

FIGURE 1. a) Scheme of the experimental setup, b) video image (symmetric temperature and potential distribution), c) video image (asymmetric temperature and potential distribution), and d) horizontal and vertical forces acting on the falling particles.

obtained. The plasma potential profile and thus the electric field distribution have been derived from probe measurements. Therefore the electric force on the particles could be determined. To measure the thermophoretic force, the temperature profiles of the neutral gas have been measured with a thermo-resistor. To evaluate the ion drag, the ion velocity has been calculated employing their mobility μ_i and the electric field structure in the plasma $v_i = \mu_i E(x)$. It must be mentioned here that the equations describing the ion drag (Eq. 4, 5) are applicable under the assumption that no ion collisions occur inside the Debye sphere, which is the case of low pressure plasmas. In our case, however, the ion mean free path is small comparable to the Debye length.

Two different experimental situations have been studied here. First, the behavior of falling particles through an unperturbed plasma, characterized by its natural radially symmetric electric potential and temperature distribution, has been analyzed. In the experiment, it can be seen that the particles move radially outward with increasing velocity (see Fig. 1b). This effect becomes stronger with increasing plasma power.

In the second case, the temperature distribution has been altered using a heat source on one of the observation windows (left side of the Fig. 1c). The probe measurements revealed that also the electric potential distribution has been perturbed in this case due to the use of large metallic parts very close to the discharge.

Fig. 2 shows the forces acting on the dust particles in the symmetric case. Therefore, we have chosen radial coordinates with the origin on the discharge axis. The radial acceleration and friction force from the radially outward moving particles are shown in Fig. 3a. Additionally, the inward electric field force is shown together with the total inward force. This total force in Fig. 3a is supposed to be balanced by the ion drag and the thermophoresis. It is seen that the sum of the measured thermophoresis and ion drag is slightly smaller than the total measured inward force. The small discrepancy between these two total forces might lie in the applicability of the ion drag equations (Eq.4,5) for our experimental conditions. For these measurements, a ten-fold increased ion drag is not compatible with our results. The same analysis has been done for the asymmetric

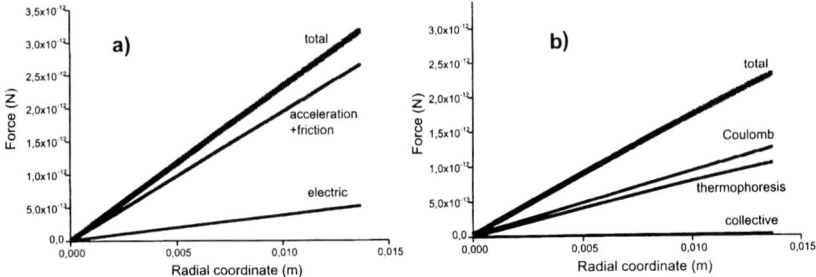

FIGURE 2. Symmetrical temperature and potential distribution case; a) measured force, from the particle trajectories analysis, taking into account the friction and the electric force, and b) theoretical force (thermophoresis and ion drag).

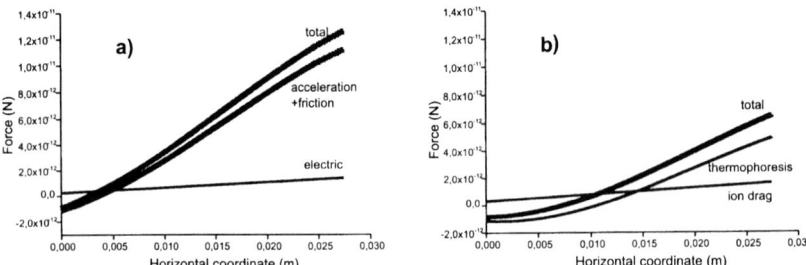

FIGURE 3. Asymmetrical temperature and potential distribution case; a) measured force, from the particle trajectories analysis, taking into account the friction and the electric force, and b) theoretical force (thermophoresis and ion drag).

experimental case. Here, also plasma regions have been observed with opposing ion drag and thermophoresis. Again, the total measured force summarizing the inward acting forces is slightly larger than the sum between the measured thermophoretic force and the ion drag. Concluding, for both cases analyzed here, the ion drag and thermophoresis have been experimentally measured, showing a quite good agreement with the theoretical predictions, and emphasizing the applicability limits of the actual models describing these two forces for our experimental conditions.

Financial assistance from INTAS (97-755) is gratefully acknowledged.

REFERENCES

1. Morfill, G. E., Thomas, H. M., Konopka, U., Rothermel, H., Zuzic, M., Ivlev, A., and Goree, J., *Phys. Rev. Lett.*, **83**, 1598–1601 (1999).
2. Akdim, M. R., and Goedheer, W. J., *Phys. Rev. E*, **65**, 015401 (2002).
3. Morfill, G., and Tsytovich, V. N., *Phys. Plasmas*, **9**, 4–16 (2002).
4. Goree, J., Morfill, G. E., Tsytovich, V. N., and Vladimirov, S. V., *Phys. Rev. E*, **59**, 7055–7067 (1999).
5. Barnes, M. S., Keller, J. H., Forster, J. C., O'Neill, J. A., and Coultas, D. K., *Phys. Rev. Lett.*, **68**, 313–316 (1992).
6. Perrin, J., Molinás-Mata, P., and Belenguer, P., *J. Phys. D: Appl. Phys.*, **27**, 2499–2507 (1994).

SECTION 5: SUMMARY AND PROGNOSIS

A Synopsis of Recent Theoretical Developments in Dusty Plasma Physics

D.A. Mendis

Department of Electrical and Computer Engineering, University of California, San Diego, La Jolla, CA 92093, USA

Abstract. A brief summary is presented of recent developments in the study of naturally occurring dusty plasmas and in theory, together with a short discussion of some outstanding problems.

INTRODUCTION

At the conclusion of this highly successful Third International Conference on the Physics of Dusty Plasmas, I have the challenging yet pleasant task of making a few summary comments, confining myself to observations and theory of naturally occurring (i.e. essentially cosmic) dusty plasmas. Laboratory dusty plasmas will be similarly treated in a companion summary paper.

The progress as well as the outstanding problems in the area of cosmic dusty plasmas, in a variety of regions, was reviewed in 3 comprehensive overview talks. Oral and poster presentations of ongoing research in the area covered a wide range of environments (including the terrestrial polar mesosphere, planetary magnetospheres and interstellar clouds) and a wide range of topics (including grain charging and growth, grain transport, and collective processes in dusty plasmas including self-gravitating ones).

In this summary, which is brief by necessity, it is impossible to do justice to all the interesting results that were presented. In any case, they will be discussed in accompanying papers in the Proceedings. So here I will confine myself to a few general comments on the progress and the outstanding problems as I see them, as well as some prognostications.

I will begin with an enumeration of the various cosmic dusty plasmas environments that have been considered so far. As can be seen in Tab. 1, these environments span a wide range from planetary rings to supernova shells with a wide range of plasma and dust parameters. Furthermore (as pointed out by Professor M. Horanyi in his overview of dusty plasmas in the solar system), there are also a variety of naturally occurring dusty plasmas. These include (a) positively and/or negatively charged dust immersed in a mution and electron plasma (as in comets and planetary rings), (b) positively charged dust and electrons (as on the moon and asteroid surfaces) and (c) positively and negatively charged dust (as in sand storms and volcanic eruptions).

I will next comment on a few specific topics.

TABLE 1. Some sample values

Environment	n_e (cm^{-3})	T eV	n_d (cm^{-3})	a	n_n (cm^{-3})	$|Z_D|$	d/λ_D	Γ_d
Saturn's E-ring	10	10-100	10^{-7}	1	1	$\sim 10^{-4}$	0.1	$< 10^{-2}$
Saturn's F-ring	10	10-100	<10	1	–	$\sim 10 - 10^{-2}$	$< 10^{-3}$	$< 10^{-4}$
Saturn's spokes	$0.1 - 10^2$	2	1	1	–	~ 10	$< 10^{-2}$	~ 1
Red Giant Photospheres	2×10^2	0.1	2	0.01	5×10^8	1	$\sim 6 \times 10^{-2}$	$\sim 2 \times 10^{-6}$
Halley's comet Inside ionopause Outside ionopause	$10^3 - 10^4$ $10^2 - 10^3$	< 0.1 ~ 1	10^{-3} $10^{-8} - 10^{-7}$	0.1-10 0.01-10	10^{10} –	$10 - 10^3$ $20 - 2 \times 10^4$	> 1 > 10	$\ll 1$ $\ll 1$
Noctilucent clouds	10^3	0.013	$10 - 10^3$	0.01-0.1	10^{14}	$1 \sim 80$	0.2	$2 \times 10^{-4} - 0.5$
Interstellar molecular clouds	10^{-3}	0.001	10^{-7}	0.2	10^4	?	0.3	$< 10^{-5}$
Zodiacal dust disc (IAU)	5	10	10^{-12}	10	–	10^4	5	$\sim 10^{-6}$
Solar F-corona	5×10^5	80	10^{-7}	0.3	–	$\sim 5 \times 10^2$	10	$\sim 10^{-8}$
Planetary Neblae (em)	10^4	1	10^{-7}	0.2	–	$\sim 3 \times 10^2$	30	$< 10^{-9}$
Supernovae shells	10^3	0.2	10	0.01	–	20	$\sim 5 \times 10^{-2}$	$\sim 10^{-3}$

Cosmic Dust-Laden Plasmas

GRAIN CHARGING

Central to all the novel phenomena manifested by dusty plasmas (cosmic or laboratory) is the electrostatic charging of the dust due to collection or emission of electrons and ions by the dust. Yet even the simplest case of the charging of a small isolated grain in a thermal plasma is not yet completely resolved, although great progress has been made in recent times as discussed by Martin Lampe in his topical review on the subject. The limitations of the so-called 'orbit motion limited' (OML) calculation of the collection current of the attractive species, due to the presence of 'potential barriers' to some positive energy ions, has been discussed by numerous authors. While it was recently shown by Lampe that the OML approach was a very good approximation in the collisionless case when $a/\lambda_D \ll 1$ (a being the grain radius and λ_D is the Debye shielding distance), what is more interesting are the consequences of ion trapping and untrapping near the grain due to charge exchange in collisional plasmas of the type that typically occur both in space and in the laboratory. In this case it has been established that while the trapped ions dominate the shielding cloud near the grain, the trapping and de-trapping of these in a steady state, leads to a substantial increase of the ion current. So, it seems that plasma kinetics of the charging of an isolated grain in a thermal plasma is now well understood due to very recent work. There is, however, another aspect of grain charging that has so far received much less attention. These are the important electrical properties of the grain such as the sticking coefficients of electrons and ions (s_e and s_i). While it is customary to assume that both s_i and s_e are close to unity, this is certainly not the case for s_e, particularly in the case when the grains are small (i.e. $a < 0.01\,\mu\text{m}$). The incoming electrons have to loose their kinetic energy by interaction with surface phonons and plasmons before they can stick to the surface. Available calculations, which show that s_e is already significantly less than unity, are not applicable to small grains, not only because multiple collisions with surface phonons are required for the electron to loose its kinetic energy, but also because the more energetic electrons in the Maxwellian tail can tunnel right through the grain. So the electron collection current could be significantly decreased. Clearly these issues need to be addressed.

Grains in the cosmic environment are not necessarily smooth spheres, so the role of grain size, shape and surface roughness needs to be addressed. Interesting work as already been done on the role of grain size on secondary electron emission. Theoretical studies predicting highly enhanced yields when the grain size is comparable to the penetration depth of the primaries have been verified by laboratory observations. Laboratory studies have also shown that the photo-electric yield increases dramatically when the grain size is comparable to the skin depth, yet existing theoretical estimates of the enhanced yields, using semi-classical approaches fall well short of the observations. Since cosmic grains, (including VSG's with $a < 0.01\,\mu\text{m}$) are often immersed in a background UV radiation field, there is clearly a need to resolve this problem. The role of electric field emission in limiting the surface charge of grains with significant surface roughness also needs further study, particularly experimental.

DYNAMICS

The role of electrodynamic forces in the orbital dynamics and orbital evolution of charged dust in planetary and cometary magnetospheres has now been recognized for some time. The Voyager spacecraft observations of Saturn's rings in early nineteen eighties as well as spacecraft observations of comet Giacobini-Zinner in the mid nineteen eighties gave this area a major boost, since several unusual observed phenomena could only be explained by a proper combination of gravitational and electrodynamic forces. The 'unexpected' observations of quasi-periodic collimated high speed streams of fine dust from Jupiter by the Ulysses and Galileo spacecrafts, which were nicely explained using a gravitoelectrodynamic model of dust injected into the Jovian magnetosphere by its volcanic satellite, Io, further boosted this area of study. All of this was discussed in the comprehensive review of dusty plasmas in the solar system by Mihaly Horanyi. Even more exciting are the predictions of Professor Horanyi and his co-workers, such as the existence of an entirely new class of nonequatorial 'halo' orbits of charged dust in the Saturnian magnetosphere. Their existence could be verified during the Cassini mission to Saturn in 2004. Another prediction that could be verified during the mission is that of Ove Havnes, viz the existence of Mach cones, mediated by the dust acoustic wave, within Saturn's rings. While it is very gratifying when otherwise puzzling observations have natural gravitoelectrodynamic explanations, it is even healthier for the field when predictions are made that could be verified in the near future. In that connection, the Rosetta mission to a comet later this decade would also provide an excellent opportunity for verifiable predictions. In fact, the first verifiable prediction of the role of dust charging in the dynamics of dust in a solar system body concerned the spatial distribution of dust of various sizes in the tail of comet Giacobini-Zinner, which was supported by in-situ observations during the 1985 encounter by the ICE spacecraft. Sporadic brightness variations of comet Halley at large heliocentric distances (< 10 AU) during its last apparition has been attributed to electrostatic levitation of surface dust, predominantly from the night side during the encounter of high speed solarwind streams by the bare cometary nucleus. This proposal is clearly verifiable during the distant encounter of the target comet by the Rosetta spacecraft. Comets provide excellent natural laboratories for the study of a whole range of dust-plasma interactions, with highly variable solarwind interaction modes as a comet approaches the sun. So with the Rosetta mission in mind, this is a good time to revisit the problem of dust-plasma interactions at comets together with their physical and dynamical consequences.

The comprehensive overview of dusty plasmas in the terrestrial ionosphere (particularly in the upper part of the mesosphere and the lower part of the thermosphere) by Ove Havnes has raised a multitude of questions pertaining to the physics and dynamics of charged dust, the answers to which are far from clear. The problem of meteoroid entry and flight in the ionosphere is a central one that clearly needs more study, both observations and theoretical. Very recent studies that self-consistently solve the combined equations of the continuity of mass, momentum, energy and charge, are promising, showing for instance how the same micro-meteoroid can have different charge polarity during different portions of its path, while thermionic emission of electrons from them could be a major source of electrons at higher altitudes. These studies need to be continued and refined. Also the continuing sounding rocket, radar and lidar studies of NLC's, PMSE's and meteor trails would lead to a more secure understanding of these phenomena in the future.

Finally, I would like to draw attention to high resolution images of circumstellar dust discs that have been obtained in very recent times. An excellent example, namely the dust disc surrounding the nearby star Vega (which is displayed on the cover page of news magazine Science News, vol. 161, no. 18, May 4, 2002), clearly shows 'rings, gaps, arcs, warps and clumps'. This dust disc, where planetary formation may be in progress, is intriguingly similar to the Saturnian rings system. There appears to be large quantities of very fine dust in this disc as well as others like that of the star Beta Pictoris. Also infrared images of the disc at different wave lengths lead to inferences of the spatial distribution of grains of different size. It seems to me that the morphology of these discs are now ripe for study taking into account the electrodynamical effects due to grain charging. Our present ignorance of the plasma and magnetic field environment there is certainly a drawback. It would, however, be instructive to see if one could reasonably match the detailed infrared images of the dust by adopting reasonable particles and fields models and following the gravitoelectrodynamic evolution of the charged dust grain orbits as was recently done for Saturn's E-ring by Mihaly Horanyi and his coworkers.

GRAIN FORMATION

Another area that deserves further attention is the important area of grain growth in the space plasma environment. Although it may not be intuitively obvious, there is a highly intimate connection between dust and plasma, the two states of matter that are furthest removed from each other by the measure of entropy. It has now been repeatedly shown in the terrestrial laboratory, how readily dust forms in chemically reactive plasmas. At the same time that dust, which is vital for the formation of planetary systems and therefore for life itself, arises entirely in cosmic plasma environments, namely the outer atmospheres of evolved red giant stars and in the expanding outer shells of supernovae. In the latter case, one can now actually 'see' when and where the dust is formed due to simultaneous visual and infrared observations. There is a dramatic increase in the infrared emission, presumably due to the rapid formation of dust in the expanding shell, precisely at the time when the visual brightness of the central star decreases due to obscuration by this newly formed dust.

In this connection laboratory studies can greatly enhance our understanding of dust formation in various cosmic environments. For instance recent studies, which emphasize the thermodynamic similarity between sooting hydrocarbon flames and the outer atmospheres of carbon stars, seems to resolve the longstanding difficulty in understanding the prolific rate of dust formation seen in the stars. The continuing studies of grain growth in laboratory plasmas would no doubt enhance our understanding of dust formation in the cosmic environment, too.

CONCLUDING REMARKS

There are several other important topics, such as the crucial role of dust in the ionization equilibria of cosmic dusty plasma, that I have not touched upon in this brief summary. I will conclude with a few very general remarks. First and foremost, the field of dusty plasma physics overall (both cosmic and laboratory) is in excellent health, and continuing to grow vigorously as seen by the growth of publications in Table 2. While the impetus to the most recent and dramatic growth phase was the fabrication in the laboratory of dust Coulomb crystals, the early boosts came from in-situ observations of cosmic plasmas (see Table 2). While the richness and complexity of the field, with numerous unresolved and challenging problems, will continue to stimulate the field ensuing its continued rapid growth into the foreseeable future, several forthcoming space missions (e.g. Cassini and Rosetta) will provide important observational boosts to the study of cosmic dusty plasmas. The fascinating high resolution photographs of dusty cosmic regions, obtained recently by the Hubble Space Telescope as well as powerful ground based telescopes where new stellar and planetary formations are taking place in plain sight (the so-called cosmic nurseries and circumstellar dust discs) are only just beginning to underscore the importance of dusty plasma phenomena in these vital astrophysical regions.

GROWTH OF PUBLICATIONS IN DUSTY PLASMAS

Period	Number of Papers	
1970–1972	22	
1973–1975	17	
1976–1978	32	
1979–1981	64	← Voyager/Saturn (Gravitoelectrodynamics)
1982–1984	74	
1985–1987	142	← Halley Encounter
1988–1990	182	← Collective Effects/Plasma Processing
1991–1993	255	
1994–1996	542	← Dust Coulomb Crystals
1997–1999	689	
2000–2002	702(>900)	

Table 2

* 'Inspec' Database (Dust, plasma)
* The first number in the last row is the number of publications for the first 28 months of that 3-year period. The number in parenthesis is the linear extrapolation for the full 36-month period.

A Synopsis of Recent Experimental Developments in Complex (Dusty) Plasma Physics

G.E. Morfill

Max-Planck-Institut für extraterrestrische Physik, 85740, Garching, Germany
and
Centre for Interdisciplinary Plasma Science, Garching, Germany

Abstract. A brief summary is presented of recent developments in laboratory studies of complex (dusty) plasmas, both in terrestrial laboratories and under microgravity conditions, together with a short discussion of possible future developments.

INTRODUCTION

The young research field of "complex dusty plasmas" has shown a dramatic growth and development in the last eight years. The publication rate has increased by a factor 10 and, more important, the number of institutes involved in the field has grown from a mere handful in 1994 to over 100 today. The field comprises diverse research topics — in the area of "dusty plasmas" there are environmental research (atmospheric, mesospheric phenomena), pollution and combustion, solar system research (zodiacal cloud, cometary science, planetary rings), astrophysics (interstellar clouds, dust formation in stellar envelopes, star and planet formation), industrial dusty plasmas (particle growth, coating, deposition, dust cleaning technology) and so on. In the area of "complex plasmas" — where the microparticles may be regarded truly as one plasma component — as opposed to being a contaminant, a plasma source or absorber, a catalyst for chemical reactions or for transferring the effects of (otherwise possibly unimportant) forces onto the plasma (e.g. gravity, radiation pressure, etc.) — there has been tremendous progress, too. It is this research area, which is the topic of this summary paper. The contribution is divided into four parts — laboratory research, research under microgravity conditions, theory, and, last but not least, the future developments.

LABORATORY RESEARCH

The remarkable aspects of fundamental plasma research are:

1. making (one component of) plasmas visible at the most elementary (the kinetic) level,
2. investigating the previously unexplored field of strongly coupled plasmas, including liquid and crystalline states;
3. observing plasma processes in "slow motion" with unprecedented resolution (the dust plasma frequency is about 10 Hz),

4. investigating complex many-particle systems under conditions of very little damping.

These aspects have precipitated a boom in this field — understandably — because a plethora of new phenomena have suddenly become accessible for experimental research. The essential elements for complex plasma research are:

1. the plasma chamber
2. the diagnostics
3. the manipulation techniques
4. the microspheres

Plasma Chambers

Different plasma chambers are being used, some are modified existing reactors and others are special designs. They range from radiofrequency (capacitatively or inductively coupled) devices, d.c. discharges of different designs to combustion plasmas. The innovation appears almost boundless, with different geometries, electrode designs, grids for plasma control, operating conditions, etc. The surprising commonality is that complex plasmas can be produced in all of these systems, different research conditions can be achieved and different processes can be studied. We will come back to this later.

Microparticle Component

Regarding the "new" plasma component — the microparticles — again a whole range of possibilities are being explored. In the laboratory the major constraint is provided by gravity — if the microparticles are too large they cannot be supported and a complex plasma is impossible. Nevertheless, there is tremendous ingenuity in overcoming this. Small (sub)micron-sized particles can be suspended in the pre-sheath of r.f. discharges, for instance, and can be used for plasma diagnostics (as a tiny in-situ probe), for the study of strong coupling phenomena and for complex plasma investigations. Larger particles can be employed to progressively investigate the plasma sheath properties at different locations from the electrode and for the study of (essentially) two-dimensional strongly coupled complex plasmas. Large microballoons can be used to investigate interparticle forces in the limit where the particle size is about the same as the Debye length, and thus complement existing collision studies carried out with smaller solid particles. Needle-shaped particles can be employed to investigate systems with rotational as well as translational degrees of freedom. Last, but not least, a new technology employing the thermophoretic force can be used to (partially) offset gravity and perform limited experiments in the bulk plasma. This technique is interesting in its own right, but in addition it can be very useful in supplementing parabolic flight tests to improve the design of microgravity experiments.

Diagnostics

The standard technique for diagnostics is laser light illumination of the microparticles and detection of their position using CCD cameras. By fanning out the laser beam one can thus monitor a 2-dimensional sheet of the complex plasma at a speed (or rate) given by the speed of the camera. The crystalline state of complex plasmas can be investigated in 3 dimensions, by scanning through the system. This does not work for the liquid or disordered states, because the identification of particles becomes a major difficulty the longer the sweeping (or scanning) time lasts. The latter is limited by the technology employed in moving the laser sheet (rotating mirrors or tilting systems as used in laser light shows), the light intensity and the optical design for focussing the cameras. In addition to these "digital techniques" being developed (where the particle is either visible in the laser light, or it is not) there are also "analog techniques", which employ thick light sheets of different colour and identify the third dimension by the colour ratio — which, when calibrated, can be translated into a distance. Other methods currently investigated employ luminescent particles (which can be used to measure the particle temperature), stroboscopic flashing of the illuminating laser (which provides a high resolution position and velocity measurement), velocimetry for identifying particle trajectories and first attempts at using colours generally to position particles accurately in three dimensions. Generally, we may say that the techniques are still quite elementary, but the degree of subtlety has increased considerably in the last few years.

Manipulation

One of the strengths of the complex plasma research is the possibility to conduct active experiments. The microparticles can be manipulated in a number of ways — using laser light pressure, modulating the voltage on the electrodes, using electron or ion beams, with ultraviolet light, thermophoresis, etc. Magnetic fields provide another means for particle manipulation. This can be either direct — via the $\underline{E} \times \underline{B}$ force — or indirect via ion drag, where the ions respond to the $\underline{E} \times \underline{B}$ force in the first place. This may lead to unexpected effects, depending on position and the coupling strength in the complex plasmas. Another interesting technique is the analysis of the phonon spectrum in plasma crystals and global modes, the latter especially for finite, small, Coulomb clusters. Last, but not least, is the "momentum technique", where either a continuous flow of neutral gas or sudden "puffs" are used to cause special situations or perturbations. This enables studies of the coupling of sound waves onto complex plasmas, rotational (angular) momentum transfer, etc.

MICROGRAVITY RESEARCH

Research under microgravity conditions is largely concentrated on the first natural science experiment to be placed in the International Space Station (ISS) — PKE-Nefedov. PKE stands for **P**lasma **K**ristall **E**xperiment, and it was named after Anatoli

Nefedov, the great Russian scientist and co-PI of PKE, who died suddenly in January 2001, before the experiment was transferred to the ISS. It is a joint project between the Max-Planck Institut für extraterrestrische Physik-Center for Interdisciplinary Plasma Science, and the Russian Academy Institute for High Energy Density (IHED), Moscow.

Additional microgravity work is now being conducted on parabolic flights both in the US and in Japan.

The major new results obtained by PKE-Nefedov so far concern the nature of complex plasma interfaces, the de-charging of complex plasma clouds, the microstructure of shock waves, the stress-free formation of plasma crystal domains, anomalous (runaway) particle coagulation and the coalescence of complex plasma clouds, to name but a few. Further work, in collaboration with GREMI (Orleans), involved particle growth experiments using PKE-Nefedov (a task for which it was not designed originally) during the Andromede mission. These examples may highlight the importance of microgravity research for this new field of complex plasmas. On Earth, experiments have to be conducted in conditions where particle levitation against gravity works — i.e. under strong electrostatic fields, such as produced in the plasma edge sheaths. In space, through the absence of gravity and by conducting experiments in the bulk plasma (and not in the sheath) it is possible to achieve a degree of homogeneity and a reduction of the influence of bulk forces by about two orders of magnitude compared to what is possible in the laboratory on the ground. This allows the study of thermodynamically interesting processes near local thermal equilibrium under conditions where stored energy is only a minor factor.

THEORY

Theory has blossomed in the past six or eight years, in part leading the experimental work, in part lagging it. Great advances have been made in the development of a kinetic theory that incorporates particle mass and charge as variables. Such a theory is difficult to develop, because the governing equations are mostly nonlinear and stochastic. For instance, particle charge fluctuations can be considered as stochastic, however, the charges also depend on the particle number density.

Numerically, the approaches using molecular dynamic (MD) techniques with "dressed" Yukawa particles have been extremely useful, e.g. for predicting the liquid / solid phase transition. They are not able, so far, to study the propagation of condensation fronts, however, and other dynamical processes. More sophisticated codes are being developed now, so that it will be interesting to see the progress.

A great strength of this field is the plethora of analytical (or semi-analytical) research that is possible. This is easy to note in the vast field of "waves" — dispersion relations, mode couplings, damping mechanisms, new wave types including surface, shear, bending and compressional waves, etc. continue being explored. Nonlinear phenomena like solitons are beginning to be studied in greater detail and the same goes for shocks — all these both at the hydrodynamic and the kinetic levels.

Another exciting area is the field of strong coupling phenomena. Investigation of Coulomb clusters using normal or global modes, both experimentally and theoretically, is a particularly convincing synthesis of physics, with phonon spectral analysis and instabilities rounding this off.

Even in the field of single particles, or in the collision experiments between two particles, lies an enormous potential. Using these particles as plasma microprobes to investigate the plasma sheath and the interparticle forces (e.g. perhaps finding the elusive "shadow force") is of great fundamental significance and should be pursued further.

FUTURE DEVELOPMENTS

Under microgravity the development in the next decade promises to be very exciting. The Plasma Kristall Experiment — PKE-Nefedov — which is still operative and performing well on the International Space Station (ISS) will be replaced in 2004 with another joint Russian/German instrument, PK-3plus. This will be based on the same r.f. discharge technology as PKE-Nefedov, however, there will be an improved electrode design, particle injectors and diagnostics. Two years later, in 2006, it is planned to launch another PK-laboratory to the ISS, PK-4. This will consist of a d.c. plasma chamber design (as pioneered at IHED, Moscow) with high resolution diagnostics to study liquid plasmas and in particular the phase transition to the crystal state. In 2007, according to current plans, "IMPF" — the International Microgravity Plasma Facility — will be brought to the ISS. This laboratory will provide research opportunities under microgravity conditions for complex plasma studies for about 10 years.

In the laboratory it is very difficult to predict what will happen. Since the first publications of the discovery of the plasma crystals in 1994, over 100 laboratories world-wide have taken up this field — some with experimental programs, some with theoretical research. The publication rate in this field has grown from a handful per year to about one per day during this time, with no sign that this is slowing down. Under such circumstances innovation abounds and new ideas thrive. Presumably, future directions will explore the regime of strongly coupled, strongly magnetised plasmas, bigger reactors will be commissioned that are able to handle many billions of particles and the physics this entails, paramagnetic particles — which interact via dipole forces will be studied in specially designed magnetised plasmas, etc.

Perhaps it is best to summarise my summary by simply stating: "I would not have been able to predict the status of this field, as it is now, back in 1994 — and I am not going to attempt to do the impossible now. Instead, I just plan to enjoy the progress and try to contribute a little myself."

Author Index

A

Afanas'ev, V., 135
Agarwal, A. K., 261, 265
Albrecht, J., 255
Amatucci, W. E., 243
Amemiya, H., 373
Amiranashvili, S. G., 463
Amroun, M., 269
Annaratone, B. M., 91
Annou, R., 269
Asano, K., 277
Avinash, K., 121

B

Bano, G., 152
Berndt, J., 305
Bharuthram, R., 442, 483
Bhattacharjee, A., 121
Bingham, R., 126, 450
Bouchoule, A., 45, 135
Boufendi, L., 45, 135
Břeň, D., 329, 353

C

Čadež, V. M., 251
Carrier, M. J., 414
Cermak, I., 378
Chaudhuri, T. K., 337
Cheung, F. M. H., 139, 325
Chu, H.-Y., 53
Chutov, Y. I., 144, 273, 277
Colwell, J. E., 235, 438
Cramer, N. F., 148, 255, 281, 475

D

Dahiya, R. P., 188
Das, N., 285
de Assis, A., 479
Djebli, M., 269

E

Doyle, T. B., 174
Dubinin, E., 220

E

Ebeling, W., 317
Efremov, V. P., 390

F

Fontcuberta i Morral, A., 45
Fortov, V. E., 3, 91, 135, 196, 247, 289, 293, 390, 394, 471

G

Galka, T., 297
Gallagher, A., 152
Galvão, R., 479
Ganguli, G., 157, 357
Gavrishchaka, V., 357
Gebauer, G., 297
Ghosh, S., 337
Goedheer, W. J., 273
Golub', A. P., 204, 386
Gorbov, D., 333
Goree, J., 200, 418
Goswami, K. S., 285
Goswami, R., 357
Gousein-zade, N. G., 463
Gupta, M. R., 337

H

Hagl, T., 135
Haigneré, C., 135
Havnes, O., 13, 454
Hellberg, M. A., 301
Hertzberg, M. P., 475
Hippler, R., 170, 333
Hong, S., 305
Horányi, M., 22, 235, 309, 313, 402, 438
Howard, J. E., 313

Hu, S., 121
Huet, S., 45

I

I, L., 53
Ignatov, A. M., 162, 317
Ikezawa, S., 184
Ivanov, A. I., 135
Ivlev, A. V., 91, 341, 349, 386

J

Jacobs, G., 321
James, B. W., 139, 148, 255, 325, 467
Jouanny, M., 45
Jovanović, D., 166
Joyce, G., 157, 357

K

Kaizr, V., 329, 353
Katoh, K., 184
Kaw, P. K., 231
Kersten, H., 170, 333
Khan, K., 426
Khan, M., 337
Kharchenko, A. V., 45
Khrapak, A. G., 394
Khrapak, S. A., 341
Klindworth, M., 345
Klumov, B. A., 349
Kong, M., 192
Konopka, U., 345
Kozeev, K., 135
Krauss, C. E., 309
Kravchenko, O. Y., 273, 277
Kroesen, G. M. W., 188
Kulhánek, P., 329, 353

L

Lai, Y.-J., 53
Lakhina, G. S., 442
Lampe, M., 59, 157, 357
Lavrov, O. A., 273

Lebib, S., 45
Lehtinen, K. E. J., 414
Li, F., 361
Lipaev, A., 135
Losseva, T. V., 204, 386

M

Maiorov, S. A., 255, 317, 410
Mamun, A. A., 83
Matthews, A. P., 301
McKenzie, J. F., 174, 220
Melzer, A., 180, 345, 495
Mendis, D. A., 501
Mikikian, M., 45, 135
Mirza, A. M., 426
Mitchell, C. J., 313
Mitchell, L. W., 398
Molotkov, V. I., 135, 390, 394
Morfill, G. E., 91, 110, 135, 243, 341, 345, 349, 365, 386, 487, 507
Müller, W.-C., 365

N

Nakamura, Y., 184, 212, 369, 373
Nascimento, I., 479
Nefedov, A. P., 91, 135, 196, 390, 394
Nemecek, Z., 378, 382
Nonaka, S., 184
Nosenko, V., 200, 418

O

Ohno, N., 277
Ono, K., 446

P

Paeva, G. V., 188
Partoens, B., 192
Pašek, J., 329, 353
Pavlu, J., 378, 382
Peeters, F. M., 192
Petrov, O. F., 135, 196, 247, 394, 471

Piel, A., 200, 345, 495
Pillay, S. R., 442
Popel, S. I., 204, 386
Poustylnik, M. Y., 390, 394
Pramanik, J., 231
Prasad, G., 231, 261, 265
Prior, N. J., 398
Pshenychnyj, A. F., 277

Q

Quaas, M., 170
Quinn, R. A., 91

R

Raadu, M. A., 422, 426, 430
Resendes, D. P., 239
Richterova, I., 378
Robertson, S., 208, 235, 309, 402, 438
Roca i Cabarrocas, P., 45
Rothermel, H., 135
Rozsa, K., 152

S

Safrankova, J., 378, 382
Saitou, Y., 212
Sakawa, Y., 434
Salimullah, M., 216
Samarian, A. A., 139, 148, 255, 325, 398, 406, 410, 467
Sarkar, S., 337
Sato, N., 66
Sauer, K., 220
Schram, P. P. J. M., 224
Schweigert, I. V., 74, 414, 418
Schweigert, V. A., 74, 418
Sedlmayr, E., 32
Semenov, Y. P., 135
Sen, A., 231
Setsuhara, Y., 446
Shafiq, M., 422, 426, 430
Shinohe, K., 434
Shoji, T., 434
Shukla, P. K., 83, 166, 239
Sickafoose, A. A., 235, 438

Singh, S. V., 442
Smirnov, R. D., 277
Sorasio, G., 239
Sternovsky, Z., 208, 402
Stoffels, E., 188
Stoffels, W. W., 188

T

Takahashi, K., 446
Takamura, S., 277
Tendler, M., 479
Teng, L.-W., 53
Thieme, H. G., 170, 333
Thomas, H. M., 91, 135, 341
Thomas Jr., E., 243
Tomita, H., 434
Torchinsky, V. M., 390, 394
Trigger, S. A., 224, 317
Tsypin, V. S., 479
Tsytovich, V. N., 110, 126, 369, 450, 454, 458, 463
Tu, P.-S., 53

U

Usachev, A. D., 289, 293

V

Vaulina, O. S., 196, 247, 406, 467, 471
Veeresha, B. M., 231
Velyhan, A., 382
Verheest, F., 251, 281, 301, 321
Vladimirov, S. V., 148, 255, 281, 410, 471, 475, 479

W

Wardle, M., 281
Watanabe, K., 458
Wiese, R., 333

Winter, J., 297, 305
Wulff, H., 170

Y

Yadav, L. L., 483
Yaroshenko, V. V., 321, 487
Yokota, T., 491

Z

Zachariah, M. R., 414
Zafiu, C., 495
Zagorodny, A. G., 224
Zeiler, A., 365
Zobnin, A. V., 289, 293